PUMPS AND PUMPING

STUDIES IN MECHANICAL ENGINEERING

Volume 1   Analysis, Synthesis and Design of Hydraulic Servosystems and Pipelines (Viersma)
Volume 2   Interordering: A New Method of Component Orientation (Den Hamer)
Volume 3   Combustion Systems of High-Speed Piston I.C. Engines (Kowalewicz)
Volume 4   Mechanical Conveyors for Bulk Solids (Colijn)
Volume 5   Railway Traction (Andrews)
Volume 6   Pumps and Pumping (Ionel)

STUDIES IN MECHANICAL ENGINEERING, 6

PUMPS AND PUMPING

with particular reference to variable-duty pumps

ION I. IONEL

Consulting Engineer, Designer-in-Chief,
Romproiect Co., Bucharest, Romania

Elsevier
Amsterdam–Oxford–New York–Tokyo, 1986

Distribution of this book is being handled by the following publishers:

for the U.S.A. and Canada

ELSEVIER SCIENCE PUBLISHING COMPANY, INC.

52 Vanderbilt Avenue,
New York, NY 10017

for the East European countries, China, Northern Korea, Cuba, Vietnam, and Mongolia

EDITURA TEHNICĂ

Piaţa Scînteii nr. 1,
79738 Bucharest, Romania

for all remaining areas

ELSEVIER SCIENCE PUBLISHERS,

25, Sara Burgerhartstraat
P.O. Box 211, 1000 AE Amsterdam, The Netherlands

Library of Congress Cataloging-in-Publication Data

Ionel, Ion I.
  "Pumps and Pumping"

  (Studies in mechanical engineering; v. 6)
  English version of: "Pompe şi instalaţii de pompare".
  Bibliography: p.
  Includes index.
  1. Pumping machinery I. Title. II. Series.
TJ900.I5913 1986   621.6   85-27478
ISBN 0—444—99528—5

With 555 illustrations and 25 tables

©EDITURA TEHNICĂ, 1986

©Translation: VALENTINA ILEA and ION I. IONEL, 1986

All rights reserved. No part of this publication may be reproduced, stored in a retrieval system, or transmitted in any form or by any means (electronic, mechanical, photocopying, recording, or otherwise) without prior written permission of the copyright owner.

PRINTED IN ROMANIA

Foreword

The work "Pumps and Pumping" responds to two major concerns of our contemporary society, namely, the increasing demand for water and the increasing limitations on energy use. In this context, the presentation of pumps and pumping highlights the use of variable-duty pumps as a technical solution to the above problems.

Enjoying an impressive background as Head of Projects and Consulting Engineer at the "Proiect-București" and "Romproiect" Institutes, Division Head at the "Enterprise for Water Supply and Sewage" in Bucharest, and author of several other scientific handbooks and numerous articles in specialized journals, the author has succeeded in offering a valuable synthesis of the concerned subject by means of lavish documentation.

The work gives the reader detailed analyses and instructions regarding pumps and the pumping-system components, as well as their behaviour under various operating conditions. Furthermore, it deals with the best schematic arrangements of design for pumping installations, required by various typically practical purposes such as pressure-boosting stations for high-rise and long-line systems; distribution and transmission pump stations; boiler supply; cooling and recirculation; pressurizing of land irrigation systems, all accompanied by their operation diagrams.

As to the scientific and technical significance of the book, worth mentioning are the analytical display of the turbo-pump and system characteristics, the derivation of the analytical expression of the speed-torque characteristic of a single turbo-pump, having in view main static and dynamic system characteristics, the adjustable-speed drive with electronic control function attained in the stator and rotor winding, the basic principles and methods of the automatically controlled pumps and water systems, both for obtaining a maximum efficiency and for reducing surges and water hammer damages.

The work focusses on the main problems relating to research, design and exploitation of pump installations, representing thus a priority in this field at the date of its publication.

The pump and pumping problems are solved algorithmically, and mathematics is used at a level that is easily understood by graduates of higher technical establishments and engineering colleges, who may also find many guidelines for their analytical schedule.

The book is graphically illustrated by numerous tables and diagrams. The practical applications, suggested and solved, serve not only to illustrate the theory, but also to add to the completeness of the text

Foreword

with its lucid and rigorous presentation, facilitating a better understanding of the concepts and solutions to the problems.

The book is addressed to experts in hydraulics and energetics who work on and develop studies of design, installment, and exploitation of hydraulic and/or power systems that use/supply water or other fluids. At the same time, it is an excellent guide for the designers of drinking- and/or industrial-water supply systems, as well as of reclamation works or irrigation networks. In many of its chapters, Hydraulic and/or Power Research Institutes will find evolved subjects on newly developed equipment. Besides, the manufacturers of pumps, electric motors, and converters will find strong reasons to switch their main production towards variable-speed pumping groups in response to the ever increasing consumer demand.

Congratulations to Dipl. Eng. Ion I. Ionel, — distinguished graduate of the Hydrotechnical Faculty in Jassy, — for this valuable and up-to-date work which will enrich the literature in this field.

Prof. Eng. Valeriu Blidaru, D. Sc.,

Hydrotechnical Department,
Polytechnical Institute of High Education in Jassy,
Romania.

Contents

Foreword 5

1. BASIC EQUATIONS OF HYDRAULICS AS APPLIED TO THE STUDY OF TURBOPUMPS

1.1 Differential equations of fluid motion for turbopumps 13
   1.1.1 Equations of ideal fluid motion 13
   1.1.2 Motion equations of real fluids (the Navier-Stokes equations) . . . 24
   1.1.3 General criteria of hydraulic similitude, as applied to turbopumps . 25
1.2 Integral equations of fluid motion as applied to turbopumps 28
   1.2.1 Bernoulli's equation for the relative motion 28
   1.2.2 The action of a stream on a moving rigid surface 30
   1.2.3 Fundamental equations of radial-flow pumps 32
   1.2.4 Fundamental equations of axial-flow pumps 39
   1.2.5 Energetic interpretation of the discharge head 50
1.3 Theory of fluid potential motion as applied to turbopumps 51
   1.3.1 Fluid potential motion around a profile 51
   1.3.2 Estimation of fluid pressure force on the profile 58
   1.3.3 Typical profiles and geometrical features 61
   1.3.4 The plane potential motion within the area of a profile network . . 67
   1.3.5 Axially-symmetrical potential motion around the careenage 70
References 170

2. PUMP PERFORMANCE PARAMETERS AND PUMP CHARACTERISTIC CURVES

2.1 Introductory concepts 71
   2.1.1 Definition and classifications 71
   2.1.2 Pressure units 72
   2.1.3 Specific energies and heads 73
   2.1.4 Unit quantities 74
2.2 Main performance parameters of turbopumps 75
   2.2.1 Pump flow rate 75
   2.2.2 Pumping head 76
   2.2.3 Pump power 77
   2.2.4 Pump efficiency 79
2.3 Similitude relationships applied to turbopumps 79
   2.3.1 Similitude concepts of turbomachine theory 79
   2.3.2 The expression of turbopump specific speed 81
   2.3.3 The relationship between the dimensionless specific speed n_s, and the double-unit dimensional quantities 86
   2.3.4 Relationships between Q, H, P and η 87
   2.3.5 Effects of specific speed on turbopump geometry and efficiency . . . 88
2.4 Hydrodynamics of turbopump impeller 90
   2.4.1 Outlining the turbopump wirling energy 90
   2.4.2 Basic concepts of the theory of turbopump impellers 91

    2.4.3 Hydrodynamics of radial-flow pump 93
    2.4.4 Hydrodynamics of axial-flow pumps 98
2.5 Estimation of turbopump performance characteristics 102
2.6 Turbopump performance characteristics curves 108
    2.6.1 Definitions and classifications 108
    2.6.2 The pumping head characteristic curve $H(Q)$ 112
    2.6.3 The power characteristic capacity curve $P(Q)$ 118
    2.6.4 The efficiency capacity characteristic curve $\eta(Q)$ 120
2.7 Characteristics of turbopumps operating in complex conditions . . . 121
    2.7.1 Pump duty 121
    2.7.2 Braking duty 122
    2.7.3 Turbine duty 124
References 127

3. OPERATION AND CONTROL OF TURBOPUMPS CONNECTED TO PUMPING SYSTEMS

3.1 Operation of turbopumps connected to pumping systems 128
    3.1.1 Determination of the pumping-plant duty point 128
    3.1.2 Stable and unstable turbopump operation 133
    3.1.3 Turbopump-to-system connection configurations 135
    3.1.4 Connection of turbopumps according to the shape of their characteristic curves 138
    3.1.5 Restriction criteria for the number of turbopumps that can be parallel-connected 141
3.2 Flow control of turbopumps connected to the system 142
    3.2.1 Shifting the duty point. Control range 142
    3.2.2 Control actions based on changes in the system characteristic 143
    3.2.3 Control actions based on changes in the turbopump characteristic 147
3.3 Estimating the pumping plant energetic indices 155
    3.3.1 Estimating the pumping special energy 155
    3.3.2 Estimating the global efficiency of the pumping plant 157
References 158

4. MECHANISMS OF CAVITATION AND CAVITATION DAMAGE IN TURBOPUMPS

4.1 Cavitation development 159
    4.1.1 Definition, effects and classification 159
    4.1.2 Cavitation coefficient 161
    4.1.3 Cavitation origins 164
    4.1.4 Dynamics of cavitation buldges 169
    4.1.5 Cavitational destruction and protection methods 174
    4.1.6 Resistance of materials to cavitational destruction 178
    4.1.7 Methods of cavitation localizing and detection 181
    4.1.8 Attenuation of cavitational destruction 181
4.2 Equations and curves characteristic of cavitation 182
    4.2.1 Defining equations of cavitation occurrance conditions . . . 182
References 187

5. HEADS, HEAD-LOSSES AND SYSTEM-HEAD CURVES IN FLOW THROUGH PUMPING SYSTEM

5.1 Elementary flow theory for the study of pipes 189
    5.1.1 Flow conditions 189
    5.1.2 Flow velocity 190
    5.1.3 Flow stabilization length 191
    5.1.4 Viscosity 191
    5.1.5 Roughness 192
    5.1.6 Reynolds number 194

5.2 Calculation of linear head losses	194
5.2.1 The friction coefficient	194
5.2.2 Calculation methods	202
5.3 Calculation of local head losses	216
5.3.1 Local resistance coefficient	216
5.3.2 Calculation methods of local head losses	232
5.4 Pumping lines calculation	235
5.4.1 Calculation of the system pumping head	235
5.4.2 Calculation of the economical diameter for discharge pipe	241
References	243

6. WATER HAMMER THEORY AND COMPUTATION MECHANICS

6.1 Basic principles of calculation technique	245
6.1.1 Water hammer phenomenon	245
6.1.2 Water hammer general equations	246
6.1.3 Basic notations and general relationships	247
6.1.4 Water hammer development	253
6.1.5 Classification of computation methods	255
6.2 Approximate computation methods based on formulae	255
6.2.1 Computation of gravity-flow pipeline	256
6.2.2 Computation of pumping flow-pipelines	257
6.3 Approximate computation methods based on graphs	260
6.3.1 Computation methods used with pumping installations having check valve at the pump	260
6.3.2 Computation methods used for pumping installations without check valve at the pump	273
6.4 Exact computation methods based on analytical solving of water hammer equations	285
6.4.1 The method of gradual elimination of water hammer functions	285
6.4.2 The method of characteristics in solving water hammer equations	286
6.5 Exact computation method based on direct integration of water hammer differential equations	298
6.5.1 Physical (sonic) waves method	299
6.5.2 Computational waves method	299
References	318

7. AUTOMATED CONTROL OF PUMPING SYSTEM BY HYDRAULIC CONTROL VALVES

7.1 Basic design analysis of hydraulic control valves used in pumping system	319
7.1.1 The control valve as a component of the automated system	319
7.1.2 Operation principle of a hydraulic control system	325
7.1.3 Static behaviour of control valves	329
7.1.4 Dynamic behaviour of control valves	346
7.1.5 Choice of control valves	349
7.2 The control of system operation by corrective regulating valves	353
7.2.1 Controls to regulate presure	353
7.2.2 Controls for flow-rate maintaining	363
7.2.3 Water-level regulations	371
7.3 Control of water hammer by protective automatic valves	378
7.3.1 Effect of valve operation on waterhammer	378
7.3.2 Classification of water hammer controls	384
7.3.3 Reverse flow prevention controls	385
7.3.4 Controls associated with pump starting and stopping	387
7.3.5 Controls for discharge at overpressures	397
7.3.6 Controls for discharge at underpressures	401
7.3.7 Controls for discharge at power failure	411
References	416

8. AUTOMATED CONTROL OF PUMPS

- 8.1 Hydraulic accumulator theory and computation methods 417
  - 8.1.1 Hydromechanical accumulators 417
  - 8.1.2 Hydropneumatic accumulators 421
- 8.2 On-off pump controls 436
  - 8.2.1 Pressure switch on-off controls 436
  - 8.2.2 On-off regulation by flowmeters 445
  - 8.2.3 On-off regulation by levelmeters 458
- 8.3. Modulating pump controls 461
- References 469

9. SPEED TORQUE CHARACTERISTIC CURVES OF TURBOPUMPS

- 9.1 System characteristic effect on turbopump characteristics 470
- 9.2 Application of similarity equations to turbopumps 475
- 9.3 Turbopumps load torque computation methods 478
- 9.4 Transitory regimes occurring at turbopumps starting 481
  - 9.4.1 Turbopumps starting against a closed discharge valve 481
  - 9.4.2 Turbopump starting against a check valve 484
  - 9.4.3 Turbopump starting against an open discharge valve 485
  - 9.4.4 Turbopump starting in reverse running 486
  - 9.4.5 Turbopumps starting provided with breaking syphon 491
- 9.5 Conclusions on turbopumps behaviour features 492
- References 497

10. SPEED CONTROL OF TURBOPUMPS RUN BY MEANS OF VARIABLE VOLTAGE D. C. MOTORS

- 10.1 Adjustable electric drive systems using commutator d.c. motors supplied by phase-controlled converters 499
  - 10.1.1 Operation principles of phase-controlled converters (controlled rectifiers) . 499
  - 10.1.2 Behaviour of d.c. motors supplied by phase-controlled converters . . 510
  - 10.1.3 Speed control of commutator d.c. motors by phase-controlled converters . 512
- 10.2 Adjustable drives with d.c. motors without a commutator supplied by load-controlled converters 517
- References 523

11. SPEED CONTROL OF TURBOPUMPS RUN BY MEANS OF STATOR CONTROLLED ASYNCHRONOUS MOTORS

- 11.1 Adjustable electric drives by varying the supply voltage 524
  - 11.1.1 Speed control through variation of stator 524
  - 11.1.2 Speed control of asynchronous motors through thyristorized variators . . 528
- 11.2 Adjustable electric drives based on varing frequency of the supply source . . 541
  - 11.2.1 Frequency control of asynchronous motors 541
  - 11.2.2 Speed control by varying frequency with static converters 545
- References 562

12. SPEED CONTROL OF TURBOPUMPS MAKING USE OF ROTOR-CIRCUIT CONTROLLED ASYNCHRONOUS MOTORS

- 12.1 Operation principle of asynchronous cascades 563
  - 12.1.1 Speed control by insertion of an additional e.m.f. in the rotor circuit . . 564
  - 12.1.2 Special actuating states of asynchronous cascade-connected motors . . 566

12.2. Adjustable-Speed Drives by Asynchronous Cascade Connection 571
    12.2.1 The Asynchronous Electrical Cascade 571
    12.2.2 The Asynchronous Electromechanical Cascade 590

13. ELECTRICAL DRIVE OF TURBOPUMPS RUN BY MEANS OF VARIABLE-SPEED TRANSMISSION MECHANISMS   599

13.1 The eddy-current coupling 599
    13.1.1 Description and basic principle 599
    13.1.2 Computation of eddy-current coupling 601
    13.1.3 Load characteristic curves 605
    13.1.4 Speed control 607
    13.1.5 Range of utilization 608
    13.1.6 System appreciation 609
13.2 Fluid coupling 609
    13.2.1 Basic principle 609
    13.2.2 Fluid coupling computation 611
    13.2.3 Characteristic curves 613
    13.2.4 Description of types 615
References 617

14. TECHNIQUES FOR MATCHING PUMPS TO ECONOMIC REQUIREMENTS   618

14.1 Pumps application to facilities of populated centres 619
    14.1.1 Pressure-boosting installations for water supply to multi-storey buildings   619
    14.1.2 Distribution waterworks for direct boosting in pump-closed systems . . 646
    14.1.3 Transmission waterworks for pumping in tank-open systems 654
14.2 Pumps applications to industrial zones facilities 666
    14.2.1 Boiled-fed pumping plants 666
    14.2.2 Condenser-cooling pumping plants 675
    14.2.3 Circulating pumping installations for thermal systems 683
14.3 Applications of pumps to land reclamation systems 693
    14.3.1 Land irrigation pumping installations 693
    14.3.2 Land irrigation pressure-boosting installations 698
    14.3.3 Land drainage pumping installations 708
References 712
Index 713

1
Basic Equations of Hydraulics as Applied to the Study of Turbopumps

1.1 Differential Equations of Fluid Motion for Turbopumps

1.1.1 Equations of Ideal Fluid Motion

Euler's equations. The concept of perfect, or ideal fluids was devised to assist understanding of the behaviour and characteristics of real fluids. A perfect fluid is one that cannot be compressed and has no viscosity. The behaviour of a real fluid is close to that of the ideal in turbopump hydraulics, and within certain admitted approximations, the cumbersome equations of real fluid motion can be replaced by the simpler ones derived for ideal fluids.

Equations of fluid motion are derived from Newton's second law of mechanics, starting with a fluid element in the shape of a parallelepiped whose faces are normal to the co-ordinate axes $Oxyz$ while the edges are dx, dy, dz

$$\Sigma \vec{F} = m\vec{a} \tag{1.1}$$

where $\Sigma \vec{F}$ represents the sum of forces acting on the element of mass m and acceleration a.

When projected on the Ox axis, equation (1.1) becomes (Fig. 1.1)

$$\Sigma F_x = ma_x \tag{1.2}$$

On the same axis, the component of acceleration will be thus

$$a_x = \frac{dv_x}{dt} = \frac{\partial v_x}{\partial t} + v_x \frac{\partial v_x}{\partial x} + v_y \frac{\partial v_x}{\partial y} + v_z \frac{\partial v_x}{\partial z} \tag{1.3}$$

where v_x, v_y, v_z are the velocity projections on the three co-ordinate axes and t is the time.

Mass m of the element is given by

$$m = \rho\, dx\, dy\, dz \tag{1.4}$$

where ρ is the specific mass, or density.

For ideal fluids, force components $\Sigma \vec{F}$ include only pressure and mass forces. And, inasmuch as pressure forces are normal to the surface they act upon, only pressure forces acting on $ABCD$ and $A'B'C'D'$ faces have non-zero projections on the Ox axis.

Let x, y, z be the co-ordinates of point M in the centre of the parallelepiped, and let p be the pressure there. Pressure on face $ABCD$ can be expressed as $p - \dfrac{1}{2}\dfrac{\partial p}{\partial x}\,\mathrm{d}x$ and on face $A'B'C'D'$ as:

$$p + \frac{1}{2}\frac{\partial p}{\partial x}\,\mathrm{d}x.$$

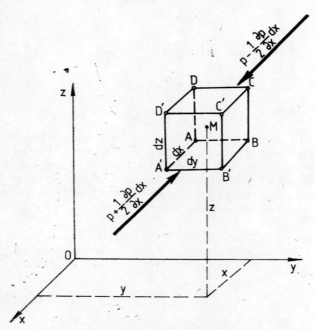

Fig. 1.1. Diagrammatic representation of the dynamical balance of a fluid element.

Pressure forces on $ABCD$ and $A'B'C'D'$ faces become

$$\mathrm{d}P_1 = \left(p - \frac{1}{2}\frac{\partial p}{\partial x}\,\mathrm{d}x\right)\mathrm{d}y\,\mathrm{d}z \quad \text{and} \quad \mathrm{d}P_2 = \left(p + \frac{1}{2}\frac{\partial p}{\partial x}\,\mathrm{d}x\right)\mathrm{d}y\,\mathrm{d}z$$

respectively. The algebraic summation of these two forces will yield the resultant of pressure forces acting along axis Ox, thus

$$\mathrm{d}P = \mathrm{d}P_1 - \mathrm{d}P_2 = -\frac{\partial p}{\partial x}\,\mathrm{d}x\,\mathrm{d}y\,\mathrm{d}z \tag{1.5}$$

The mass force is proportional to the element mass. For the projection of the mass force on the Ox axis, we can write

$$f_x = \rho X\,\mathrm{d}x\,\mathrm{d}y\,\mathrm{d}z \tag{1.6}$$

where X is the unit mass force having the dimensions of acceleration.

Consequently we can write

$$\Sigma F_x = \left(\rho X - \frac{\partial p}{\partial x}\right)\mathrm{d}x\,\mathrm{d}y\,\mathrm{d}z$$

and

$$\left(PX - \frac{\partial p}{\partial x}\right) dx\, dy\, dz = \rho \left(\frac{\partial v_x}{\partial t} + v_x \frac{\partial v_x}{\partial x} + v_y \frac{\partial v_x}{\partial y} + v_z \frac{\partial v_x}{\partial z}\right) dx\, dy\, dz$$

Writing the same relations for the components along axes Oy and Oz, we obtain Euler's system of differential equations for the fluid motion

$$\left.\begin{array}{l} X - \dfrac{1}{\rho} \dfrac{\partial p}{\partial x} = \dfrac{\partial v_x}{\partial t} + v_x \dfrac{\partial v_x}{\partial x} + v_y \dfrac{\partial v_x}{\partial y} + v_z \dfrac{\partial v_x}{\partial z} \\[6pt] Y - \dfrac{1}{\rho} \dfrac{\partial p}{\partial y} = \dfrac{\partial v_y}{\partial t} + v_x \dfrac{\partial v_y}{\partial x} + v_y \dfrac{\partial v_y}{\partial y} + v_z \dfrac{\partial v_y}{\partial z} \\[6pt] Z - \dfrac{1}{\rho} \dfrac{\partial p}{\partial z} = \dfrac{\partial v_z}{\partial t} + v_x \dfrac{\partial v_z}{\partial x} + v_y \dfrac{\partial v_z}{\partial y} + v_z \dfrac{\partial v_z}{\partial z} \end{array}\right\} \quad (1.7)$$

To equation (1.7), we now add the continuity equation which, for incompressible liquids (ρ = constant), has the expression

$$\frac{\partial v_x}{\partial x} + \frac{\partial v_y}{\partial y} + \frac{\partial v_z}{\partial z} = 0 \qquad (1.8)$$

Equation (1.8) can be derived simply enough by applying the principle of mass conservation. The difference between the amount of fluid flowing in and out of a given volume (which is fixed in space) during a given time interval dt is equal to the change in the fluid mass within the volume considered, during the same time interval dt. Thus if the stream velocity is v at point M, then, during the time interval dt, an amount of fluid $\left(v_x - \dfrac{1}{2}\dfrac{\partial v_x}{\partial x} dx\right) dy\, dz\, dt$ will flow in through face $ABCD$ while an amount $\left(v_x + \dfrac{1}{2}\dfrac{\partial v_x}{\partial x} dx\right) dy\, dz\, dt$ will flow out of face $A'B'C'D'$. The difference between the two amounts of fluid is

$$\left(v_x + \frac{1}{2}\frac{\partial v_x}{\partial x} dx\right) dy\, dz\, dt - \left(v_x - \frac{1}{2}\frac{\partial v_x}{\partial x} dx\right) dy\, dz\, dt = \frac{\partial v_x}{\partial x} dx\, dy\, dz\, dt$$

Likewise, we obtain the differences between the amounts of fluid flowing in and out through the faces normal to axes Oy and Oz as $\dfrac{\partial v_x}{\partial y} dx\, dy\, dz\, dt$ and $\dfrac{\partial v_z}{\partial z} dx\, dy\, dz\, dt$.

The sum of these differences is, of course, zero, since the fluid is incompressible. Summing up and dividing by $dx\, dy\, dz\, dt$, we obtain equation (1.8).

We now have four equations with the four unknowns v_x, v_y, v_z and p. To integrate these equations, one has to know the original conditions, that is the values of the unknowns at time $t = t_0$, as well as the limit conditions, i.e. the velocity or pressure values over certain surfaces that restrict the fluid motion. The most common limit conditions are the stream velocity vector, which is tangent to the rigid walls that restrict the fluid motion, and the pressure, which is constant and known at the free surface.

The Gromek—Lamb transformation and cases of integration of differential equations for the motion of ideal fluids. Consider the equation of fluid motion in direction Ox

$$X - \frac{1}{\rho}\frac{\partial p}{\partial x} = \frac{\partial v_x}{\partial t} + v_x\frac{\partial v_x}{\partial x} + v_y\frac{\partial v_x}{\partial y} + v_z\frac{\partial v_x}{\partial z}$$

and subtract from its two members the expression

$$\frac{\partial}{\partial x}\left(\frac{v^2}{2}\right) = \frac{\partial}{\partial x}\left(\frac{v_x^2 + v_y^2 + v_z^2}{2}\right) = v_x\frac{\partial v_x}{\partial x} + v_y\frac{\partial v_y}{\partial x} + v_z\frac{\partial v_z}{\partial x} \quad (1.9)$$

then we obtain

$$X - \frac{1}{\rho}\frac{\partial p}{\partial x} - \frac{\partial}{\partial x}\left(\frac{v^2}{2}\right) = \frac{\partial v_x}{\partial t} + v_y\left(\frac{\partial v_x}{\partial y} - \frac{\partial v_y}{\partial x}\right) + v_z\left(\frac{\partial v_x}{\partial z} - \frac{\partial v_z}{\partial x}\right) \quad (1.10)$$

where $\frac{\partial v_y}{\partial x} - \frac{\partial v_x}{\partial y} = 2\omega_z$ and $\frac{\partial v_x}{\partial z} - \frac{\partial v_z}{\partial x} = 2\omega_y$, represent the components of whirl rot $\vec{v}$ (or curl $\vec{v}$) on axes Oz and Oy.

If mass forces derive from a potential

$$X = -\frac{\partial U}{\partial x}; \quad Y = -\frac{\partial U}{\partial y}; \quad Z = -\frac{\partial U}{\partial z} \quad (1.11)$$

as is the case with turbopumps, equation (1.7) becomes

$$-\frac{\partial}{\partial x}\left(U + \frac{p}{\rho} + \frac{v^2}{2}\right) = \frac{\partial v_x}{\partial t} + 2v_z\omega_y - 2v_y\omega_z$$

Applying the same reasoning to the other two equations (1.7) we eventually obtain the Gromek—Lamb form of the equations for the motion of ideal fluids

$$\left. \begin{aligned} -\frac{\partial}{\partial y}\left(U + \frac{p}{\rho} + \frac{v^2}{2}\right) &= \frac{\partial v_x}{\partial t} + 2(v_z\omega_y - v_y\omega_z) \\ -\frac{\partial}{\partial y}\left(U + \frac{p}{\rho} + \frac{v^2}{2}\right) &= \frac{\partial v_y}{\partial t} + 2(v_x\omega_z - v_z\omega_x) \\ -\frac{\partial}{\partial z}\left(U + \frac{p}{\rho} + \frac{v^2}{2}\right) &= \frac{\partial v_z}{\partial t} + 2(v_y\omega_x - v_x\omega_y) \end{aligned} \right\} \quad (1.12)$$

In the case of potential motions, fluid elements do not take part in the rotational motion and hence $\omega_x = \omega_y = \omega_z = 0$, so that the stream velocity derives from a potential φ:

$$v_x = \frac{\partial \varphi}{\partial x} \; ; \; v_y = \frac{\partial \varphi}{\partial y} \; ; \; v_z = \frac{\partial \varphi}{\partial z} \tag{1.13}$$

Therefore, we can write

$$\frac{\partial v_x}{\partial t} = \frac{\partial^2 \varphi}{\varphi x \partial t} = \frac{\partial}{\partial x}\left(\frac{\partial \varphi}{\partial t}\right); \frac{\partial v_y}{\partial t} = \frac{\partial}{\partial y}\left(\frac{\partial \varphi}{\partial t}\right)$$

and

$$\frac{\partial v_z}{\partial t} = \frac{\partial}{\partial z}\left(\frac{\partial \varphi}{\partial t}\right)$$

while equations (1.12) can be written in the simple form

$$\left.\begin{array}{l}\dfrac{\partial}{\partial x}\left(U + \dfrac{p}{\rho} + \dfrac{v^2}{2} + \dfrac{\partial \varphi}{\partial t}\right) = 0 \\[4pt] \dfrac{\partial}{\partial y}\left(U + \dfrac{p}{\rho} + \dfrac{v^2}{2} + \dfrac{\partial \varphi}{\partial t}\right) = 0 \\[4pt] \dfrac{\partial}{\partial z}\left(U + \dfrac{p}{\rho} + \dfrac{v^2}{2} + \dfrac{\partial \varphi}{\partial t}\right) = 0\end{array}\right\} \tag{1.14}$$

which is the same as:

$$U + \frac{p}{\rho} + \frac{v^2}{2} + \frac{\partial \varphi}{\partial t} = C(t) \tag{1.15}$$

where $C(t)$ is a function of time only, and not of variables x, y, or z.

When mass forces are only weight, $U = gz$ and equation (1.15) becomes

$$z + \frac{p}{\rho g} + \frac{v^2}{2g} + \frac{\partial \varphi}{\partial t} = C(t) \tag{1.16}$$

where g is the gravitational acceleration.

Equations (1.15) and (1.16) — are called Lagrange's equations.

When the potential motion is also steady state we can use Bernoulli's equation

$$U + \frac{p}{\rho} + \frac{v^2}{2} = \text{constant} \tag{1.17}$$

or

$$z + \frac{p}{\rho g} + \frac{v^2}{2g} = \text{constant} \tag{1.18}$$

where the constant has the same value for all the fluid mass in motion.

Basic Equations of the Hydraulics as Applied to the Study of Turbopumps

For steady state motion, terms $\dfrac{\partial v_x}{\partial t}, \dfrac{\partial v_y}{\partial t}, \dfrac{\partial v_z}{\partial t}$ are cancelled out, and thus the system of equations (1.12) becomes

$$-d\left(U + \frac{p}{\rho} + \frac{v^2}{2}\right) = 2(\omega_y v_z - \omega_z v_y)\,dx + 2(\omega_z v_x - \omega_x v_z)\,dy +$$
$$+ 2(\omega_x v_y - \omega_y v_x)dz$$

or

$$-d\left(U + \frac{p}{\rho} + \frac{v^2}{2}\right) = 2\begin{vmatrix} dx & dy & dz \\ \omega_x & \omega_y & \omega_z \\ v_x & v_y & v_z \end{vmatrix} = 2\Delta \tag{1.19}$$

The total differential is zero when determinant "Δ" is zero; this will happen when one of the lines has zero terms or when the terms of two lines are proportional. When "Δ" = 0, Bernoulli's expression results

$$U + \frac{p}{\rho} + \frac{v^2}{2} = C \tag{1.20}$$

where C is an integration constant.

Conditions for $\Delta = 0$ are met in the following cases:

Potential motions

$$\omega_x = \omega_y = \omega_z = 0$$

In this case, constant C (1.20) has the same value throughout the whole fluid mass in motion.

Along the stream lines for which we can write the following equations

$$\frac{dx}{v_x} = \frac{dy}{v_y} = \frac{dz}{v_z} \tag{1.21}$$

Note that constant C in equation (1.20) will take values that differ from one stream line to another.

Along the whirl lines, for which the following equations can be written

$$\frac{dx}{\omega_x} = \frac{dy}{\omega_y} = \frac{dz}{\omega_z}, \tag{1.22}$$

constant C will take values that differ from one whirl line to another.

Helical motions are characterized by the proportionality between the velocity and the whirl vectors

$$\frac{\omega_x}{v_x} = \frac{\omega_y}{v_y} = \frac{\omega_z}{v_z} = \lambda(x, y, z) \tag{1.23}$$

where $\lambda(x, y, z)$ is the proportionality factor; in this case, the constant C in equation (1.20) is the same throughout the whole fluid mass in motion, as far as relations (1.23) are observed.

Intrinsic equations of real fluid motion. Let us consider a fluid particle travelling on its path and being, at a moment t, at point M a distance s from the origin O (Fig. 1.2); let us further consider a right-angled trihedron t, n, ν, with apex in point M, while t is directed along the tangent to the path, n is directed along the main normal to the path and ν, along the binormal; then the particle velocity can be written

$$\vec{v} = v\vec{\tau} \tag{1.24}$$

where $\vec{\tau}$ is the vector tangent to the path.

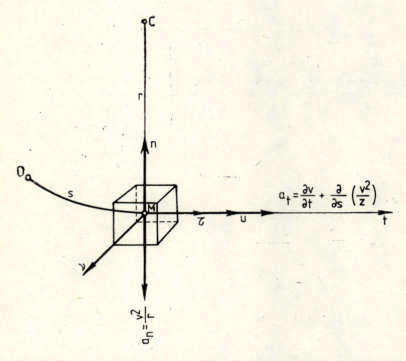

Fig. 1.2. Notation for deriving the intrinsic equations of fluid motion.

The components of acceleration along the three co-ordinate axes are determined by differentiating relation (1.24) with respect to time

$$\frac{d\vec{v}}{d_t} = \vec{\tau}\frac{dv}{d_t} + v\frac{d\vec{\tau}}{d_t} \tag{1.25}$$

where

$$\frac{dv}{dt} = \frac{\partial v}{\partial t} + v\frac{\partial v}{\partial s} = \frac{\partial v}{\partial t} + \frac{\partial}{\partial s}\left(\frac{v^2}{2}\right) \tag{1.26}$$

and
$$\frac{d\vec{\tau}}{dt} = \frac{d\vec{\tau}}{ds} \times \frac{ds}{dt} = \frac{\vec{n}}{r} \times v \tag{1.27}$$

where r is the radius of path curvature in point M, while $\vec{n}$ is the vector along the path normal, directed towards the centre of the curvature. Denoting by a_t, a_n, a_ν the projections of acceleration on the three axes, it follows

$$a_t = \frac{\partial v}{\partial t} + \frac{\partial}{\partial s}\left(\frac{v^2}{2}\right) \tag{1.28}$$

$$a_n = \frac{v^2}{r} \tag{1.29}$$

$$a_\nu = 0 \tag{1.30}$$

The equations of motion, as referred to these co-ordinate axes are called *intrinsic equations of motion* and have the expressions

$$\left. \begin{array}{l} f_t - \dfrac{1}{\rho}\dfrac{\partial p}{\partial s} = \dfrac{\partial v}{\partial t} + \dfrac{\partial}{\partial_s}\times\left(\dfrac{v^2}{2}\right) \\[6pt] f_n - \dfrac{1}{\rho}\dfrac{\partial p}{\partial n} = \dfrac{v^2}{r} \\[6pt] f_\nu - \dfrac{1}{\rho}\dfrac{p}{\partial \nu} = 0 \end{array} \right\} \tag{1.31}$$

For liquids in a gravitational field, after dividing by g, equations (1.31) become

$$-\frac{\partial}{\partial s}\left(z + \frac{p}{\rho g} + \frac{v^2}{2g}\right) = \frac{1}{g}\frac{\partial v}{\partial t} \tag{1.32}$$

$$-\frac{\partial}{\partial n}\left(z + \frac{p}{\rho g}\right) = \frac{v^2}{gr} \tag{1.33}$$

$$-\frac{\partial}{\partial \nu}\left(z + \frac{p}{\rho g}\right) = 0 \tag{1.34}$$

Equation (1.34) shows that along the binormal direction, the pressure change follows the law of hydrostatics. Along the direction of the normal, pressure changes are expressed by equation (1.33), which shows that as we move away from the curvature centre, the pressure increases in proportion to $\rho \dfrac{v^2}{r}$.

Along the tangent direction, the pressure varies according to Bernoulli's law

$$z + \frac{p}{\rho g} + \frac{v^2}{2g} + \frac{1}{g}\times \int_a^b \frac{\partial v}{\partial t}\,ds = \text{constant} \tag{1.35}$$

where a and b are points along the path.

Transcription of Euler's equations to cylindrical co-ordinates. In some problems of turbo-machine hydraulics, it is more convenient to refer the fluid motion to a system of cylindrical co-ordinates.

In cylindrical co-ordinates, the position of point M is determined by the distance z from plane xOy, by the distance r from axis Oz, and by the angle θ subtended by radius r with plane xOz (Figure 1.3). Between the Cartesian co-ordinates and the new ones, we can write the relations

$$z = z; \; x = r \cos \theta; \; y = r \sin \theta \tag{1.36}$$

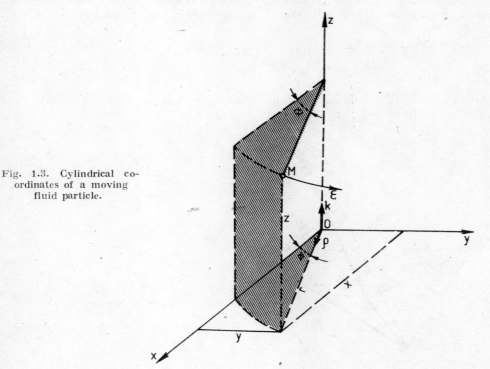

Fig. 1.3. Cylindrical co-ordinates of a moving fluid particle.

In cylindrical co-ordinates, velocity projections become

$$v_z = \frac{dz}{dt}; \; v_r = \frac{dr}{dt}; \; v_\theta = r\frac{d\theta}{dt} \tag{1.37}$$

Acceleration is found by differentiating the velocity vector with respect to time. The expression of the velocity vector is

$$\vec{v} = \vec{k}\frac{dz}{dt} + \vec{\rho}\frac{dr}{dt} + \vec{\varepsilon}r\frac{d\theta}{dt} \tag{1.38}$$

where $\vec{k}$, $\vec{\rho}$, and $\vec{\varepsilon}$ are unit vectors on the direction z, r and θ.

It follows then

$$\frac{d\vec{v}}{dt} = \vec{k}\frac{d^2z}{dt^2} + \vec{\rho}\frac{d^2r}{dt^2} + \vec{\varepsilon}\frac{d}{dt}\left(r \times \frac{d\theta}{dt}\right) + \frac{dz}{dt} \times \frac{d\vec{k}}{dt} +$$

$$+ \frac{dr}{dt} \times \frac{d\vec{\rho}}{dt} + \frac{d\theta}{dt} \times \frac{d\vec{\varepsilon}}{dt} \qquad (1.39)$$

Derivatives of unit vectors $\vec{k}$, $\vec{\rho}$, $\vec{\varepsilon}$ with respect to time will be found by knowing that while k is fixed, unit vectors $\vec{\rho}$ and $\vec{\varepsilon}$ can rotate; it follows that

$$\frac{d\vec{k}}{dt} = 0 \; ; \; \frac{d\vec{\rho}}{dt} = \vec{\varepsilon}\frac{d\theta}{dt} \quad \text{and} \quad \frac{d\vec{\varepsilon}}{dt} = -\vec{\rho}\frac{d\theta}{dt} \qquad (1.40)$$

because the $d\vec{\rho}$ variation has the same direction as vector $\vec{\varepsilon}$, while the $d\vec{\varepsilon}$ variation is contrary to vector $\vec{\rho}$ (Fig. 1.4).

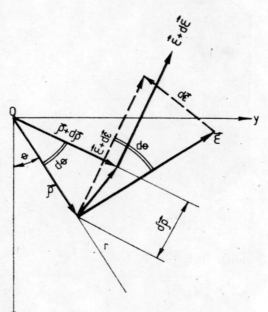

Fig. 1.4. Vector derivatives with respect to time.

Substituting relation (1.40) in (1.39) and taking into account (1.37), it results in

$$\frac{d\vec{v}}{dt} = \vec{k}\frac{dv_z}{dt} + \vec{\rho}\left(\frac{dv_r}{dt} - \frac{v_\theta^2}{r}\right) + \vec{\varepsilon}\frac{1}{r}\frac{d(r v_\theta)}{dt} \qquad (1.41)$$

Thus, the motion equations along the three axes of cylindrical co-ordinates take the form

$$\left.\begin{aligned} f_z - \frac{1}{\rho} \times \frac{\partial p}{d_z} &= \frac{dv_z}{dt} \\ f_r - \frac{1}{\rho} \times \frac{\partial p}{\partial r} &= \frac{dv_r}{dt} - \frac{v_\theta^2}{r} \\ f_\theta - \frac{1}{\rho} \times \frac{\partial p}{r \partial \theta} &= \frac{1}{r} \times \frac{d(rv_\theta)}{dt} \end{aligned}\right\} \quad (1.42)$$

where the derivatives of velocity components have the expressions

$$\left.\begin{aligned} \frac{dv_z}{dt} &= \frac{\partial v_z}{\partial t} + v_z \frac{\partial v_z}{\partial z} + v_r \frac{\partial v_z}{\partial r} + v_\theta \frac{1}{r} \frac{\partial v_z}{\partial \theta} \\ \frac{dv_r}{dt} &= \frac{\partial v_r}{\partial t} + v_z \frac{\partial v_r}{\partial z} + v_r \frac{\partial v_r}{\partial r} + \frac{v_\theta}{r} \frac{\partial v_r}{\partial \theta} \\ \frac{d(rv_\theta)}{dt} &= r \frac{\partial v_\theta}{\partial t} + rv_z \frac{\partial v_\theta}{\partial z} + rv_r \frac{\partial v_\theta}{\partial r} + v_\theta \frac{\partial v_\theta}{\partial \theta} + v_\theta v_r \end{aligned}\right\} \quad (1.43)$$

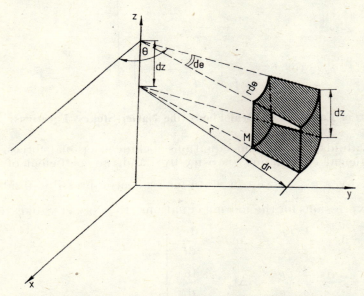

Fig. 1.5. Notation for deriving the continuity equation in cylindrical co-ordinates.

In this reference system, the continuity equation is directly derived by applying the mass conservation principle.

Taking an elemental volume with the edges dz, dr and $rd\theta$, the differences between the amount of inflowing and outflowing fluid in unit time in the elemental volume fixed in space along directions $z-r-A$ will be (Figure 1.5)

$$\left.\begin{aligned} r\,d\theta\,dr v_z - r\,d\theta\,dr\left(v_z + \frac{\partial v_z}{\partial z}dz\right) &= -d\theta\,dr\,\frac{\partial(v_z r)}{\partial z}dz \\ r\,d\theta\,dz v_r - (r+dr)\,d\theta\,dz\left(v_r + \frac{\partial v_r}{\partial r}r\right) &= -d\theta\,dz\,\frac{\partial(v_r r)r}{\partial r}dr \\ d_r z v_\theta - dr\,dz\left(v_\theta + \frac{\partial v_\theta}{\partial \theta}d\theta\right) &= -dr\,d\theta\,\frac{\partial(v_\theta r)}{r\,\partial \theta}d\theta \end{aligned}\right\}$$

Because the fluid is considered to be incompressible, the continuity condition results in the simple form

$$\frac{\partial(v_r r)}{\partial r} + \frac{\partial(v_\theta r)}{r\,\partial \theta} + \frac{\partial(v_z r)}{\partial z} = 0 \qquad (1.44)$$

In cylindrical co-ordinates, the velocity vector (curl) components, rot $\vec{v}$, along the three co-ordinate axes, have the expressions

$$\left.\begin{aligned} 2\omega_r = \operatorname{rot}\vec{v}_r &= \frac{1}{r}\frac{\partial v_z}{r\,\partial \theta} - \frac{\partial v_\theta}{\partial z} \\ 2\omega_\theta = \operatorname{rot}\vec{v}_\theta &= \frac{\partial v_r}{\partial z} \frac{\partial v_z}{\partial r} \\ 2\omega_z = \operatorname{rot}\vec{v}_z &= \frac{1}{r}\left[\frac{\partial(r v_\theta)}{\partial r} - \frac{\partial v_r}{\partial \theta}\right] \end{aligned}\right\} \qquad (1.45)$$

1.1.2 Motion Equations of Real Fluids (the Navier-Stokes Equations)

For real fluids, the motion equations include frictional forces. Denoting the coefficient of dynamic viscosity by μ and the coefficient of kinematic viscosity by $\nu = \dfrac{\mu}{\rho}$, we can write for real incompressible fluids, the following expressions for the motion equations (the Navier-Stokes equations)

$$\left.\begin{aligned} X - \frac{1}{\rho}\frac{\partial p}{x} + \nu \Delta v_x &= \frac{d v_x}{dt} \\ Y - \frac{1}{\rho}\frac{\partial p}{\partial y} + \nu \Delta v_y &= \frac{d v_y}{dt} \\ Z - \frac{1}{\rho}\frac{\partial p}{\partial z} + \nu \Delta v_z &= \frac{d v_z}{dt} \end{aligned}\right\} \qquad (1.46)$$

where $\Delta = \dfrac{\partial^2}{\partial x^2} + \dfrac{\partial^2}{\partial y^2} + \dfrac{\partial^2}{\partial z^2}$ is Laplace's operator.

It is worth noticing that in the presence of a potential motion, the terms containing the viscosity effect disappear from equation (1.46). Indeed, for potential motions we have

$$\nu \Delta v_x = \nu \frac{\partial^2}{\partial x^2}\left(\frac{\partial \varphi}{\partial x}\right) + \nu \frac{\partial^2}{\partial y^2}\left(\frac{\partial \varphi}{\partial x}\right) + \nu \frac{\partial^2}{\partial z^2}\left(\frac{\partial \varphi}{\partial x}\right) = \nu \left(\frac{\partial^3 \varphi}{\partial x^3} + \frac{\partial^3 \varphi}{\partial x \partial y^2} + \frac{\partial^3 \varphi}{\mathrm{d}x \partial z^2}\right) = \nu \frac{\partial}{\partial x}\left(\frac{\partial v_x}{\partial x} + \frac{\partial v_y}{\partial y} + \frac{\partial v_z}{\partial z}\right) = 0$$

and likewise $\Delta v_y = \Delta v_z = 0$.

This fact justifies the substitution of ideal fluids for real ones in those ranges of motion where the particles are not drawn into a rotational motion; this is a case frequently applied to the study of pump casings, if the effects of the impeller blades are neglected.

1.1.3 General Criteria of Hydraulic Similitude, as Applied to Turbopumps

According to the general theory of similitude, two phenomena are similar if they have the same nature, and are described by the same equation.

Therefore, to obtain the expression for all hydraulic similitude criteria, all that is needed is to consider that for two similar motions, equations (1.46) are identical.

Between two quantities corresponding to the two motions, there are constant scale coefficients in homologous points, thus

$$\alpha_1 = \frac{l_1}{l_2} \ ; \ \alpha_v = \frac{v_1}{v_2} \ ; \ \alpha_\rho = \frac{\rho_1}{\rho_2} \ldots$$

If for motion *1* we write the system of equations (1.46), then for motion *2* the following system will result

$$\alpha_f X - \frac{\alpha \partial p}{\alpha_\rho \alpha_l} \times \frac{1}{\rho}\frac{\partial p}{\partial x} + \frac{\alpha_v \cdot \alpha_v}{\alpha_l^2}\nu \Delta v_x = \frac{\alpha_v}{\alpha_t}\frac{\mathrm{d}v_x}{\mathrm{d}t} \qquad (1.46/1)$$

Dividing by $\dfrac{\alpha^2 v}{\alpha_l}$ and putting the identity condition between the two systems of equations (1.46) and (1.46/1), the hydraulic similitude criteria result

$$\frac{\alpha_l}{\alpha_v \alpha_t} = 1 \text{ or } \mathrm{Sh} = \frac{l}{vt} = \text{ditto} \qquad (1.47)$$

This is the Strouchal criterion or the homochrony, characteristic only to non-steady state motions

$$\frac{\alpha_f \alpha_l}{\alpha_v^2} = 1 \text{ or (for } f = g\text{) Fr} = \frac{v^2}{gl} = \text{ditto} \qquad (1.48)$$

This is the Fourde criterion, characteristic of all motions where gravity is the driving force

$$\frac{\alpha \delta_p}{\alpha_\rho \alpha_v^2} = 1 \quad \text{or} \quad \text{Eu} = \frac{\delta_p}{\rho v^2} = \text{ditto.} \tag{1.49}$$

This is Euler's criterion and its presence implies the existence of pressure as a driving force.

$$\frac{\alpha_v \alpha_v}{\alpha_l^2} = 1 \quad \text{or} \quad \text{Re} = \frac{vl}{\nu} = \text{ditto.} \tag{1.50}$$

This is the Reynolds criterion and is characteristic of fluid motions where frictional forces are important.

In the case of turbopumps and fluids flowing through pressurized pipes, the determining forces are the pressure, frictional and inertial forces. As a consequence, the criteria to be taken into consideration are Eu, Re and Sh.

When the motion is in steady state, the Sh criterion is not important any more. When frictional forces are negligible with respect to the other forces, or when the motion is self-modelling according to criterion Re, criterion Eu becomes dominant.

In some cases, still another criterion is important in the study of hydraulic machines, i.e. the Kármán criterion

$$\text{Ka} = \frac{v_i' v_j'}{\bar{v}^2} \tag{1.51}$$

where v_i' and v_j' are the pulsating components of velocity along axes i and j, while $\bar{v}$ is the local mean velocity. This criterion characterizes the fluid turbulence due to factors that are external to the range where the motion is analyzed, and pertains to determinant criteria, that is, it should be considered as a limit condition.

The above criteria become specific in the case of turbopumps. Let us first consider two characteristic velocities that can vary independently from one another and are given by the limit conditions below:

— the peripheral velocity u is proportional to nD, where D is the impeller diameter and n the speed; and

— the average velocity v at the impeller outlet is proportional to $\frac{Q}{D^2}$ (with axial-flow pumps, $v = \frac{4Q}{\pi D^2}$, while with radial-flow pumps, $v = \frac{Q}{D^2} \times \frac{D}{\pi b}$, where b is the width of the radial outlet of the stream channel).

Two pumps are similar if the cumulative conditions for the geometric, kinematic, and dynamic similitude are met.

Geometric similitude requires that ratio $\dfrac{D}{b}$ be the same for both pumps, and hence:

$$\frac{D}{b} = K = \text{constant}$$

while kinematic similitude is met when for both pumps

$$\frac{v}{u} \sim \frac{Q}{nD^3} = q_s = \text{constant} \tag{1.52}$$

Dynamic similitude is satisfied when criterion Eu = constant applies, i.e.

$$\frac{\partial p}{\rho v^2} = \frac{gh}{n^2 D^2} = h_s = \text{constant} \tag{1.53}$$

where $\rho g H$ is the pressure force as given by the head of the discharged liquid. Quantity q_s is called the *specific flow*, and h_s is called the *specific head*. For two similar turbopumps, these two quantities are constant (by definition).

Relation (1.53) can be also written in the form

$$n' = \frac{nD}{\sqrt{gH}} = \frac{1}{\sqrt{h_s}} = \text{constant} \tag{1.53/1}$$

where n' is equal to the double unit dimensionless speed.

Eliminating speed n between equations (1.52) and (1.53) results in

$$n = \frac{\sqrt{gH}}{D\sqrt{h_s}} \; ; \; q_s = \frac{Q\sqrt{h_s}}{D^2\sqrt{gH}}$$

Thus we obtain the double unit dimensionless flow rate

$$Q' = \frac{Q}{D^2\sqrt{gH}} = \frac{q_s}{\sqrt{h_s}} = \text{constant} \tag{1.54}$$

Eliminating diameter D between equations (1.52) and (1.53) results in

$$n'_q = \frac{nQ^{1/2}}{(gH)^{3/4}} = \frac{q_s^{1/2}}{h^{3/4}} = \text{constant} \tag{1.55}$$

which is called the dimensionless *specific speed* when expressed as a function of the rate of flow.

When expressed as a function of power $P = \rho Q H$, the dimensionless specific speed has the expression

$$n'_s = \frac{n}{\rho^{1/2} gH} \sqrt{\frac{P}{\sqrt{gH}}} \tag{1.56}$$

Traditionally, Q, n_q' and n_s' are used as dimensional quantities, bearing in mind that $g =$ constant, $\rho =$ constant, if the two pumps under consideration handle the same liquid (this will be further dealt within Chapter 2).

1.2 Integral Equations of Fluid Motion as Applied to Turbopumps

1.2.1 Bernoulli's Equation for the Relative Motion

Consider the motion of a fluid within a curvilinear channel that rotates uniformly around an axis which, for simplifying purposes, we suppose to be vertical (Fig. 1.6).

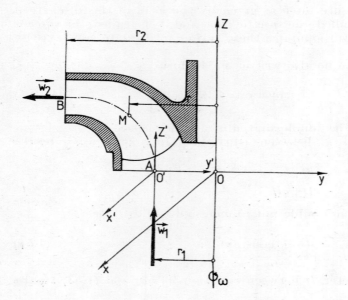

Fig. 1.6. Diagrammatic representation of fluid relative motion.

Such a motion is characteristic of turbines and pumps. Let us further assume that the fluid flows in direction AB, while the angular velocity of the rotational motion is constant i.e. $\omega =$ constant.

Denoting the absolute velocity of the fluid particles along the AB direction by c, the relative velocity of the fluid particles with respect to a system of axes that are tied to the channel by $\vec{w}$, and the transport through the channel by $\vec{u}$, then $\vec{c} = \vec{u} + \vec{w}$.

Now, if we define a system of co-ordinates $Oxyz$ for the channel in rotational motion so that the vertical axis around which the rotation takes place at a velocity $\vec{w}$ is Oz, while axes Ox and Oy are tied to the channel, then Bernoulli's equation (1.20) holds true on any stream line

$$U = \frac{p}{\rho} + \frac{w^2}{2} = C \tag{1.57}$$

where v is the velocity w of the relative motion.

In this case, among the mass forces acting upon the fluid particle in relative motion, we should consider both the gravitational and inertial forces resulting from the rotational motion of the channel. Along the AB line, for any point M located at distance $r = \sqrt{x^2 + y^2}$ from axis Oz, the forces will be

$$f_x = \omega^2 x; \; f_y = \omega^2 y; \; f_z = -g \tag{1.58}$$

and

$$-\frac{\partial U}{\partial x} = \omega^2 x; \; -\frac{\partial U}{\partial y} = \omega^2 y; \; -\frac{\partial U}{\partial z} = -g \tag{1.59}$$

because the unit inertial force is $\omega^2 \vec{r}$.

It follows that

$$\left. \begin{array}{l} \mathrm{d}U = \dfrac{\partial U}{\partial x}\,\mathrm{d}x + \dfrac{\partial U}{\partial y}\,\mathrm{d}y + \dfrac{\partial U}{\partial z} = -\omega^2 x\,\mathrm{d}x - \omega^2\mathrm{d}y + g\,\mathrm{d}z \\[6pt] U = -\dfrac{\omega^2}{2}(x^2 + y^2) + gz + C \end{array} \right\} \tag{1.60}$$

or

$$U = -\frac{\omega^2 r^2}{2} + gz + C = -\frac{u^2}{2} + gz + C$$

where C is a constant of integration.

For two points, 1 and 2, along the AB line, Bernoulli's equation (1.57) takes the form

$$z_1 + \frac{p_1}{\rho g} + \frac{w_1^2 - u_1^2}{2g} = z_2 + \frac{p_2}{\rho g} + \frac{w_2^2 - u_2^2}{2g} \tag{1.61}$$

When load losses h_t produced between A and B are also taken into consideration, then Bernoulli's equation for a stream line through a channel in relative motion becomes

$$z_1 + \frac{p_1}{\rho g} + \frac{w_1^2 - u_1^2}{2g} = z_2 + \frac{p_2}{\rho g} + \frac{w_2^2 - u_2^2}{2g} + h_t \tag{1.62}$$

For the streaming fluid tube of a finite length, equation (1.62) should be multiplied by $\rho g dQ$ and integrated over the whole length of the jet. Then, dividing by $\rho g Q$, we obtain Bernoulli's equation, still of the (1.62) form, except that to each term a correlation coefficient is attached, which is generally close to unity. For this reason, equation (1.62) is assumed to be true even for finite lengths of jets in relative motion.

When no transport motion is ($u = 0$) present while the motion is gradually varied within the length sections *1* and *2*, Bernoulli's equation takes its usual form

$$z_1 + \frac{p_1}{\rho g} + \frac{\alpha_1 v_1^2}{2g} = z_2 + \frac{p_2}{\rho g} + \frac{\alpha_2 v_2^2}{2g} + h_t \qquad (1.63)$$

where v_1 and v_2 are the mean stream velocities through the length section, and α_1 and α_2 the Coriolis correction coefficients, whose usual values are

$$\alpha \simeq 1.05 - 1.10 \qquad (1.64)$$

1.2.2 The Action of a Stream on a Moving Rigid Surface

Let us assume that a curved surface *1—2*, in a rectilinear and uniform motion of velocity u, is traversed by a fluid vein whose relative velocities (with respect to the solid surface) are w_1 and w_2 at the inlet and outlet respectively (see Fig. 1.7). It is further assumed that the fluid motion is steady.

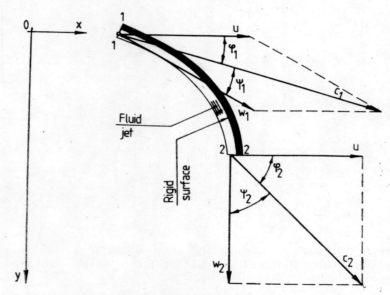

Fig. 1.7. Action of a fluid jet on a curved rigid surface.

Let Q be the rate of fluid flow through any given length, permanently adherent to the curved surface *1—2*; for the quantity of fluid in motion, we can write the equation

$$\beta_1 \rho Q \vec{w}_1 - \beta_2 \rho Q \vec{w}_2 + \vec{G} + \vec{P}_1 + \vec{P}_2 + \vec{R} = 0 \qquad (1.65)$$

where $\vec{R}$ is the reaction of the curved surface; β is the Boussinesq correction coefficient ($\beta \approx 1.02 - 1.03$), $\vec{G}$ is the fluid weight between two length sections, $1-1$ and $2-2$ at the inlet and at the outlet respectively; and P_1 and P_2 the pressure forces in sections $1-1$ and $2-2$ respectively.

When the fluid vein in the two length sections $1-1$ and $2-2$ is in contact with the atmospheric pressure and the fluid weight is negligible, relation (1.65) becomes

$$\beta_1 \rho Q \vec{w}_1 - \beta_2 \rho Q \vec{w}_2 + \vec{R} = 0 \tag{1.66}$$

The expression for the fluid action on the curved surface is

$$\vec{F} = -\vec{R} = \beta_1 \rho Q \vec{w}_1 - \beta_2 \rho Q \vec{w}_2 \tag{1.67}$$

Denoting by φ and ψ the angles subtended by the absolute velocity $\vec{c}$ with the transport velocity $\vec{u}$ and with the relative velocity $\vec{w}$, the expression for the projections along direction u (i.e. direction Ox) and along the normal to u (direction Oy) of equation (1.67) can be written

$$\left. \begin{array}{l} F_x = \rho Q \left[\beta_1 w_1 \cos(\varphi_1 + \psi_1) - \beta_2 w_2 \cos(\varphi_2 + \psi_2) \right] \\ F_y = \rho Q \left[\beta_1 w_1 \sin(\varphi_1 + \psi_1) - \beta_2 w_2 \sin(\varphi_2 + \psi_2) \right] \end{array} \right\} \tag{1.68}$$

Flow rate Q is determined by the relation

$$Q = w\omega \tag{1.69}$$

where ω is the cross-sectional area of the fluid vein.

If the surface is a plane and the fluid vein is normal to it, we can write for the action of the fluid vein upon that surface the expression (Fig. 1.8)

$$F = \beta_1 \rho Q w \tag{1.70}$$

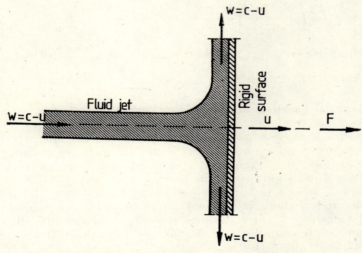

Fig. 1.8. Action of a fluid jet on a flat rigid surface.

The power of the fluid vein acting upon a rigid surface in rectilinear uniform motion has the expression

$$P = \vec{F} \times \vec{u} \tag{1.71}$$

When the surface is flat and normal to the direction of the fluid vein, relation (1.71) becomes

$$\left. \begin{array}{l} P = \beta_1 \rho Q w u \\ P = \beta_1 \rho \omega w^2 u = \beta_1 \rho \omega (c-u)^2 u \end{array} \right\} \tag{1.72}$$

Denoting the theoretical power of the fluid vein by P_t, as defined by relation

$$P_t = \alpha \rho \omega \frac{c^3}{2} \tag{1.73}$$

where α is the Coriolis coefficient, the efficiency of the fluid vein acting upon a flat surface in rectilinear uniform motion will be

$$\eta = \frac{P}{P_t} = \frac{2\beta}{\alpha} \left(1 - \frac{u}{c}\right)^2 \frac{u}{c} \tag{1.74}$$

and has its maximal value for $u = \dfrac{c}{3}$.

Now, assuming that the rigid surfaces are succeeding one another (e.g. blades mounted on an impeller), we can take it that the fluid vein acts upon one blade with mass $\rho \omega c$ instead of mass $\rho \omega w$. Consequently, in relations (1.67) and (1.68) $Q = \omega c$, and equations (1.72) and (1.74) become

$$P = \beta_1 \rho \, \omega c (c-u) u = \beta_1 \rho \omega c \, (cu - u^2) \tag{1.75}$$

$$\eta = \frac{2\beta}{\alpha} \left(1 - \frac{u}{c}\right) \frac{u}{c} \tag{1.76}$$

In such a case, $\eta = \eta_{max}$ will be obtained for $u = \dfrac{c}{2}$, which should be borne in mind with hydraulic machines.

1.2.3 Fundamental Equations of Radial-Flow Pumps

Any radial-flow pump has as main components the impeller (*1*), mounted on a shaft (*2*), rotated by a motor (Fig. 1.9). The impeller includes the surface of revolution; it is grooved to form blades (*4*) which feature either cylindrical or double-curvature surfaces.

The liquid sucked in through connection (*5*) is carried through the impeller channels, where its velocity and pressure are increased, then through the casing (*6*), also divided into fixed blades by channels (*7*), then into the volute chamber (*8*) and further into the diffuser connected to the discharge pipe. In the stator and the volute chamber, most of the kinetic energy is converted into potential energy.

Thus, by centrifuging towards the volute chamber the fluid sucked in through the suction pipe, the pump converts the mechanical power into hydraulic pressure and kinetic power.

Now, let us assume that within the pump, the fluid moves without friction (no friction against the rigid walls) along paths that are parallel to the impeller blades. Theoretically, this condition can be met by allowing an infinite number of blades of zero thickness.

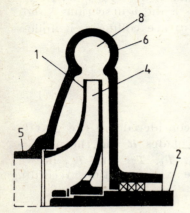

Fig. 1.9. Longitudinal cross-section through a radial-flow pump.

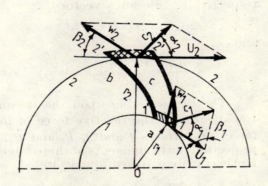

Fig. 1.10. Diagrammatic representation for deriving Euler's equation.

Let us further denote by u the peripheral velocity of the impeller normal to the impeller radius; by w, the fluid relative velocity; and by c the absolute velocity. The $1-1$ section is considered to be the inlet section, while $2-2$ is the outlet section from the impeller (Fig. 1.10). For the fluid between these two sections, we apply the theorem of the momentum; according to this theorem, the time derivative of the moment, as referred to one axis of motion of a system of material points, is equat to the sum of moments of forces acting on the system considered and referred to the same axis.

During time dt, the fluid between two consecutive blades of the impeller moves from position $1-2$ over to position $1'-2'$. Taking three given volumes, a, b, c along the streaming fluid between two blades, we can write for the change with reference to the axis of revolution (the impeller axis) in the moment of the fluid motion during time dt the expression

$$d\vec{\mathscr{L}} = [\vec{\mathscr{L}}(c) + \vec{\mathscr{L}}(b)]_{t+dt} - [\vec{\mathscr{L}}(c) + \vec{\mathscr{L}}(a)]_t \qquad (1.77)$$

where $\vec{\mathscr{L}}$ denotes the moment of the fluid motion within volumes a, b and c as referred to the axis of revolution. Since the motion is in steady state, it follows that

$$[\vec{\mathscr{L}}(c)]_{t+dt} = [\vec{\mathscr{L}}(c)]_t \qquad (1.78)$$

and consequently

$$d\vec{\mathscr{L}} = \vec{\mathscr{L}}(b) - \vec{\mathscr{L}}(a)$$

In b, the momentum is therefore $\vec{C}(b) = \rho Q \beta_2 \vec{c}_2 dt$, while the moment of this angular momentum with respect to the axis of rotation is $\vec{\mathscr{L}}(b) = \rho Q_i \beta_2 dt \vec{c}_2 \times \vec{r}_2$ and $\mathscr{L}(b) = \rho Q_i \beta_2 dt c_2 r_2 \cos \alpha_2$, where Q_i is the stream line rate of flow. Likewise, $\vec{\mathscr{L}}(a) = \rho Q_i \beta_1 \vec{c}_1 \vec{r}_1$ and $\mathscr{L}(a) = \rho Q_i \beta_1 dt c_1 r_1 \cos \alpha_1$; α_1 and α_2 being the angles subtended by vectors $\vec{c}$ and $\vec{u}$ in sections 1 and 2 respectively. Because vectors $\vec{\mathscr{L}}(b)$ and $\vec{\mathscr{L}}(a)$ are collinear, it follows

$$d\mathscr{L} = \mathscr{L}(b) - \mathscr{L}(a)$$

and

$$\frac{d\mathscr{L}}{d_t} = \rho Q_i (\beta_2 c_2 r_2 \cos \alpha_2 - \beta_1 c_1 r_1 \cos \alpha_1) \tag{1.79}$$

The forces acting upon the stream line considered are the gravitational force $\vec{G}$, the reactive force of impeller blades $\vec{R}$ and the pressure forces on surfaces 1 and 2, $\vec{P}_1$ and $\vec{P}_2$. Let T_G, T_R, T_{p1} and T_{p2} be the projection of the moments of these forces on the axis of rotation. According to the theorem stated above, we can write

$$\frac{d\mathscr{L}}{d_t} = \Sigma T = T_G + T_R + T_{p1} + T_{p2} \tag{1.80}$$

The moment of the force active on the blades, M_F, is equal to moment M_R but of opposite sign. Consequently

$$T_F = T_G + T_{p1} + T_{p2} + \rho Q_i (\beta_2 c_2 r_2 \cos \alpha_2 - \beta_1 c_1 r_1 \cos \alpha_1) \tag{1.81}$$

This is the fundamental equation of radial-flow pumps, as derived by Euler.

Assuming that P_1 and P_2 are about normal to surfaces $1-1$ and $2-2$, it follows that $\mathscr{L}_{p1} = \mathscr{L}_{p2} = 0$. If all n channels of the pumps are taken into consideration, $\mathscr{L}_G = 0$, and for symmetry reasons, it follows that

$$T_F = \rho Q (\beta_2 c_2 r_2 \cos \alpha_2 - \beta_1 c_1 r_1 \cos \alpha_1) \tag{1.82}$$

where $Q = \Sigma Q_i$ is the pump output.

The power at the impeller shaft is denoted by P and has the expression

$$P = T_F \omega = \rho Q (\beta_2 c_2 r_2 \cos \alpha_2 - \beta_1 c_1 r_1 \cos \alpha_1) \tag{1.83}$$

or

$$P = \rho Q (\beta_2 c_{2u} u_2 - \beta_1 c_{1u} u_1) \tag{1.84}$$

where $\omega = \dfrac{u}{r}$, is the angular velocity, and $c_u = c \cos \alpha$ is the projection of the absolute velocity in the direction of peripheral velocity.

On the other hand, if considered as a hydraulic generator, the pump power has the expression

$$P = \rho Q H_{t_\infty} \quad (1.85)$$

where H_{t_∞} represents the theoretical pumping head for the case when the number of blades is infinite. From equations (1.83) and (1.85) it follows that

$$H_{t_\infty} = \frac{\beta_2 c_{2u} u_2 - \beta_1 c_{1u} u_1}{g} \quad (1.86)$$

Velocity triangles in sections *1* and *2* are shown in Fig. 1.10. Component c_m along the direction of the absolute velocity radius is called the meridian velocity $c_m = c \cos\alpha$.

Components c_u and c_m are essential in the determination of pump operational characteristics. Component c_u is a factor of the pumping head expression (1.86), while component c_m is a factor of the output expression $Q_t = \pi D b c_m$, where D is the impeller diameter and b the height of the channel of the flowing fluid.

The impeller with an infinite number of zero-thickness blades is called an ideal impeller, and equation (1.86) represents the fundamental equation of radial-flow pumps with an ideal impeller. Actually, the blade thickness is different from zero, and their number is finite.

For a finite number of blades, within the space between the blades a whirl is developed, $\vec{\omega}_z$, whose sign is opposite to that of the pump rotation speed (Fig. 1.11, a—c).

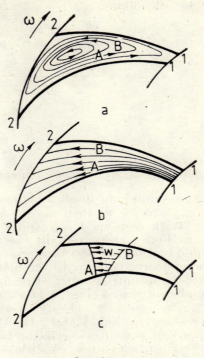
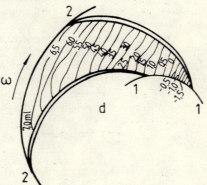

Fig. 1.11. Induced whirls and velocity and pressure distributions around the blades of a radial pump

To explain the generation of this whirl, all that is needed is to estimate vector rot $\vec{w}$ of the relative motion in cylindrical co-ordinates (r, θ, z). When expressed as a function of vector $\vec{c}$ components, vector $\vec{w}$ components have the expression: $w_r = c_r$; $w_\theta = c_u \omega_r$ and $w_z = c_z$. From equation

(1.45), it follows that

$$\left.\begin{aligned}
2\omega_r &= \frac{\partial w_z}{r\partial\theta} - \frac{\partial w_\theta}{\partial z} = \frac{\partial c_z}{r\partial\theta} - \frac{\partial c_u}{\partial z} = 2\omega_{ra} \\
2\omega_\theta &= \frac{\partial w_r}{\partial z} - \frac{\partial w_z}{\partial r} = \frac{\partial c_r}{\partial z} - \frac{\partial c_z}{\partial r} = 2\omega_{\theta a} \\
2\omega_z &= \frac{\partial (r\omega_\theta)}{r\partial r} - \frac{\partial \omega_r}{r\partial\theta} = \frac{1}{r}\frac{\partial (rc_u)}{r\partial r} - \frac{\partial c_r}{r\partial\theta} - 2\omega = 2\omega_{za} - 2\omega
\end{aligned}\right\} \quad (1.87)$$

where ω_{ra}, $\omega_{\theta a}$, ω_{za} are components of vector $\frac{1}{2}\text{rot }\vec{c}$.

Assuming an absolute potential motion results in $\text{rot }\vec{c} = 0$ or $\omega_{ra} = \omega_{\theta a} = \omega_{za} = 0$, then $\omega_z = -\omega$.

With sections $1-1$ and $2-2$ closed, stream lines would appear as in Fig. 1.11,a. Actually, sections $1-1$ and $2-2$ being open, the liquid motion can best be represented by graphically summing the velocity vectors corresponding to the motion of an ideal liquid within an impeller with an infinite number of blades, and within an impeller with a finite number of blades (Figs. 1.11,b and 1.11,c).

The resulting velocity over face A of the blade is lower than that over face B (Fig. 1.11,d). On the other hand (according to Bernoulli's equation), pressures exerted on face B will be higher than those on face A. In certain operation conditions, it may be that the pressure exerted on face A is nil or even negative, in which case a whirl area will be developed.

Owing to the whirl generated in the space between the blades, as the stream lines come out of the impeller, they will be deflected in a direction opposite to that of the rotational motion, while as they go into the impeller, they will be deflected in the direction of the rotational motion [1.5], [1.10]. The peripheral components of the absolute velocity will have the expressions

$$\left.\begin{aligned} c'_{2u} &= c_{2u} - k_2 u_2 \\ c'_{1u} &= c_{1u} + k_1 u_1 \end{aligned}\right\} \quad (1.88)$$

while the velocity triangles will suffer changes as shown in Fig. 1.12.

The expression for the theoretical discharge head of the pump with a finite number of blades is similar to relation (1.86)

$$H_t = \frac{c'_{2u} u_2 - c'_{1u} u_1}{g} \quad (1.89)$$

where Boussinesq's coefficients are taken at unit values.

With equation (1.88), we obtain

$$H_t = H_{t\infty} - \frac{k_2 u_2^2 + k_1 u_1^2}{g} = k_n H_{t\infty} \quad (1.90)$$

where

$$k_n = 1 - \frac{k_2 u_2^2 + k_1^2 u_1^2}{gH_{t_\infty}} < 1$$

is the coeficient of the decrease in the theoretical discharge head as a result of the finite number of blades.

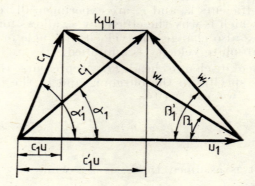

Fig. 1.12. Velocity triangles for an impeller with a finite number of blades, "Z"

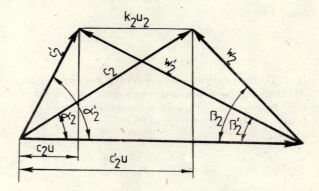

Various formulae have been suggested for the determination of coefficient k_n. To a first approximation, we can use Proskura's expression [1.8]

$$k_n = \frac{2\sigma}{n} \frac{1}{1 - \left(\frac{D_1}{D_2}\right)^2} \tag{1.91}$$

where n is the number of blades and D_1 and D_2 the internal diameter (ID) and external diameter (OD) of the impeller respectively, while $\sigma = (0.55 - 0.65) + 0.6 \sin \beta_2$.

The actual discharge head is lower than the theoretical one owing to internal pressure losses of the pump. Denoting these pressure losses by h_f, we can write for the actual discharge head the following expression

$$H = H_t - h_f = H_t\left(1 - \frac{h_f}{H_t}\right) = \eta_h H_t = k_n \eta_h H_{t\infty} \qquad (1.92)$$

where $\eta_h < 1$ is the hydraulic efficiency.

Coefficients k_n and η_n are experimentally determined for each typical pump, which is why the efficiency η_h allows for the blade thickness. Blade thickness also changes the velocity triangle in that the radial component of the absolute velocity is amplified.

Since discharge head (H) represents the difference between the energies specific to the outgoing and to the incoming flow of the pump impeller, we can write

$$H = \frac{1}{\rho g}(p_2 - p_1) + \frac{1}{2g}(c_2^2 - c_1^2) = h_{p\infty} + h_{c\infty} \qquad (1.93)$$

where it is assumed that $z_1 = z_2$, while $h_{p\infty} = \frac{1}{\rho g}(p_2 - p_1)$ denotes the static head and $h_{c\infty} = \frac{1}{2g}(c_2^2 - c_1^2)$ is the dynamic head.

The ratio

$$\mu = \frac{h_{p\infty}}{h_{c\infty}} \qquad (1.94)$$

is called the impeller *reaction coefficient* and characterizes the structure of the latter from an energetic point of view. For radial-flow pumps, the *reaction coefficient* essentially depends on the shape of the impeller blades. In current practice

$$\alpha_1 = \frac{\pi}{2}, \quad c_{1u} = 0 \text{ and } c_{1m} = c_{2m} \qquad (1.95)$$

and, ignoring friction, it follows that

$$\left.\begin{array}{l} h_{c\infty} = \dfrac{c_{2u}^2}{2g} \; ; \quad H = \dfrac{c_{2u}u_2}{g} \\[2mm] h_{p\infty} = H - h_{c\infty} = \dfrac{c_{2u}}{2g}(2u_2 - c_{2u}) \text{ and } \mu = 1 - \dfrac{c_{2u}}{2u_2} \end{array}\right\} \qquad (1.96)$$

Two limit cases are to be emphasized: $\mu = 1$ and $\mu = 0$.

When $\mu = 0$, the whole energy received is converted into potential energy, since $u_2 \neq \infty$ and hence $c_{2u} = 0$: $\alpha_2 = \dfrac{\pi}{2}$ and $h_{c\infty} = h_{p\infty} = H_t = 0$.

An impeller with a pure reaction is therefore not a reasonable proposition because it fails to build up a discharge head. Indeed, to obtain such a discharge head, the condition is: $c_{2u} > 0$ or $c_{2u} > c_{1u}$, and thus

$$\beta_2 > \beta_{2min} = \arctan \frac{c_{2m}}{u_2} \tag{1.97}$$

(If $\beta_2 < \beta_{2min}$, $\mu > 1$, $H_t < 0$, $c_{2u} < 0$ and the pump operates as a hydraulic turbine).

When $\mu = 1$, the whole energy received by the pump is converted into kinetic energy, that is the impeller has a pure action. In such a case, $c_{2u} = -2u_2$, $H_t = h_{c\infty} = \dfrac{2u_2^2}{g}$ and $h_{p\infty} = 0$.

Such a pump is the cause of important pressure losses, because an important part of the kinetic energy should be converted into potential energy in the stator and in the diffuser. Therefore, the condition is

$$\beta_2 < \beta_{2max} = -\arctan \frac{c_{2m}}{u_2} = \pi - \beta_{2min} \tag{1.98}$$

1.2.4 Fundamental Equations of Axial-flow Pumps

With axial pumps, both the impeller (*1*) and the casing (*2*) are mounted within a tube-shaped body, (*3*), thus making the general direction of liquid flow parallel to the axis of revolution (Fig. 1.13, *a*).

The impeller (*1*) with its blades (*4*) can be likened to a ship's propeller and for this reason, axial-flow pumps are also called propeller pumps.

Taking two cylindrical cross-sections through the impeller at distances r and $r + dr$ (Fig. 1.13, *b*) and assuming further that we develop one of these sections in the plane, we obtain the network of plane profiles (Fig. 1.13, *c*) situated at a distance $t = 2\pi r/n$ from one another, where n is the number of blades (Fig. 1.13). Thus, the flow around the blades between two coaxial cylinders with radii r and $r + dr$ respectively can be substituted by the plane motion of the liquid around an infinite network of profiles.

Here we shall introduce the concept of profile chord, which represents the profile length l, and incidence angle α, denoting the angle α subtended by the upstream direction of stream at infinity with the chord (axis) direction (Fig. 1.16).

The forces acting on the pump blades are determined by estimating the forces active on one profile of the network, which is done by applying the theorem of momentum. By choosing a control surface limited by two stream lines, *ad* and *bc*, situated at a distance t from one another, and by two parallel lines, *ab* and *cd*, of lengths t (Fig. 1.14), and neglecting the frictional forces acting upon the control surface as well as the weight of the liquid inside the control surface, the equation for the momentum will have the expression

$$m(\vec{w}_2 - \vec{w}_1) = \vec{P} - \vec{F} \tag{1.99}$$

where $\vec{F}$ is the action exerted by the liquid upon the blade (equal but of opposite sign to the blade reaction); $\vec{P}$ is the vector of pressure forces acting upon the control surface; and m is the mass passing in unit time

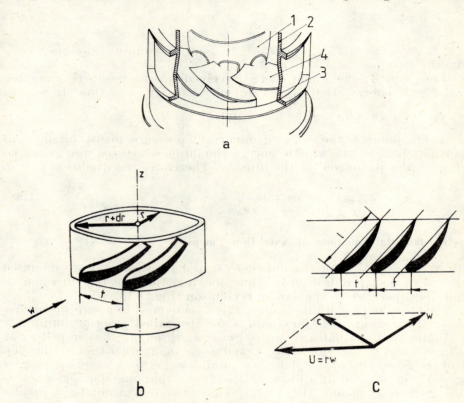

Fig. 1.13. Profile network for an axial pump

over the control surface. Denoting by w_z the velocity component along axis Oz, normal to lines ab and cd, it follows that

$$w_z = w_1 \sin \beta_1 = w_2 \sin \beta_2 = w_{1z} = w_{2z} \tag{1.100}$$

$$m = \rho t w_z \tag{1.101}$$

where ρ is the specific mass (the profile spread being taken as unity).
Projecting equation (1.99) along axes Ou and Oz results in

$$\left. \begin{array}{l} \rho t w_z (w_2 \cos \beta_2 - w_1 \cos \beta_1)\, \mathrm{d}r = F_u \\ \rho t w_z (w_2 \sin \beta_2 - w_1 \sin \beta_1) = (p_1 - p_2)\, t - F_z \end{array} \right\}$$

or, with equation (1.100)

$$F_u = \rho\, t w_z (w_1 \cos \beta_1 - w_2 \cos \beta_2) \tag{1.102}$$

and
$$F_z = (p_2 - p_1)t \tag{1.103}$$

The pressure difference $p_2 - p_1$ can be expressed by means of Bernoulli's equation

$$p_2 - p_1 = \rho \frac{w_1^2 - w_2^2}{2} - \rho g h_f \tag{1.104}$$

where h_f denotes the pressure losses between ab and bd sections.

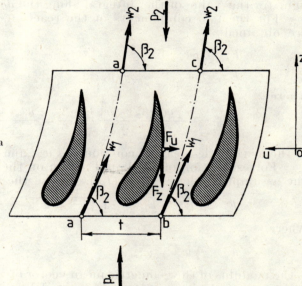

Fig. 1.14. Control surface for a profile network.

Relation (1.103) therefore becomes:
$$F_z = \rho \frac{w_1^2 - w_2^2}{2} t - \rho g h_f t \tag{1.105}$$

Making the due transformations:
$$\left.\begin{array}{l} w_1^2 = w_1^2(\cos^2\beta_1 + \sin^2\beta_1)\,;\ w_2^2 = w_2^2(\cos^2\beta_2 + \sin^2\beta_2) \\ w_1^2 - w_2^2 = w_1^2\cos^2\beta_1 - w_2^2\cos^2\beta_2 + w_1^2\sin^2\beta_1 - w_2^2\sin^2\beta_2 - \\ -w_1^2\cos\beta_1 - w_2^2\cos^2\beta_2 = (w_1\cos\beta_1 - w_2\cos\beta_2)(w_1\cos\beta_1 + \\ + w_2\cos\beta_2) \end{array}\right\}$$

It follows that
$$F_z = \rho t(w_1\cos\beta_1 - w_2\cos\beta_2)\frac{w_1\cos\beta_1 + w_2\cos\beta_2}{2} - h_f t \tag{1.106}$$

Velocity circulation is defined as integral $\Gamma = \oint \vec{v} \cdot d\vec{s}$ estimated for a closed random profile and assuming that the positive sign of circulation according to conditions set forth in Fig. 1.14 is anticlockwise (v is the velocity vector while ds is the directed curve elements). Velocity circulation on the $ab - cd$ profile will thus result as

$$\Gamma = [w_1 \cos \beta_1 + w_2 \cos(\pi - \beta)]t \text{ or } \Gamma = (w_1 \cos \beta_1 - w_2 \cos \beta_2)t \quad (1.107)$$

since the value of the integral along the bc line is equal, but of opposite sign, to the value of the integral along the da line.

For the two components of the reaction, the following expressions are obtained

$$F_u = \rho w_z \cdot \Gamma \quad (1.108)$$

$$F_z = \rho w_{mu} \Gamma - \rho g h_f t \quad (1.109)$$

where

$$w_{mu} = \frac{w_2 \cos \beta_2 + w_1 \cos \beta_1}{2} = \frac{w_{2u} + w_{1u}}{2} \quad (1.110)$$

is the arithmetic mean of components w_{2u} and w_{1u}.

For estimation of the liquid action on the blade input section, with the assumption that $h_f = 0$ (F_{tz}), we use the expression

$$F_t = \sqrt{F_u^2 + F_{tz}^2} = \rho \Gamma \sqrt{w_z^2 + w_{um}^2} = \rho \Gamma w_m \quad (1.111)$$

where

$$w_m = \sqrt{w_z^2 + w_{um}^2} \quad (1.112)$$

is the modulus of the geometric mean vector of vectors $\vec{w}_1$ and $\vec{w}_2$ (Fig. 1.15) and of the velocity at infinity, denoted by w_∞; F_t is the action of the liquid on the blade, assuming that friction is negligible; and F_{tz} is the component of this action along axis Oz.

It is easy to show that vector $\vec{F}_t$ is normal to vector $\vec{w}_m$

$$\tan \beta_m = \frac{w_z}{w_{mu}}$$

and

$$\tan \beta_m = \frac{F_u}{F_{tz}} = \frac{\rho_t w_z \Gamma}{\rho\, t w_{mu} \Gamma} = \frac{w_z}{w_{mu}}$$

Hence:

$$\varphi = \beta_m \quad (1.113)$$

as shown in Fig. 1.16. Relations (1.111) and (1.113) were derived by Jukowski and define the lift in the case of a profile situated within a network, for an ideal fluid.

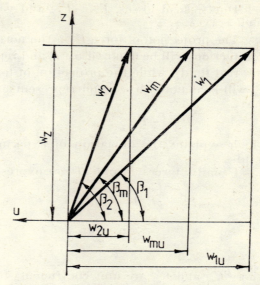

Fig. 1.15. Geometric mean of w_1 and w_2 vectors.

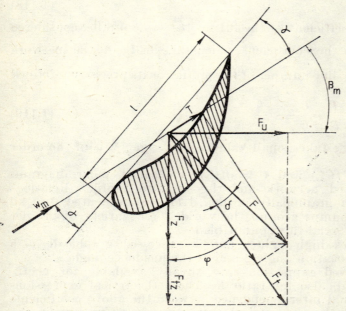

Fig. 1.16. Forces acting on a profile.

Due to load losses, $F_z < F_{tz}$, and vector $\vec{F}$ subtends at axis Oz an angle $\varphi + \delta > \varphi$.

The projection of force $\vec{F}$ on the normal to velocity $\vec{w}_m$ (that is on the $\vec{F}_t$ direction) will be denoted by P and represents the lift (in the conditions of a real fluid), while the projection of force $\vec{F}$ on the direction of velocity $\vec{w}_m$ will be denoted by $\vec{T}$ and represents the frontal resistance. It follows that

$$|\vec{P}| < |\vec{F}_t| \tag{1.114}$$

which explains the deviation of experimental data from the theoretical results.

Usually, forces P and T are expressed in the form:

$$P = \rho C_{yr} l \frac{w_m^2}{2} \tag{1.115}$$

$$T = \rho C_{xr} l \frac{w_m^2}{2} \tag{1.116}$$

where C_{yr} and C_{xr} are unit coefficients for the lift and resistance, for an outline (profile) situated in the network.

As a general rule, these coefficients are estimated with relations

$$C_{yr} = \left(\frac{C_{yr}}{C_y}\right) C_y = K C_y \tag{1.117}$$

$$C_{xr} = \varepsilon C_{yr} \tag{1.118}$$

where C_y is the unit coefficient of the lift for the same profile, considered separately; $K = \dfrac{C_{yr}}{C_y}$ is the coefficient which takes care of the network effects; and ε is the quality number of the profile, or its precision, defined as:

$$\varepsilon = \tan \delta = \frac{T}{P} = \frac{C_{xr}}{C_{yr}} \tag{1.119}$$

As a general rule, ε takes small values since angle δ is of the order of 1° or 2°.

Relations $C_{yr} = f(C_{xr})$ and $C_y = f(C_x)$ are called polar diagrams for both the integrated network and the separate profile. Relations $C_y = f(C_x)$ are shown graphically in Fig. 1.17, as presented by all constructors of such pumps around the world, for a variety of profiles used in the design of axial-flow pump blades.

The ratio K takes high values and is expressed graphically as a function of t/l, of the location angle β and of the incidence angle α.

Theoretically derived expressions are already available for coefficients C_y and C_{yr}, and also for ratio K. These theoretical expressions emphasize the functional interdependence between the above coefficients on the one hand, and the geometrical characteristics of profiles in the network and the physical characteristics of the fluid on the other. However, these expressions should be corrected by empirical coefficients.

As an example, Fig. 1.18 represents ratio $K = K\left(\beta, \dfrac{t}{l}\right)$ for $\alpha = 2.5°$ for the case of a flat plate (blade) in a network [1.8], [1.9] while Fig. 1.19 represents the same ratio for two hydrodynamical profiles $K = K\left(\dfrac{t}{l}\right)$, both theoretical and experimental, with $\alpha = 0°$ and $\beta =$ constant [1.9]

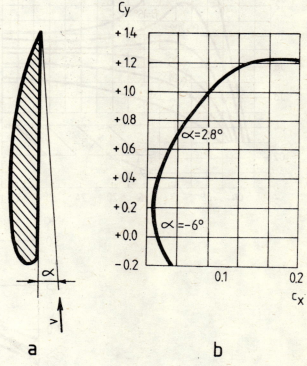

Fig. 1.17. Polar diagram of separate profile:

a — incidence angle; b — diagrammatical representation of function $C_y = f(C_x)$.

On the same diagram of Fig. 1.19 are shown curves $K_0 = K_0\left(\dfrac{t}{l}\right)$ for the flat plate (blade) in a network, as given by theoretical expressions.

It is worth mentioning that for angles of incidence of zero lift of the separate profile, ratio K takes values of $\pm \infty$, which further increases the lack of accuracy in the calculations. This is why two coefficients were recently suggested to characterize the network effects, one to represent the relative increase in the lift of the separate profile

$$x = \dfrac{dC_{yr}}{d\alpha}\bigg|_{\alpha=0} = 0 : \dfrac{dC_y}{d\alpha} = \bigg|_{\alpha=0} 0 \qquad (1.120)$$

Basic Equations of the Hydraulics as Applied to the Study of Turbopumps

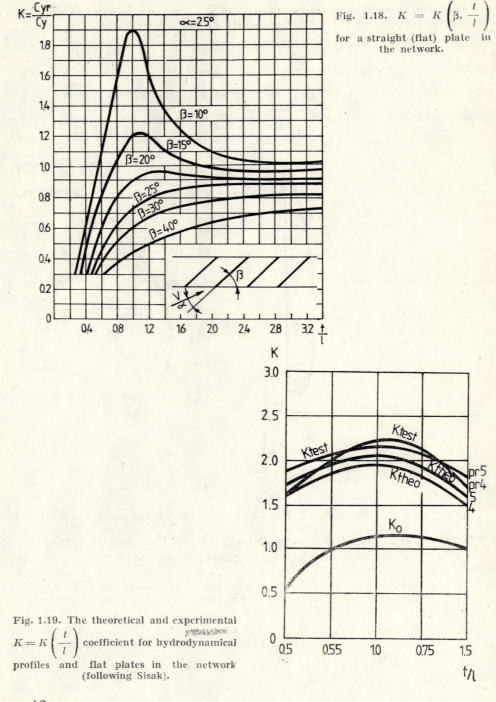

Fig. 1.18. $K = K\left(\beta, \dfrac{t}{l}\right)$ for a straight (flat) plate in the network.

Fig. 1.19. The theoretical and experimental $K = K\left(\dfrac{t}{l}\right)$ coefficient for hydrodynamical profiles and flat plates in the network (following Sisak).

46

and the other $C_{yr}|_{\alpha=0}$, both coefficients being functions of $\frac{t}{l}$ and β. Curves x and $C_{yr}|_{\alpha=0}$ were determined experimentally at the Laboratory for Hydraulics of the Polytechnical Institute in Timișoara (Romania) for some Carafoli profiles with a rounded peak [1.7] and for the MHT-3 profile [1.9], (Fig. 1.20).

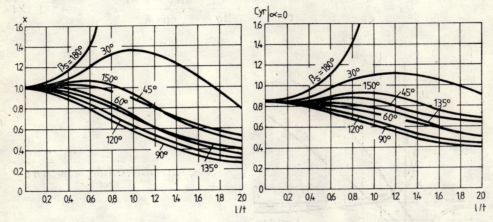

Fig. 1.20. Diagrams for the determination of x and $C_{yr}|\alpha=0|$ coefficient for the MHT-3 profile (Centre of Hydraulical Research, Timișoara).

However, some authors think that coefficient C_{yr} can be derived by calculation, and even suggest mathematical expressions to this effect. Thus, Schlichting suggests

$$C_{yr} = C_{yr}|_{\alpha=0} + 2\pi X \alpha \qquad (1.121)$$

and Proskura

$$C_{yr} = 2\pi K_0 \frac{\Delta}{1 - \frac{\Delta}{\tan \beta}} + 2\pi \frac{K_0}{1 - \frac{\Delta}{\tan \beta}} \alpha \qquad (1.122)$$

where α is the angle of incidence, K_0 is the coefficient accounting for the network effects in the case of flat plates (blades), while Δ is the incidence angle at a zero lift of the profile in the network.

More accurate values for these coefficients are obtained from the experimental data derived by processing characteristic curves, such as those illustrated in Fig. 1.21 for the MHT-1 profile [1.9].

During pump operation, the mechanical work done per unit time by force e is expressed as

$$e = uF_u = uF \cos\left(\frac{\pi}{2} - \varphi - \delta\right) \qquad (1-123)$$

or

$$e = uF \sin(\beta_m + \delta) \qquad (1-123a)$$

Now, if $dQ = n c_z t dr$ is the flow rate for a square cross-section described by cylinders of radii r and $r + dr$, and H_t is the theoretical pumping head, the energy per unit time impressed by the pump on the liquid flowing at a rate dQ is $dE_r = \rho H dQ/\eta_h$, where η_h represents the hydraulic efficiency. From relation $dE = dE_r$, it follows that

$$H_t = \frac{H}{\eta_h} = C_{yr} \times \frac{l}{t} \times \frac{u}{c_z} \times \frac{w_m^2}{2g} \times \frac{\sin(\beta + \delta)}{\cos \delta} \tag{1.124}$$

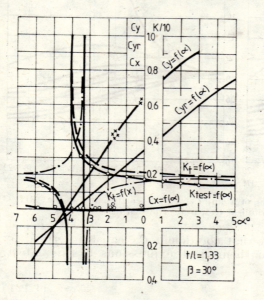

Fig. 1.21. Characteristics of the $MHT-1$ profile.

or

$$C_{yr} \frac{l}{t} = \frac{H}{\eta_h} \times \frac{c_z}{u} \times \frac{2g}{w_m^2} \times \frac{\cos \delta}{\sin(\beta + \partial)} \tag{1.124/1}$$

where c_z is the component of the absolute velocity along axis Oz ($c_z = w_z$).

Expression (1.124) represents another form of the basic equation derived for the axial-flow pump. Euler's general equation (1.86) still holds true; of course, with $u_1 = u_2 = u$, it follows that

$$H_t = \frac{u(c_{2u} - c_{1u})}{g} \tag{1.125}$$

Assuming that H_t is constant along radius r of the impeller, and using relation $u = \frac{2\pi n}{60} r$, where n is the impeller speed, equation (1.125) becomes for the extreme values of the radius

$$r_i(c_{2u} - c_{1u})_i = r_e(c_{2u} - c_{1u})_e \tag{1.126}$$

Taking $c_{1u} \approx 0$

$$r_i c_{2u\ i} = r_e c_{2u\ e}. \tag{1.127}$$

But $c = \sqrt{c_z^2 + c_u^2}$ and, inasmuch as for all practical purposes c_z does not depend on r, it follows that for $r_i < r_e$ the result will be $c_{2u\ i} > c_{2u\ e}$ and $c_{2i} > c_{2e}$; hence, the absolute velocity is higher around the hub and consequently the pressure, by Bernoulli's expression, will be lower near the hub. This explains why cavitation phenomena develop first in the area of the impeller hub.

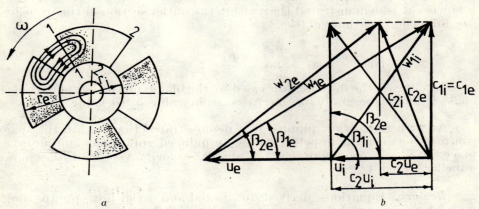

Fig. 1.22. Velocity triangles and the induced whirl for axial-flow pumps.

To keep constant the axial component of the absolute velocity c_z between angles β_i and β_e (in the hub area and at the periphery, respectively), the following relation should exist (Fig. 1.22,b).

$$\beta_i > \beta_e \tag{1.128}$$

while the inlet and outlet pitches of the blade, i.e. h_1 and h_2 respectively, should be independent of the radius. Indeed, from the velocity triangles at the impeller inlet it follows that

$$\frac{c_z}{u_i} = \operatorname{tg}\beta_{1i} \text{ and } \frac{c_z}{u_e} = \operatorname{tg}\beta_{1e} \tag{1.129}$$

and, by definition, pitches p_{1i} and p_{1e} have the expressions

$$\frac{p_{1i}}{\pi D_i} = \operatorname{tg}\beta_{1i} \ ; \ \frac{p_{1e}}{\pi D_e} = \operatorname{tg}\beta_{1e} \tag{1.130}$$

As a consequence

$$c_z = u_i \frac{p_{1i}}{D_i} = u_e \frac{p_e}{\pi D_e}$$

Now, since
$$u_i = \frac{\pi D_i n}{60} \quad \text{and} \quad u_e = \frac{\pi D_e n}{60}$$
it follows that
$$p_{1i} \frac{n}{60} = p_{1e} \frac{n}{60} = \frac{p_1 n}{60} = p_{1s} \tag{1.131}$$

where $p_{1s} = p_1 \omega$ is the product of the impeller pitch in the inlet cross-section and the impeller angular velocity in the outlet cross-section. Likewise, it is demonstrated that within the outlet section of the impeller, the following relations exist

$$p_{2i} \frac{n}{60} = p_{2e} \frac{n}{60} = \frac{p_2 n}{60} = p_{2s} \tag{1.132}$$

To increase the impulse impressed by the blades on the liquid particles, angle β should be gradually increased from value β_1 to value β_2. Hence, $p_{2s} > p_{1s}$.

As with radial-flow pumps, it is demonstrated that within the space confined between the impeller blades, an induced rotational motion takes place, cu $\vec{\omega}_z = -\vec{\omega}$, where $\vec{\omega}$ is the angular velocity vector of the pump impeller (Fig. 1.22,a).

Remark. Equations derived for radial and axial-flow pumps hold true for mixed, axial-flow ultra-slow and radial-flow ultra-fast pumps too, allowing for the geometric characteristics specific to such pumps.

1.2.5 Energetic Interpretation of the Discharge Head

Let us consider Bernoulli's equation which does not take into account the pressure losses. Replacing in this equation the relative velocity w by the absolute velocity c and by the peripheral velocity u according to relation

$$w^2 = u^2 + c^2 - 2u c_u \tag{1.133}$$

results in

$$z_1 + \frac{p_1}{\rho g} + \frac{c_1^2}{2g} = z_2 + \frac{p_2}{\rho g} + \frac{c_2^2}{2g} - \frac{u_2 c_{u2} - u_1 c_{u1}}{g} \tag{1.134}$$

With equation (1.86), where Boussinesq's coefficient is taken as unity, equation (1.134) becomes

$$H_{t\infty} = \left(z_2 + \frac{p_2}{\rho g} + \frac{c_2^2}{2g} \right) - \left(z_1 + \frac{p_1}{\rho g} + \frac{c_1^2}{2g} \right) \tag{1.135}$$

This means that the pumping head H_t represents the specific energy given up by the pump to the stream.

If the sections for which Bernoulli's equation is derived are upstream from the suction pipe at the pump inlet and downstream from the discharge pipe at the pump outlet, and if the pressure losses between the two reference sections are also introduced, Bernoulli's equation will have the expression

$$z_1 + \frac{p_1}{\rho g} + \frac{\alpha_1 v_1^2}{2g} = z_2 + \frac{p_2}{\rho g} + \frac{\alpha_2 v_2^2}{2g} + h_z - H \qquad (1.136)$$

where v_1 and v_2 are the mean velocities, α_1 and α_2 the Coriolis coefficients, and H the actual discharge head.

1.3 Theory of Fluid Potential Motion as Applied to Turbopumps

Fluid motion may be considered as a potential notion if the frictional forces are negligible in comparasion with the inertial forces, that is if the fluid particles do not take part in the rotational motion. Frictional forces are due to physical properties of the fluids, namely their viscosity and adherence to the surfaces of solid bodies they contact when in motion. Close to the surface of solid bodies, a limit layer is developed where the motion is not potential any more. In that area where the streams break away from the surface of solid bodies, whirls are generated and frictional forces are again very important. Outside the limit layer and the whirls, where the effects of the solid body walls are negligible, the motion takes on a potential character and hence within this range the theory of fluid potential motion can be applied. There are cases when the thickness of the limit layer and of the whirls is negligible in comparison with the stream cross-section area and consequently, to a good approximation, the motion may be considered as a potential one along the whole range covered by the moving fluid.

In most cases applying to the hydrodynamic study of turbopumps, fluid can be assumed to be in potential motion, since the limit layer thickness is comparatively small and the whirls are avoided (at least in optimal operating conditions) by the very profile of solid bodies in contact with the fluid.

1.3.1 Fluid Potential Motion Around a Profile

For a better understanding of the interaction between the stream and the pump impeller blades, let us consider the fluid motion around a profile.

It was experimentally demonstrated that a profile placed in a plane-parallel stream experienced two forces, being the resistance force T parallel to the stream direction, and lift P, normal to the stream direction. Assuming a *perfect* fluid, free of whirls, computations lead to D'Alembert's well known paradox, according to which both the lift and the resistance are nil. Consequently, if it is desired to explain and estimate the resistance and lift forces, real fluids should be considered instead.

With real fluids, a limit layer is developed at the profile surface wherein the frictional forces are high and the motion is rotative (with whirls). The resistance force T is given by the sum of all frictional forces at the contact surface, while the lift P is given by the sum of all whirls produced inside the limit layer (Fig. 1.23).

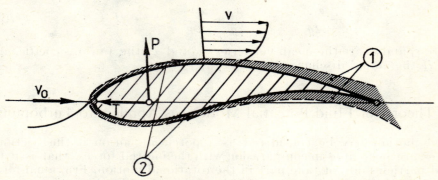

Fig. 1.23. Profile flow diagram within the profile zone area:
1 — limit layer; *2* — frictional force

The method of Jukowski [1.4] has the merit of substituting for the sum of effects of all the whirls produced inside the limit layer (and thus within the range covered by the moving fluid) the effect of one single whirl, placed inside the profile (and thus outside the range covered by the moving fluid). In this way he was able to maintain the assumption of the ideal fluid, and it was possible to develop the whole theory of interaction between profiles and the fluid in potential motion.

The profile outline is a stream line, and the fluid motion around the profile resulting by summing up a rectilinear, parallel potential motion and a whirl motion whose centre is inside the profile is called *the potential motion with circulation* „Γ".

The potential motion with fluid circulation around any profile may be obtained by isogonal transformation of the potential motion with circulation around a circular cylinder. There are, therefore, two problems to be solved, that:

— of the potential motion with fluid circulation around a cylinder of a circular cross-section, and that

— of the isogonal transformation of the cylinder into any profile.

The potential motion with fluid circulation around a cylinder is described by the complex potential

$$f(z) = v_0 \left(z + \frac{R^2}{z} \right) + \frac{\Gamma}{2\pi i} \ln \frac{z}{R} \tag{1.137}$$

where R is the radius of the cylinder with centre at the origin of the co-ordinate axes in plane z; v_0 is the velocity, at infinity, of the fluid, subtending at the Ox axis an angle $\alpha = \arctan \dfrac{v_{0y}}{v_{0x}}$; while Γ is the circulation around the cylinder.

This motion results from the superposition of three motions: a plane parallel one, at an angle $\alpha = \arctan \dfrac{v_{0y}}{v_{0x}}$ with respect to axis Ox which is defined by the complex potential $f_1(z) = v_0 z$; a motion described by the complex potential $f_2(z) = v_0 \dfrac{R^2}{z}$ and due to a double source; and a motion due to a whirl, the complex potential of which is $f_3(z) = \dfrac{\Gamma}{2\pi i} \ln \dfrac{z}{R}$.

Velocity circulation Γ along the closed curve C is given by integral

$$\Gamma = \oint_C \vec{v} \times \mathrm{d}\vec{r} = \oint_C v_s \mathrm{d}s \qquad (1.138)$$

and represents the mechanical work of the velocity vector along the closed contour C. The projection of this vector on the tangent at curve C is denoted by v_s, while $\mathrm{d}s$ is the elemental arc of curve C.

Of course, with plane motions, $\vec{v} = v_x \vec{i} + v_y \vec{j}$ and $\mathrm{d}\vec{r} = \vec{i}\,\mathrm{d}x + \vec{j}\,\mathrm{d}y$, and relation (1.138) becomes

$$\Gamma = \oint_C v_x \mathrm{d}y + v_y \mathrm{d}x \qquad (1.139)$$

For the complex potential (1.137), the critical points where velocity is zero result from equation

$$\frac{\mathrm{d}f}{\mathrm{d}z} = v_0 \left(1 - \frac{R^2}{z^2}\right) + \frac{\Gamma}{2\pi i z} = 0 \qquad (1.140)$$

from which

$$z = \left| \frac{\Gamma}{4\pi v_0} \right| i \pm \sqrt{R^2 - \frac{\Gamma^2}{16\pi^2 v_0^2}} \qquad (1.141)$$

If $\dfrac{\Gamma}{4\pi v_0} \leqslant R$, it follows that $|z| = R$ and the two points are

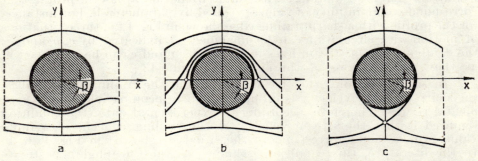

Fig. 1.24 Potential motion with circulation around a cylinder (following Jukowski).

distinct and placed on the cylinder (Fig. 1.24,*a*). The argument of the critical point is obtained from the relation

$$\sin \beta = \frac{\Gamma}{4\pi R v_0} \qquad (1.142)$$

For $\Gamma = 0$, the two critical points are situated at the intersection of the cylinder with axis Ox (Fig. 1.24,*b*). For $\frac{\Gamma}{4\pi v_0} > R$, one of the critical points is placed outside the cylinder (Fig. 1.24,*c*).

Therefore, an infinity of possible motions is obtained for the circulation of a fluid around a cylinder of radius R, as a function of quantity Γ.

Let us assume a contour C in plane z (Fig. 1.25), and further that the function which gives an isogonal representation of the outside of contour C in plane z, external to the circle of radius R in plane t, is $t(z)$. The complex potential of the fluid motion around contour C is given by the function.

$$f(z) = v_0 \left[t(z) + \frac{R^2}{t(z)} \right] + \frac{\Gamma}{2\pi i} l_n t(z) \qquad (1.143)$$

By means of function $t(z)$, the stream lines in plane z around contour C are transformed into stream lines in plane t around the circle of radius R.

Point $t = \infty$ is common to all stream lines. Putting the condition that $t(\infty) = \infty$, we obtain in plane z a motion that maintains this property. If we further put the condition that $\frac{dt(\infty)}{dz} = 1$, then it follows that at infinity, the velocity of the stream will be

$$\frac{df}{dz} = \left| \frac{df}{dt} \right|_{\substack{z=\infty \\ t=\infty}} \times \frac{dt}{dz}_{z=\infty} = v_0 \times 1 = v_0 \qquad (1.144)$$

Hence, function $f(z)$, as derived from relation (1.141), leads to an infinity of potential motions, with fluid circulation around a profile that corresponds to an infinity of values yielded by parameter Γ. For instance, for the profile with a sharp trailing edge shown in Fig. 1.26, three situations of motion with circulation can be imagined as indicated by *a*, *b* and *c*; the configuration of stream lines in situations *a* and *c* can be most varied, as they are functions of the circulation value Γ.

In such situations, the stream goes round the trailing edge, and at the sharp peak the velocity is infinite, which is incompatible with experimental results. Therefore, according to experimental data, the condition required is that the stream velocity at the trailing edge be of a finite value, which would correspond to the situation shown in Fig. 1.26, *b*. In the literature, this condition is known as the Cheaplighin postulate [1.2], [1.4].

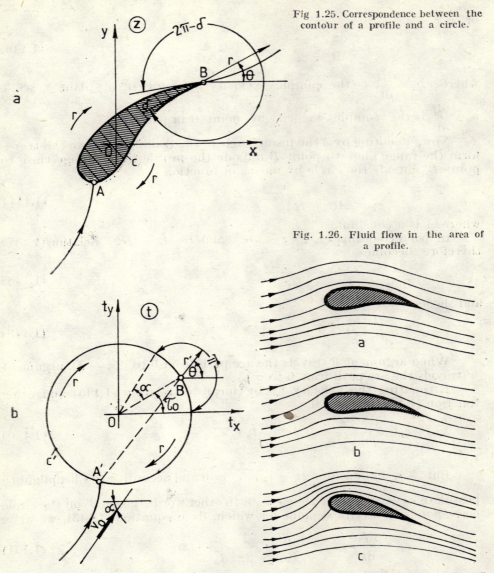

Fig 1.25. Correspondence between the contour of a profile and a circle.

Fig. 1.26. Fluid flow in the area of a profile.

Denoting by B the point that defines the trailing edge of profile C in plane z (see Fig. 1.25), and by B' the corresponding point on circle C' resulting from an isogonal transformation of profile C of the form

$$t = t(z)$$

we can write

$$\left.\frac{df}{dt}\right|_{B'} = \left.\frac{df}{dz}\right|_{B} \times \frac{dz}{dt} \qquad (1.145)$$

or
$$v_{B'} = v_B \cdot \frac{dz}{dt} \tag{1.146}$$

where $v_{B'} = \frac{df}{dt}$ is the complex velocity at point B' in plane t, while $v_B = \frac{df}{dz}$ is the complex velocity at point B in plane z.

Now, denoting by δ the inside angle of the trailing edge, we can transform the range close to point B outside the profile into a range close to point B' outside the circle by means of function

$$z - z_B = A(t - t_{B'})^{\frac{2\pi - \delta}{\pi}} \tag{1.147}$$

where A is a constant.

In polar co-ordinates, $z - z_B = re^{i\theta}$ and $t - t_{B'} = re^{i\theta'}$. Relation (1.147) therefore becomes

$$re^{i\theta} = Ar'^{\frac{2\pi - \delta}{\pi}} \times e^{i\frac{2\pi - \delta}{\pi}\theta'} \tag{1.148}$$

and hence

$$\theta = \frac{2\pi - \delta}{\pi}\theta' \tag{1.149}$$

When argument θ travels the arc path from θ to $(2\pi - \delta)$, argument θ' travels the arc path from 0 to π.

Estimating the velocity $v_{B'}$ by means of relation (1.145) and using function (1.147), results in

$$v_{B'} = v_B \frac{2\pi - \delta}{\pi} A(t - t_{B'})^{\frac{2\pi - \delta}{\delta}} \tag{1.150}$$

But $\delta < \pi$ and $\lim_{t \to t_{n'}} (t - t_{B'})^{\frac{\pi - \delta}{\pi}} = 0$ and according to Cheaplighin's postulate $v_B \neq \infty$, and hence $v_{B'} = 0$. In other words, point B' on the circle corresponds to a critical point for which, from equation (1.143), we write

$$v_B = \frac{df}{dt} = v_0 - \frac{v_0 R^2}{t_{B'}^2} + \frac{\Gamma}{2\pi i} \frac{1}{t_{B'}} = 0 \tag{1.151}$$

Now, putting

$$t_{E'} = Re^{i\tau_0} \;;\; v_0 = |v_0|e^{i\alpha} \tag{1.152}$$

where τ_0 is the argument of point B' while α is the angle subtended by the velocity at infinity with axis Oz or with axis Ot (the incidence angle) results in

$$\Gamma = -4\pi R |v_0| \frac{e^{i(\alpha - \tau_0)} - e^{-i(\alpha - \tau_0)}}{2i}$$

or

$$\Gamma = -4\pi R |v_0| \sin(\alpha - \tau_0) \tag{1.153}$$

As shown in Fig. 1.26, $\alpha > \tau_0$, since straight line $A'B'$ is parallel with the velocity of the stream at infinity. Therefore, $\Gamma < 0$. If $\alpha = \tau_0$, then $\Gamma = 0$ and the axis thus defined is called the *axis of nil circulation*. Putting $\beta = \alpha - \tau_0$, where is the angle between the stream direction at infinity and radius OB', relation (1.153) can also be written as

$$\Gamma = -4\pi R |v_0| \sin \beta \tag{1.154}$$

It is worth mentioning that the circulation value around a profile does not change through an isogonal transformation. Indeed, by definition, circulation Γ has the expression (1.138) or (1.139).

$$\Gamma = \oint_C v_s \, ds = \oint_C v_x dx + v_y dy$$

Starting from relation (1.139), circulation Γ can be also written as

$$\Gamma = \operatorname{Re} \oint_C f'(z) \, dz = \operatorname{Re} \oint_C (v_x - iv_y)(d_x + id_y) =$$
$$= \operatorname{Re} \oint_C v_x dx + v_y dy + i(v_x dy - v_y dx) = \oint_C v_x ds \tag{1.155}$$

However, on a closed curve surrounding the profile, the flow is defined by the integral: $\oint_C v_x dy - v_y d_x = \oint_C d\psi = 0$ and therefore

$$\Gamma = \oint_C f'(z) \, dz \tag{1.155/1}$$

Substituting for t the variable $t(z)$, that is the variable by means of which the motion around the contour profile is transformed into curve C' in plane t, integral (1.155/1) becomes

$$\Gamma = \oint_C f'_z[z(t)] \frac{dz}{dt} dt = \oint_{C'} f'_t(t) \, dt \tag{1.156}$$

because

$$f'_z[z(t)] = f'_t \times \frac{dt}{dz}$$

At the same time, it is important to note that, inasmuch as $f'(z)$ is an analytical function both outside the profile and along its contour, curve C or C' can take any shape.

1.3.2 Estimation of the Fluid Pressure Force on the Profile

Let us consider the motion of a fluid around a profile C the velocity of this motion at infinity is v_0, and is directed along axis Ox (Fig. 1.27). Neglecting the fluid weight, the pressure at any point can be estimated with relation

$$p = A - \frac{\rho}{2}|v|^2 \qquad (1.157)$$

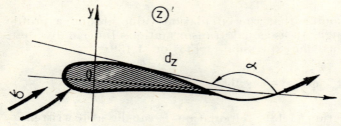

Fig. 1.27. Diagrammatic representation for estimating the pressure force exerted by the fluid on the profile

where A is a constant, ρ is the specific mass, and $|v|$ is the modulus of the stream velocity at the point considered.

Since the pressure is normal to the profile outline, the pressure force on the length section dz will have the expression

$$dP = \left(A - \frac{\rho}{2}|v|^2\right) i \, dz \qquad (1.158)$$

(vector $i \times dz$ is normal to vector dz).

The pressure force, active along the whole contour C, has the expression

$$P = \oint_C \left(A - \frac{\rho}{2}|v|^2\right) i \, dz = -\frac{i\rho}{2}\oint_C |v|^2 \, dz \qquad (1.159)$$

because $\int_C A \, dz = 0$ (contour C being a closed one).

If $f(z)$ is the complex potential of the fluid motion around profile C, then the complex velocity will be $w = v_x - iv_y = f'(z)$, the velocity modulus will be $|v| = f'(z)$, and the velocity vector will be $\vec{v} = \overline{f'(z)}$, where $\overline{f'(z)}$ is the conjugate of $f'(z)$. Hence

$$P = \frac{i\rho}{2}\oint_C |f'(z)|^2 \, dz \qquad (1.160)$$

Inasmuch as along contour C the velocity is directed along the tangent to C, we can write

$$\text{Arg } dz = \text{Arg } \overline{f'(z)} = \alpha$$

and
$$\text{Arg } f'(z) = -\alpha$$

It follows therefore
$$\text{Arg } \{[f'(z)]^2 \, dz\} = -2\alpha + \alpha = -\alpha$$
and
$$\text{Arg } |f'(z)| \, dz = \alpha$$

Therefore, in any point on the C contour quantities $|f'(z)|^2 dz$ and $[f'(z)]^2 dz$ are conjugate, having the same modulus, while the arguments are equal but of opposite sign.

By passing in equation (1.160) to conjugates, one gets the first formula of Blasius—Cheaplighin

$$\bar{P} = \frac{i\rho}{2} \int_C [f'(z)]^2 \, dz \tag{1.161}$$

Likewise, the resulting moment of pressure forces about the origin is derived by estimating first the elemental moment

$$dM = \text{Re } (iz \, dP)$$

then, by integration

$$M = -\text{Re } \frac{\rho}{2} \int_C z \left(\frac{df}{dz}\right)^2 dz \tag{1.161/1}$$

which is the second equation of Blasius—Cheaplighin.

Function $f'(z)$ is analytical and can be developed as a series around point $z = \infty$

$$f'(z) = v_0 + \frac{c_{-1}}{z} + \frac{c_{-2}}{z^2} + \ldots \tag{1.162}$$

where $v_0 = v_{x0} - iv_{y0}$ is the complex velocity at infinity.

Choosing a contour $\mathscr{L}$ which encloses contour C of the profile, we can write

$$\oint_{\mathscr{L}} f'(z) dz = \oint_{\mathscr{L}} (v_x - iv_y)(dx + idy) =$$
$$= \oint_{\mathscr{L}} v_x dx + v_y dy + i \oint_{\mathscr{L}} v_x dy - v_y dx = \Gamma + iQ \tag{1.163}$$

where Γ is the circulation on contour C, while $Q = 0$ is the flow of the velocity vector through contour C.

On the other hand, from relation (1.161) it follows

$$c_{-1} = \frac{1}{2\pi i} f'(z) \, dz \tag{1.164}$$

and, comparing (1.163) with (1.164), results in

$$c_{-1} = \frac{\Gamma}{2\pi i} \qquad (1.165)$$

Squaring relation (1.162) and adequately grouping the terms, one can write

$$[f'(z)]^2 = v_0^2 + \frac{2v_0 c_{-1}}{z} + \frac{c_{-1}^2 + 2v_0 c_{-2}}{z^2} + \ldots \qquad (1.166)$$

and hence it follows that

$$\int [f'(z)]^2 dz = 2v_0 c_{-1} \times 2\pi i = 2v_0 \Gamma \qquad (1.167)$$

The first Blasius—Cheaplighin equation becomes

$$\overline{P} = \rho i v_0 \Gamma = i\rho (v_{x0} - i v_{y0}) \Gamma \qquad (1.168)$$

or else, passing to conjugate quantities, we get Jukowski's theorem

$$\vec{P} = P_x + i P_y = -i\rho (v_{x0} + i v_{y0}) \Gamma = -i\rho \vec{v}_0 \Gamma \qquad (1.169)$$

where $\vec{v}_0$ is the velocity vector.

From relation (1.169) it follows that *if* $\Gamma < 0$, i.e. if the circulation is in a clockwise direction, the direction of the pressure force is obtained by turning vector v_0 by an angle $\frac{\pi}{2}$ in an anti-clockwise direction; and *if* $\Gamma > 0$, the direction of force P is found by turning vector $\vec{v}_0$ by an angle $\frac{\pi}{2}$ in a clockwise direction.

But the pressure force P is actually the lift, so according to Jukowski's theorem, the amount of lift for a certain profile is given by product $\rho \cdot v_0 \Gamma$, while the direction is found by turning vector $\vec{v}_0$ by an angle $\frac{\pi}{2}$, in a direction opposite to the circulation.

Circulation Γ is given in relation (1.153), and consequently one can write for the lift

$$P = 4\pi \rho\, v_0^2 R \sin(\alpha - \tau_0) \qquad (1.170)$$

or

$$P = 4\pi \rho\, v_0^2 R \sin \beta \qquad (1.170/1)$$

where α is the angle subtended by the velocity direction at infinity with the Ox axis (the angle of incidence); τ_0 is the argument of the critical point B'; and $\beta = \alpha - \tau_0$ is the angle between the velocity direction at infinity and the direction of the vector radius of the critical point (Fig. 1.25).

As was to be expected for an ideal fluid, the motion with circulation around a profile only permits the determination of lift. For the resistance force, parallel with the stream velocity, the zero value is found, according to the above-mentioned D'Alembert principle.

For the lift and the resistance, it is a common practice to use the expressions

$$P = \frac{\rho}{2} C_y v_0^2 \tau b \qquad (1.171)$$

$$\Gamma = \frac{\rho}{2} C_x v_0^2 \tau b \qquad (1.172)$$

analogous to (1.115) and (1.116) respectively, where C_y and C_x are the lift and resistance coefficients respectively, τ is the profile chord, and b is the profile span. Coefficient C_y may be derived theoretically (from

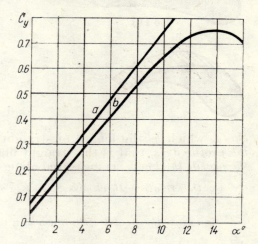

Fig. 1.28. Diagrammatic representation of function $Cy = f(\alpha)$:
  a — theoretical; b — experimental.

Jukowski's formula for profiles with infinite spread), or else by experimental means. Both C_y and C_x depend on the incidence angle α, that is the angle subtended by the flow velocity at infinity with the profile axis. For each value of the angle, the points in the $C_x - C_y$ co-ordinate plane describe a curve. Figure 1.28 is an illustration of the $C_y = f(\alpha)$ diagram, established both experimentally and through calculations. As can be seen, the lift determined theoretically is higher than that determined experimentally, as also corroborated by relation (1.114).

1.3.3 Typical Profiles and Geometrical Features

Any profile is characterized by a *leading edge* (the rounded, front side) and a *trailing edge* (the sharp, rear side). According to their trailing edges, profiles fall into the following classification:
  — sharp-tip, or Jukowski profiles (Fig. 1.29, *a*);
  — dihedral-tip, or Kármán—Trefftz profiles, (Fig. 1.29, *b*);
  — round-tip, or Carafoli profiles (Fig. 1.29, *c*).

The upper side of the profile is called the back of the blade (extrados), while the lower side is called the underside of the blade (or intrados). Depending on their upper and lower side curvatures, profiles can take a variety of geometrical shapes. To help in characterizing the geometrical shape of the profile, the following concepts are introduced: *the profile*;

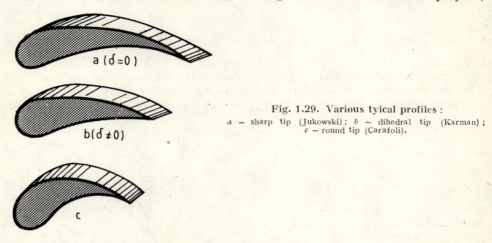

Fig. 1.29. Various tyical profiles:
a — sharp tip (Jukowski); *b* — dihedral tip (Karman); *c* — round tip (Carafoli).

chord, which is the length of straight segment AB and is denoted by l *the profile axis*, that is the straight line passing through points A and B; *the profile thickness*, as measured along the normal to the axis and denoted by e; *the relative thickness*, $\dfrac{e}{l} = \varepsilon$; *the maximum relative thickness* $\dfrac{e_{max}}{l} = \varepsilon_{max}$; *the profile framework*, i. e. the mean thickness line (Fig.1.30). As a common practice, extrados and intrados lines are given in a co-ordinate system.

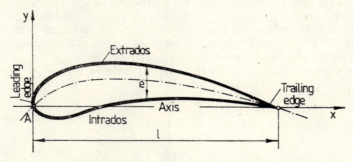

Fig. 1.30. Geometrical elements of a profile.

In any study of the various typical profiles, either the direct or the reverse problem has to be solved. The direct problem consists of the determination of the complex potential of motion for a given profile, while the reverse problem consists of the determination of the profile according to a complex potential function.

The direct problem is the most important in practice but causes considerable difficulties owing to the very intricate calculations involved. In the literature, various methods are suggested for the solution of the direct problem. It seems that numerical methods based on electronic computers offer the best prospects.

The reverse problem is easier. It is solved either by direct isogonal transformations, more accessible to engineering computations, or by integral equations in the complex range.

In what follows, the author will illustrate the method of isogonal transformations for the Jukowski profile and will derive the complex potential functions for the Kármán—Trefftz and Carafoli profiles. The Jukowski profile is obtained by transforming circle K' in plane t, into contour K in plane z, by means of function

$$z = \left(t + \frac{a^2}{t}\right) \tag{1.173}$$

where a is a constant whose meaning is readily apparent.

Consider in plane t a circle, called the basic circle, that intersects axis Ot_x in two points, at distances $-a$ and $+a$ respectively, from the origin. In Fig. 1.31,b, the centre of the basic circle is on the Ot_x axis. Function (1.173) will transform this circle from plane t into a segment of straight line with its ends at points $-2a$ and $+2a$ respectively, in plane z (Fig. 1.31,a). Indeed, by putting $t = ae^{i\theta}$, one finds

$$z = a(e^{i\theta} + e^{-i\theta}) = 2a \cos \theta \tag{1.174}$$

and consequently: $Im\, z = 0$.

A circle C_1' in plane t, concentric with the basic circle, corresponds in plane z to an ellipse C_1, having the focal points at $-2a$ and $+2a$. Indeed, putting $t = re^{i\theta}$, $(r > a)$, results in

$$z = \left(re^{i\theta} + \frac{a^2}{r}e^{-i\theta}\right)$$

whence

$$x = \left(r + \frac{a^2}{r}\right) \cos \theta$$

$$y = \left(r - \frac{a^2}{r}\right) \sin \theta$$

$$\frac{x^2}{\left(r + \frac{a^2}{r}\right)^2} + \frac{y^2}{\left(r - \frac{a^2}{r}\right)^2} = 1$$

The Jukowski profile is obtained in plane z by transforming, with the help of function (1.173), a circle K' (called the generating circle), in plane t,

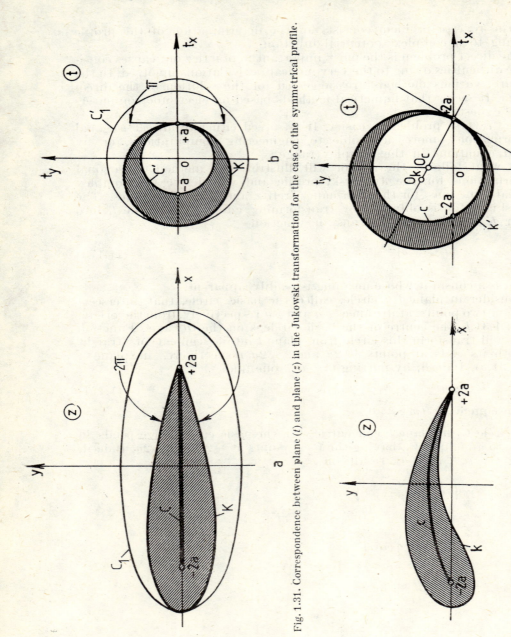

Fig. 1.31. Correspondence between plane (*t*) and plane (*z*) in the Jukowski transformation for the case of the symmetrical profile.

Fig. 1.32. Correspondence between plane (*t*) and plane (*z*) in the Jukowski transformation for the case of an asymmetrical profile.

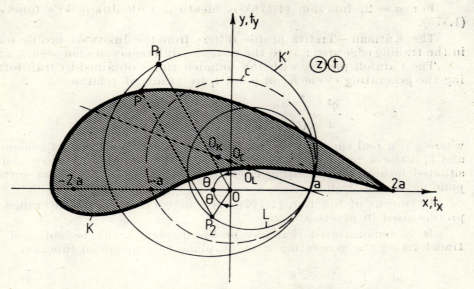

Fig. 1.33. Diagrammatical correspondence of the Jukowski profile.

from having its centre on the abscissa axis and being tangent to circles C' and C'_1. The profile thus obtained is symmetrical to the axis and has the peak angle $\delta = 0$. Indeed, function (1.173) can be also written in the form

$$\frac{z-2a}{z+2a} = \left(\frac{t-a}{t+a}\right)^2 \tag{1.175}$$

from which it follows that in points $t = \pm a$, the angles in plane t are doubled in plane z, in points $z = \pm 2a$; thus angle π of the generating circle in point $t = +a$ becomes 2π for contour K in the corresponding point $K'z = +2a$.

The asymmetrical Jukowski's profile is found with the same function (1.173), if the basic circle has its centre on the co-ordinate axis and passes through points $t = \pm a$, while the generating circle K' passes through point $t = \pm a$ and is tangent at the basic circle (Fig. 1.32).

The Jukowski profile can be constructed point by point, either analytically or graphically. The graphical construction is obtained by geometrically summing up transformation $z = t$ and transformation through reversion $z = \dfrac{a^2}{t}$ (Fig. 1.33).

The profile with the dihedral tip (Kármán—Trefftz) is found by applying to the generating circle K' in plane t the transformation

$$\frac{z-\sigma a}{z+\sigma a} = \left(\frac{t-a}{t+a}\right)^\sigma \tag{1.176}$$

with $\sigma = 2 - \dfrac{\delta}{\pi}$, where δ is the peak angle of the profile.

For $\sigma = 2$, function (1.176) is identical with Jukowski's function (1.175).

The Kármán—Trefftz profile differs from the Jukowski profile only in the trailing edge area; as to the rest, the differences are not significant.

The Carafoli profile, with the rounded tip, is obtained by transforming the generating circle K' in point t, by means of relation

$$z = t + \frac{q^2}{t} + \frac{q_n}{(t - \lambda_i)^n} \tag{1.177}$$

where q is a real constant, q_n is a complex constant, n is a real constant, and λ_i is the complex root that cancels out the complex velocity and is situated inside the generating circle, yet very close to the point corresponding to the trailing edge.

By means of function (1.177) one can cover a very wide range of profiles used in practice [1.2].

It is demonstrated that any profile in plane z can be obtained by transforming the generating circle in plane t by means of function

$$z = t + \frac{a_1}{t} + \frac{a_2}{t^2} + \ldots \tag{1.178}$$

But also function (1.176) can be written in the form (1.178)

$$z = t + \frac{\sigma^2 - 1}{3} \frac{q^2}{t} + \ldots \tag{1.179}$$

Denoting by O the origin of co-ordinate axes Oxy for vector OB, the relation resulting from (1.178) can be written

$$\overline{OB} = a + \frac{a_1}{a} + \frac{a_2}{a^2} + \ldots$$

Profiles used in practice are slightly curved and of a comparatively small thickness. In such a case (Fig. 1.25) segment OB represents half of the profile chord and relation (1.178) is limited to the first two terms. Thus

$$\overline{OB} = 2a = \frac{l}{2} \approx 2R$$

where l is the profile chord and R is the radius of the generating circle. Therefore

$$R \approx \frac{1}{4} \; ; \; P = \rho\pi/v_0^2 \sin \beta \approx \rho\pi l \beta v_0^2 \tag{1.180}$$

where it is assumed that $\beta \approx \sin \beta$. Comparing (1.171) with (1.180) results in

$$C_y = 2\pi\beta \tag{1.181}$$

Experimentally, one can get

$$C_y = 5.4 \, \beta \qquad (1.182)$$

according to the graph shown in Fig. 1.28.

1.3.4 The Plane Potential Motion Within the Area of a Profiles Network

Theoretical study of linear and circular profile networks has concerned many hydraulic engineers and mathematicians during the last 50 years or so. Classic contributions to the solution of this problem were made by Romanian scientists as well, including D. Pavel, E. Carafoli, Caius Iacob, I. Anton, S. Hâncu, V. Burchiu etc.

The following presentation, which is by no means exhaustive, illustrates the method of isogonal transformations, as applied to the study of linear [1.8] and circular [1.2], [1.12] profile networks.

Networks of linear profiles. Consider in plane z a network of rectilinear profiles, as illustrated in Fig. 1.34. The fluid motion in the rectilinear network corresponds to the fluid motion along the real axis $O\zeta$ in plane ζ, generated by a system of sources and whirls placed at points ζ_a and ζ_b and in their corresponding points with respect to $O\zeta$, $\bar{\zeta}_a$ and $\bar{\zeta}_b$.

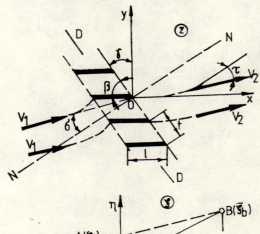

Fig. 1.34. Network of rectilinear profiles in the z plane and its transformation in the ζ plane.

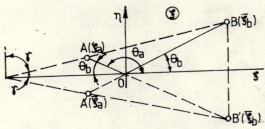

The transformation function has the expression

$$z = \frac{t}{2\pi} \left[e^{i\gamma} \ln \frac{\zeta - \zeta_a}{\zeta - \zeta_b} + e^{-i\gamma} \ln \frac{\zeta - \bar{\zeta}_a}{\zeta - \bar{\zeta}_b} \right] \qquad (1.183)$$

while between the quantities shown in the figure we can write the relationships

$$\gamma = \delta = \frac{\pi}{2} \; ; \; \theta_a + \theta_b = \pi \; ; \; \frac{r_a}{r_b} = \frac{\sin(\theta_b - \delta)}{\sin(\theta_b + \delta_b)} \qquad (1.184)$$

$$\frac{\pi l}{t \cos \gamma}(\pi - 2\theta_b)\operatorname{tg}\gamma = \frac{\sin(\theta_a - \delta)}{\sin(\theta_a + \delta)} \qquad (1.185)$$

It is readily seen that with the notation used in Section 1.2.4, one can write $\beta_1 = \frac{\pi}{2} - \sigma$ and $\beta_2 = \frac{\pi}{2} - \tau$, while the velocity at infinity v_m, is the geometric mean of velocities v_1 and v_2.

The network is characterized by the blade length, l, the pitch t, and the set angle $\gamma = \frac{\pi}{2} - \delta$. From relation (1.184), θ_a and $\frac{r_a}{r_b}$ are found as a function of θ_b, and from relation (1.185) θ_b is found as a function of ratio $\frac{l}{t}$. All the quantities can therefore be expressed as a function of the two parameters γ and $\frac{l}{t}$. Given angle σ, angle τ is found from relation

$$\operatorname{tg}\tau = \operatorname{tg}\sigma + \left(1 - \frac{r_a}{r_b}\right) \frac{\sin(\gamma - \sigma)}{\cos\gamma \times \cos\sigma} \qquad (1.186)$$

and the angle φ subtended by vector $\vec{F}_t$ (the action of the liquid on the blade) with direction NN (Oz as in Section 1.2.4)

$$\cot\varphi = \frac{1}{2}(\operatorname{tg}\tau + \operatorname{tg}\sigma) \qquad (1.187)$$

Velocity v_m results from relation

$$v_m = \frac{\cos\sigma}{\sin\varphi} \qquad (1.188)$$

Component F_u of force F_t along direction $D-D$ of the network for one blade has the expression

$$F_u = \rho \, t(v_1 \cos\sigma)^2 \, (\operatorname{tg}\tau - \operatorname{tg}\sigma) \qquad (1.189)$$

while the value of force F_t has the expression

$$F_t = F_u \sqrt{1 + \frac{1}{4}(\operatorname{tg}\tau + \operatorname{tg}\sigma)^2} \qquad (1.190)$$

Having now the force F_t, one can find the coefficient of unit lift of the blade in the network, C_{yr}, and ratio $k = \frac{C_{yr}}{C_y}$ where C_y is the coefficient of unit lift of the separate blade relation (1.121). Some considerations on the agreement with experimental data were presented in Section 1.2.4.

Networks of circular profiles. Function (1.183) transforms the plane z into plane ζ and the Sörensen function

$$Z = \left[\left(\frac{\zeta - \zeta_a}{\zeta - \zeta_b}\right)\left(\frac{\zeta - \bar{\zeta}_a}{\zeta - \bar{\zeta}_b}\right) e^{-2i\gamma}\right]^{\frac{1}{n}} \tag{1.191}$$

transforms the plane ζ into plane Z. The n spiral blades in plane Z are represented along axis ζ, and the network of plane profiles in plane z is transformed into a network of circular profiles following logarithmic spirals, in plane Z, (Fig. 1.35).

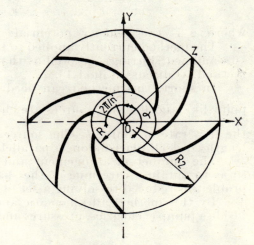

Fig. 1.35. Network of circular profiles.

The complex potential of motion in plane ζ has the expression:

$$f(\zeta) = -\frac{Q}{2\pi n} \ln \frac{(\zeta - \zeta_a)(\zeta - \bar{\zeta}_b)}{(\zeta - \zeta_b)(\zeta - \bar{\zeta}_b)} - \frac{i\Gamma_0}{2\pi} \ln \frac{\zeta - \zeta_a}{\zeta - \zeta_b} -$$

$$- \frac{i(\Gamma_0 + \Gamma)}{2\pi} \ln \frac{\zeta - \bar{\zeta}_b}{\zeta - \zeta_b} +$$

$$+ \frac{1}{2\pi} \int_{-\infty}^{\infty} q(\lambda) \ln(\zeta - \lambda) \, d\lambda \tag{1.192}$$

where Q is the output of the pump with n blades of a span equal to unity, $q(\lambda)$, Γ_0 and Γ; these quantities have to be estimated in the actual operating conditions of the pump.

One author [1.12] suggests the determination of v_{ref}, which is equivalent to the velocity at infinity v_m in the plane profile network and corresponds to the circular profile network

$$v_{ref} = \frac{1}{4\pi R_0} \sqrt{4Q^2 + [2\Gamma_0 + (n-1)\Gamma]^2} \tag{1.193}$$

1.3.5 The Axially-Symmetrical Potential Motion around the Careenage

Neglecting the existence of blades, the liquid motion inside most axial-flow pumps might be considered, with a few acceptable errors, as an axially symmetric potential motion. The pump hub is a closed careenage (streamlined skirting) where $\psi = 0$, and the external tube is the surface where $\psi = Q$, with Q being the pump output.

For the velocity components, one can write the expressions

$$v_z = \frac{\partial \varphi}{\partial z} = \frac{1}{r}\frac{\partial \psi}{\partial r} \ ; \ v_r = \frac{\partial \varphi}{\partial r} = -\frac{1}{r}\frac{\partial \psi}{\partial z} \ ; \ v_\theta = 0 \qquad (1.194)$$

where φ is the velocity potential.

The method currently applied to trace the outline of the hub careenage (streamlined skirting), as well as that of the external tube, is the source method, as discussed in [1.1].

In short, this method can be described as follows. A number of n point-like and line-like sources are chosen so that $\sum_{k=1}^{n} Q_k = 0$, with Q_k being the flow rate of source k. The sources are placed along a straight line of length l, itself placed along a parallel, or other form of stream.

The number n, arangement and intensity Q_k of sources are chosen so as to obtain a careening of the shape and size desired. Generally, the problem is reduced to solving a set of systems of equation s.

By this method all kinematic and dynamic characteristics of stream inside a pump (velocities, pressures and stream lines) may be found.

REFERENCES

1.1 BURCHIU, V. *Contribuții la hidrodinamica și optimizarea parametrilor turbinelor elicoidale*. Doctor Thesis, Bucharest, 1970.
1.2 CARAFOLI, E., OROVEANU, T. *Mecanica fluidelor*. Academy Publishing House, Bucharest, 1952.
1.3 HÂNCU, S. *Modelarea hidraulică in curenți de aer sub presiune*. Academy Publishing House, 1967.
1.4 LOITIANSKY, L. G. *Mehanika jidkosti i gaza*. Moscow, 1950.
1.5 PAVEL, D. *Mașini hidraulice*. Didactic and Pedagogic Publishing House, Bucharest 1955.
1.6 PAVEL, D. HANCU, S. *Utilaje hidromecanice pentru sisteme de imbunătățiri funciare, stații de pompare*. Ceres Publishing House, Bucharest, 1974.
1.7 POPA, O. *Rețele de profile Carafoli*. Doctor Thesis, I. P. Timișoara, 1961.
1.8 PROSKURA, G. F. *Gydrodinamika gydromashin*. Mashinostroenie, Moscow, 1954.
1.9 SISAK, E. *Considerații asupra coeficientului de influență a rețelelor de profile hidrodinamice de curbură redusă*. Papers of the Conference on Hydraulic Machines, I. P. Timișoara, 1964.
1.10 STEPANOFF, A. I. *Pompes centrifuge et pompes hélices*. Dunod Publishing House, Paris, 1961.
1.11 STEFAN, I. MIHALACHE, C. *Elemente de hidraulică și mașini hidraulice*. Military Academy Publishing House, Bucharest, 1970.
1.12 ZIDARU, G. *Contribuții la hidrodinamica rețelelor circulare de profile cu aplicație la turbo-mașini*. Doctor Thesis, Bucharest, 1970.

2

Pump Performance Parameters and Pump Characteristic Curves

2.1 Introductory Concepts

2.1.1 Definition and Classifications

Definitions. Turbopumps are hydrodynamic, swirling-motion machines generating hydraulic energy which, owing to the mechanical energy contributed from outside, is added to the kinetic and potential energy of the liquid flowing through them. Such machines carry the liquid as a result of the whirling energy exchange. Energy transfer occurs through the rotational motion of the liquid driven by the active member of the machine, called *the impeller*. The energetic process takes place in two successive stages. Firstly, due to the rotational motion, a growth in the liquid kinetic energy is produced and secondly, the liquid is passed through several channels of variable cross-sections where the kinetic energy is converted into pressure energy. The body of channels that carries out this process is called *the casing*, and is an integral part of the turbopump. In the dual definition of turbopumps as hydrodynamic, swirling-motion machines, the first part refers to the kinetic-to-pressure energy conversion processes achieved by the machine and the concept of swirling motion refers to the rotational motion and particularly to the fact that the flow of the fluid streaming through the rotor is similar to a whirl.

Classification. All turbopumps are classified as radial-flow, mixed-flow or axial-flow (helical) pumps. These designations denote the liquid flow through the turbopump impeller (during the process of increase in the hydraulic energy), which is, respectively, in radial, diagonal (cross) or axial directions, referred to the axis of rotation. Thus in radial-flow pumps, the liquid flow is normal to the axis and the swirling-energy exchange occurs owing to the centrifugal effects on the liquid, while in axial-flow pumps the liquid flows parallel to the rotational axis and the energy transfer is due to the exclusive action of the swirling energy, manifest in the form of ,,Γ'' flows around the blades. With mixed-flow pumps, which are a compromise between the two above, the liquid flow is in a half-radial direction, or half-axial with respect to the axis of rotation. In principle, all turbopumps include an impeller and a casing both having the same energetic functions, but each of the above three categories of pumps is of a different structural design.

2.1.2 Pressure Units

Pressure can be expressed in relative or absolute units, with differences being caused by the atmospheric pressure and by the height above sea level. For pressure, there are three different kinds of units, namely *atmospheric, relative and absolute*. This can be better understood by looking at Fig. 2.1.

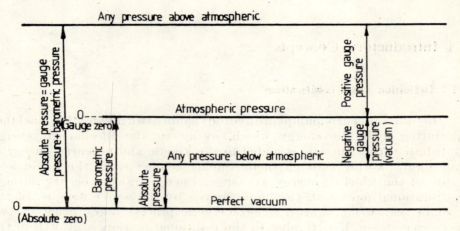

Fig. 2.1. Diagrammatic representation of pressure symbols.

The atmospheric (barometric) pressure, denoted by p_{bar}, is the pressure built up in any place on the surface of the earth by its gaseous envelope. Its value varies with height above sea level (altitude) and local climate. Thus, when turbopumps are mounted at heights well above sea level (taken as a reference level), due consideration should be given to the drop in the atmospheric pressure.

The relative (gauge) pressure, denoted by p_{rel}, is the presure as measured with respect to the atmospheric one. It can take positive values, when it is greater than the atmospheric pressure, or negative values, when it is lower than the atmospheric pressure; in the last case, the relative pressure is called *vacuum or underpressure* (negative gauge pressure). The relative pressure is the most frequently used pressure measuring unit in turbopump applications. Thus, the great majority of pressure measuring and control instruments read either positive or negative pressure values; the zero point on the instrument scale then corresponds to the atmospheric pressure.

The absolute pressure, denoted by p_{abs}, is the pressure above the absolute zero (i.e. perfect vacuum). It therefore represents the sum of atmospheric and relative pressure values

$$p_{abs} = p_{bar} + p_{rel} \tag{2.1}$$

The zero point on the instrument scale corresponds to *absolute* or *perfect vacuum*.

Systems of pressure units. According to the international system of units (SI), pressure p is expressed in $\left[\dfrac{N}{m^2}\right]$, which unit is used in the case of low pressures. In practice, to express high pressure values, multiples of the former units are used, such as $\left[\dfrac{daN}{cm^2}\right]$ and $\left[\dfrac{daN}{mm^2}\right]$. The relations between these units, as expressed in the international system (SI) and in practice, are given in Table 2.1.

Table 2.1 **The relations between pressure units (1 Pa = Pascal = 1 N/m² = 1 kg/m s²)**

Units of measurement	Pa = N/m²	N/mm²	daN/cm²	daN/mm²	kgf/cm²	kgf/mm²
1 N/m²	1	10^{-6}	10^{-5}	10^{-7}	$0.102 \cdot 10^{-4}$	$0.102 \cdot 10^{-6}$
1 N/mm²	1 000 000	1	10	0.1	10.2	0.102
1 daN/cm²	100 000	0.1	1	0.01	1.02	0.0102
1 daN/mm²	10 000 000	10	100	1	102	1.02
1 kgf/cm²	98 000	0.0981	0.981	0.00981	1	1.01
1 kgf/mm²	9 810 000	9.81	98.1	0.981	100	1

2.1.3 Specific Energies and Heads

Consider a streaming liquid particle situated at a height z with respect to a reference plane, moving at a velocity v under a pressure p; this particle will have several typical specific energies by its unit weight (neglecting the friction). All these energies represent the terms included within Bernoulli's equation

$$H = H_p + H_c = \left(z + \frac{p}{\rho g}\right) + \frac{v^2}{2g} = \text{constant} \qquad (2.2)$$

The energetic meaning of terms in the above relation is as follows:

z — the positional specific energy;

$\dfrac{p}{\rho g}$ — pressure specific energy;

$H_p = z + \dfrac{p}{\rho g}$ — potential specific energy;

$H_c = \dfrac{v^2}{2g}$ — kinetic specific energy;

H — total specific energy.

The geometric meaning of the same terms is as follows:

z — geodetic head of the relevant point;

$\dfrac{p}{\rho g}$ — piezometric head;

$H_p = z + \dfrac{p}{\rho g}$ — piezometric pressure;

$H_c = \dfrac{v^2}{2g}$ — kinetic head (velocity head);

H — hydrodynamic head (load).

Equation (2.2) is an energy balance relation expressing the law of energy conservation in the case of an ideal liquid. Therefore, any exchange of specific energy can take place without any change in the total specific energy.

For turbopump casings, when pressure losses are also taken into consideration, equation (2.2) is written

$$\frac{p}{\rho g} + \frac{v^2}{2g} + z + h = \text{constant} \tag{2.3}$$

The first terms are heads (m) of the potential and kinetic pressure, while h represents the head losses.

For the turbopump impeller, $\dfrac{v^2}{2g}$ is replaced by $\dfrac{w^2 - u^2}{2g}$ and thus result relations of the form

$$\frac{w^2 - u^2}{2g} + \frac{p}{\rho g} + z + h = \text{constant} \tag{2.4}$$

$$\frac{w_1^2 - u_1^2}{2g} + \frac{p_1}{\rho g} + z_1 + h_1 = \frac{w_2^2 - u_2^2}{2g} + \frac{p_2}{\rho g} + z_2 + h_2 \tag{2.5}$$

2.1.4. Unit Quantities

For turbopumps, the unit quantities (reduced to a head of 1 m, an impeller diameter of 1 m and 1 HP) are as follows:

$n_q = nQ^{0.5}/H^{0.5}$ — the specific speed;
$Q'_1 = Q/D^2\sqrt{H}$ — unit flow rate;
$P'_1 = P/D^2 H^{3/2}$ — unit power;
$n'_1 = nD/\sqrt{H}$ — unit speed.

Out of these four quantities, specific speed n_q is indicative of the pump type and represents the main classification criterion (see Table 2.2).

Table 2.2 The classification of turbopumps as a function of specific speed

Pump type	n_q	$n_s \approx 4q$	k
Vane-flow field	4—12	16—48	0.08—0.23
Radial-flow field	8—45	32—180	0.15—0.85
Mixed-flow field	40—160	160—540	0.75—1.90
Axial-flow field	100—300	400—1200	1.90—5.60

In this table:

$$n_q = n \times \frac{Q_{opt}^{1/2}}{H_{opt}^{3/4}} \; ; \quad n_s = n \frac{Q_{opt}^{1/2}}{H_{opt}^{5/4}} \; ; \quad k = 2\pi \times \frac{n \times Q_{opt}^{1/2}}{(g \times H_{opt})^{3/4}}.$$

2.2 Main Performance Parameters of Turbopumps

2.2.1 Pump Capacity

Pump capacity (Q), or flow rate, is the liquid volume flowing through the pump per unit time

$$Q = \frac{V}{t} \tag{2.6}$$

As a rule, high flow rates are measured in m³/s or cu m³/day, and low rates in dm³/s or m²/h. The conversion from one measuring unit to another is given in Table 2.3.

Table 2.3.a **The relations between flow rate units as expressed in : m³/h, dm³/s, dm³/min, m³/s**

Flow rate	m³/h	dm³/s	dm³/min	m³/s
1 m³/h	1	0.278	16.67	2.778×10^{-4}
1 (dm³/s)	3.600	1	60	1×10^{-3}
1 (dm³/min)	0.060	0.01667	1	1.667×10^{-5}

Table 2.3.b **The relations between flow rate units as expressed in : dm³/s, m³/h, m³/day**

dm³/s	m³/h	m³/day	m³/h	m³/day	dm³/s	m³/day	dm³/s	m³/h
1	2	3	4	5	6	7	8	9
1	3.6	86	1	24	0.28	1	0.0115	0.042
2	7.2	173	2	48	0.56	2	0.0231	0.083
3	10.8	259	3	72	0.83	3	0.0347	0.125
4	14.4	346	4	96	1.11	4	0.0462	0.167
5	18.0	432	5	120	1.39	5	0.0578	0.208
6	21.6	518	6	144	1.67	6	0.0694	0.250
7	25.2	605	7	168	1.94	7	0.0810	0.292
8	28.8	691	8	192	2.22	8	0.0925	0.333
9	32.4	778	9	216	2.50	9	0.1041	0.375
10	36.0	864	10	240	2.78	10	0.1157	0.417
12	43.2	1 037	12	288	3.33	12	0.1388	0.500
14	50.4	1 210	14	336	3.89	14	7.1620	0.583
16	57.6	1 382	16	384	4.44	16	0.1851	0.667
18	64.8	1 555	18	432	5.00	18	0.2083	0.750
20	72.0	1 720	20	480	5.56	20	0.2314	0.833
25	90.0	2 160	25	600	6.94	25	0.2893	1.052
30	108.0	2 592	30	720	8.33	30	0.3472	1.250
35	126.0	3 024	35	840	9.72	35	0.4051	1.458
40	144.0	3 456	40	960	11.11	40	0.4629	1.667
45	162.0	3 888	45	1 080	12.50	45	0.5208	1.875
50	180.0	4 320	50	1 200	13.89	50	0.5787	2.083
55	198.0	4 752	55	1 320	15.28	55	0.6365	2.292
60	216.0	5 184	60	1 440	16.67	60	0.6944	2.500
65	234.0	5 616	65	1 560	18.06	65	0.7523	2.708

Table 2.3,b Continuation

1	2	3	4	5	6	7	8	9
70	252.0	6 048	70	1 680	19.44	70	0.8101	2.917
75	270.0	6 480	75	1 800	20.83	75	0.8680	3.125
80	288.0	6 912	80	1 920	22.22	80	0.9259	3.333
85	306.0	7 344	85	2 040	23.61	85	0.9838	3.542
90	324.0	7 776	90	2 160	25.00	90	1.0416	3.750
95	342.0	8 208	95	2 280	26.39	95	1.0995	3.958
100	360.0	8 640	100	2 400	27.78	100	1.1574	4.167
110	396.0	9 504	110	2 640	30.56	110	1.2731	4.583
120	432.0	10 368	120	2 880	33.33	120	1.3888	5.000
130	468.0	11 232	130	3 120	36.11	130	1.5046	5.417
140	504.0	12 096	140	3 360	38.89	140	1.6203	5.839
150	540.0	12 960	150	3 600	41.67	150	1.7361	6.250
160	576.0	13 824	160	3 840	44.44	160	1.8518	6.667
170	612.0	14 688	170	4 080	47.22	170	1.9675	7.083
180	648.0	15 532	180	4 320	50.00	180	2.0833	7.500
190	684.0	16 416	190	4 560	52.78	190	2.1990	7.917
200	720.0	17 280	200	4 800	55.56	200	2.3148	8.333
300	1 080.0	25 920	300	7 200	83.33	300	3.4722	12.500
400	1 440.0	34 560	400	9 600	111.11	400	4.6296	16.667
500	1 800.0	43 200	500	12 000	138.89	500	5.7870	20.833
600	2 160.6	51 840	600	14 400	166.67	600	6.9444	25.000
700	2 520.0	60 480	700	16 800	194.44	700	8.1018	29.167
800	2 880.0	69 120	800	19 200	222.22	800	9.2592	33.333
900	3 240.0	77 760	900	21 600	250.00	900	10.4166	37.500
1 000	3 600.0	86 400	1 000	24 000	277.78	1 000	11.5740	41.667

2.2.2 Pumping Head

Pumping head (H), or the pump load, represents the increase in useful energy content ($H = E_{ds} - E_{sc}$) of the pumped liquid, as referred to the unit weight $\left[\dfrac{\text{daNm}}{\text{daN}}\right]$; hence the measuring unit will be the meter.

Equation of the pumping head. To establish the pumping head of a turbopump, only the head (pressure) losses within it are taken into consideration, that is those produced from the suction to the discharge connections. Hydraulic losses in the pipes external to the pump are ignored because they have no relation whatever with the pumping head, in much the same way as there is no relationship whatever between the voltage across an electric generator and the voltage drops in the external leads connected to it.

According to Fig. 2.2 and bearing in mind Bernoulli's equation, on the suction side, the energy of one Newton of liquid is

$$E_{sc} = z_{sc} + \frac{p_{sc}}{\rho g} + \frac{v_{sc}^2}{2g} \tag{2.7}$$

Likewise, on the discharge side, the energy will be

$$E_{ds} = z_{ds} + \frac{p_{ds}}{\rho g} + \frac{v_{ds}^2}{2g} \tag{2.8}$$

Subtracting, and noting $z = z_{ds} - z_{sc}$, the equation for the pumping head will result

$$H = E_{ds} - E_{sc} = z + \frac{p_{ds} - p_{sc}}{\rho g} + \frac{v_{ds}^2 - v_{sc}^2}{2g} \qquad (2.9)$$

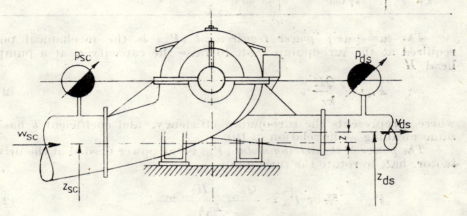

Fig. 2.2. Notation for deriving the pumping head.

where p_{sc} — the suction pressure as measured with respect to the pump shaft (if the pump shaft is above the suction head term $-p_{sc}$ becomes $+p_{sc}$);

p_{ds} — the discharge pressure, as measured with respect to the same reference plane;

v_{sc}, v_{ds} — velocity of the liquid through the suction and discharge connections respectively, as measured at the points where the pressure gauges are connected to the line.

2.2.3 Pump Power

Pump power (P) represents the mechanical work effected by such a machine in unit time $\left[\dfrac{\text{daNm}}{\text{s}}\right]$ to raise the liquid volume corresponding to capacity $Q[\text{m}^3/\text{s}]$ at a pumping head $H[\text{m}]$.

There are three distinct power concepts, as follows:

The turbopump useful power (P_{us}) represents the hydraulic power transmitted to the cycled fluid as it passes through the turbopump impeller

$$P_{us} = \frac{\rho Q H}{k} \qquad (2.10)$$

where k is a coefficient whose value is given in Table 2.4 and depends on the units of measurement chosen to express quantities ρ, Q, H and P.

Table 2.4 **The value of coefficient k**

P	ρ	Q	H	k
HP	kg/dm³	l/s	m	75
kW	kg/dm³	l/s	m	102
HP	kg/dm³	m³/h	m	270
kW	kg/dm³	m³/h	m	367

The turbopump power requirement (P_{rq}) is the mechanical power required at the turbopump shaft to raise the capacity Q at a pumping head H

$$P_{rq} = \frac{\rho QH}{k\eta} \tag{2.11}$$

where η represents the turbopump efficiency, and coefficient k has the same meaning as in relation (2.10).

The power of the driving motor (P_m) is the power needed at the driving motor shaft to rotate the turbopump

$$P_m = \frac{P_{rq}}{\eta'} \text{ or } P_m = \frac{Q_{P_{rq\,max}} H_{P_{rq\,max}}}{k\eta\eta'} k_1 \tag{2.12}$$

where η' is the speed reducer (gear) efficiency; when there is direct coupling between turbopump and the driving motor $\eta' = 1$:

$Q_{P_{rq\,max}}, H_{P_{rq\,max}}$ — the pumping capacity or flow rate and head, respectively, corresponding to the highest power requirements, taken from the performance characteristic curves of the turbopump, in actual operating conditions; as a rule, with radial and axial-flow pumps, the maximum power requirements (as taken from the characteristic curve $P(Q)$ of the turbopump) will be found at the boundaries of the flow rate range, ΔQ:

k_1 — a practical safety coefficient (see Table 2.5), introduced to take care of the increase in driving motor power which can occur in order to avoid overloads; this coefficient arises from inaccuracy in the characteristic curves for the turbopump and pumping network, and from additional power losses caused by defective seals at the turbopump shaft, the reducer gear, or the physical properties of the cycled liquid, etc.

Table 2.5 **The safety coefficient k_1 as a function of the motor power P_m**

P_m, kW	1.5	1.5—4	4—7.5	7.5—40	40—100	100—200	≥ 200
k_1	1.50	1.25	1.20	1.15	1.10	1.08	1.05

2.2.4 Pump Efficiency

The hydraulic and mechanical performance of a turbopump is expressed by its efficiency, which is defined as the ratio of the useful power (P_{us}) to the power requirements (P_{rq})

$$\eta = \frac{P_{us}}{P_{rq}} = \frac{\rho Q H}{k P_{rq}} \tag{2.13}$$

where k takes a value shown in Table (2.4), depending on the units chosen to express parameters ρ, Q, H and P, while $\eta = \eta_h \eta_v \eta_m = 0.70 - 0.92$ represents the global or gross efficiency of the turbopump, i.e. the hydraulic efficiency (η_h) $\times$ the volume efficiency (η_v) $\times$ the mechanical efficiency (η_m).

The hydraulic efficiency $\eta_h = \Delta H/H = 0.72 - 0.95$ includes hydraulic losses within the pump, caused by the friction of the liquid against its mechanical components.

The volume efficiency $\Delta Q/Q = 0.90 - 0.99$ represents flow-rate losses caused because not all of the cycled liquid reaches the discharge side of the pump ($\Delta Q = Q_{sc} - Q$).

The mechanical efficiency $\eta_m = \dfrac{\Delta P}{P} = 0.85 - 0.99$ includes losses caused by mechanical friction that may occur within the turbopump.

Global efficiency enables users to compare turbopump performances, while the partial efficiencies are used by constructors in design studies.

2.3 Similitude Relationships Applied to Turbopumps

2.3.1 Similitude Concepts of Turbomachine Theory

Water motion within a hydraulic machine is of a complex character, and thus the integration of equations describing the fluid motion is not possible. One has resort instead to simplified hydrodynamical equations, as well as to the construction of models (simulation); experimental studies carried out on such models give some indications of the performance of such machines when operated in industrial conditions. The results obtained on small-scale "mock-ups" can be applied to large turbopumps only if certain similitude conditions, dealt with in some detail in Chapter 1, are satisfied, viz: geometrical similitude $\lambda = \dfrac{L}{L_m}$; kinematic similitude λ, and $\tau = \dfrac{t}{t_m}$; dynamic similitude λ, τ and $\varphi = \dfrac{F}{F_m}$, where λ, τ, φ are dimensionless similitude parameters for length (λ), time (τ) and force (φ) respectively, while "m" denotes "mock up", as distinct from the actual machine, which has no subscript.

In Chapter 1, the similitude criteria were defined as follows:

— Reynolds $\quad \mathrm{Re} = \dfrac{\mu D}{\nu}$;

— Froude $\quad \mathrm{Fr} = \dfrac{c^2}{gL}$;

— Euler $\quad \mathrm{Eu} = \dfrac{p}{\rho c^2}$;

— Mach $\quad \mathrm{Ma} = \dfrac{c}{a}$;

— Weber $\quad \mathrm{We} = \dfrac{\rho c^2 L}{\tau_s}$;

— Strouhal $\quad \mathrm{Sh} = \dfrac{ct}{L}$,

where: p — pressure; L — length; g — gravity acceleration; ν — kinematic coefficient of the viscosity; c — absolute velocity; a — acceleration of sound propagation; τ_s — superficial stress (tension).

Owing to the complexity of intervening phenomena in turbopump modelling, these three similitude criteria are not used directly, but combinations of them are used instead. When the simulation conditions do not allow the simultaneous application of several criteria, the criterion that represents the most important physical property is taken into consideration, and adequate calculations are made.

In the operation of a pump, the pressure, viscosity and inertial forces predominate, so that criteria Re and Sh or Eu and Sh will have to be simultaneously obeyed too.

The resulting quantity is specific speed n_s, which represents the similitude criterion for all turbomachines.

Starting from $\mathrm{Eu} = \dfrac{p}{\rho c^2}$ and $\mathrm{Sh} = \dfrac{ct}{L}$ and noting pressure $p = \rho g H$, flow rate $Q = c.S$ and gravity $g = \rho g/g$, we get $\mathrm{Eu} = \dfrac{gH}{c^2}$, whence

$$c = \left(\dfrac{\rho g H}{\rho g \, \mathrm{Eu}}\right)^{1/2} = \left(\dfrac{gH}{\mathrm{Eu}}\right)^{1/2}.$$

If we denote by $Q_2' = \dfrac{S}{D_2^2}\left(\dfrac{g}{\mathrm{Eu}}\right)^{1/2} = \dfrac{Q}{D_2^2 \sqrt{H}} = \dfrac{cS}{D_2^2 \sqrt{H}} = \left(\dfrac{Hg}{\mathrm{Eu}}\right)^{1/2} \times$

$$\times \dfrac{S}{D_2^2 \sqrt{H}} \qquad (2.14)$$

the two dimensional unit flow rate, then

$$Q_2' = \dfrac{Q}{D_2^2 \sqrt{H}} \left[\dfrac{\mathrm{m}^{0.5}}{\mathrm{s}}\right] \qquad (2.15)$$

In the case of turbopumps, length L is the corresponding diameter at the impeller output D_2, and $t = \dfrac{1}{n}$, knowing that speed n(rpm) has the dimensions of angular velocity $|\omega| = |s^{-1}|$.

From relation $\mathrm{Sh} = \dfrac{ct}{L} = \dfrac{c}{nD_2}$, we derive $n = \dfrac{c}{D_2 \mathrm{Sh}}$ and bearing in mind the expression of $c = \left(\dfrac{gH}{\mathrm{Eu}}\right)^{1/2}$ as derived before, the speed will be

$$n = \frac{c}{D_2 \mathrm{Sh}} = \left(\frac{gH}{\mathrm{Eu}}\right)^{1/2} \cdot \frac{1}{D_2 \mathrm{Sh}} = \frac{\sqrt{g}}{\mathrm{Sh}\sqrt{\mathrm{Eu}}} \cdot \frac{\sqrt{H}}{D_2} = n_2' \frac{\sqrt{H}}{D_2}$$

Here, the dimensional double-unit speed is denoted $n_2' = \dfrac{\sqrt{g}}{\mathrm{Sh}\sqrt{\mathrm{Eu}}}$, so that

$$n_2' = \frac{nD_2}{\sqrt{H}} \left[\frac{\mathrm{m}^{0.5}}{\mathrm{s}}\right] \tag{2.16}$$

Now, considering the efficiencies of a "mock-up" pump with D_{2m} and an industrial one with $D_2 \neq D_{2m}$, we can state that $\eta_{hm} \neq \eta_h$, being in fact

$$\eta_h = 1 - (1 - \eta_{hm})\left[\frac{D_{2m}}{D_2}\right]^{0.25} \tag{2.17}$$

Consequently, except for the case when the efficiency is to be determined, the net drop $H_n = \dfrac{H}{\eta_h}$ is used in the calculations, so that for $\eta_h \neq 1$, the dimensional double-unit quantities result:

$$Q_2' = \frac{Q}{D_2^2}\left(\frac{\eta_h}{H}\right)^{1/2}; \quad n_2' = \frac{nD_2}{H^{0.5}} \cdot (\eta_h)^{0.5} \tag{2.18}$$

2.3.2 The Expression of Turbopump Specific Speed

Dimensional specific speed n_s as referred to power. According to Coriolis, $\bar{c} = \bar{u} + \bar{w}$, so that absolute velocity c is to be found in its rotational $u = r\omega = \pi D n/60$, and relative w, components. But c, u and w can be expressed as a function of head H, according to Torricelli

$$c = k_c\sqrt{2gh}, \quad u = k_u\sqrt{2gH}, \quad w = k_w\sqrt{2gH}$$

where for $2gH = 1$ m, coefficients k_c, k_u, k_w become specific speeds.

Considering the cylindrical surface $S_2 = \psi \pi D_2 b_2$ of the water as it leaves the pump impeller at a velocity $c_2 = k_{c_2}\sqrt{2gH}$, we can write for the flow rate.

$$\left.\begin{array}{l} Q = S_2 C_2 = \psi\pi D_2 b_2 k_{c_2}\sqrt{2gH} \text{ with } b_2 = \lambda_2 D_2 \\ Q = (\pi\sqrt{2g}\psi\, k_{c_2}\lambda_2)\sqrt{H} D_2^2 = \dfrac{\sqrt{H}}{a^2} D_2^2 \end{array}\right\}$$

Machine constants were concentrated, $\frac{1}{\alpha^2}\pi\sqrt{2g}\,\psi k_{c_3}\,\lambda_2$, from which the diameter results

$$D_2 = a\sqrt{\frac{Q}{\sqrt{H}}} \tag{2.19}$$

Pump suction capacity is indicated by $\sqrt{\frac{Q}{\sqrt{H}}} = \sqrt{Q_1}$, where the simple unit flow rate $Q_1 = \frac{Q}{\sqrt{H}}$, hence $Q_1 = Q$ for $H = 1$ m.

Pump dimensional specific speed is established by starting from the definition of coefficient $k_{u2} = \frac{u_2}{\sqrt{2gH}}$ assigned to the peripheral velocity $u_2 = \pi D_2 n/60$, where the following substitutions are successively made: $D_2 = a\sqrt{\frac{Q}{\sqrt{H}}}$ and $Q = \frac{P \times \eta_p}{13.33\,H}$ (from the expression for the power). Therefore

$$ku_2 = \frac{u_2}{\sqrt{2gH}} = \frac{D_2 n}{60\sqrt{2gH}} = \frac{\pi n a \sqrt{\frac{Q}{\sqrt{H}}}}{60\sqrt{2gH}} \left(\frac{\pi a \sqrt{\eta_P}}{60\sqrt{2g} \times \sqrt{13.33}}\right) \frac{n}{H}\sqrt{\frac{P}{\sqrt{H}}}$$

Putting the energetic parameters n, H and P on one side, and the constants a, k_u, η_P and numerical coefficients on the other one can write

$$\frac{n}{H}\sqrt{\frac{P}{\sqrt{H}}} = 309\,\frac{ku_2}{a\sqrt{\eta}}$$

The expression

$$n_s = \frac{n}{H}\sqrt{\frac{P}{\sqrt{H}}} = n \times p^{0.5} H^{-1.25} \tag{2.20}$$

is called *the dimensional specific speed as referred to the power*; meaning the speed (rpm) of an ideal turbopump geometrically similar to the real one, and so dimensioned as to discharge the water at a unit pumping head $H = 1$ m when driven at the unit power $P = 1$ HP.

When put in the form of relation (2.20), the specific speed is a dimensional quantity

$$|n_s| = \left|\frac{1}{S} \times \frac{\text{da N} \cdot \text{m}^{1/2}}{S} \times \frac{1}{\text{m}^{5/4}}\right| = \left|\frac{\text{da N}^{1/2}}{\text{m}^{1/4} S^{1/2}}\right|$$

To speed up calculations of the specific speed the diagram shown in Fig. 2.3. was drawn up, based on the expressions

$$n_s = \frac{3.65\,n}{60} \times \frac{Q_{opt}^{1/2}}{H_{opt}^{3/4}} \tag{2.21}$$

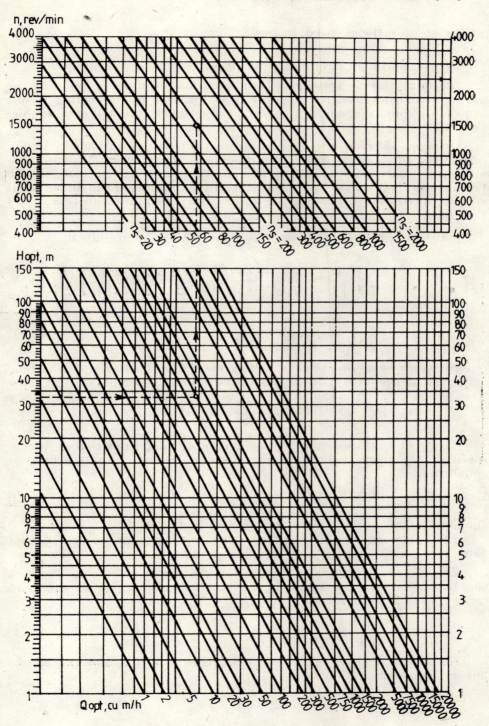

Fig. 2.3. Diagrammatical determination of specific speed n_s.

where n — rotational speed, in rpm;
Q_{opt} — optimum flow rate, in cu.m/min;
H_{opt} — optimum pumping head, in m.

Remark. Specific speed n_s is estimated for the working point at the optimal efficiency, and refers to the pumping head corresponding to one impeller of a multi-stage turbopump, or else to only the half of a double-flow turbopump.

Dimensional specific speed as a function of flow rate n_q. For any pump, estimating the specific speed n_s presupposes that power P, and hence efficiency η, are known beforehand. For the pump design analysis, the parameters H, Q and n, are specified, and for this reason the dimensional specific speed is used in the form n_q, i.e. expresseed as a function of the flow rate.

Substituting in expression (2.20) $P = 13.33 \dfrac{QH}{\eta}$ results in

$$n_s = \frac{n}{H} \sqrt{\frac{P}{\sqrt{H}}} = \frac{n}{H} \sqrt{\frac{13.33\, QH}{\eta \sqrt{H}}} = \sqrt{\frac{13.33}{\eta}} \times \frac{n}{\sqrt{H}} \sqrt{\frac{Q}{\sqrt{H}}}$$

The expression

$$n_q = \frac{n}{\sqrt{H}} \sqrt{\frac{Q}{\sqrt{H}}} = nQ^{1/2} H^{-3/4} \qquad (2.22)$$

represents *the pump dimensional specific speed as a function of the flowrate*. When put in the form (2.22), specific speed is a dimensional quantity, for

$$|n_q| = \left| \frac{1}{S} \times \frac{(m^3)^{1/2}}{S} \times \frac{1}{m^{3/4}} \right| = |m^{3/4} S^{-3/2}|.$$

To speed up the calculations involved in the determination of specific speed n_q, the diagram shown in Fig. 2.4 was drawn up, on the basis of equation

$$n_q = \frac{n}{60} \times \frac{Q_{opt}^{1/2}}{H_{opt}^{3/4}} \qquad (2.23)$$

the notation being the same as in equation (2.21).

In the determination of specific speed n_q, the same remark as that made for specific speed n_s is to be borne in mind.

Equivalence between dimensional specific speeds n_s and n_q. Between the two specific speeds n_s and n_q, one can write, according to equations (2.20) and (2.23), the relation

$$n_s = \frac{3.65}{\sqrt{\eta}} n_q \qquad (2.24)$$

Since with modern pumps $\eta > 0.83$, then $n_s \approx 4\, n_q$ and this can be put in the dimensional form

$$|n_q| = |\, m^{3/4} S^{-3/2}\,|$$

Similitude Relationships Applied to Turbopumps

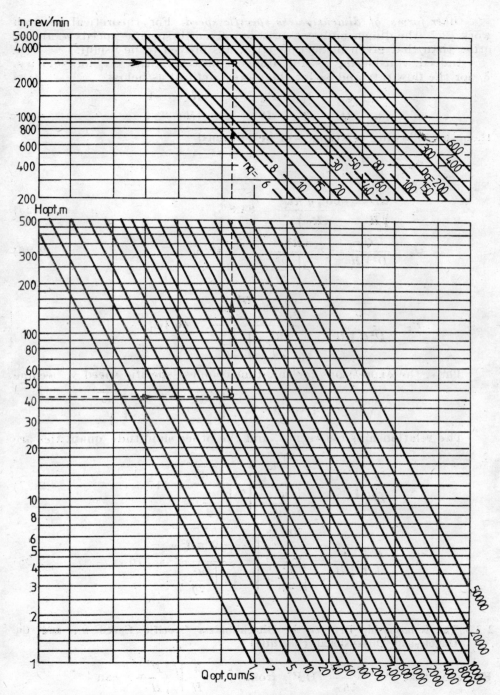

Fig. 2.4. Diagrammatic representation of specific speed n_q

Other forms of dimensionless specific speed. For theoretical research work [2.7] the dimensionless expression given to specific speeds n_s and n_q other than that given in Section 1.1.3, has proved to be useful.

Indeed, considering as unit dimensionless quantities ν_1 for velocity, δ_1 for the flow rate and ω_1 for the power, defined as below

$$\nu_1 = \frac{n_m}{n} \; ; \; \frac{Q_m}{Q} = \frac{\delta_1}{\delta} \; ; \; \frac{P_m}{P} = \frac{\omega_1}{\omega} \; ; \; \frac{H_m}{H} = \frac{1}{\mu}$$

then the following relations will be obtained

$$\frac{n_m}{n} = \sqrt{\frac{H_m}{H}} \; ; \; \nu_1 = \frac{1}{\sqrt{\mu}}$$

$$n_2' = \frac{nD}{\sqrt{H}} = \frac{60\sqrt{g}\,R\omega}{\pi R \omega \sqrt{\mu}} = 84.8\,\nu_1 \tag{2.25}$$

$$Q_2' = \frac{Q}{D^2\sqrt{H}} = \frac{\pi R^2 \mu \delta}{D^2 \sqrt{\frac{u^2}{2g}\mu}} = 3.48\,\delta_1 \tag{2.26}$$

$$P_2' = \frac{P}{D^2 H \sqrt{H}} = \frac{\frac{\rho g}{2g} u^3 \pi \frac{D^2}{4} \omega}{D^2 \frac{u^2}{2g} \mu \sqrt{\frac{u^2}{2g}\mu}} = 46.4\,\omega_1$$

Thus, the expressions for the dimensionless specific speed ν_s become

$$\nu_s = \nu_1 \sqrt{\delta_1} = \nu_1 \sqrt{\omega_1}\,; \; \nu_s = \frac{1}{\sqrt{\mu}} \sqrt{\frac{\delta}{\sqrt{\mu}}}\,; \; \nu_s = \frac{1}{\mu} \sqrt{\frac{\omega}{\sqrt{\mu}}} \cdot$$

The relationships between ν_s and the other similitude quantities are

$$n_q = \frac{n}{\sqrt{H}} \sqrt{\frac{Q}{\sqrt{H}}} = \frac{\frac{30}{\pi}\omega}{\sqrt{\frac{u^2}{2g}\mu}} \sqrt{\frac{\pi R^2 u \delta}{\sqrt{\frac{u^2}{2g}\mu}}} = 117.5\,\nu_s \tag{2.27}$$

$$n_s = \frac{n}{H} \sqrt{\frac{P}{\sqrt{H}}} = \frac{\frac{30}{\pi}\omega}{\sqrt{\frac{u^2}{2g}\mu}} \sqrt{\frac{\frac{\rho}{2} u^3 \pi R^2 \omega}{75 \sqrt{\frac{u^2}{2g}\mu}}} = 576\,\nu_s \tag{2.28}$$

2.3.3 The Relationship Between the Dimensionless Specific Speed n_s, and the Double-unit Dimensional Quantities

Knowing that $P = \frac{\rho HQ}{75\eta}$ [HP], from $n_s = \frac{n}{H}\sqrt{\frac{P}{\sqrt{H}}}$ we can write

$$n_s = \frac{n}{H}\sqrt{\frac{\rho\, QH}{75\eta\sqrt{H}}} = \frac{nD_2}{\sqrt{H}}\sqrt{\frac{\rho}{75\eta}\frac{Q}{D_2^2\sqrt{H}}}$$

or

$$n_s = 3.65\, n_2' Q_2' \frac{1}{\sqrt{\eta}} \tag{2.29}$$

Whether expressed as n_s or as n_q, dimensional specific speed is a very important quantity in the theory, design analysis, building and operation of turbopumps, because it :
— offers a classification of turbopumps, being the most synthetic criterion ;
— intervenes in all similitude calculations, being the parameter of turbopump complex similitude ;
— serves for turbopump standardization ;
— permits the analytical calculation of costs and weights of all kinds of turbopumps ;
— is applied to construct the characteristics and the operational topographies of turbopumps ;
— is used in turbopump dimensional analysis, since all pump sizes depend on n_s.

2.3.4 Relationships Between Q, H, P and η

Case of identical specific speeds : $n_s = n_{sm}$. Knowing the geometric similitude parameter, $\lambda = \dfrac{L}{L_m}$, velocities are expressed, as $\dfrac{L}{t}$ and time t, by the inverse of the rotational speed

$$|n| = \left|\frac{l}{t}\right| = |S^{-1}|$$

and thus we can write the relations

$$\frac{Q}{Q_m} = \frac{L^2 c\, \eta_v}{L_m^2 c_m \eta_{vm}} = \lambda^3 \frac{n}{n_m} \times \frac{\eta_v}{\eta_{vm}} \tag{2.30}$$

$$\frac{H}{H_m} = \lambda^2 \left(\frac{n}{n_m}\right)^2 \frac{\eta_h}{\eta_{hm}} \quad \text{with} \quad H = H_t \times \eta_h \tag{2.31}$$

$$\frac{P}{P_m} = \lambda^5 \left(\frac{n}{n_m}\right)^3 \frac{\eta_{mec.m}}{\eta_{mec}} \tag{2.32}$$

Efficiency similitude is obtained from

$$\eta_h = 1 - \frac{\Sigma h_r}{H_t} \quad \text{with} \quad \frac{\Sigma h_r}{H_t} = \frac{Q}{\mathrm{Re}^\alpha D_2^\beta}$$

$$\mathrm{Re} = \frac{cD}{\nu} = k_u \sqrt{2gH}\, \frac{D}{\nu} \quad \text{so that}$$

$$\text{Re} = 4.43\, k_u H^{0.5} \frac{D_2}{\nu} \text{ for the pump and}$$

$$\text{Re}_m = 4.43\, k_u H_m^{0.5} \frac{D_{2m}}{\nu_m} \text{ for the "mock-up" pump}$$

Finally

$$\frac{1 - \eta_h}{1 - \eta_{h\,mock-up}} = \left(\frac{H_m}{H}\right)^{0.5\alpha} \times \left(\frac{D_{2m}}{D_2}\right)^{\alpha+\beta} \times \left(\frac{\nu}{\nu_m}\right)^{\alpha}.$$

Now, choosing [2.4] $\alpha = 0.2$ and $\beta = 0.05$, we get the efficiency similitude relation in the form given by Moody

$$\eta_h = 1(1 - \eta_{hm})\left(\frac{\nu}{\nu_m}\right)^{0.2}\left(\frac{H_m}{H}\right)^{0.1}\left(\frac{D_{2m}}{D^2}\right)^{0.25} \tag{2.33}$$

Hence the global pump efficiency will be

$$\eta = \eta_h \eta_m \eta_v \text{ where } \eta_m \eta_v = \left(0.97 - \frac{0.2}{P_m^{0.25}}\right)$$

Case of non-identical specific speeds: $n_s \neq n_{sm}$. From the already known relations, $n_s = n \times P^{1/2} H^{-5/4}$

$$P = 13.33\, QH\, \frac{1}{\eta}\, [\text{HP}]; \quad \eta = \eta_{mec}\eta_v\eta_h; \quad \eta_{mv} = \eta_m\eta_v = (0.97 - 0.2\, P^{-1/4})$$

and we find $D = \lambda D_m$, where λ has the value

$$\lambda = \frac{n_m}{n}\left(\frac{n_s}{n_{s_m}}\right)^{2/3}\left(\frac{H}{H_m}\right)^{1/2}\left(\frac{\eta_{vm}}{\eta_{mv}}\right)^{1/2} \tag{2.34}$$

Relations (2.30) — (2.34) allow the results of studies carried out on "mock-up" pumps to be transferred onto real pumps performing in actual production conditions, and also to work out a self-similitude concept of the pump performing in varying Q, H, n and η conditions, this case having a direct importance in practice.

2.3.5 Effects of Specific Speed on Turbopump Geometry and Efficiency

Effects of specific speed on impeller geometry. Specific speed has no physically practical value, but is used to characterize a certain typical pump.

If the geometrical parameters of a pump impeller are denoted by: D_0 — diameter of the pump suction connexion; D_1 — diameter of the impeller blade inlet; D_2 — blade output diameter; b_2 — width of flow channels (between the impeller disks) at the impeller blade outlet; then the shapes of impellers in the meridian plane (Fig. 2.5), as well as their geometrical parameters (Fig. 2.6), can be determined as a function of specific speed n_q.

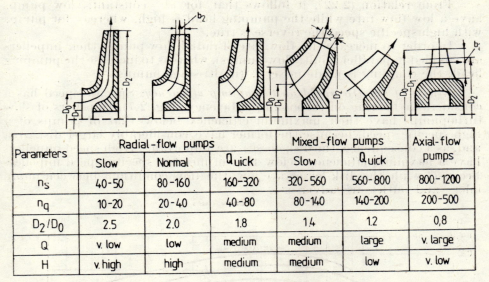

Parameters	Radial-flow pumps			Mixed-flow pumps		Axial-flow pumps
	Slow	Normal	Quick	Slow	Quick	
n_s	40-50	80-160	160-320	320-560	560-800	800-1200
n_q	10-20	20-40	40-80	80-140	140-200	200-500
D_2/D_0	2.5	2.0	1.8	1.4	1.2	0,8
Q	v. low	low	medium	medium	large	v. large
H	v. high	high	medium	medium	low	v. low

Fig. 2.5. Geometries of turbopump impellers (in a meridian plane), as a function of specific speeds n_s and n_q.

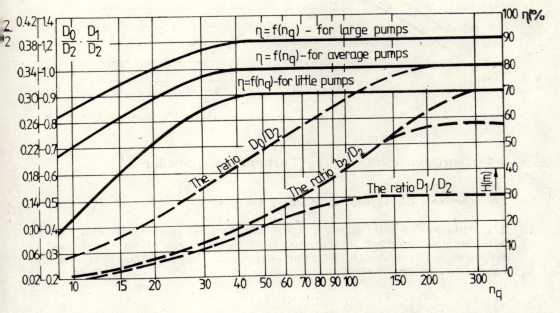

Fig. 2.6. Change in the overall sizes of turbopump impeller geometry as a function of specific speed n_q.

From relation (2.22), it follows that for $n =$ constant, slow pumps have a low flow-rate while the pumping head is high, whereas for pumps with high specific speed the reverse is true.

In order to increase the flow rate of radial-flow pumps their impellers are coupled in parallel (multi-flow pumps), whereas to increase the pumping head, the coupling is made in series (multi-stage pumps).

Effects of specific speed on turbopump efficiency. Specific speed has a considerable bearing on turbopump efficiency (Fig. 2.7). Impellers of slow turbopumps have their maximum efficiency below that of pumps of a high specific speed, because the former have considerably larger disk area and, as a consequence, the disk losses are important. High speed impellers have a maximum efficiency below that of medium specific speed impellers because, while the disk losses are low, the flow conditions between pump inlet and outlet are adverse.

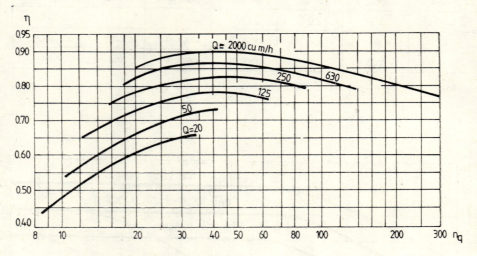

Fig. 2.7. Effect of specific speed n_q on turbopump efficiency η.

2.4 Hydrodynamics of the Turbopump Impeller

2.4.1 Outlining the Turbopump Whirling Energy

The equation of fluid motion (1.46) was derived in the first chapter. In what follows, another interpretation of the concept of pumping head is given, so as to outline the whirling energy. In a vectorial form, the motion equation can be written

$$\overline{\frac{dc}{d_t}} + \overline{\frac{\nabla p}{\rho}} = \bar{f} + \overline{\frac{dF_t}{dm}} \tag{2.35}$$

Integrating by one arc element, $d\bar{r}$, results in

$$\int_1^2 \frac{\partial \bar{c}}{\partial t} d\bar{r} + \frac{c_2^2 - c_1^2}{2} + 2\int_1^2 \bar{\omega} \times \bar{c} \times d\bar{r} + \int_1^2 \frac{dp}{\rho} + U_2 - U_1 = \int_1^2 \frac{d\overline{F}_t}{dm} d\bar{r} \quad (2.36)$$

which is the substantial derivative of velocity c, of the form

$$\frac{d\bar{c}}{dt} = \frac{\partial \bar{c}}{\partial t} + (\bar{c}\nabla)\bar{c} = \frac{\partial \bar{c}}{\partial t} + \nabla\left(\frac{c^2}{2}\right) + (\nabla \times \bar{c}) \times \bar{c} \quad (2.37)$$

and, according to Helmholtz, $\nabla \times \bar{c} = 2\bar{\omega}_{whirl}$. Since $U = gz + ct$ is the potential and $\rho(X, Y, Z) = $ constant, we may write:

$$U_2 - U_1 = g(Z_2 - Z_1) \text{ and } \int_1^2 \frac{dp}{\rho} = \frac{p_2 - p_1}{\rho}. \quad (2.38)$$

Thus equation (2.36) becomes

$$\frac{p_2 - p_1}{\rho g} + \frac{c_2^2 - c_1^2}{2g} + (Z_2 - Z_1) = -\frac{1}{g}\int_1^2 \frac{\partial c}{\partial t} d\bar{r} - \frac{2}{g}\int_1^2 \bar{\omega} \times \bar{c} \, d\bar{r} - h_{1,2} \quad (2.39)$$

where

$$\int_1^2 \frac{d\overline{F}_t}{dm} d\bar{r} = -g \times h_{1,2}$$

represents the term for the lost mechanical work.

The first two terms of the second side of equation (2.39) emphasize both the accelerating (or decelerating) and the whirling mechanical work which, when summed up, yield the useful mechanical work of the turbopump.

2.4.2 Basic Concepts of the Theory of Turbopump Impellers

As shown in Chapter 1 and referring to Figs. 2.8 and 2.9, Bernoulli's equation has the expression

$$\frac{p_1}{\rho g} + \frac{W_1^2}{2g} + \frac{U_1}{g} = \frac{p_2}{\rho g} + \frac{W_2^2}{2g} + \frac{U_2}{g} \quad (2.40)$$

since

$$U = -\frac{r^2\omega^2}{2} + gz + ct \text{ (with } r^2\omega^2 = u^2\text{)}$$

According to the Pythagora's generalized theorem

$$W^2 = u^2 + c^2 - 2uc\cos\alpha; \quad W^2 - u^2 = c^2 - 2uc_u, \text{ where } c \times \cos\alpha = c_u \quad (2.41)$$

and equation (2.40) becomes (2.42), i.e. the same as (1.14)

$$\frac{p_1}{\rho g} + \frac{c_1^2}{2g} + Z_1 = \frac{p_2}{\rho g} + \frac{c_2^2}{2g} + Z_2 - \frac{c_{u2}u_2 - c_{u1}u_1}{g} \quad (2.42)$$

The last term of equation (2.42) represents the energy exchange that takes place at the inlet *1* and at the outlet *2* of the impeller blade; it is called the *"theoretic head for an infinity of blades"* and has the same expression as (1.43), i.e.

$$H_{t\infty} = \frac{c_{u2}u_2 - c_{u1}u_1}{g} \tag{2.43}$$

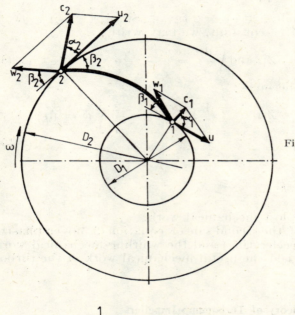

Fig. 2.8. Flow kinematics and overall sizes of turbopump impellers.

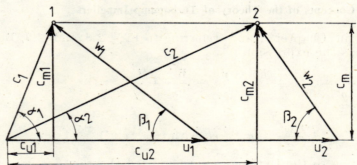

Fig. 2.9. Speed triangles at the blade inlet and outlet for an ideal impeller $(z = \infty, s = 0)$.

A comparison of equations (2.43) and (2.42) will show that

$$H_{t\infty} = -\frac{2}{g} \int_1^2 \bar{\omega} \times \bar{c} \times d\bar{r} \tag{2.44}$$

Now, the expression

$$H_{t\infty} = \frac{c_2^2 - c_1^2}{2g} + \frac{u_2^2 - u_1^2}{2g} + \frac{W_2^2 - W_1^2}{2g} \qquad (2.45)$$

will be taken as the basic equation in the study of turbopumps.

Relation (2.43) can be also derived by applying the method of the kinetic moment of impulse forces, as shown in Chapter 1, or else the method of potential motions, as will be shown elsewhere.

The expression of the pump hydraulic power, $P_h = \rho Q H_{t\infty} = T_h \omega$, with $H_{t\infty}$ from equation (2.43), is used to derive the hydraulic moment (Euler), which is similar to expression (1.81)

$$T_h = \frac{P_h}{\omega} = \frac{Q}{g}(c_{u2} \times r_2 - c_{u1} \times r_1) \qquad (2.46)$$

Taking the expression for the circulation (1.138) we get: at the impeller inlet $\Gamma_{1\infty} = 2\pi c_{u1} r_1$, while at the impeller outlet, $\Gamma_{2\infty} = 2\pi c_{u2} r_2$. Since $\Gamma_{p\infty} = \dfrac{(\Gamma_2 - \Gamma_1)_\infty}{z}$ is the circulation (flow) around one of the z blades of the impeller, we find

$$z \Gamma_{p\infty} = 2\pi (c_{u2} r_2 - c_{u1} r_1)$$

Substituting $(c_{u2} r_2 - c_{u1} r_1) = g H_{t\infty}$ from relation (2.43), results in:

$$H_{t\infty} = \frac{z \Gamma_{p\infty}}{2\pi g}$$

$$P_h = \rho g Q H_{t\infty} = \frac{\rho g Q z}{2\pi g} \Gamma_{p\infty} = \frac{\rho g Q}{2\pi g}(\Gamma_{2\infty} - \Gamma_{1\infty}) \qquad (2.47)$$

2.4.3 Hydrodynamics of Radial-Flow Pumps

Bearing in mind the real flow process, the finite number of blades (z) and their thickness (s), a mathematical model should be chosen for this study. We shall proceed to estimation of the velocity triangles in steps, as follows.

The ideal impeller, with $z = \infty$ and $s = 0$. Using suffixes 1 and 2 for the impeller blade inlet and outlet respectively, the velocity diagram for the ideal impeller results as shown in Fig. 2.9, where the following notation (as given in the literature) is used throughout

α_1, α_2 — angles between c and $u = r\omega = \pi Dn/60$
β_1, β_2 — relative angles between w and u

$$\left. \begin{array}{l} c_{m1} = c_1 \sin \alpha_1 = w_1 \sin \beta_1; \quad c_{u1} = c_1 \cos \alpha_1 \\ c_{m2} = c_2 \sin \alpha_2 = w_2 \sin \beta_2; \quad c_{u2} = c_2 \cos \alpha_2 \\ \bar{c} = \bar{c}_u + \bar{c}_m \end{array} \right\} \qquad (2.48)$$

Components c_{m1} and c_{m2} are used to estimate the flow rate Q of the pump, and components c_{u1} and c_{u2} to estimate the discharge head $H_{t\infty}$ (from relation 2.43) for $z = \infty$.

The semi-real impeller with $z \neq \infty$ and $s = 0$. The semi-real impeller produces a non-uniform velocity owing to the relative whirl, the blades being taken as partial, solid boundaries. Whirl $\overline{\omega}_{rel}$ is constant, axial and of the opposite sign to angular velocity ω_{shaft}, as shown in expression (1.87)

$$\overline{\omega}_{relative} = -\overline{k}\omega \tag{2.49}$$

The relative whirl, $\overline{\omega}_{rel}$, impresses a relative velocity called "induced", $\delta\overline{w}$, that is added to the ideal impeller relative velocity $\overline{w}_\infty$, so that

$$\overline{w} = \overline{w}_\infty + \delta\overline{w} \tag{2.50}$$

According to Coriolis, $\overline{c} = \overline{u} + \overline{w}$, which permits the construction of velocity triangles for the semi-real impeller.

In [2.5] the average velocities induced by the relative whirl around the closed boundaries were estimated and the results, as projections along $\overline{u}$ direction, were given by expressions

$$\left.\begin{array}{l} w_{u1-\text{average}} = \dfrac{\omega t_1}{2} \sin \beta_1 \\[6pt] w_{u2-\text{average}} = \dfrac{\omega t_2}{2} \sin \beta_2 \end{array}\right\} \tag{2.51}$$

where $t_1 = \dfrac{\pi D_1}{Z}$ and $t_2 = \dfrac{\pi D_2}{Z}$ are the pitches.

With these specifications, the corrected velocity diagram is of the form shown in Fig. 2.10, where component c_{u2} drops in value by $n_2 u_2$, thus reducing the discharge head.

The real impeller (with $z \neq \infty$ and $s \neq 0$). Because of blade thickness ($s \neq 0$), the cross-section will be restricted just after the rotor inlet and before its outlet, and since $Q = $ constant, the meridian velocities c_{m1}, c_{m2} will increase.

Using the notations of Fig. 2.11, *a*, and bearing in mind the continuity equation we can write for the corrected meridian velocity

$$c'_m = c_m \frac{2\pi rb}{(2\pi r - Z\Delta t)b} = \frac{c_m}{\varepsilon} \tag{2.52}$$

where the contraction $\varepsilon < 1$, owing to the fact that $\Delta t = \dfrac{s}{\sin \beta}$, so that

$$c'_m = c_m + \Delta c_m = c_{m_1} \frac{1}{1 - \dfrac{Zs}{2\pi r \sin \beta}} \tag{2.53}$$

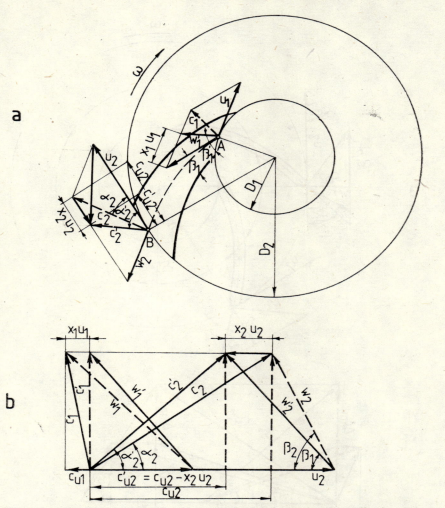

Fig. 2.10. Speed triangles at the blade inlet and outlet for the semi-real impeller ($z \neq \infty, s = 0$).

where c_m — velocity of the ideal impeller;
        z — impeller blades;
        r — current radius ($D_{1/2} < r < D_{2/2}$);
        s — blade thickness;
        β — angle of relative velocity, between $\bar{w}$ and $\bar{u}$.

Velocity diagram for the real impeller will result from Fig. 2.11 b. The introduction of the model of the semi-real impeller makes $H_{t\infty}$ simulate H_t, and hence, according to the already known expressions (1.86) and (1.89):

— for the ideal impeller, $H_{t\infty} = \dfrac{c_{u2}u_2 - c_{u1}u_1}{g}$ (2.54)

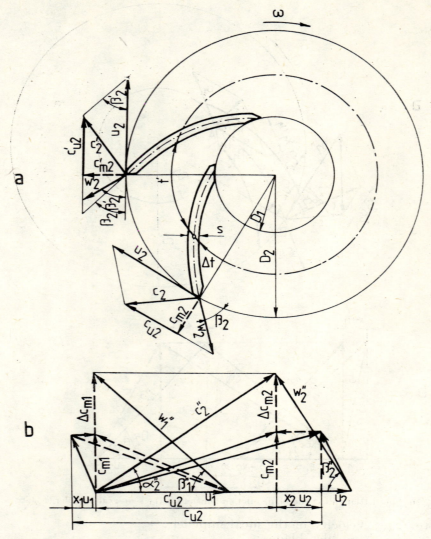

Fig. 2.11. Velocity triangles at the blade inlet and outlet for the real impeller ($z \neq \infty$ $s \neq 0$).

— for the semi-real impeller $H_t = \dfrac{c'_{u2}u_2 - c'_{u1}u_1}{g}$ (2.55)

Now, introducing the induced efficiency k_i as being the same as k_n, according to the notation used in relation (1.90)

$$k_i = \frac{H_t}{H_{t\infty}} = \frac{c'_{u2}u_2 - c'_{u1}u_1}{c_{u2}u_2 - c_{u1}u_1}$$ (2.56)

The finite blades $z \neq \infty$ lead to a pumping head reduced by δH_t, so that $k_i = \dfrac{H_{t\infty} - \delta H_t}{H_{t\infty}} = 1 - \dfrac{\delta H_t}{H_{t\infty}}$, the values varying between 0.6 and 0.8, for various parameters.

Considering in relations (2.54) and (2.55) the orthogonal input $\alpha_1 = 90°$, $c_{u1} = 0$ and $c_{u2} - c'_{u2} = x_2 u_2$, we can write

$$H_{t\infty} - H_t = \frac{u_2}{g}(c_{u2} - c'_{u2}) = \frac{\eta_2 u_2^2}{g} \tag{2.57}$$

Now, introducing in the above relation the parameter p to say that $H_{t\infty}$ is greater than H_t

$$H_{t\infty} = (1+p)H_t \text{ or } H_t = \frac{1}{1+p}H_{t\infty}; \text{ results } k_i = \frac{1}{1+p}.$$

Terms p and k_i were estimated by several authors (see relation (1.91)) and the results were broadly of the form

$$k_i = \frac{1 - \dfrac{t_2 \omega}{2 c_{u2}} \sin \beta_2}{1 + \dfrac{r_1}{r_2} \dfrac{t_1 \omega}{2 c_{u2}} \sin \beta_1} \tag{2.58}$$

According to more recent studies, k_i and p are explained by the effects of blade interactions within the profile network.

From relation (2.47), the whirl interpretation of k_i could be

$$k_i = \frac{\Gamma_p}{\Gamma_{p\infty}} \text{ and } \frac{1}{1+p} = \frac{\Gamma_p}{\Gamma_{p\infty}} \text{ or } p = \frac{\Gamma_{p\infty}}{\Gamma_p} - 1 \tag{2.59}$$

that is to say that on the profile, taken separately in the network, are acting the hydrodynamic effects of the $(z-1)$ profiles of the network, and this is exactly what the influence coefficient k_i stands for. Coefficient p, that reduces the discharge from $H_{t\infty}$ down to H_t, also expressed the fact that $z \neq \infty$ reduces the centrifugal effect because of the stream lines being bent as they pass through the impeller. Other relations, more applicable than equation (2.58), show the dependence of k_i on D_2/D_1 and β_2, so that in the first stage of the design analysis, the values given in Table 2.6 can be used.

Table 2.6 **The coefficient k_i as a function of D_2/D_1 and β_2 for $z=8$ blades**

$\dfrac{D_2}{D_1}$	Pumps without vaned diffuser			Pumps with vaned diffuser		
	β_2					
	20°	30°	40°	20°	30°	40°
1.5	0.705	0.635	0.600	0.780	0.730	0.685
2.0	0.740	0.700	0.670	0.820	0.780	0.750
3.0	0.760	0.730	0.706	0.850	0.810	0.780

As shown by expression (1.29), the real discharge head is, in the last resort

$$H = \eta_h k_i H_{t\infty} = \eta_h H_t \tag{2.60}$$

and is lower than head H_t at $z \neq \infty$ owing to load losses $h_r = k_r Q^2$, which reduce the hydraulic efficiency of the pump

$$\eta_h = 1 - \frac{k_r Q^2}{H} < 1 \tag{2.61}$$

In actual practice, pumps with $\eta_h = 0.9 - 0.96$ and having large D_2 were built.

2.4.4 Hydrodynamics of Axial-Flow Pumps

The high velocity of a flow with respect to H leads to a small number of blades z, and also to hydrodynamical shapes offering a minimal resistance.

The three-dimensional motion of water is analytically studied by means of simplifying hypotheses, which are subsequently checked by laboratory tests conducted on "mock-ups".

The approximately cylindrical flowing surfaces of the impeller intersect its blades and by display in the plane area a co-axial cylinder of radius r, along $2\pi r$ will appear the z profiles that build up the profile network considered in Section 1.2.4.

Thus, for $r=1$, (i.e. the unit span), the bearing forces, $F_z = \rho w_{u.med.} \Gamma_p$; and the tangential forces $F_u = \rho w_z \cdot \Gamma_p$ were obtained. For an elemental span dr, these forces are

$$\left. \begin{array}{l} \mathrm{d}F_z = w_{u\,med.} \Gamma_p \times \mathrm{d}r \\ \mathrm{d}F_u = \rho\, w_z\, \Gamma_p \times \mathrm{d}r \end{array} \right\} \tag{2.62}$$

The torque of the force about the rotational axis, from the hub to the periphery, i.e. $F_u = \int_{ri}^{ro} \mathrm{d}F_u$, yields the resisting torque T_r for one blade of the impeller ($z = 1$); hence, for z blades, the total torque will be (taking $\eta_h = 1$)

$$T_r = z \times \rho\, w_z \int_{ri}^{re} \Gamma_p r \times \mathrm{d}r = \rho w_z Z \cdot \Gamma_p \frac{r_o^2 - r_i^2}{2}$$

or else:

$$T_r = \rho \times H_t Q \frac{1}{\omega} \tag{2.63}$$

But

$$\eta h \neq 1, \text{ so that } T_r = \frac{\rho Q H}{\omega \eta_h} \tag{2.64}$$

Hydrodynamics of the Turbopump Impeller

For the pure axial pump, $u_1 = u_2 = r\omega = \dfrac{\pi D n}{60}$ and, inasmuch as for the z blades each Γ_p processes the circulation differentials between the impeller input Γ_1 and output Γ_2, it follows that

$$z\Gamma_p = \Gamma_2 - \Gamma_1 \tag{2.65}$$

The relations for the hydraulic torque T_h and the circulation Γ_p are [2.4], [2.5]

$$T_h = \frac{\rho g Q}{g \eta_h} r(c_{u2} - c_{u1}) \parallel \frac{\rho g Q}{2\pi g \eta^u} \times z \times \Gamma_p \tag{2.66}$$

$$\Gamma_p = 2\pi g \frac{H}{\eta_h \omega Z} = 589 \frac{H}{n z \eta_h} \tag{2.67}$$

Starting from relations (1.115) and (1.116) and bearing in mind (1.119) as well as Fig. 2.12, for the two components of hydrodynamic force dF_u and dF_z that act upon the blade element of span dr, one can write the expressions

$$dF_u = dP \sin \beta_m + dT \cos \beta_m \tag{2.68}$$

$$dF_z = dP \cos \beta_m - dT \sin \beta_m \tag{2.69}$$

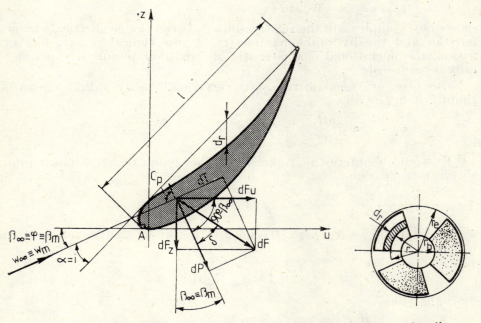

Fig. 2.12. Notation for the hydrodynamical study of the axial — flow pump impeller blade: $\tan \delta = T/P = C_{xr}/C_{yr}$ = profile fineness;
$C_p = f(C_T)$ = $C_y = f(C_x)$ = polar coordinates; $T = dx$, $P = dy$.

or else

$$dF_u = \frac{\rho}{2} W_m^2 \, l \, dr \, C_{yr}(\sin \beta_m + \varepsilon \cos \beta_m) \tag{2.70}$$

$$dF_z = \frac{\rho}{2} W_m^2 \, l \, dr \, C_{yr}(\cos \beta_m - \varepsilon \sin \beta_m) \tag{2.71}$$

from which, after transformation, we can write for the force the expression

$$dF_u = \frac{\rho}{2} W_m^2 l \times C_{yr} \frac{\sin(\beta_m + \delta)}{\cos \delta} \, dr \tag{2.72}$$

that determines the pumping head, and the force

$$dF_z = \frac{\rho}{2} W_m^2 \, l \cdot C_{yr} \frac{\cos(\beta_m + \delta)}{\cos \delta} \, dr \tag{2.73}$$

which will serve to estimate the axial thrust.

The fundamental equation of axial-flow pumps, as derived in Chapter 1 in its two forms (1.124/1)

$$H_t = \frac{C_{yr}}{2g} \frac{l}{t} \frac{u}{c_z} W_m^2 \frac{\sin(\beta_m + \delta)}{\cos \delta} \tag{2.74}$$

$$\frac{C_{yr} \cdot 1}{t \cdot \cos \delta} = \frac{2gH_t \cdot c_z}{u W_n^2 \sin(\beta_m + \delta)} \tag{2.74/1}$$

shows how complex are the relationships between the geometrical characteristics and the hydraulic parameters of this typical pump; for this reason, the operational characteristic of axial-flow pumps is experimentally established.

Because, as a general rule, δ is very small, relation (2.74/1) can be simplified in the form

$$C_{yr} \frac{l}{t} \approx \frac{2gH_t}{u W_m}$$

It would be interesting to know the behaviour of axial-flow pumps, which can be done by writing

$$\frac{\sin(\beta_m + \delta)}{\cos \delta} = \sin \beta_m \left(1 + \frac{\mathrm{tg}\,\delta}{\mathrm{tg}\,\beta_m}\right)$$

$u = k_u \sqrt{2gH}$, $W_m = k_{wm} \times \sqrt{2gH}$ and $c = W_m \sin \beta_m$ and it follows from equation (2.74)

$$H_t = \frac{C_{yr}}{t} l \times k_u \times k_{wm} \left(1 + \frac{\mathrm{tg}\,\delta}{\mathrm{tg}\,\beta_m}\right) H$$

and therefore

$$\eta_h = \frac{H}{H_t} = \frac{t}{C_{yr} \times 1 \times k_u \times k_{wm} \left(1 + \frac{\mathrm{tg}\,\delta}{\mathrm{tg}\,\beta_m}\right)} \tag{2.75}$$

Hydrodynamics of the Turbopump Impeller

If we call the head losses in the pump impeller h_{r1}, and those in the casing h_{r2}, the hydraulic efficiency of the pump will be

$$\eta_h = 1 - \frac{h_{r1} + h_{r2}}{H_t} = 1 - (\eta_{h_1} + \eta_{h_2}) \qquad (2.75/1)$$

Since the energy losses due to hydraulic friction might be estimated as the mechanical work of force (d_T), it follows that

$$\rho g \times dQ \times h_{r1} = Z \times W_m \times dT = Z \times W_m dP \times \tan \delta$$

and eventually

$$h_{r1} = \frac{C_{yr}}{2g} \frac{l}{t} W_m^2 \frac{\tan}{\sin \beta_m} \qquad (2.76)$$

If relation (2.75/1) is used, the result will be

$$\eta_{h1} = \frac{\sin \delta}{\sin(\beta_m + \delta)} \frac{W_m}{u} \qquad (2.77)$$

Assuming $h_{r_2} = \zeta \dfrac{Cu_2^2}{2g}$ with $\zeta = 0.15 - 0.25$, the general expression for the hydraulic efficiency of axial-flow pumps results in the form

$$\eta_h = 1 - \frac{\sin \delta}{\sin (\beta_m + \delta)} \times \frac{W_m}{u} - \zeta \frac{Cu_2^2}{2gH_t} = f(r) \qquad (2.75/2)$$

As can be seen from relation (2.75/2), the hydraulic efficiency differs from one blade profile to another, so that the impeller average efficiency will be

$$\eta_{h\,med} = \frac{1}{r_o^2 - r_i^2} \int_{r_i}^{r_o} \eta_h(r) \times r \times dr \qquad (2.78)$$

The diagram is made with angle $\beta_{2i} = 90°$ and $\alpha_i = 90°$, whatever the radius, as shown in Fig. 2.13. It is worth mentioning that angles β_2 are smaller than $90°$ and reach this value only at the limit — $\beta_{2i} = 90°$.

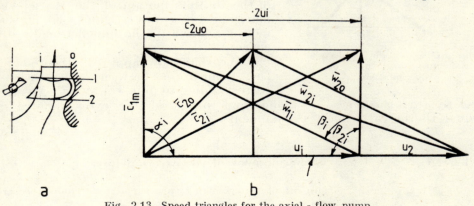

Fig. 2.13. Speed triangles for the axial - flow pump.

2.5 Estimation of Turbopump Performance Characteristics

The working member of any pump is the impeller (one or several). It is mounted on a shaft and driven by the motor coupled to the pump. On radial-flow pumps, the impeller is built up out of surfaces of revolution and is divided into channels by fixed blades whose surfaces are either cylindrical or have a double curvature. The fluid is led through these channels and, according to Bernoulli's equation, increases its pressure and motion (kinetic) energy; once past the channels, the fluid is led through the stator, where most of the motion (kinetic) energy is converted into position (potential) energy.

The motion of an elementary fluid particle within the strict boundaries set by the rotor of a radial flow-pump (Fig. 2.14) will now be considered. Let us assume that the liquid moves freely inside the pump (without friction), following paths that are parallel to the pump impeller blades. This condition can only be achieved if the blades have no thickness and are infinite in number. Now, let us further assume that the liquid motion is laminar and continuous. The fluid particle entering the impeller at point A moves, under the action of centrifugal forces, in a translation motion from the centre towards the periphery, at a relative velocity; at the same time, it acquires a rotational motion, along with the impeller, at an angular velocity ω, and a linear velocity (the peripheral velocity) u, both velocities growing as the particle moves away from the impeller centre. The particle arriving at the impeller periphery (point C) is thrown into the stator at the absolute velocity c_2. The trajectory in space of the absolute motion of the particle is determined by curve AC, and that of the motion relative to the impeller blades by curve AB.

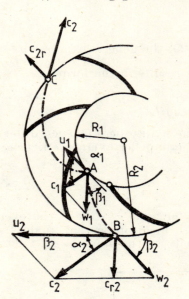

Fig. 2.14. Notation for deriving the turbopump fundamental operational equations.

We shall now apply the theorem of the momentum between the sections corresponding to points A and B. According to this theorem, the momentum torque of the external force is equal to the variation in the moment

$$T = mc_2 r_2 \cos \alpha_2 - mc_1 r_1 \cos \alpha_1 \qquad (2.79)$$

where $T = \dfrac{P}{\omega} t = \dfrac{A}{\omega}$ is the moment of the external force;

A — the mechanical work produced by the external force.

It is obvious that mechanical work A leads to an increased energy of particle motion, that is further converted into pressure energy. Thus, neglecting the losses in the turbomachine, the theoretical pumping head H_{theor} is related to the mechanical work by the expression

$$A = mgH_{theor} \text{ or } T = \frac{mgH_{theor}}{\omega} \tag{2.80}$$

Now, using relations (2.79) and (2.80) and bearing in mind that $r = u$, we can write for the theoretical pumping head in the case of an impeller with an infinite number of blades

$$H_{theor} = \frac{u_2 c_2 \cos \alpha_2 - u_1 c_1 \cos \alpha_1}{g}$$

And, because in practice the liquid enters the impeller at an angle $\alpha_1 = 90°$, we can say that

$$H_{theor} = \frac{u_2 c_2 \cos \alpha_2}{g} \tag{2.81}$$

Actually, the real pumping head of a turbopump is lower than the theoretical head because pressure losses occur and, at the same time, the velocities are non-uniformly scattered across the two side-surfaces of the impeller (this phenomenon being caused by the recycling of one part of the fluid). As a consequence, a pressure coefficient k_h is introduced, whose expression is

$$k_h = \eta_h k_c \frac{c_2 \cos \alpha_2}{u_2}$$

where η_h is the hydraulic efficiency, while k_c is the recycling coefficient.

Turbopump design parameters (i.e. the impeller geometry, the number of blades etc.) are determining factors for the pressure coefficient. Now, if in relation (2.81) we include this coefficient, which is always lower than unity, the expression for the pumping head will be

$$H = k_h \frac{u_2^2}{g} - \frac{k_h r_2^2}{g} w^2 \tag{2.82}$$

As can be seen from this relation, *the pumping head developed by the turbopump is proportional to the square of the angular rotational speed.*

The theoretical capacity of the turbopump Q_{theor}, is obtained by multiplying the impeller output cross-sectional area by the radial velocity component of the fluid streaming out of the impeller (see Fig. 2.14). Thus

$$Q_{theor} = \pi d_2 \, b c_{2r} \tag{2.83}$$

where b denotes the width of the turbopump impeller.

By means of the velocity triangle at the outlet (see Fig. 2.14), we can write the relation

$$c_2 \cos \alpha_2 = u_2 - c_{2r} \cot \beta_2$$

Substituting this in relation (2.81) and introducing c_{2r} from (2.83), an idealized relation is obtained relating the pumping head and the capacity, i.e. the theoretical characteristic $H(Q)$ of the turbopump

$$H_{theor} = \frac{u_2^2}{g} - \frac{u_2}{g} \times \frac{\cot \beta_2}{\pi d_2 b} Q_{theor}$$

If we now express the linear velocity u_2 as a function of the rotation speed n, in rpm

$$u_2 = \frac{\pi d_2 n}{60}$$

and we use the notation

$$A_1 = \frac{\pi^2 d_2^2}{3\,600\,g} \; ; \; B_1 = \frac{\cot \beta_2}{60\,g\,b}$$

the result will be

$$H_{theor} = A_1 n^2 - B_1 n Q_{theor}. \tag{2.84}$$

For a turbopump with several stages (that is, with series-coupled impellers) and, hence, summed-up pumping heads, equation (2.84) becomes

$$\frac{H_{theor}}{i} = A_1 n^2 - B_1 n Q_{theor} \tag{2.84 a}$$

For a turbopump with double-flow, (that is with parallel-coupled impellers), and, hence, summed-up flow rates, equation (2.84) becomes

$$H_{theor} = A_1 n^2 - B_1 n \frac{Q_{theor}}{i}$$

where i denotes the number of series or parallel-coupled impellers.

The last equation expresses a linear relationship between the pumping head and the capacity. It has an idealized character because, like relation (2.81), it excludes the losses due to recycling processes, which are actually independent of the discharge pipe flow rate. Allowing for these losses, which could be expressed through a component $C_1 n^2$, results in

$$H'_{theor} = A_1 n^2 - B_1 n Q - C_1 n^2 = (A_1 - C_1) n^2 - B_1 n Q$$

Likewise, it must be remembered that relation (2.81) was derived for the optimal performance conditions of the turbopump, i.e. for a rated flow of Q_r. When $Q < Q_r$, the fluid stream (flow) comes off the blade back-wall, while when $Q > Q_r$, the fluid stream comes off the blade front-wall. Pressure losses due to the fluid stream pull-off have vital influence on the turbopump speed-torque characteristic.

If in relation (2.84), which is the expression for the theoretical characteristic $H(Q)$ of the turbopump, due consideration is given to losses outlined in Fig. 2.15 that take place inside the turbopump, then the real characteristic $H(Q)$ is obtained as a second order expression

$$H = A_2 n^2 + B_2 nQ + C_2 Q^2 \tag{2.85}$$

By multiplying both terms of equation (2.84) by $\rho g Q$, one gets relation (2.86) that expresses the dependence between power and capacity, i.e. the $P(Q)$ characteristic of the turbopump, thus

$$P' = A_3 n^2 Q - B_3 n Q^2 \tag{2.86}$$

where P' is the power and Q the external capacity.

This relation does not take into consideration either the recycling or the mechanical losses, for neither of them is a function of the external capacity of the turbopump. Indeed, the recycling losses are produced as a consequence of the parasite recycling, between the rotor and the stator, of one portion of the flow, while the mechanical losses are the consequence

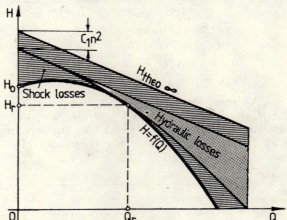

Fig. 2.15. Turbopump head-capacity characteristic.

of friction between bearings and seals, and between the rotor disk and the cycled fluid. If one takes these losses into consideration, along with the fact that they are proportional to the third power of the speed, the equation of the real characteristic of the turbopump power requirements as a function of the capacity results in the form

$$P = A_3 n^2 Q - B_3 n Q^2 + D_3 n^3 \tag{2.86,a}$$

If the turbopump rotational speed is constant ($n = $ constant), the above relation takes the following particular form

$$P = A_3' Q + B_3' Q^2 + D_3 \tag{2.86,b}$$

Likewise, in the case of relation (2.85) the useful power of the turbopump is determined

$$P_u = A_2' n^2 Q + B_2' n Q^2 + C_2' Q^3 \tag{2.87}$$

When the rotational speed of the turbopump is constant ($n = $ constant), relation (2.87) takes the particular form

$$P_u = A_2'' Q + B_3'' Q^2 - C_2'' Q^3 \tag{2.88}$$

Now, referring the useful power P_u to the power requirements P results in

$$\eta = \frac{\varrho Q H}{102 P} \tag{2.89}$$

Substituting the power requirements as expressed by relation (2.86,a) in (2.89) results in

$$\eta = \frac{\rho}{102} \times \frac{QH}{A_3 n^2 Q - B_3 nQ^2 + D_3 n^3} \qquad (2.90)$$

Taking account of the pumping head H as expressed by relation (2.85), one can write for the real characteristic of the efficiency, as a function of the capacity, the relation

$$\eta = \frac{\rho}{102} \cdot \frac{A_2 n^2 Q + B_2 nQ^2 + C_2 Q^3}{A_3 n^2 Q - B_3 nQ^2 + D_3 n^3} \qquad (2.90,a)$$

When the turbopump speed is constant ($n = $ constant), the above relation can be written

$$\eta = \frac{A_2'' Q + B_2'' Q^2 + C_2'' Q^3}{A_3' Q - B_3' Q^2 + D_3'} \qquad (2.90,b)$$

As seen from relation (2.90,b) and from the relevant diagram (Fig. 2.16), the efficiency curve $\eta(Q)$ shows a clear maximum which determines the rational range of use of the turbopump, that is the range where Q and H take values (for $n = $ constant) for which the efficiency is high enough (see Fig. 2.17).

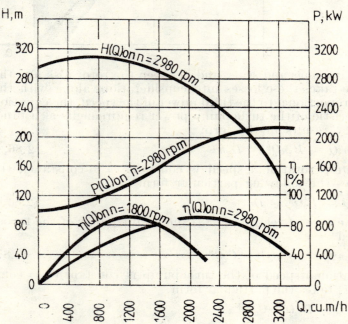

Fig. 2.16. Concrete characteristics of a radial-flow pump.

The analytical determination of the performance characteristics of real turbopumps requires first the determination of coefficients A, B, C and D, by means of at least three points taken from the characteristics supplied by the constructors of such machines (see Fig. 2.16).

Estimation of Turbopump Performance Characteristic

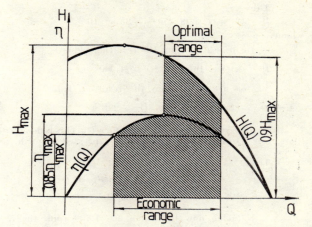

Fig. 2.17. Range of rational use for a radial-flow pump.

Example 1.1. Analytical determination of the $\eta(Q)$ characteristic. Construct, for a turbopump whose rated speed is $n_R = 2.980$ rpm, the function $\eta(Q)$ for a speed $n = 0.6\, n_R = 1.800$ rpm.

Solution. Find first characteristics $H(Q)$ and $P(Q)$, then characteristic $\eta(Q)$.

(1) *Determination of characteristic $H(Q)$.* This is done with relation (2.85); coefficients A_2, B_2 and C_2 included within this relation are determined for three points, namely

$Q = 0$ m³/h; $H = 300$ m; $P = 1.000$ kW
$Q = 600$ m³/h; $H = 310$ m; $P = 1.100$ kW
$Q = 2200$ m³/h; $H = 260$ m; $P = 1.880$ kW

Coefficient A_2 is found thus

$$A_2 n^2 = 300\ ;\quad A_2 = \frac{300}{2980^2} = 33.6 \cdot 10^{-6}$$

Substituting in relation (2.85) the value obtained for coefficient A_2, as well as values Q and H for the other two points, results in

$C_2 = -22 \times 10^{-6}$; $B_2 n = 0.03$; $B_2 = 10 \times 10^{-6}$

Thus:
$$H = (33.6\, n^2 + 10nQ - 26\, Q^2) \times 10^{-6}$$

and it follows that

— for $n = 2.980$ rpm the resulting value is
$$H = 300 + 30 \times 10^{-3} Q - 22 \times 10^{-6} Q^2$$

— for $n = 1.800$ rpm the resulting value is
$$H = 110 + 18 \times 10^{-3} Q - 22 \times 10^{-6} Q^2$$

(2) *Determination of characteristic $P(Q)$.* This is done with relation (2.86, a). Coefficients A_3, B_3 and D_3 in the relation are also determined for three points, namely

$$D_3 n^3 = 1000\ ;\quad D_3 = \frac{1000}{2980^3} = 38 \cdot 10^{-5}$$

$B_3 n = 145 \times 10^{-6}$; $B_3 = -48.5 \times 10^{-9}$

$A_3 n^2 = 0.1$; $A_3 = 11 \times 10^{-9}$

Thus
$$P = (11\, n^2 Q + 48.5\, nQ^2 + 38\, n^3)\, 10^{-9}$$
whence:
— for $n = 2.980$ rpm the resulting value is
$$P = 0.1 Q + 145 \times 10^{-6} Q^2 + 1.000$$
— for $n = 1.800$ rpm the resulting value is
$$P = 0.036 Q + 87 \times 10^{-6} Q^2 + 220$$

(3) *Determination of characteristic* $\eta(Q)$. This is done with relation (2.90, a) for $\rho = 1\,\text{kg/dm}^3$. Flow rate Q taken from Fig. 2.16 is expressed in m³/h, so a factor 3.600 should be introduced.
Thus
$$\eta = \frac{\rho}{102 \times 3600} \cdot \frac{A_2 n^2 Q + B_2 nQ^2 + C_2 Q^3}{A_3 n^2 Q - B_3 nQ^2 + D_3 n^3}$$

Introducing now the value $n = 1.800$ rpm. the result is
$$\eta = \frac{1}{102 \times 3.6} \cdot \frac{110 Q + 18 \times 10^{-3} Q^2 - 22 \times 10^{-6} Q^3}{0.036 Q + 87 \times 10^{-6} Q^2 + 220}$$

The $\eta(Q)$ characteristic for $n = 1.800$ rpm is shown in Fig. 2.16 (in broken lines). It will be seen that the maximum efficiency corresponds to a speed $n_R = 0.6$; it has the value 0.87, and occurs for $Q = 1.200$ m³/h.

Analysis of application 1.1 shows that the maximum efficiency does not change with change in the turbopump speed, which in turn proportionally affects the capacity of the pump. This is explained by the structure of relation (2.90,a), which shows that whereas the turbopump capacity varies proportionally with the speed, its efficiency does not change.

The shape of characteristics $H(Q)$, $P(Q)$ and $\eta(Q)$ depends on the turbopump construction specifications, which determine the values of coefficients A, B, C and D.

2.6. Turbopump Performance Characteristic Curves

2.6.1 Definitions and Classifications

Definitions. The performance of any turbopump can be expressed through a relation between the hydraulic quantities (Q, H) and the mechanical ones (P, n, η), this relation being in the form of function $f(Q, H, n, \eta) = 0$. This function can be shown graphically in a Cartesian system of co-ordinates and the resulting curve is called the *pump characteristic*. With this end in view, we choose among the variables the two having the greatest importance in the study of the operating conditions, while the other quantities are considered as parametrical. We thus obtain a function of two variables, called the turbopump characteristic. Its graphical representation, for a given range of variation in the values of the parameters, is called the characteristic curve.

The various possible characteristic curves are identified by the chosen characteristic quantity or by their purpose. Thus we have: the pumping head characteristic curve, $H = f(Q)$; the power characteristic curve, $P = f(Q)$; the efficiency characteristic curve, $\eta = f(Q)$ etc. A complete characteristic includes all these separate curves.

A turbopump is designed for optimal performance at parameters H and Q and at a constant speed n. But it is equally important to know its performance at other parameters, H_x, Q_x, and at a variable speed, n_x.

The performance characteristic curves are equally useful to three interested parties: the turbopump designer, for choosing the final structural solution after laboratory and actual tests; the pumping plant designer, to help him in the choice of the most adequate turbopump design for the application in view; and the user, to help him choose the best operating conditions.

Classification. There are three typical characteristic curves associated with turbopumps, namely the working, relative, and universal characteristics (or topograms).

Working characteristics. Those curves that represent the variation of one or several quantities as a function of another, taken alone and considered to be an independent variable, are called *working characteristics*. Their expression is of the form $y = f(x)$, where y is one of the characteristic quantities (the pumping head H, the power P, the efficiency η, etc.), and x is the independent variable (say, the flow rate Q).

When the speed, n, is considered to be constant, the graphical representation of relationships $H = f(Q)$, $P = f(Q)$ and $\eta = f(Q)$ will yield the working characteristics, for if $n =$ constant, then the varying quantities will be H, Q, P and η. If we now assume that one of these is an independent variable, say Q, and we represent graphically the variation of the others, i.e. H, P and η, we get the curves shown in Fig. 2.18, each called after the variable quantity it represents.

Thus curve $H(Q)$ is called the characteristic curve of the pumping head; curve $P(Q)$ is called the characteristic curve of the power; curve $\eta(Q)$, the characteristic curve of the efficiency.

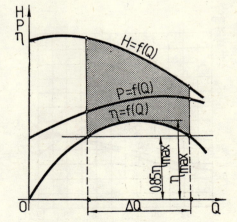

Fig. 2.18. Working characteristics.

Relative (gauge) characteristics. The need for characteristic curves, and their general use (as established through numerical analysis, or measurements carried out on real turbopumps and applied to similar dynamical turbopumps), have led to the conclusion that dimensionless relative quantities should be used as variables. All turbopump characteristic curves that can be drawn up using as variables the percentage values of the relevant quantities (percentages of the reference values, of which the most frequently are the rated values), represent *the relative (gauge) characteristics*. They allow comparisons to be made among various typical turbopumps. For illustration purposes, Fig. 2.19 represents the relative characteristics of various typical current turbopumps. As shown, the pumps have various specific speeds, which affords an easy comparison from the aspect of the main quantity variation; as also shown (Fig. 2.19,b) the lowest power variation (as a function of the flow rate) is found with mixed-flow pumps ($n_s = 280$), and the highest with radial-flow ($n_s = 42$) and axial-flow ($n_s = 700$) pumps.

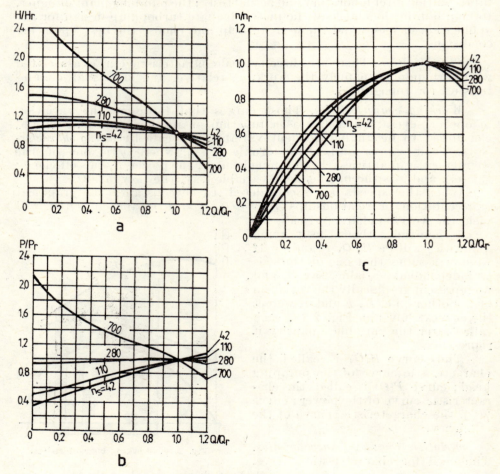

Fig. 2.19. Relative characteristics.

Universal characteristics (*topograms*). When the working characteristics of a turbopump, drawn for variable speeds or for several blade angles, and the topographic curves of iso-efficiencies η/η_{max} (practically constant), appear on the same diagram, this representation is called the *universal characteristic*; when this is expressed in relative values, it is called the *turbopump topogram*.

A turbopump topogram is drawn by representing, in a system of co-ordinates Q_x, H_x, functions of the form $H_x = f(Q_x n)$ (where the rotational speed n is a parameter), and by joining co-ordinates of the points of equal efficiency (see Fig. 2.20). When similitude is perfect, the iso-effi-

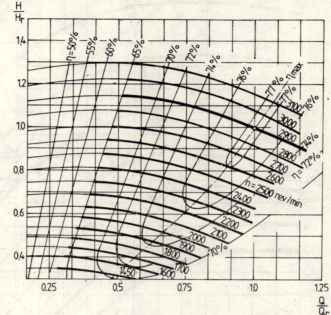

Fig. 2.20 Universsal characteristic of a turbopump.

ciency curves are parabolae. Because at low speeds efficiency η drops (the $Re = uD/\mu$ number is small), and also because at high speeds and flow rates cavitation phenomena might develop, the iso-efficiency curves are closed both at the bottom and at the top (that is towards Q_{min} and Q_{max}) and so resemble ellipses.

Using as co-ordinates the dimensionless values H/H_r, Q/Q_r, η/η_r and n/n_r, we obtain the performance topogram of the turbopump shown in Fig. 2.21.

All points on the same curve, say curve $\eta/\eta_{max} = 0.9$, show that efficiency $\eta = 0.9\ \eta_{max}$ is constant and equal to 90% of the turbopump maximum efficiency.

The same turbopump can operate at various flow rates, pumping heads and speeds; however, the economic performance at maximum efficiency is obtained only for those Q, H and n values that lead to point A (the optimal performance point). Curve $I - I'$ represents the main axis of topographic curves η/η_{max} = constant, and those operating conditions that best maintain the optimal efficiency are to be found along this curve.

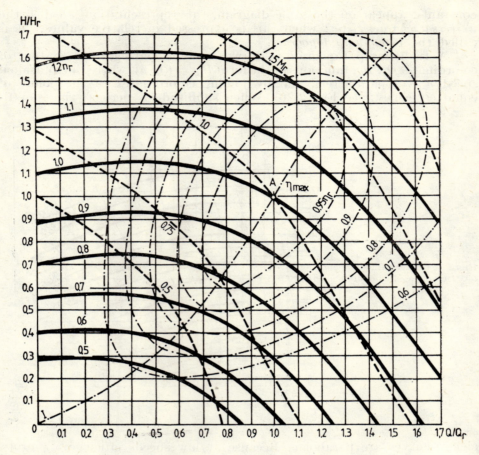

Fig. 2. 21. Performance topogram of the turbopump.

2.6.2 The Pumping Head-capacity Characteristic Curve $H(Q)$

Effects of outlet angle β_2. Assuming the water inlet into the pump is orthogonal, hence $\alpha_1 = 90°$ and $c_{u1} = 0$, the theoretical head, as computed before for $z = \infty$ and $s = 0$, is from Section 2.4

$$H_{t\infty} = c_{u2} \frac{u_2}{g} \tag{2.91}$$

For the same inlet conditions, three typical impellers are to be distinguished, according to whether $\beta_2 \lessgtr 90°$. The speed diagram at the outlet for each typical impeller, having the same u_2 and c_{m2} since n and Q are the same, is shown in Fig. 2.22.

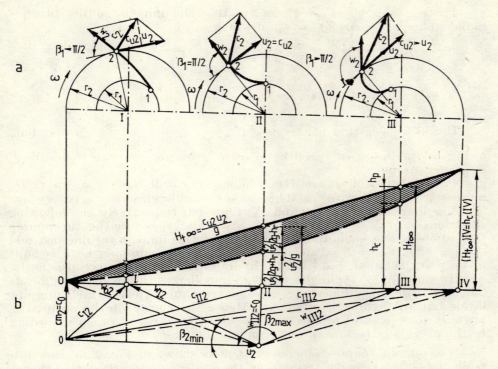

Fig. 2.22. Blade shape in a parallel plane (*a*) and the velocity triangles at the outlet (*b*) for pump impellers with $\beta \gtreqless 90°$.

Putting $\beta_2 < 90°$ (Case I), which is the case most frequently met with high-efficiency radial-flow turbopumps, results in

$$H_{t\infty} = h_{p\infty} + h_{c\infty} = c_{u2}\frac{u_2}{g} \qquad (2.92)$$

where

$$h_{p\infty} = \frac{u_2^2 - u_1^2}{2g} + \frac{w_1^2 + w_2^2}{2g} - h_r \text{ is the manometric term, while}$$

$$h_{c\infty} = \frac{c_2^2 - c_1^2}{2g} + h_r \text{ is the kinematic one.}$$

The linear variation of $H_{t\infty}$ with speed c_{u2}, for which $h_{p\infty} = 0.58\, H_{t\infty}$ and $t_{c\infty} = 0.42\, H_{t\infty}$ is shown in Fig. 2.22,b. Angle β_2 being small, c_{u2} is small too, hence $H_{t\infty}$ is low; but with c_2 being small, losses $h_r = kc_2^2$ are small too, and therefore the hydraulic efficiency η_h of the impeller will be high.

113

Putting $\beta_2 = 90°$ (Case II) means that the impeller outlet blade is radial and $c_{u2} = c_2 \cos 90° = u_2$, in which case

$$h_{e\infty} = \frac{c_{u2}^2}{2g} + h_r = \frac{u_2^2}{2g} + h_r$$

$$h_{p\infty} = \frac{u_2^2}{2g} - h_r$$

It will be seen that in this case $H_{t\infty} = h_{p\infty} + h_{e\infty} = \frac{u_2^2}{g}$ grows but, owing to the increased speed c_2, losses h_r grow too and the hydraulic η_h efficiency drops.

Putting $\beta_2 > 90°$ (Case III), the impeller will have $c_{u2} > u_2$; $H_{t\infty}$ increases very much with c_2 and hence the hydraulic efficiency η_h is very low.

The limit cases (Cases 0 and IV) represent respectively the following situations: (a) $h_{p\infty} = 0$; $h_{e\infty} = 0$; $H_{t\infty} = 0$, and therefore the turbopump fails to operate as such and becomes, below this limit, a turbine instead; (b) $h_{p\infty} = 0$; $h_{e\infty} = H_{t\infty}$, and therefore the whole pumping head is built up only by the kinetic head and the hydraulic efficiency drops consistently.

To sum up, low β_2 angles result in impellers with low pumping heads but high hydraulic efficiency, whereas large β_2 angles lead to high pumping heads but low hydraulic efficiency. Hence the choice of the optimal β_2 angle is a conjectural problem, to be solved from case to case according to the prevailing technical and economic factors.

The internal characteristic $H = f(Q)$. As shown in Fig. 2.23, flow-rate Q changes the velocity triangle at the outlet, so, for $Q_x < Q$, velocity $Cm_{2x} < Cm_2$ is obtained. And, inasmuch as β_2 (hence the directions of w_{2x}) is kept unchaged, relation (2.92) yields, for n = constant

$$H_{t\infty} = \frac{u_2}{g} Cu_{2x} = \frac{u_2}{g} \left(u_2 - \frac{\cot \beta_2}{\pi D_2 b_2} Q_x \right) \tag{2.93}$$

because $cu_{2x} = u_2 - Cm_{2x} \cot \beta_2$, and $Cm_{2x} = \frac{Q_x}{\pi D_2 b_2}$. Expression (2.33) represents the linear dependence of $H_{t\infty} = f(Q_x)$ in the shape of a straight line (see Fig. 2.24), that is parallel to or inclined with respect to Q_x according to whether β_2 is equal to, larger or smaller than 90°.

For no-load running conditions, $Q_x = 0$, and hence the discharge valve is closed, and for any β_2 we can write

$$H_{t\infty} = \frac{u_2^2}{g} \tag{2.94}$$

Now, supposing we have to deal with the semi-real impeller for which $z \neq \infty$ and $s = 0$, instead of $H_{t\infty}$ (the theoretical pumping head $H_{tx} < H_{t\infty x}$), we can write

$$H_{tx} = \frac{1}{(1+p)} H_{t\infty x} = \frac{u_2}{g(1+p)} \left[u_2 - \frac{\cot \beta_2}{\pi D_2 b_2} Q_x \right] \tag{2.95}$$

Theory of Fluid Potential Motion as Applied to Turbopumps

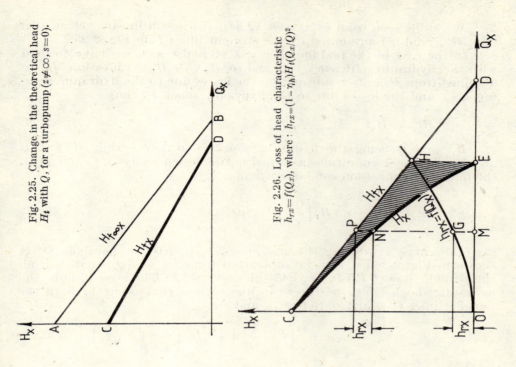

Fig. 2.25. Change in the theoretical head H_t with Q, for a turbopump ($z \neq \infty, s=0$).

Fig. 2.26. Loss of head characteristic $h_{rx} = f(Q_x)$, where: $h_{rx} = (1-\eta_{th})H_t(Q_x/Q)^2$.

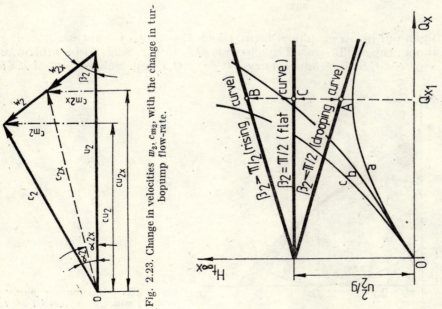

Fig. 2.23. Change in velocities w_2, c_{m2}, with the change in turbopump flow-rate.

Fig. 2.24. Pumping head and power characteristic curves for the ideal impeller.

115

As can be seen from expression (2.35), H_{tx} is still linear, yet smaller than $H_{t\infty x}$ and is represented by the straight line CD in Fig. 2.25.

In the case of the real impeller, $z \neq \infty$ and $s \neq 0$, we have to deal with the hydraulic efficiency (η_h) and so $H_x = \eta_h H_{tx}$. Elevation H_x is obtained from H_{tx}, after subtracting the losses due to the hydraulic resistance h_{rx} and the losses due to the hydraulic shock h_s, thus

$$H_x = H_{tx} - h_{rx} - h_s \tag{2.96}$$

H_{tx} and Q being the hydraulic shock conditions, while H_t and Q are the non-shock conditions assumed in the design analysis, i.e. (H_d, Q_d), the following expression can be written for h_{rx}

$$h_{rx} = (1 - \eta_h) H_t \left(\frac{Q_x}{Q}\right)^2 = f(Q_x) \tag{2.97}$$

The curve expressing the load losses, $h_{rx} = f(Q)$, is parabola OGH shown in Fig. 2.26; it is to be subtracted from $H_{tx} = f(Q)$, that is from the straight line $CPHD$, to yield the curve CNE. The losses due to the hydraulic shock are to be traced to dissipations that occur at both the impeller inlet (h_{s1}) and outlet (h_{s2}), and so

$$h_s = h_{s1} + h_{s2} = \frac{\varphi}{2g}\left(1 - \frac{Q_x}{Q}\right)^2 \left[u_1 + \left(\frac{u_2^2}{1+p} \cdot \frac{D_2}{D_3}\right)^2\right] \tag{2.98}$$

where

$$\varphi = 0.3 + 0.6 \frac{\beta_2}{90°}$$

When put into a graphical form, shock $h_s = f(Q_x)$ is expressed through a parabola; this is illustrated by curve $h_{s0}Q_sG$ in Fig. 2.27, to be subtracted from curve CHE to produce curve $H_x = f(Q_x)$, represented by $AMSR$.

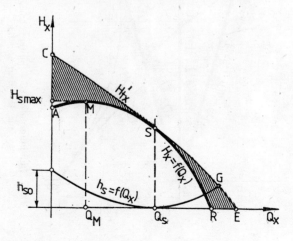

Fig. 2.27. Change in the pumping head H_x with Q_x (the $AMSR$ curve).

The analytical form of relationship $H_x = f(Q_x)$ is inferred from equation (2.96), where h_{rs} and h_s are replaced by expressions (2.97) and (2.98). Since $\mu_1 = \pi D_1 \dfrac{n}{60}$ and $\mu_2 = \pi D_2 \dfrac{n}{60}$, after regrouping we obtain for the relationship $H_x = f(Q_x, n)$ the expression

$$H_x = k_1 n_x^2 + 2 k_2 n_x Q_x - k_3 Q_x^2 \tag{2.99,a}$$

or else, for $n = $ constant

$$H_x = C_1 + C_2 Q_x - k_3 Q_x^2$$

with

$$C_1 = k_1 n^2 \,;\, C_2 = 2 k_2 n \tag{2.99, b}$$

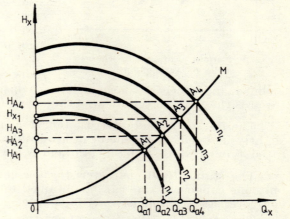

Fig. 2.28. $H = f(Q_x)$ characteristic for a turbopump whose speed n is chosen as a parameter.

Coefficients k_1, k_2, k_3 represent constant values for a certain turbopump, while equation (2.99, a) represents the curve of the pumping head, or the internal characteristic of the turbopump for a constant speed.

For a speed different from the first one (for instance the rated speed) another parabola is obtained, congruent with the first one, as shown in Fig. 2.28; for n_1, n_2, n_3, n_4, other parabolae will be drawn having the vertices A_1, A_2, A_3, A_4 lined up along a reverse parabola (concave upwards), OM. Knowing the characteristic curve for a certain speed (n_1), the curve of the pumping head can be drawn for any other (n_2). To this end, we take the highest point (A_1) on the parabola, corresponding to (n_1), then we draw parabola OA_1M, with ordinate H_x as a centre line. Then we take curve $H_{x1} A_1 n_1$, in a translation motion until vertex (A_1) reaches (A_2), the co-ordinates being:

$$Q_{a2} = Q_{a1} \frac{n_2}{n_1} \,;\, H_{A2} = H_{A1} \left(\frac{n_2}{n_1}\right)^2 \tag{2.100}$$

The congruence law holds true over the whole performance range of the turbopump, except for cavitation conditions.

Shape of head-capacity, $H(Q)$ *characteristic*. This characteristic can be of several kinds (Fig. 2.29): rising, drooping, steep, flat, stable or unstable.

The rising $H(Q)$ characteristic (curve a) is that curve for which the head rises continuously as the capacity is descreased; it has the shape of a parabola whose vertex lies on the pumping head axis.

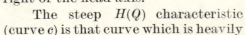

Fig. 2.29. Various head-capacity characteristics for turbopumps.

The drooping $H(Q)$ characteristic (curve b) is that curve for which the dead head [the intersection point between characteristic $H(Q)$ and the pumping head axis determines the dead head, H_0] is lower than the head corresponding to other capacities; this curve has the shape of a parabola with vertex shifted to the right of the head axis.

The steep $H(Q)$ characteristic (curve c) is that curve which is heavily inclined, and hence the head increases considerably from the rated flow towards the zero; the gradient of this curve is greater than or equal to 0.5.

The flat $H(Q)$ characteristic (curve d) is that curve which is only slightly inclined, and hence the head only slightly increases from the rated flow towards zero; its gradient is lower than or equal to 0.1.

The stable $H(Q)$ characteristic (see curve a) is that curve for which only one capacity value can be obtained for any value of the head: this characteristic is implicitly a rising curve.

The unstable $H(Q)$ characteristic (see curve e) is that curve which generates the same head for two or more capacities. Such a characteristic is common with axial-flow pumps.

The main cause of differences between the characteristics of axial-flow and radial-flow pumps is the existence of parasite flows (streams) within the impeller channels.

2.6.3 The Power-Capacity Characteristic Curve $P(Q)$

Writing the expression for the ideal impeller power ($z = \infty$, $s = 0$) in the form $P_{t\infty} = \rho g Q H_t$ and making the substitution $H_{t\infty} = A - B' \cot \beta_2 Q_x$ results in

$$P_{t\infty} = A Q_x - B Q_x^2 \qquad (2.101)$$

which shows that the power varies parabolically or linearly with the capacity according to whether $\beta_2 \lessgtr 90°$, as illustrated in Fig. 2.30.

But since the theoretical power of the semi-real impeller ($z \neq \infty$, $s = 0$) has the expression $P_t = \sum P_{t\infty}$, and $\sum \Delta P_p$ represents the power lost through hydraulic, volumetric and mechanical losses, it follows that for $\beta_2 < 90°$ the power curves shown in Fig. 2.31 result, according to the following relation

$$P_p = P_t + \Sigma \Delta P_p = \frac{P_t}{\eta_p} \qquad (2.102)$$

It should be mentioned [as shown on (Fig. 2.31)] that $\Sigma \Delta P$ is minimum for the designed capacity (i.e. the rated flow), while for $Q = 0$, the resulting power will be $P = P_0 \neq 0$, which is called *shut-off power*.

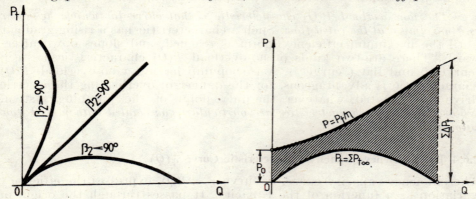

Fig. 2.30. Change in the turbopump power as a function of flow rate.

Fig. 2.31. Power characteristics for $\beta_2 < 90°$.

Again in Fig. 2.30, one can see that for $\beta_2 > 90°$ (the case of an impeller with blades bent in a forward direction, which is very rare), the power grows very rapidly with the flow rate. This is an adverse condition for the driving motor and must be borne in mind even for pumps fitted with normal impellers (whose blades are bent backwards), because after repairs or overhauls any phase-reversal in the electric drive brings about a reversal in the pump rotation direction which, for the case of the impeller turning in the reverse direction, means $\beta_2 > 90°$. The danger is the higher, the greater the opening of the discharge valve.

According to their shape, characteristic curves $P(Q)$ may be with or without overload (Fig. 2.32).

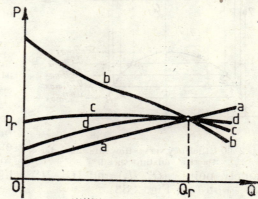

Fig. 2.32. Various power-capacity characteristics.

The overload $P(Q)$ characteristic is that curve for which the power increases with flows higher or lower than the rated flow. There are two kinds of overload characteristics: rising and drooping. The rising $P(Q)$ characteristic is associated with radial-flow pumps (curve *a*), and the drooping characteristic with axial-flow pumps (curve *b*). With such turbopumps, the power of the electric drive will be chosen as a function of the value of the maximum power requirements over the ΔQ flow range for which the turbopump under consideration was chosen.

The non-overload $P(Q)$ characteristic is that curve for which the power is a maximum at the rated flow. Such a characteristic has a rising gradient until the maximum efficiency point is reached, and slopes down afterwards. There are two kinds of no-overload $P(Q)$ characteristics: steep (curve *d*) and flat (curve *c*). A turbopump having a no-overload $P(Q)$ characteristic is advantageous, for the danger of overloading the electric drive is eliminated, whatever the duty point of the pumping system. *No-overload $P(Q)$ characteristics are particularly associated with mixed-flow pumps* ($n_s \approx 280$).

2.6.4 The Efficiency-Capacity Characteristic Curve $\eta(Q)$

The efficiency-capacity curve, or curve $\eta = f(Q)$, expresses the efficiency variation as a function of the capacity. It passes through the origin of the system of co-ordinate axes $\eta - Q$, reaches a maximum value at the design rated flow, then slopes down to zero (for $H = 0$). This curve shows a very flat maximum for radial-flow pumps, and for this reason such machines offer a very wide flow range; whereas with axial-flow pumps the curve vertex is sharper, which means that their flow range is more restricted.

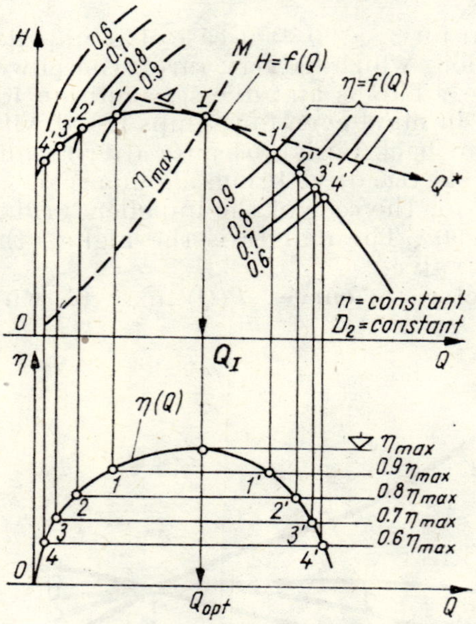

By putting the turbopump efficiency in the form $\eta_h = 1 - \dfrac{\Sigma h_r}{H_t}$ with $h_r = f(Q^2)$, the curve $\eta_h = f(Q)$ of a turbopump operating at a constant speed can be represented as shown in Fig. 2.33.

It was theoretically demonstrated, and experimentally corroborated, that on the $H = f(Q)$ curve there are pairs of points for which the efficiencies are equal, namely points (Q_1, H_1) and (Q'_1, H'_1) which have the efficiencies $\eta_1 = \eta'_1$ and so on (see Fig. 2.33).

Fig. 2.33. Efficiency versus flow rate characteristic at a constant speed.

As already shown, the global efficiency of a turbopump is $\eta = \eta_m \eta_v \eta_h$, with η_v and η_m being little affected by the speed change (from 3% to 5%). Thus, by changing the turbopump speed curves, different efficiency curves are obtained (Fig. 2.34), for which η_{max} corresponds to speed n_2 (optimal speed).

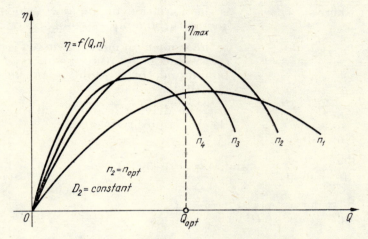

Fig. 2.34. Efficiency versus flow rate characteristics at a variable speed.

2.7 Characteristics of Turbopumps Operating in Complex Conditions

All hydraulic machines (including tubopumps) are reversible, either converting hydraulic into mechanical energy, or vice-versa. In what follows, turbopump performance will be considered for three characteristic duties, namely as a pump, as a brake and as a turbine.

2.7.1 Pump Duty

In pumping duty, the turbopump carries the flowrate Q_P, corresponding to a pumping head $H_p = H$ and to an efficiency η_p, so that at the shaft the following power is developed

$$P_p = \frac{\rho Q_p H_p}{102 \, \eta_p} \qquad (2.103)$$

On starting the turbopump, the electric drive supplies an active starting torque higher than the resistive one, and so the rotational speed increases and reaches the rated value after a certain time t_s called the "starting time".

The dynamic equation of the combined rotational members is expressed in the form

$$I \frac{d\omega}{dt} = M_m - M_r \tag{2.104}$$

where $\dfrac{d\omega}{dt}$ — is the variation of angular velocity of the motor-turbopump assembly;

$I = mi^2 = \dfrac{GD^2}{4g}$ — is the moment of inertia (where i is the gyration radius of rotating members of the turbopump-motor assembly, transmission included);

T_m, T_r — are the driving (motor) and the resisting (load) torques respectively.

Now, since

$$\frac{d\omega}{dt} = \frac{\pi}{30} \frac{d_n}{d_t} \quad \text{and, because } 30 \times 4 \times g/\pi = 375, \text{ it follows}$$

$$\frac{dn}{dt} = 375 \frac{M_m - M_r}{GD^2} \tag{2.105}$$

Thus the time needed for starting, i.e. for the speed to rise from 0 to rated speed n_r, will be

$$t_a = \frac{GD^2}{375} \int_0^{n_r} \frac{dn}{M_m - M_r} \tag{2.106}$$

2.7.2 Braking Duty

When the electricity supply to the turbopump drive is switched off, the driving torque drops and so does the rotational speed of the motor-turbopump assembly. The time interval taken for the speed to fall from n_r to zero is called the *braking time* (t_b) and is to be borne in mind when water hammers are to be analysed.

In order to find the speed variation with time for the braking duty, one of the methods below can be applied.

The constant efficiency method. When the electricity supply is switched off, $T_m = 0$, and the equation of motion becomes

$$i \frac{d\omega}{dt} = -M_r = \frac{P}{\omega} = -\frac{30}{\pi} \frac{P}{n} \tag{2.107}$$

If for the standard performance conditions we take parameters Q_0, H_0, P_0, η_0 and $n_0 = n_R$, then, according to the similitude laws, it follows

$$Q = Q_0 \left(\frac{n}{n_0}\right) \text{ and } H = H_0 \left(\frac{n}{n_0}\right)^2 \tag{2.108}$$

so that
$$P = \frac{\rho Q_0 H_0}{\eta} \left(\frac{n}{n_0}\right)^3 \qquad (2.109)$$

But since $\dfrac{dn}{dt} = \dfrac{30}{\pi} \dfrac{d\omega}{d_t}$, and putting $k = \left(\dfrac{30}{\pi}\right)^2 \dfrac{\rho}{I\eta} Q_0 H_0$, (where $\eta = \eta_0 =$ constant), the result will be

$$\frac{dn}{n^2} = -k \frac{dt}{n_0^3} \qquad (2.110)$$

The function $n = f(t)$, being the variation of speed with time, is obtained by integration, using an integration constant $C = \dfrac{1}{n_0}$ so chosen that, for $t = 0$, the result is $n = n_0$ ($\omega = \omega_0$)

$$n = \frac{n_0}{1 + \dfrac{k}{n_0^2} t} \quad \text{or} \quad n = \frac{n_0}{1 + \dfrac{P_0}{I \omega_0^2} t} \qquad (2.111)$$

From relation (2.111) we get $n = 0$ for $t_b = t = \infty$; the expression is therefore an approximate one, because actually $t_b = 5 - 15$ sec.

The variable efficiency method. As known, the performance efficiency of any pump varies with the speed, and hence $\eta = f(n)$; thus, a new expression results for $n = f(t)$.

Considering that

$$\eta = 1 - \eta_0) \left(\frac{n_0}{n}\right)^{1/5} \qquad (2.112)$$

as well as equation (2.110), we get after substitution

$$\frac{dn}{d_t} = -k_1 \frac{n_0^2}{1 - (1 - \eta_0) \dfrac{n_0}{n}^{1/5}} \qquad (2.113)$$

with

$$k_1 = \left(\frac{30}{\pi}\right)^2 \frac{\rho Q_0 H_0}{I n_0^3}$$

Simplifying equation (2.113) by writing $A = (I - \eta_0) n_0^{1/5}$, results in

$$\frac{dn}{d_t} = -k_1 \frac{n^{11/5}}{n^{1/5} - A} \qquad (2.114)$$

or

$$\frac{dt}{dn} = -\frac{1}{k_1 n^2} + \frac{A}{k_1} \frac{1}{n^{11/5}} \qquad (2.115)$$

Thence by integration

$$t = \frac{1}{k_1 n} + \frac{5A}{k_1} \int n^{-6/5} + C_1 \qquad (2.116)$$

where $C_1 = \frac{5}{6} \frac{A}{k_1} n_0^{-6/5} - \frac{1}{k_1 n_0}$, as established by the limit conditions $(t = 0, n = n_0)$. After substitution

$$t = \frac{1}{k_1 n_0}\left(\frac{n_0}{n} - 1\right) - \frac{5A}{6k_1} \frac{1}{n_0^{6/5}} \left[\left(\frac{n}{n_0}\right)^{-6/5} - 1\right] \qquad (2.117)$$

or finally,

$$t = \frac{I\omega_0}{P_0 \eta_0}\left(\frac{n_0}{n} - 1\right) - 0.823\,(1 - \eta_0)\frac{I\omega_0^2}{P_0 \eta_0}\left[\left(\frac{n}{n_0}\right)^{-6/5} - 1\right] \qquad (2.118)$$

which shows a more rapid braking action of the pumping unit.

On pumping units with a high moment of inertia (GD^2), hence having a long braking time (t_b) and having no check valve on the discharge pipe, flow reversal $(-Q)$ can start at a time $t < t_b$; in such a case, the turbopump operates in the second braking mode.

The finite difference method. When written in finite differences, the motion equation (2.107) becomes

$$n_2 - n_1 = -\frac{450}{\pi^2 I}\left(\frac{Q_1 H_1}{n_1 \eta_1} + \frac{Q_2 H_2}{n_2 \eta_2}\right)\Delta t \qquad (2.119)$$

or

$$n_2 - n_1 = -k(m_1 + m_2)\Delta t \qquad (2.120)$$

where $k = 450\,\rho/\pi^2 I$ and $m = QH/n\eta$.

Assuming that the operating values are n_0 and m_0, then equation (2.120) can also be written in the form

$$\alpha_2 - \alpha_1 = -k_1(\beta_1 + \beta_2)\Delta t \qquad (2.121)$$

where $\alpha = \frac{n}{n_0}$, $\beta = \frac{m}{m_0}$, $k_1 = \frac{km_0}{n_0}$

By means of equation (2.120) or (2.121) we construct by successive trials the turbopump curve, $n = f(t)$. Thus, knowing n_1 and m_1, at a time t_1, Δt is increased and n_2 is computed by accepting initially that $m_1 = m_2$. Then, m_2 is determined from the topogram of the relevant pump, and n_2 is computed to a second approximation, and so on.

2.7.3 Turbine Duty

After $t > t_b$, the turbopump being without a check valve, it begins to work in a turbine mode. Indeed, the liquid will flow through the turbopump in a reverse direction, $(-Q)$, under a head H_t and at a velocity n_t. If the turbopump were coupled to an electric generator this would produce the useful mechanical power

$$P_t = \frac{\rho\, Q_t H_t}{102\, \eta_t} \tag{2.122}$$

where η_t is the global efficiency of the pump running in turbine duty;
Q_t — the capacity as a turbine;
$H_t = H_g - \sum h_r$ — the net available head.

When in turbine mode, the turbopump starts racing in no-load running. More often than not, this has detrimental consequences upon its rotating members because the centrifugal forces are increased by the increased rotational speed. To avoid this phenomenon, and also to protect the unit against the adverse effects of water hammers in the discharge pipe, a check valve is mounted on the latter. However, for specific speeds within the interval $n_s = 35-55$, the check valve can be omitted. The racing phenomenon is more dangerous for the electric motor of the turbopump, since this motor can only afford overspeeds of 20 per cent above the rated speed.

The topogram of a turbopump (as expressed in relative quantities as against the Q_0, H_0, η_0, n_0 values). for the three operating duties described above, is illustrated in Fig. 2.35.

The operational characteristic curves for the three duties mentioned before are drawn by the constructor, on the basis of laboratory tests. There are, however, some reports about tentative theoretical determinations of the operational characteristic curves of turbopumps, in the four quadrants; these theoretical determinations all start from the Euler's torque equation and use relationships borrowed from the impeller dynamics. Considering head H as a parameter and applying the above methods we find for $n = f(Q)$

$$\left. \begin{array}{l} n = \dfrac{-BQ^2 \pm \sqrt{B^2Q^2 + 4A(KQ^2 + H)}}{2A} \\[2mm] n' = \dfrac{-BQ^2 \pm \sqrt{B^2Q^2 - 4A(KQ^2 - H)}}{2A} \end{array} \right\} \tag{2.123}$$

where $K = \dfrac{1-\eta_h}{Q^2}$; $A = \dfrac{\pi^2}{60^2 g}(D_2^2 - D_1^2)$

and $B = -\dfrac{1}{60 g b_2 \operatorname{tg} \beta_2} + \dfrac{1}{60\, g b_1 \operatorname{tg} \beta_1}$.

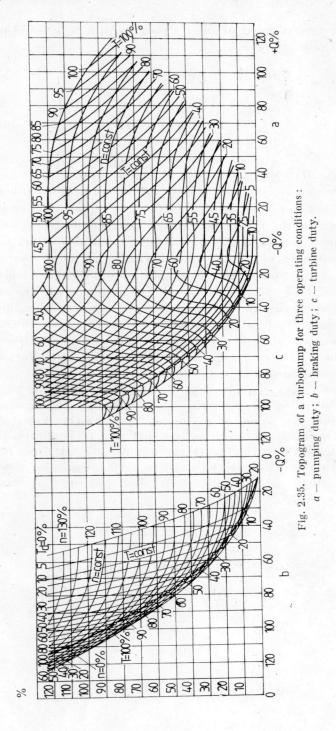

Fig. 2.35. Topogram of a turbopump for three operating conditions:
a — pumping duty; *b* — braking duty; *c* — turbine duty.

REFERENCES

2.1. KIRILOV, I. I. *Teoryia turbomashiny*. Mashinostroienie, Leningrad, 1972.
2.2. LOMAKIN, A. A. *Tzentrobejnye i oseviie nasosi*, Mashinostroienie, Moscow, 1966.
2.3. PAVEL, D. HÂNCU, S. *Utilaje hidromecanice pentru sistemele de îmbunătățiri funciare, stații de pompare*. Ceres Publishing House, 1974.
2.4. PAVEL, D. *Mașini hidraulice*. Energy Publishing House, Bucharest, 1955.
2.5. PFLEIDERER, C. *Strommungs Machienen*. Springer Verlag, Berlin, 1952.
2.6. PFLEIDERER, C. *Die Kreiselpumpen für Flussikeiten und Gase*. Springer Verlag, Berlin, 1961.
2.7. TASCA, D., ZIDARU, G. *Contribuții la studiul similitudinii turbomașinilor, mărimi unitare adimensionale*. I. P. Bucharest, 1965.

3

Operation and Control of Turbopumps Connected to Pumping System

3.1 Operation of Turbopumps Connected to Pumping System

3.1.1 Determination of the Pumping-plant Duty Point

Suppose a turbopump is connected to a pumping system with the purpose of carrying a flow-rate Q at a pumping head H, in order to cover the total pumping head H_t needed by the system (H_t being the sum of static, H_{st}, and dynamic, H_{dy} heads). Operation becomes possible when the head achieved by the turbopump equals that needed by the pumping system.

Figure 3.1 represents, in a system of $H - Q$ co-ordinates, the pump characteristic, $H(Q)$, and the system characteristic, $H_s(Q)$. The duty point of the pumping plant, comprising the turbopump and the system, lies where the manometric head, supplied by the pump, reaches the system pumping head, that is at point A. This point was obtained graphically at the intersection of the two characteristic curves, that is of the turbopump and of the pumping system. Co-ordinates of point A can be analytically derived by solving the system of equations

$$\begin{cases} H_t = H_{st} + RQ_x^2 \\ H_x = C_1 + C_2 Q_x - C_3 Q_x^2 \end{cases} \tag{3.1}$$

and finding Q_x by solving the following further equation

$$C_1 - H_{st} + C_2 Q_x - (h_3 - R)Q_x^2 = 0 \tag{3.2}$$

Capacity Q corresponding to point A should, in the ideal case, be coincident with the capacity corresponding to the maximal turbopump efficiency (η_{max}). In Fig. 3.1 $\eta_A \neq \eta_{max}$, since only a particular choice of the turbopumps, leading to $Q = Q_A$, can lead to $\eta_A = \eta_{max}$.

When the design analysis, carried out to establish the parameters of a pumping system, is based on uncertain assumptions (for instance, as a result of poor choice of the frictional coefficient, or of the introduction of additional hydraulic losses), a real characteristic $H_{sI}(Q)$ will result for the system which is different from that given by the analysis, $H_{sII}(Q)$, and consequently a real duty point will result that will be shifted from the latter and will thus yield a differential flow rate ΔQ (Fig. 3.2). This difference is smaller, the steeper the characteristic curve of the turbopump connected to the pumping system [curve $H_1(Q)$] — for which the difference will be ΔQ_1. When the characteristic is flat [curve $H_2(Q)$], the difference will be bigger i.e. ΔQ_2.

When the system only includes one turbopump and one pumping pipe, finding the duty point is a simple matter. When, however, the system includes several turbopumps and several pipes, all having different characteristics, finding the duty point is more complicated. In what follows, it will be shown how to find the duty point for a variety of cases currently met with in practice.

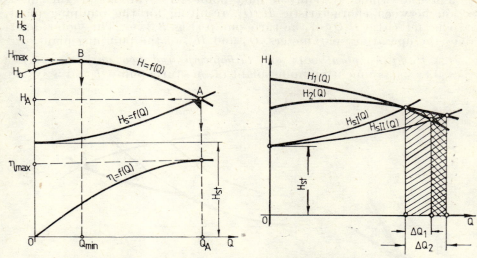

Fig. 3.1. Pump and system characteristics [$H(Q)$ and $H_s(Q)$].

Fig. 3.2. Shift of the real operational point as against the estimated point.

(1) *Pumping plants with one turbopump and two series-connected pipes.* Consider a pumping plant including a turbopump *1* and a system comprising pipe sections *I* and *II* of different diameters, yet series-connected and operating under a static pumping head of H_{st} (Fig. 3.3, *a*). Curves

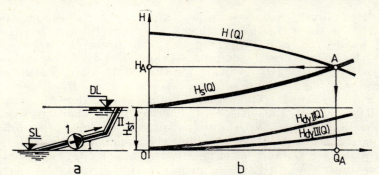

Fig. 3.3. Operational point of a pumping plant including one turbopump and two series-connected pipe sections:
a — process flow diagram; *b* — diagrammatic representation.

$H_{dyI}(Q)$ and $H_{dyII}(Q)$ represent the characteristics of dynamic pumping heads of pipe sections *I* and *II*. The characteristic curve $H_s(Q)$ resulting for the system will be found by the summation of ordinates of characteristic curves $H_{dyI}(Q)$ and $H_{dyII}(Q)$, and of the ordinate of the static head characteristic H_{st} = constant, for the same outputs. Thus

$$H_t = H_{st} + H_{dyI} + H_{dyII} \tag{3.3}$$

For the whole system, the duty point (A) will lie at the intersection between characteristic $H_s(Q)$, resulting for the pumping plant, and characteristic $H(Q)$ of the turbopump (Fig. 3.3, b). This point determines the operating parameters, Q_A and H_A, of the pumping plant.

(2) *Pumping plant with one turbopump and two parallel-connected pipes.* Let us assume a pumping plant of one turbopump *1* and a system

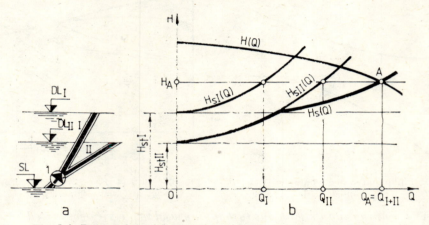

Fig. 3.4. Duty point of a pumping plant including one turbopump and two parallel-connected pipe sections:
a — process flow diagram; *b* — diagrammatic representation.

including two parallel-connected pipes *I* and *II*, that discharge the liquid carried at two different heads (levels), as determined by the static pumping heads H_{stI} and H_{stII} (Fig. 3.4, *a*). Curves $H_{rI}(Q)$ and $H_{rII}(Q)$ represent the characteristics of pipe sections *I* and *II* respectively. The characteristic curve $H_s(Q)$ resulting for the whole plant will be obtained by summing up the abscissae of characteristic curves $H_{sI}(Q)$ and $H_{sII}(Q)$ for the same output. The duty point A for the whole pumping plant will appear at the intersection between the characteristic drawn for the pumping system $H_s(Q)$ and that drawn for the turbopump $H(Q)$. As previously, this point determines the operating parameters of the pumping plant, i.e. Q_A and H_A (Fig. 3.4, *b*). Capacity Q_A represents at the same time both the turbopump output and the sum of capacities Q_I and Q_{II}.

The partial capacities carried through the separate pipes *I* and *II* are found at the intersection of a horizontal line through point *A* and the curves $H_{sI}(Q)$ and $H_{sII}(Q)$.

(3) *Pumping plant with two parallel-connected turbopumps on a common pipe.* Let us assume a pumping plant including two turbopumps,

1 and *2*, and *a* common discharge pipe *I*, delivering under a static head H_{st} (Fig. 3.5, *a*).

Curves $H_1(Q)$ and $H_2(Q)$ represent the characteristics of turbopumps *1* and *2* respectively. The resulting turbopump characteristic $H(Q)$ (for the parallel duty) is found by summing up the abscissae of characteristics $H_1(Q)$ and $H_2(Q)$ for the same ordinates. For the whole pumping plant

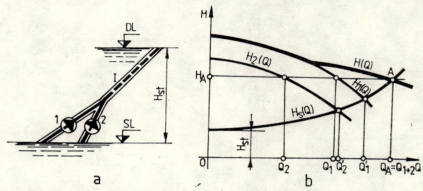

Fig. 3.5. Duty point of a pumping plant including two parallel-connected pumps discharging on a common pipe :
a — process flow diagram; *b* — diagrammatic representation.

the duty point *A* lies at the intersection of the curve resulting for the turbopumps, $H(Q)$, and of characteristic $H_s(Q)$ resulting for the pumping system. This point determines the operating parameters of the pumping plant, i.e. Q_A and H_A.

At the intersection of a horizontal straight line through point *A* with curves $H_1(Q)$ and $H_2(Q)$, the duty points corresponding to each separate pump will be found, and implicitly their partial capacities Q_1 and Q_2. It will be seen that these capacities are lower than Q_1 and Q_2 drawn for the separate operation of each turbopump, taken alone. And they are the lower, the flatter the turbopump characteristics $H(Q)$ and the steeper the pumping system characteristic, $H_s(Q)$.

(4) *Pumping plant with two turbopumps connected in parallel through two pipe sections, to a common pipe.* Consider two turbopumps, *1* and *2*, parallel-connected and delivering through two separate pipe sections *I* and *II* until point *B*, where they are jointed into a single common pipe *III* (Fig. 3.6, *a*). The turbopump characteristics, as reduced at point *B*, are $H_{1\,red}(Q)$ and $H_{2\,red}(Q)$. These curves were drawn as the difference between the ordinates of turbopump characteristics $H_1(Q)$ and $H_2(Q)$ respectively, and the ordinates of the separate pipe section characteristics, i.e. $H_{sI}(Q)$ and $H_{sII}(Q)$ respectively for the same capacities (Fig. 3.6, *b*). Thus

$$H_{1\,red} = H_1 - H_{tI}; \quad H_{2\,red} = H_2 - H_{tII} \tag{3.4}$$

By combining the turbopump characteristics drawn for point *B*, i.e. $H_{1\,red}(Q)$ and $H_{2\,red}(Q)$, the turbopump characteristic for the parallel operation is found. The duty point of the pumping plant, point A, will lie at the intersection of turbopump reduced characteristic $H_{red}(Q)$ and of characteristic $H_{sIII}(Q)$ resulting for the common pipe. This point determines the operating parameters of the pumping plant, i.e. Q_A and H_A (Fig. 3.6, c).

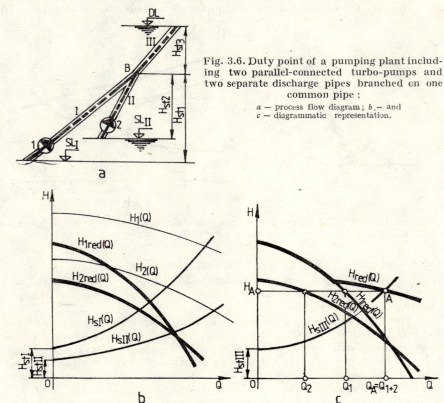

Fig. 3.6. Duty point of a pumping plant including two parallel-connected turbo-pumps and two separate discharge pipes branched on one common pipe:
a — process flow diagram; b — and
c — diagrammatic representation.

Taking a horizontal straight line through point A, the separate operational points for each separate turbopump will be found, i.e. $H_{1\,red}(Q)$ and $H_{2\,red}(Q)$ respectively, and implicitly their partial capacities, i.e. Q_1 and Q_2 respectively.

(5) *Pumping plant with two series-connected turbopumps and two separate pipes*. Let us assume two series-connected turbopumps, *1* and *2*, delivering on two separate pipes, *I* and *II*, discharging at two different levels, under static heads H_{stI} and H_{stII} (Fig. 3.7).

The case illustrated in Fig. 3.7 can be reduced to that shown in Fig. 3.4, if the reduced characteristic curve $H_{s\,II\,red}(Q)$ of the system is known. This curve results as the difference between the ordinates of the characteristic curve $H_2(Q)$ of the turbopump and ordinates of characteristic curve $H_{sII}(Q)$ of the pumping system for the same capacities (Fig. 3.7, b).

By combining the reduced characteristic curve $H_{sI}(Q)$ of pipe I with that of pipe II, $H_{sIIred}(Q)$, the characteristic curve resulting for the system is obtained as $H_s(Q)$. The duty point of turbopump 1 (point A) will lie at the intersection of this curve with the turbopump characteristic, $H_1(Q)$, (Fig. 3.7, c). This point determines the operating parameters of

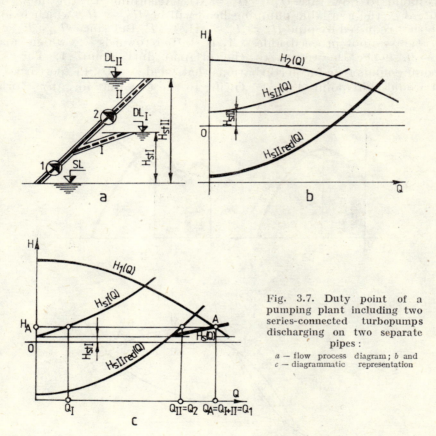

Fig. 3.7. Duty point of a pumping plant including two series-connected turbopumps discharging on two separate pipes:

a — flow process diagram; b and c — diagrammatic representation

turbopump 1, i.e. Q_A and H_A. The intersection points of a horizontal straight line through point A with curves $H_{sI}(Q)$ and $H_{sII}(Q)$ will yield the partial capacities, Q_I and Q_{II}, carried through pipes I and II. Output Q_{II} is the same as output Q_2 delivered by turbopump 2.

3.1.2 Stable and Unstable Turbopump Operation

Assuming a turbopump whose working characteristic has the shape of a parabola with apex in the first quadrant (see Fig. 3.8, c) of the system of $H-Q$ axes, and assuming further that such a pump operates at point A

as a component of a pumping system having $H_{st} < H_0$, we can state that with such pumps we can distinguish two operating zones. They are the unstable zone for $0 < Q < Q_M$ and the stable zone, for $Q_M < Q < Q_A$, where Q_A is the duty capacity. Assuming again that an external cause increases the pumped capacity by $(+\Delta Q)$, the new duty point A_1 will correspond to flow rate $Q_{A_1} = Q_A + \Delta Q$. According to the new duty point, A_{A_1}, the available pumping head will be $H_{A1} < H_A$, which is lower than that required by pipe $H_s = H_{A1} + \Delta H_1 + \Delta H$. But since $H_{A1} < H_{sA1}$, the operation cannot proceed unless A_1 is shifted towards A, which makes $\Delta Q \to 0$, hence the return to the original duty point A. The same reasoning holds true for perturbations that yield $(-\Delta Q)$, i.e. perturbations that would shift point A in A_2. Owing to the fact that for any working

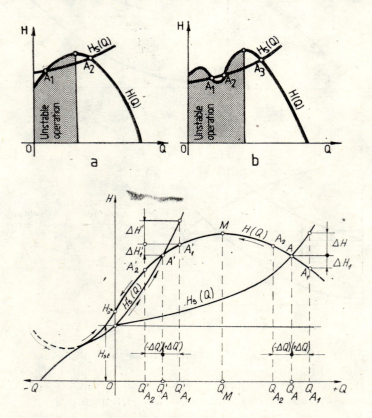

Fig. 3.8. Stable and unstable operation of a turbopump:
a — radial-flow pump; b — axial-flow pump; c — diagrammatic representation.

parameters on the right-hand branch of point M, after the perturbations disappear, the system resumes its initial operating conditions, this branch is called the *stable branch*.

Let the branch on the left-hand side of point M be $0 < Q < Q_M$, and let A' be a hypothetical duty point. When an external perturbation develops that raises the pumped capacity by $(+ \Delta Q')$, the duty point A' will lie in A'_1, corresponding to a higher capacity, i.e. $Q'_{A_1} = Q'_A + \Delta Q'$. For the new working point the pumping head available to the pump is $H'_{A1} = H_A + \Delta H'_1$, lower than that needed by the system, $H'_s = H'_A + \Delta H'_1 + \Delta H'$, and thus, in order to meet the requirements, H'_{A1} should grow and tend towards H_M. With the perturbations $(-\Delta Q')$ the situation is similar, and consequently on this branch the duty point does not resume its original position after the perturbing phenomenon is eliminated. For this reason, the branch on the left-hand of point M is called *the unstable branch*.

3.1.3 Turbopump-to-system Connection Configurations

Parallel-connection. This configuration has the main purpose of increasing the pumping plant capacity.

(1) *Parallel-connection of identical turbopumps.* Assuming $Q_1 = Q_2$ to be the turbopump capacities at a certain moment in time, and $H_1 = H_2$ to be the pumping heads, the combined output results as

$$Q_c = Q_1 + Q_2 = 2Q_1 = 2Q_2 \qquad (3.5)$$

The characteristic of this configuration is found by successively summing up the outputs of the two turbopumps for various pumping heads. In the case under consideration, the abscissae of curves K_1 and K_2 are doubled to get the characteristic of the combined structure $K_1 + K_2$ (Fig. 3.9).

When the two turbopumps are connected to a pumping system with a given characteristic, S_1, while C is the duty point of a single pump, then for the whole system the duty point will be C. In this situation, the combined capacity Q_c for the two pumps is lower than the sum of the two capacities $Q_1 + Q_2$, and capacity Q_c will depend on the shape of the system characteristic; in fact, it will be greater the smaller the slope of the system characteristic (i.e. curve S_1 with

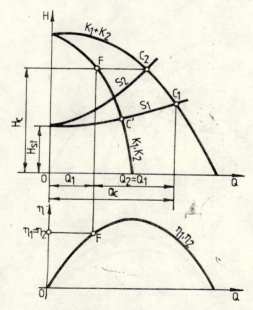

Fig. 3.9. Parallel-connection of two identical pumps.

respect to curve S_2). Likewise, it can be seen that for this connection, the pumping head of the system is increased with respect to the pumping head of a single pump ($H_c > H'_c$), and this is the more so, the steeper the slope of the system characteristic.

In conclusion, the results of the parallel-connection depend greatly on the system particulars which this configuration should match.

(2) *The efficiency of the parallel-connection* is called the ratio of the useful power of the configuration to the power requirements of the two pumps

$$\eta_{c.par.} = \frac{\rho Q_c H_c}{P_1 + P_2} \tag{3.6}$$

Because of their connection, the two turbopumps will have the duty point at the level of pumping head H_c, the output discharged by each turbopump will be $Q_f = 1/2 Q_f$, and if the total efficiency is the same for both turbopumps, then

$$P_1 = P_2 = \frac{\rho Q_f H_c}{\eta} = \frac{1}{2} \cdot \frac{\rho Q_c H_c}{\eta} \tag{3.7}$$

that is

$$\eta_{c.par.} = \frac{Q_c H_c}{\frac{1}{2} \cdot \frac{\rho Q_c H_c}{\eta} + \frac{1}{2} \frac{\rho Q_c H_c}{\eta}} = \eta \tag{3.8}$$

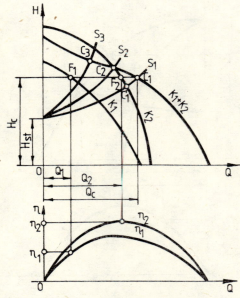

Fig. 3.10. Parallel-connection of two non-identical pumps.

Therefore, whenever two identical turbopumps are parallel-connected, the coupling efficiency will be the same as the total efficiency of the turbopump used, as shown by the duty point F (see also Fig. 3.9).

(3) *Parallel-connection of non-identical turbopumps.* This situation is illustrated in Fig. 3.10, where it can be seen that for certain performance ranges situated above the maximum pumping head of the smaller pump, the characteristic of the whole configuration lies below the characteristic of the bigger pump. This is a consequence of the fact that one of the two turbopumps operates in braking mode (that is, flybacks will develop in the smaller pump). Therefore, this typical coupling configuration is advisable only for those systems

whose S_1 and S_2 characteristics intersect the $K_1 + K_2$ characteristic on the C_2C_1D branch. Characteristics of the S_3 type are not recommended, because the flow discharged is even lower than that of a single turbopump of the configuration; in addition, one of the pumps works in the braking mode.

The efficiency of the configuration is found in the same way, that is by taking a horizontal straight line through point C at the elevation of the pumping head H_c; thus the efficiencies are found by determining each duty point of each separate turbopump on the efficiency curves (i.e. points F_1 and F_2). Hence

$$\eta_{c.par} = \frac{\rho Q_c H_c}{\dfrac{\rho Q_1 H_1}{\eta_1} + \dfrac{\rho Q_2 H_2}{\eta_2}} = \frac{Q_c}{\dfrac{Q_1}{\eta_1} + \dfrac{Q_2}{\eta_2}} \qquad (3.9)$$

In conclusion, two or more non-identical parallel-connected turbopumps perform as one single turbopump whose characteristic would be $K_1 + K_2$.

(4) *Series-connection*. The main purpose of this configuration is to increase the pumping head of the plant.

(5) *Series-connection of identical turbopumps*. The flow rate cycled through the turbopumps is the same ($Q_1 = Q_2$) due to the mass continuity principle, so in order to find the characteristic of this configuration, i.e. $K_1 + K_2$, the ordinates of the pumping head curves, K_1 and K_2, will be successively doubled (see Fig. 3.11).

The efficiency of the configuration will depend on the system characteristic shape. In contrast with the parallel-connection, the series-connection is effective when the system characteristic is of the S_1 type, that is, when the slope of the characteristic is steep.

(6) *The efficiency of the series-connection* is the ratio of the power corresponding to the effective (actual) pumping head of the configuration and of the turbopump power requirements

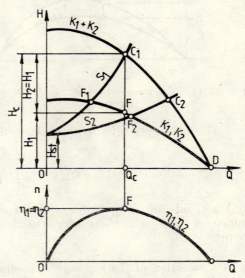

Fig. 3.11. Series-connection of two identical turbopumps.

$$\eta_{c.ser} = \frac{\rho Q_c H_c}{P_1 + P_2} \qquad (3.10)$$

and, since the capacity of the configuration, as given by the working point C, is $Q_1 = Q_2 = Q_c$ and $H_c = H_1 + H_2 = 2H_1 = 2H_2$, it follows

$$\eta_{c.ser.} = \frac{\rho Q_c H_c}{\dfrac{\rho Q_1 H_1}{\eta} + \dfrac{\rho Q_2 H_2}{\eta}} = \eta \tag{3.11}$$

that is, like the parallel-connection, the efficiency of the connection is equal to efficiency of each turbopump.

(7) *Series-connection of non-identical turbopumps*. Looking at the diagram of this typical configuration (Fig. 3.12), one can see that there is an area S_2D where it cannot be recommended, because the pumping head is lower than that developed by one single turbopump separately connected to the system. Indeed, within this area, part of the pumping head developed by the turbopump having a K_2 characteristic is used to make up for the operation on the negative branch of the pumping head curve of the other turbopump, when the pumping capacities are higher than those given by a zero pumping head.

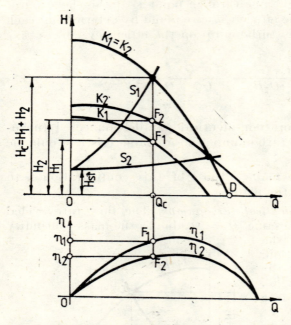

Fig. 3.12. Series-connection of two non-identical pumps.

The efficiency of the combination is given by the relation

$$\eta_{c.ser} = \frac{\rho Q_c H_c}{\dfrac{\rho Q_1 H_1}{\eta_1} + \rho \dfrac{Q_2 H_2}{\eta_2}} = \frac{H_c}{\dfrac{H_1}{\eta_1} + \dfrac{H_2}{\eta_2}} \tag{3.12}$$

and depends on the position of the working point of the configuration, and hence on the system characteristic and on the efficiencies of the two turbopumps.

3.1.4 Connection of Turbopumps According to the Shape of their Characteristic Curves

(1) *Connection of flat characteristic turbopumps*. Let us consider two separate systems of which one has a steep characteristic while the other has a flat one; to each network, two flat characteristic $H(Q)$ turbopumps are connected. Let us denote by $H_{sl}(Q)$ the curve of the steep character-

istic system and by $H_{sII}(Q)$, that of the system with a flat characteristic, while the curve for the parallel-connection of the turbopumps will be denoted by $2H_{par}(Q)$ and that of the series connection, by $2H_{ser}(Q)$. These notations refer to Figs. 3.13 and 3.14.

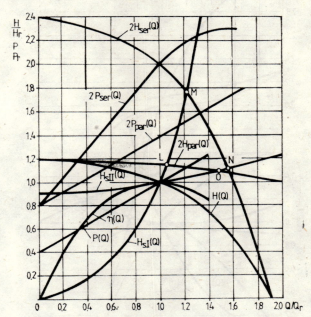

Fig. 3.13. Head and power resulting when pumps with flat characteristics are connected to the system.

Now, at the intersection of characteristic curves $H_{sI}(Q)$ and $2H_{par}(Q)$ we find the working point for the system, L, which yields the flow rate $Q_L = 1.07\ Q_r$, while at the intersection of characteristic curves $H_{sI(Q)}$ and $2H_{ser}(Q)$, we find the working point M which yields the flow rate $Q_M = 1.23\ Q_r$ (Fig. 3.13). The difference of capacities Q_M and Q_L shows a substantial gain in output, $Q = 0.16\ Q_r$, and therefore in the case of a pumping system with a steep characteristic, the series connection of turbopumps with flat characteristics is recommended.

At the intersection between characteristic curves $H_{sII}(Q)$ and $2H_{par}(Q)$ we find the working point O, yielding the capacity $Q_0 = 1.48\ Q_r$ and the power $P_0 = 1.64\ P_r$, while at the intersection of characteristic curves $H_{sII}(Q)$ and $2H_{ser}(Q)$ we find point N, which yields the capacity $Q_N = 1.57\ Q_r$ and the power $P_N = 2.3\ P_r$. The $Q_N - Q_0$ difference shows a negligible gain in output, $Q = 0.09\ Q_r$, whereas the power difference shows a substantial power gain of $P = 0.66\ P_r$. In view of the high power requirements, the resulting ratios are $P_N/Q_N = 1.47$ and $P_0/Q_0 = 1.11$, and it follows that for a system having a flat characteristic the parallel connection of turbopumps with flat characteristics is recommended.

(2) *Connection of turbopumps with a steep characteristic.* Let us consider two separate systems or networks, of which one has a steep and the other a flat characteristic; to each system, two turbopumps both having a

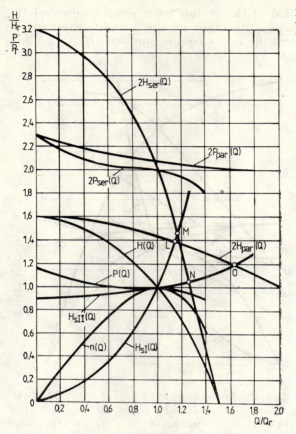

Fig. 3.14. Head and power resulting when pumps with a steep characteristic are connected to the system.

steep characteristic, $H(Q)$, are connected. The curves drawn for the pumping system and for the turbopump configurations have the same notations as in Fig. 3.13, with the same meanings. At the intersection between characteristic curves $H_{sI}(Q)$ and $2H_{par}(Q)$ we find point L, yielding capacity $Q_L = 1.14\, Q_r$ and power $P_L = 2.08\, P_r$, while at the intersection between characteristic curves $H_{sI}(Q)$ and $2H_{ser}(Q)$ we find point M, yielding capacity $Q_M = 1.16\, Q_r$ and power $P_M = 1.95\, P_r$ (Fig. 3.14). The $Q_M - Q_L$ difference shows a negligible output gain, $Q = 0.02\, Q_r$. In contrast to turbopumps with a flat characteristic (see Fig. 3.13), power P_M is lower than power P_L. This is due to the flat power characteristic, $P(Q)$, of turbopumps. Therefore, in the case of a pumping system having a steep characteristic $H_s(Q)$, the series-connection of turbopumps having steep characteristics is recommended.

The working point N, yielding capacity $Q_N = 1.26\, Q_r$, will be found at the intersection between characteristic curves $H_{sII}(Q)$ and $2H_{par}(Q)$, whereas the duty point M, yielding capacity $Q_M = 1.62\, Q_r$, will be found at the intersection between characteristic curves $H_{sII}(Q)$ and $2H_{ser}(Q)$. The $Q_N - Q_M$ difference will show a substantial gain, i.e. $\Delta Q = 0.36\, Q_r$.

Therefore, in the case of a pumping system whose characteristic curve $H_s(Q)$ is flat, the parallel-connection of turbopumps having a steep characteristic $H(Q)$ is recommended.

3.1.5 Restriction Criteria for the Number of Turbopumps that can be Parallel-connected.

The number of turbopumps that may be parallel-connected affects both the unit capacity of the turbopumps and the general capacity of the pumping plant; it is a function of the components of the system pumping head. Indeed, as can be seen in Fig. 3.15, when the system pumping head has no dynamic component the unit output will not depend on the number of turbopumps involved (curve a); in the reverse case, the unit output will be in inverse proportion to the number of turbopumps (curve b). Looking now at Fig. 3.16, it will be seen that the general capacity, delivered

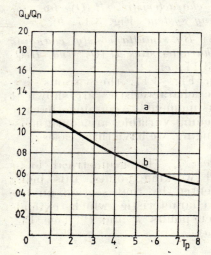

Fig. 3.15. Change in the unit output, Q_u, of a turbopump, as a function of the number of parallel-connected turbopumps, T_p, and of the nature of the pumping system.

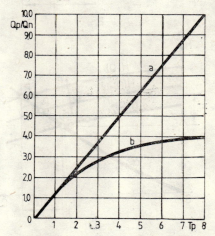

Fig. 3.16. Change in the output of the pumping system, Q_p, as a function of the number of parallel-connected turbopumps, T_p and of the nature of the pumping system.

by the pumping system, whose pumping head has no dynamic component (curve a), varies linearly with the number of pumps, while that delivered by a pumping system, whose pumping head has a dynamic component (curve b), varies nonlinearly with their number. Therefore in the case of a system whose pumping head has a dynamic component, a so-called "System saturation" is produced with increase in the number of pumps, which calls for restriction on the number of parallel-connected pumps. The maximum number of parallel-connected pumps is limited by that duty point beyond which the cost of the next additional pump connected to the system is not economically offset by the gain in performance.

3.2 Flow Control of Turbopumps Connected to the System

3.2.1 Shifting the Duty Point. Control Range

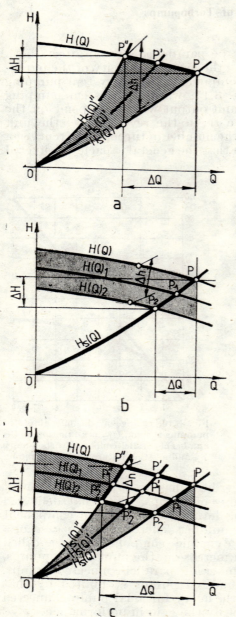

In principle, the flow rate of a pumping system can be controlled by shifting its duty point, which can be achieved in three ways:

(1) *By changing only the pumping system characteristic $H_s(Q)$, and keeping the same pump characteristic $H(Q)$*, (Fig. 3.17, a).

(2) *By changing only the pump characteristic $H(Q)$, while keeping the same characteristic $H_s(Q)$ for the pumping system* (Fig. 3.17, b).

(3) *By simultaneously changing both characteristics $H_s(Q)$ and $H(Q)$* (Fig. 3.17, c).

In Fig. 3.17, the various hydraulic characteristics are shown at various moments. Assuming that the intervening variables are continuously changing, we have three cases:

Case 1. The geometrical locus for all working points, this being represented by that section on the $H(Q)$ characteristic which extends between points P and P'' (see Fig. 3.17, a).

Case 2. The geometric locus of all working points, this being represented by that section on the $H_s(Q)$ characteristic that extends between points P and P_2 (see Fig. 3.17, b).

Case 3. The geometric locus of all operating points, this being repre-

Fig. 3.17. Control of the pumping system flow rate, by shifting the duty point:
a — by changing the pumping system characteristic, $H_s(Q)$; b — by changing the turbopump characteristic, $H(Q)$; c — by simultaneously changing both characteristics, $H_s(Q)$ and $H(Q)$.

sented by that surface that lies between points P, P'', P_2'', and P_2 (see Fig. 3.17, c).

When choosing the control procedure, several aspects should be borne in mind, particularly including economy, reliability and the automation potentialities.

As a general rule, the control action of a pumping plant extends only over the flow capacity. The major index defining the quality of the flow rate control is the range denoted by ΔQ and defined as the ratio of the minimum capacity, Q_{min} to the maximum one, Q_{max} that can be achieved through the control procedure under consideration

$$\Delta Q = \frac{Q_{min}}{Q_{max}}$$

Now, if the minimum capacity is taken as the basic quantity, equal to unity, the control range can be expressed as a ratio such as 1 : 10, 1 : 20, etc.

3.2.2 Control Actions Based on Changes in the System Characteristic

These control actions are all based on alteration of the pumping system characteristic, $H_s(Q)$, while keeping unchanged the turbopump characteristic, $H(Q)$. They are achieved through very simple means but are associated with high energy losses, and as a consequence the overall efficiency of the pumping plant is reduced (see Section 3.3.2). That is why these control actions were called "quantitative controls" in the literature.

(1) *Control action by throttling (lamination) of the fluid flow.* This is achieved by means of a valve V mounted on the turbopump discharge pipe, P (Fig. 3.18, a). By closing this valve, an additional head loss is

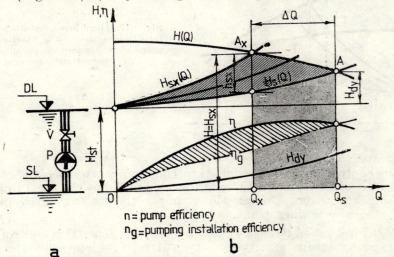

Fig. 3.18. Control through throttling of the fluid flow :
a — flow process diagram; *b* — diagrammatic representation.

artificially created in the system, which changes the system dynamic head, H_d, and thereby the dynamic characteristic of the pumping plant, $H_d(Q)$ (Fig. 3.18, b).

The hydraulic plant, including the turbopump and the pumping system, recovers its balance when the pumping head of the turbopump, H, equals that of the system H_t. Hence we can write

$$H = H_t = H_{st} + H_{dy} + h_{rx} \tag{3.13}$$

where H_{st} is the system static head; H_{dy} the system dynamic head; and h_{rx} the additional variable head loss, artificially introduced into the system by the valve V. By varying the head loss, h_{rx}, by means of the valve V, the pumping system capacity is changed over the control range ΔQ (see Fig. 3.18, b); the situation created by complete shut-off of the valve is represented by a perfectly vertical position of curve $H_{sx}(Q)$, that is perfectly overlapped on the pumping head axis, OH, for which $Q = 0$ (dead head).

The energy losses associated with the control process (through throttling) are represented by the area bordered by the turbopump characteristic $H(Q)$ on one side, and that of the system, $H_s(Q)$ on the other (see Fig. 3.18). These losses are the higher, the steeper the pump characteristic and the flatter the system characteristic (Fig. 3.19). That is why control action through throttling is recommended only for pumping systems fitted with radial-flow pumps having low specific speeds, $n_s < 100$, as the $H(Q)$ characteristic of such pumps is flat.

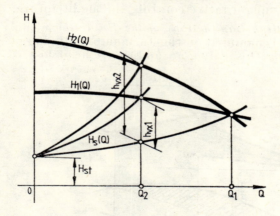

Fig. 3.19. Head losses resulting from the throttling action, as a function of the shape of turbopump and pumping system characteristic curves.

The control process through throttling offers the benefit of using structurally simple components and means, but it has the great drawback of causing high energy losses which, implicitly, lower the overall efficiency of the pumping plant.

(2) *Control action through by-passing the fluid flow.* This control action is achieved by means of an auxiliary valve, V, mounted on a pump derivable (by-pass) pipe P (Fig. 3.20, a). In this way, the discharged flow is partially returned to the suction side (i.e. to the suction tank).

Referring to Fig. 3.20 *b*, curve $H_s(Q)$ represents the characteristic of the discharge pipe, and curve $H_{der}(Q)$ that of the by-pass. By summing up the two curves, the characteristic $H'_s(Q)$ resulting for the system is found. At the intersection of this latter characteristic with the turbopump characteristic, $H(Q)$, the duty point for the pumping plant (point A_3) and its performance parameters, i.e. Q_{A3} and H_{A3}, are obtained.

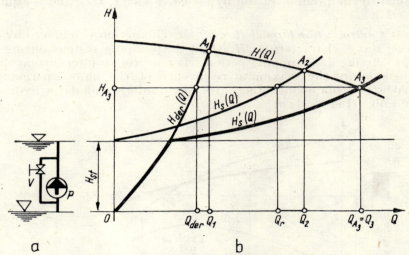

Fig. 3.20. Control through by-passing the fluid flow:
a — process flow diagram; *b* — diagrammatic representation.

Whenever the by-pass capacity control action is used, the system characteristic becomes more flat and the duty point for the whole plant is shifted to the right (that is from point A_2 towards point A_3), which thus increases the turbopump capacity (see Fig. 3.20, *b*). The resulting capacity will therefore be

$$Q_3 = Q_r + Q_{der}$$

where Q_3 is the overall flow; Q_r the discharged flow; Q_{der} the by-passed flow.

Thus, whenever using the flow-by-pass control action, turbopump capacity will be increased above the rated one, and at the same time the power requirements will be increased too, but this applies only to radial-flow pumps. To eliminate the detrimental effects of overheating the pump, whenever applying this control process it is advisable to use only axial- and mixed-flow pumps, because they have drooping and flat power characteristics, $P(Q)$, respectively.

However, the control process through flow by-pass can be applied with radial-flow pumps too, but in such cases the electric drives should be adequately oversized. In conclusion, flow by-pass control can be used in the two following cases:

— with high specific speed turbopumps, since with such machines the danger of unstable operation capacities close to zero is eliminated;

— with low specific speed turbopumps, when it is desired to eliminate overheating of the cycled liquid, since such machines operate at capacities close to zero.

The flow by-pass control action is a complex process that requires the oversizing of the hydraulic components of the pumping plant; at the same time the process is associated with high energy losses that are proportional to the product of the bypassed capacity, Q_{der}, and the pumping head H_{A3}.

(3) *Control action through flow off-set.* This action is achieved by changing the static characteristic $H_{st}(Q)$ of the pumping system during turbopumps starting and stopping periods. It requires an intermittent duty for the turbopump, the maximum recurrence of the on-off periods being limited by the compensating action of a hydraulic accumulator (hydro-pneumatic tank) (Fig. 3.21, *a*).

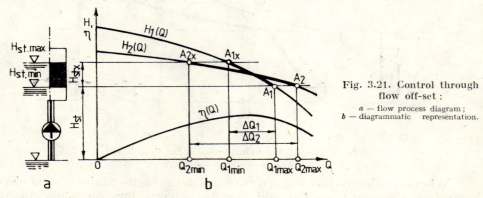

Fig. 3.21. Control through flow off-set:
a — flow process diagram;
b — diagrammatic representation.

The hydraulic accumulator is branched on the discharge side. Its purpose is to off-set the difference between the flow rate required by the system and that pumped by the turbopump as the hydraulic accumulator is filled up or drained off, the static head, H_{st}, of the pumping system is accordingly increased or lowered (Fig. 3.21, *b*).

The balance of the turbopump system hydraulic plant is reached when the pumping head of the turbopump, H, is equal to that of the pumping plant, H_t. Hence, we can write

$$H = H_t = H_{st} + H_{st\,x} \tag{3.14}$$

where H_{st} is the static head of the pumping plant, and $H_{st\,x}$, the deviation from the same, for the said plant.

With such control actions, the recommended turbopumps are those with a steep characteristic $H(Q)$ [i.e. curve $H_1(Q)$ in Fig. 3.21, *b*], because in this case the flow-rate control range ΔQ_1 is rather narrow, and hence the turbopumps will be used close to their maximum efficiency. If turbopumps with a flat characteristic $H(Q)$ are used [i.e. curve $H_2(Q)$ in Fig. 3.21, *b*], the resulting flow-rate control range, ΔQ_2, will be broad and the corresponding efficiency values, found on the $\eta(Q)$ curve, will be low as they represent extreme working points on the characteristic.

The off-set control offers some benefits (in the first place, low operating costs) but at the same time requires high investment costs (due to the exaggerated volume of the hydraulic accumulator). Since the volume of the latter is proportional to the unit output of the controlled pump, this method is recommended only for pumping systems fitted with low flow rate pumps ($Q \approx 10$ 1/s).

3.2.3 Control Actions Based on Changes in the Turbopump Characteristic

These are carried out by changing the turbopump characteristic, $H(Q)$, while keeping the same characteristic for the pumping system, $H_s(Q)$. In practice, they are achieved by means either of electric adjustable speed drives, or special mechanical gears (by altering the position of impeller blades or by adjustable inlet guide vanes). Whatever the control solution chosen, the energy losses are small and the overall efficiency of the pumping plants is considerably increased over those of unadjustable pumps (by 30 per cent); likewise, the electric power requirements are substantially reduced (by up to 25 per cent).

There are three main procedures for controlling the performance of pumping plants by acting on the turbopump characteristic: (a) by varying the turbopump rotational speed; (b) by altering the position of turbopump impeller blades; (c) by adjustable inlet guide vanes.

(1) *The control action by varying the turbopump rotational speed.* This is achieved with the help of adjustable-speed drives. Indeed, by varying the turbopump driving speed, its characteristic $H(Q)$ is changed and the flow discharged by the pump is changed in direct proportion. This method can give for the turbopumps involved artificial parabolic characteristics $H(Q)$, having the concave side upwards; such characteristics ideally overlap on the pumping system characteristic $H_s(Q)$, and result in considerably higher efficiencies for the pumping plant (some 30 per cent higher).

(2) *Quality indices of the speed control method.* The main indices that define the quality of a speed control procedure are:

a. *The speed control range.* This is noted by Δn and is defined as the ratio of the minimum to the maximum speeds (n_{min} and n_{max}) that can be achieved through the control procedure under consideration. Thus, we can write

$$\Delta n = \frac{n_{min}}{n_{max}} \qquad (3.15)$$

Now, assuming the minimum speed is the basic speed, taken at unit value, the speed control range can be expressed as a ratio, such as $1:2$, $1:5$, $1:10$ etc.

b. *The speed control coefficient.* This is denoted by k_{sc} and is defined as the ratio of the $n_r - n_{min}$ difference to the rated speed, n_r, thus

$$k_{sc} = \frac{n_r - n_{min}}{n_r} \qquad (3.16)$$

To assist in finding a relation for the speed control coefficient as a function of the pumping head, the diagram $Q - H$ was illustrated in Fig. 3.22 for a few values of the rotational speed. When the output is

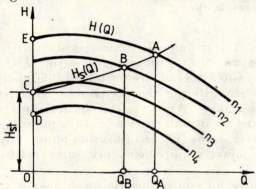

Fig. 3.22. Diagram to derive the speed control coefficient.

controlled over the range $\Delta Q = 0-100\%$ of Q_r through speed variation (which is the case illustrated in Fig. 3.22), the minimum speed ($n_{min} = n_3$) corresponds to point C, where the turbopump pumping head reaches the static head of the system, H_{st}. At this point, the turbopump capacity is zero ($Q_c = 0$). Any further speed reduction down to a value n_4 (lower than the minimum speed) is not rational, for in such a situation the turbopump pumping head is lower than the system static head, and thus the turbopump stops discharging.

Now, substituting in relation (2.85) the co-ordinates of point $C(H = H_{st}; Q = 0)$ results in

$$H_{st} = A_2 n_{min}^2 \qquad (3.17)$$

then, making the same operation for the co-ordinates of point E ($H = H_0; Q = 0$)

$$H_0 = A_2 n_r^2 \qquad (3.18)$$

where H_0 is dead head.

Figures 3.23 typify the relationships between head, capacity and efficiency at varying speeds of turbopumps.

Combining relations (3.17) and (3.18) results in

$$n_{min} = n_r \sqrt{\frac{H_{st}}{H_0}} \qquad (3.19)$$

which, when substituted in (3.16) leads to

$$k_{sc} = 1 - \sqrt{\frac{H_{st}}{H_0}} \qquad (3.20)$$

(3) *Relationship between speed control range and the shape of the pumping system characteristic.* Any pumping system can fall into one of three particular cases.

(4) *System having a static head only.* This is a relatively seldom met case in practice. However, we are often confronted with systems whose maximum

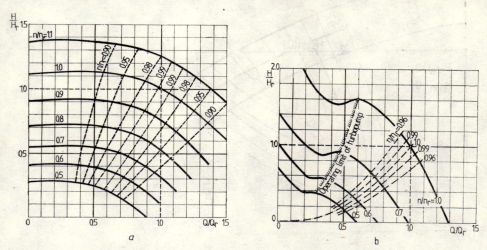

Fig. 3.23. The relationships between head, capacity and efficiency at varying speeds:

a — for radial-flow pump ($n_q = 20$); b — for axial-flow pump ($n_q = 200$).

pumping head is the static head. Such is the case of pumping systems with very short, generally high-rise networks as, for instance, pumping systems used to increase the pressure of domestic water supplies in high-rise buildings. Indeed, in order to control the flow rate, the turbopump in such cases has available only a narrow speed control range ($\Delta n \leqslant 1 : 1.25$), and within the minimum capacity range the pumping plant develops only a low efficiency, because the resultant artificial characteristic $H(Q)$ of the turbopump (shown by the bold lines in Fig. 3.24) lies far enough from the characteristic of optimal efficiency, $\eta_{opt}(Q)$.

Thus, to avoid the unstable performance of the turbopump within the range of minimum capacities, radial-flow pumps with high specific speed ($n_s > 80$) should be used; furthermore, to avoid for such pumps the low-efficiency range of operation, the control range of the pumping plants should be chosen to cover an adequate flow range ΔQ.

(5) *System having a dynamic head only.* This typical system applies in particular to recycling pumping plants. Indeed, to control the flow-rate in such a case, the turbopump should cover a wide speed control range, i.e. $\Delta n > 1 : 2$. On the other hand, the pumping plant performs at a high efficiency over the range of minimum capacities because the resultant artificial characteristic is perfectly coincident with the optimal efficiency characteristic $\eta_{opt}(Q)$ (Fig. 3.25). As a consequence, the capacity of the carrying system, ΔQ, is theoretically unlimited.

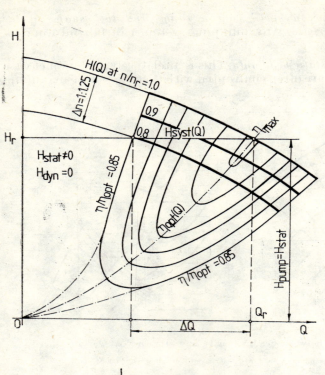

Fig. 3.24. Speed control range Δn for turbopumps operating in a pumping system with only a static head.

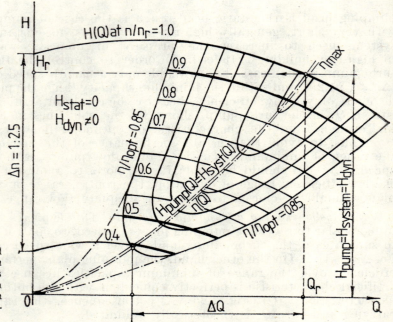

Fig. 3.25. Speed control range Δn for turbopumps operating in a pumping system with only a dynamical head.

The above operating conditions are ideal for the control of pumping plants fitted with variable-speed turbopumps, in contrast to those plants fitted with turbopumps driven at a constant speed, for which such conditions are totally inadequate.

(6) *System with both a static and a dynamic head*. This is the case generally met in practical pumping plants. To control the pumping plant capacity in such a case, the variable speed turbopump should cover a medium-width speed control range, while performing at acceptable efficiencies over the range of minimum flow rates; this is because the artificial characteristic obtained by turbopump control (see the bold line curve in Fig. 3.26) is close to that of the optimal efficiency, $\eta_{opt}(Q)$. As a consequence, over this range of capacities of the pumping plant, ΔQ is but little restricted.

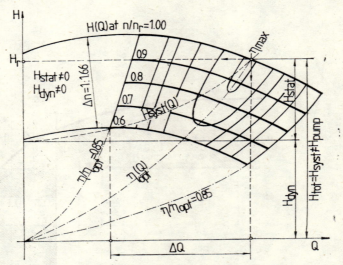

Fig. 3.26. Speed control range Δn for turbopumps operating in a pumping system with both static and dynamical heads.

(7) *Comparison with other procedures*. Compared with throttling or by-passing, the rotational speed control has no detrimental effect on the efficiency. Indeed, if the rotational speed of a turbopump is changed, say, to 90% n_r, its output drops to 90% Q_r, its pumping head to 81% H_r and the power to 73% P_r, but the efficiency keeps pace with the rated efficiency ($\eta = 100\% \; \eta_r$), as compared with $\eta = 97\% \; \eta_r$ for $Q = 90\% \; Q_r$ for throttling, and with $\eta = 93\% \; \eta_r$ for 115% Q_r for by-passing (see also Fig. 3.27, b). Thus there are considerable economic benefits of the control action based on speed variation.

(8) *Control action by altering the position of impeller blades*. This procedure is characteristic of axial-flow pumps with variable-pitch blades and is based on the alteration of performance characteristic curves through changing the impeller blades angle of tilt which, in turn, brings about a change in the profile hydraulic characteristics. The procedure is beset with difficulties in practice and is also very expensive; it is therefore seldom applied.

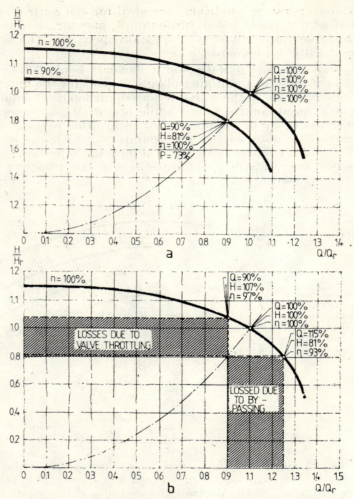

Fig. 3.27. Comparison between several flow control procedures.
a — by adjustable-speed; *b* — by throttling and by-passing the fluid flow.

The universal characteristic (topogram) of this control action is illustrated in Fig. 3.28, where δ denotes the deviation, at a certain moment, of the rotor blade incidence angle from the optimal angle which resulted from the design analysis. As can be seen from the topogram, the general shape of efficiency curves is similar to those drawn up for the speed variation method of control.

This process secures a flow rate variation from 60 per cent to 100 per cent Q_r. The results of this procedure are very good, and close to those of the speed control process. In practice, however, this solution is made rather arduous by constructional difficulties involved by the synchronous blade rotation mechanism (gear) at all working stages.

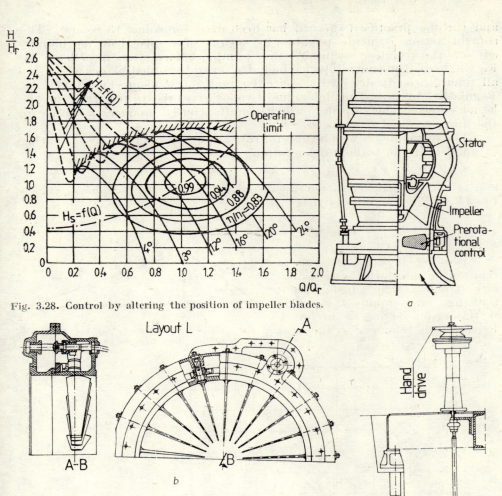

Fig. 3.28. Control by altering the position of impeller blades.

Fig. 3.29. Diagonal pump with adjustable inlet guide vanes:
a — cross-section b — guide vanes; c — drive for regulating device.

Rotor blades adjustment is currently applied to control the capacity and pumping head of large-flow-axial pumps used in land water pumping plants.

(9) *Control by adjustable inlet guide vanes*[1]. This typical control process is comparatively new with pumps, being borrowed from fan

---

[1] The first pump with adjustable inlet guide vane was designed by the English scientist Osborne Reynolds in 1875.

and turbine practice (where it has been used for some 15 years). The control action is made possible by means of a special device mounted on the suction side of the pump, before the impeller (see Fig. 3.29, a). The control device includes several profiled blades which are all simultaneously driven by a circular rack mechanism (Fig. 3.29, b, c). Regulation by adjusting inlet guide vanes gives rise to pre-whirl in the liquid entering the impeller, which simultaneously produces a change in the discharge and total head.

The principle on which this control action relies is as follows. Bearing in mind that the turbopump pumping head is

$$H_t = E_{ds} - E_{sc}$$

and taking into consideration the absolute, relative and peripheral velocities (Fig. 3.30), it follows that

$$H_t = \frac{1}{g}(u_2 c_{ou} - u_1 c_{ou}) \tag{3.21}$$

Looking at Fig. 3.29 and at relation (3.21), it will be seen that any change in angle α_0 brings about a change in component c_{ou}, that is, the term referring to the impeller input is changed while that

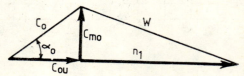

Fig. 3.30. Speed triangle for the control action illustrated in Fig. 3.29.

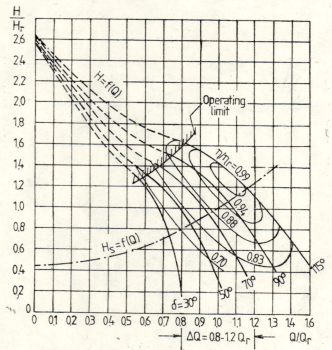

Fig. 3.31. Influence of pre-whirl on the characteristic of a mixed-flow pump.

which refers to the impeller output stays the same. Consequently, the total head H_t of the turbopump can be raised or lowered according to whether the peripheral component of the absolute velocity c_{ou} falls or rises; particularly with mixed-flow and axial-flow impellers, this peripheral component generates a pre-whirl of the fluid in the same direction as the impeller at low flow rates, and in an opposite direction, at high flow rates.

If angle α_0 is changed from $+60°$ to $-40°$ for the blades mounted up-stream of the impeller inlet of a mixed-flow pump, then the universal characteristic (topogram) shown in Fig. 3.31 results. As can be seen, pre-whirl control is recommended with mixed-flow pumps, for this process permits only narrow ranges of economic flow-rate control ($\Delta Q = 80 - 120\% \, Q_r$). This typical control process finds applications to the pumping capacity and head control of high output mixed-flow pumps used in cooling plants.

3.3 Estimating the Pumping Plant Energetic Indices

3.3.1 Estimating the Pumping Plant Specific Energy

The specific energy required by a pumping plant is symbolized by e and is described as the energy required to raise the unit flow rate Q at the pumping head H. It is given by the ratio of the turbopump power requirements P to the product of the carried mass flow rate Q, and the efficiency, η_m, of the electric motor that drives the turbopump considered

$$e = \frac{P}{Q\eta_m} \qquad (3.22)$$

By substituting $\rho Q H/\eta$ for P, we get the specific energy as a function of the turbopump pumping head, H

$$e = \frac{H}{\eta \eta_m} \qquad (3.23)$$

In the above relations, the following notation is used:
P is turbopump power requirements, in W;
Q — turbopump capacity, in cu.m/h;
H — turbopump pumping head, in m;
η — turbopump efficiency, per cent;
e — specific energy of the pumping plant, in Wh/m³;
η_m — efficiency, in per cent, of the driving electric motor.

The flow rate of the pumping plant is controlled by choosing the capacity range, ΔQ, on the abscissa axis, as a function of the electric motor available to drive the turbopump mounted in the system, depending whether the electric drive is running at variable or at constant speed. Referring to relation (3.23) and at the same time bearing in mind the diagrams shown in Fig. 3.32, we offer the following considerations [1.5]:

(1) *With turbopump driven at a constant speed, the variation range of the flow*, ΔQ, should be chosen on the descending section of the efficiency

curve $\eta(Q)$, beginning from the maximum efficiency point, η_{max}, for over this section the specific energy takes minimum values (Fig. 3.32, *a*).

(2) *With turbopumps driven at a variable speed, the variation range of the flow*, ΔQ, should be chosen on the ascending section of the efficiency curve $\eta(Q)$, ending in the maximum efficiency point, η_{max}, for over this section the specific energy takes minimum values (Fig. 3.32, *b*).

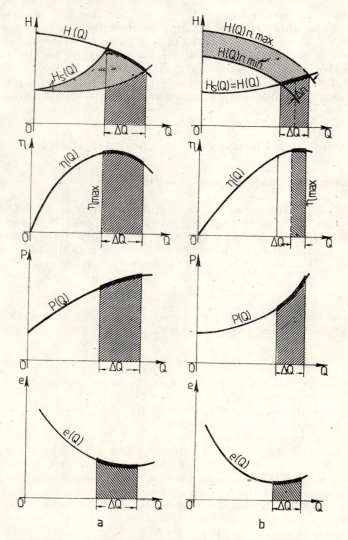

Fig. 3.32. Rational choice of the output range for minimum power expenditure :

a — for a turbopump with a constant speed; *b* — for a turbopump with a variable speed.

To distinguish among pumping plants with different heads, we may resort to a comparison index. This is the unit specific energy, and is given by the ratio of the specific energy, as defined above, to the system pumping head H_t, thus [1.4]

$$e_u = \frac{e}{H_i} \qquad (3.24)$$

where e_u is the unit specific energy, in Wh/m³.

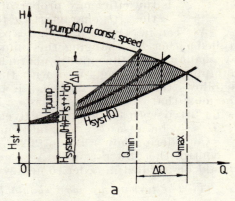

3.3.2 Estimating the Global Efficiency of the Pumping Plant

The global (total) efficiency η_g of a pumping plant is the ratio of the useful power P_u (needed to achieve the useful pumping head H_t of the system) to the power requirement P (needed to achieve the real pumping head H of the turbopump), for a given capacity Q or range of capacities ΔQ (Fig. 3.33). Hence

$$\eta_g = \frac{P_u}{P} \qquad (3.25)$$

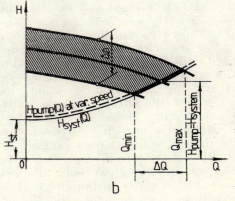

Substituting $\rho QH/k$ for the useful power P_u, and $\rho QH/\eta$ for the power requirement P, as in relations (2.10) and (2.11), gives

$$\eta_g = \eta \frac{H_t}{H} \qquad (3.26)$$

where H_t is the useful pumping head, as determined on the system characteristic $H_s(Q)$ for the given capacities Q;

H is the real pumping head (developed by the turbopump), as determined on the turbopump characteristic $H(Q)$ for the given capacities Q;

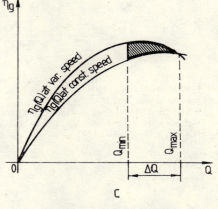

Fig. 3.33. Diagrams for estimation of the global efficiency of a pumping plant:

a — for a turbopump with a constant speed; *b* — for a turbopump with a variable speed; *c* — a comparison diagram of the global efficiency for the two cases above.

157

η is the efficiency proper of the turbopump, as determined on the $\eta(Q)$ characteristic for the given capacity Q.

As seen in Section 3.2, the capacity of a pumping plant can be controlled either by changing the pumping system characteristic $H_s(Q)$, or by changing the turbopump characteristic $H(Q)$. In the former case, the pumping plant being fitted with turbopumps driven at constant speed, it follows that $H_t/H < 1$, because as a general rule $H > H_t$ and hence the most that can be obtained is η_g. In the latter case, the pumping plant being fitted with turbopumps driven at a variable speed, it follows that $H_t/H \approx 1$, because $H \approx H_t$ and hence $\eta_g \approx \eta$ is within reach.

Considering now relation (3.26), one can see that when it is desired to control the flow rate over a given range ΔQ, the following conditions should be fulfilled in order to achieve a high efficiency [1.5]:

(1) *The turbopump used should have a high efficiency*

(2) *The turbopumps used should be capable of securing an H_t/H ratio as close to the unit value as possible.*

The two conditions above can be fulfilled as follows:

— for the first condition, the turbopumps chosen should have a high built-in efficiency, and should perform close to the point representing the optimal value of this efficiency;

— for the second condition, there are two options, namely:

(a) if the pumping plant is fitted with turbopumps driven at a constant speed, choose flat characteristics $H(Q)$ for the turbopumps; or if the turbopumps have a steep characteristic, the pumping head should be chosen within narrow ranges;

(b) if the pumping plant is fitted with turbopumps driven at a variable speed, choose parabolic artificial characteristics $H(Q)$ with up-turned concavity for the turbopumps, these characteristics being close to, and ideally even coincident with, the system characteristic $H_s(Q)$.

REFERENCES

3.1. ADDISON, H. *Centrifugal and other rotodynamic pumps*. Chapman and Hall, 1966.
3.2. HALBERG (GmBH& Co.). *Hidraulische Grundlagen für entwurf von Kreiselpumpen Anlagen* Ludwigshafen, 1972.
3.3. IONEL, I. I. *Stations de pompages à régulation débitmetrique*. La Technique de l'eau et de lassainessement, Bruxelles, No. 1, (1972).
3.4. IONEL, I. I. *Instalații de pompare reglabile*. Technical Publishing House, Bucharest, 1976.
3.5. IONEL, I. I. *Acționarea electrică a turbomașinilor*. Technical Publishing House, Bucharest, 1980.
3.6. KARASSIK, I. *Engineers' guide to centrifugal pumps*. McGraw-Hill Book Company. New York, 1964.
3.7. MESSINA, T. *Operating limits of centrifugal pumps in parallel*. Pumps, N . 6, London.

4
Mechanisms of Cavitation and Cavitation Damage in Turbopumps

4.1 Cavitation Development

4.1.1 Definition, Effects and Classification

Definition. Cavitation is characterized by the occurrence, development and sudden collapse of some cavities, filled with vapours or gases in the mass of a flowing liquid when its temperature is constant and its pressure falls to a certain critical value. The bubbles of vapour or gas, called cavities (the cavities may be bubbles, vapor-filled pockets, or combination of both), are then carried by the stream towards regions of higher than critical pressure, where the vapours condense suddenly generating strong shocks at pressures of some tens or hundreds of thousands of atmospheres, which may cause severe mechanical damage.

Cavitation is more or less similar to the phenomenon of liquids boiling. Generally, we speak of cavitation when the macroscopic discontinuities are generated by variation of hydrodynamic parameters, and of boiling when they are generated by variation of thermal flux. In both cases the growth of microscopic cavities is due to unbalance between their inner and outer pressures, but in the first case the cause is the drop of outer pressure, and in the second, the rise of inner pressure. Visually, the cavitation region appears as an amorphous, yellowish-white mass.

Effects. Cavitation has been defined mainly in terms of its effects: the change of hydrodynamic characteristics of the stream, the destruction of turbopump materials, the production of noise and, if severe enough, vibrations.

The occurrence of cavitation leads to change of the stream lines, to discontinuities in the liquid phase, to the formation of a water, vapour, and gas mixture leading to the modification of hydraulic resistances and thus to change of the characteristic curves of turbopumps. Thus, the presence of cavitation in turbopumps is expressed by the change of flow, that is by reduced efficiency. Due to the small width of the channel between two blades of radial-flow pumps, the small pressures occurring on the surface of the rotor blade secure the condition for cavitation to extend rapidly on the whole section of the channel, and thus for the sudden fall in the cavitation region of the pumping head, $H(Q)$, and of the efficiency, $\eta(Q)$, curves (Fig. 4.1, a).

In axial-flow pumps, since the channel width between two blades is relatively big, the cavitation takes only a part of the flowing section and the drop of operating characteristics is slow (Fig. 4.1, b). Sometimes in axial-flow pumps, cavitation can lead to a rise of operating characteristics during the first stage of development. This phenomenon is because the cavitational careening formed on the extrados of blade profile leads to the lift rise.

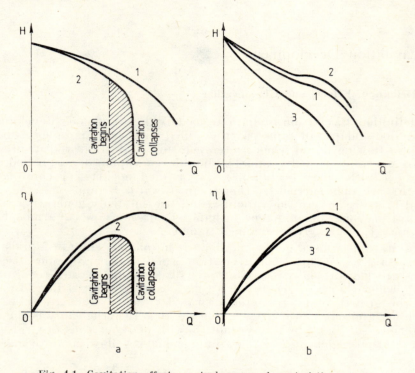

Fig. 4.1. Cavitation effects on turbopump characteristic curves:
a — at radial-flow pumps: *1* — the $H(Q)$ and $\eta(Q)$ characteristics, for normal operating conditions; *2* — the same, for cavitational conditions; *b* — at axial-flow pumps: *1* — the $H(Q)$ and $\eta(Q)$ characteristics, for normal operating conditions; *2* — the same for local cavitation conditions; *3* — the same for developed cavitation conditions.

The most spectacular effect of cavitation is the destruction of solid bodies, be they metallic or non-metallic, chemically active or inert. The destruction of the turbopump material takes place both by erosion or pitting, generated by the strong shock occurring as cavities collapse, and by corrosion, produced by the chemical action of gases released from the liquid on the metallic or non-metallic walls.

Generally, cavitation is a dynamic process and is thus accompanied by noise and severe vibrations. The noise occurs mainly as cavities collapse, and the vibrations of the liquid are transmitted to the turbopump blades and to the pumping installations in general, or to their associated constructions.

Classification. The occurrence of the first visible bubbles defines the stage of *incipient cavitation*. The association of several bubbles forms a cavitational cloud, a region characteristic of the stage of developed or *industrial cavitation*. The expansion of the cavitation region to a wider range leads to *supercavitation*. Cavitation regions can be classified as: mobile cavitation, fixed cavitation, swirling cavitation and vibratory cavitation.

Mobile cavitation occurs when the individual transitory bubbles, or the cloud of bubbles, formed in the low-pressure points of a liquid ($p_{min} = p_{cr}$), are moving with the stream, expanding at first and then collapsing suddenly.

Fixed cavitation occurs when the liquid stream is separating from the solid surface of the immersed body and shows a turbulent boiling in the separation region. The liquid near the cavitation region contains several mobile and relatively small cavities, which are growing at the upstream end of the cavitation region, retain their geometry until downstream, and then collapse and disappear.

Swirling cavitation generally occurs in the centres of the eddies, present in a stream, such as in suction tubes of pumps or around bridge piers. This type of cavitation has a shorter life cycle than the others.

Vibratory cavitation occurs on the surface of solid bodies immersed in liquids where critical pressures of vaporization are generated by an oscillating motion. This type of cavitation is present, for instance, in the cooling chambers of internal-combustion engines.

4.1.2 Cavitation coefficient

The complex nature of cavitation requires the definition of some similarity coefficients or criteria, characteristic of the cavitational stage. In order to establish such a coefficient we shall analyse the geometric and hydrodynamic parameters and the physical properties of the liquid. Worth mentioning among these parameters is the varporization pressure, p_v, at a given temperature, since we know that it is practically equal to the critical pressure, $p_{cr}(p_v = p_{cr})$ of the liquid for cavitation to begin.

In order to establish some cavitation coefficients let us study the flow around an aerofoil, Fig. 4.2. Thus, Bernoulli's equation written between A and M is

$$\frac{p}{\rho g} + \frac{v_\infty^2}{2g} + z = \frac{P_M}{\rho g} + \frac{v_M^2}{2g} + z_M + \Sigma h_{lA-M} \qquad (4.1)$$

If $z = z_M$ and the hydraulic losses are small ($h_{lA-M} = 0$)

$$\frac{p_M}{\rho g} = \frac{p_\infty}{\rho g} - \frac{v_M^2}{2g} - \frac{v_\infty^2}{2g}$$

$$\frac{p_M - p_v}{\frac{\rho}{2} v_\infty^2} = \left[\frac{p_\infty - p_v}{\frac{\rho}{2} v_\infty^2}\right] - \left[\left(\frac{v_M}{v_\infty}\right)^2 - 1\right] \qquad (4.2)$$

here

$$C_{pM} = \frac{p_M - p_\infty}{\frac{\rho}{2} v_\infty^2} = 1 - \left(\frac{v_M}{v_\infty}\right)^2. \tag{4.3}$$

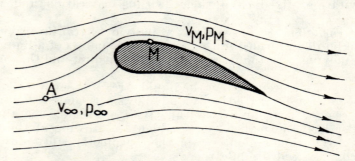

Fig. 4.2. Aspect of streamlines around a hydrodynamic profile.

If $p_M = p_{min}$ and $v_M = v_{max}$

$$\sigma = \frac{p_{min} - p_v}{\frac{\rho}{2} - v_\infty^2} = \left[\frac{p_\infty - p_v}{\frac{\rho}{2} v_\infty^2}\right] - \left[\left(\frac{v_{max}}{v_\infty}\right)^2 - 1\right]. \tag{4.4}$$

By introducing the notation

$$\sigma_{pr} = \frac{p_\infty - p_{min}}{\frac{\rho}{2} v_\infty^2} = \left(\frac{v_{max}}{v}\right)^2 - 1 \tag{4.5}$$

$$\sigma_{pr} = -(C_{p\,max}) = k_{p\,max} \tag{4.5a}$$

$$\sigma_{inst} = \frac{p_\infty - p_v}{\frac{\rho}{2} v_\infty^2} \tag{4.6}$$

we get

$$\sigma = \frac{p_{min} - p_v}{\frac{\rho}{2} v_\infty^2} = \sigma_{inst} - \sigma_{design} = \lambda - k_{p\,max} \tag{4.7}$$

This equation shows that the operation can be cavitational or normal. Thus, if

$p_{min} = p_v = p_{cr}, \quad \sigma_{inst} = \sigma_{design}$ — cavitation is incipient;
$p_{min} < p_v = p_{cr}, \quad \sigma_{inst} < \sigma_{design}$ — cavitation has developed or we have supercavitation;
$p_{min} > p_v = p_{cr}, \quad \sigma_{inst} > \sigma_{design}$ — the operation is without cavitation.

We can say that σ_{design} and σ_{inst} can be defined as cavitation coefficients, since their magnitude and ratio determine normal or cavitational operation. Thus

$$\sigma_{design} = -(C_{p\,min}) = \frac{p_\infty - p_{min}}{\frac{\rho}{2}v_\infty^2}, \text{ or } \sigma_{design} = \left(\frac{v_{max}}{v_\infty}\right)^2 - 1$$

is the cavitation coefficient of turbopump profile, which depends on its geometry, on the incidence angle α of the stream velocity, on the operating conditions, and the physical characteristics of the liquid, that is on ρ and v on Re when the hydraulic losses are also considered. The cavitation coefficient may be obtained directly from the pressure distribution curve.

The classical theory of incipient cavitation assumes that this phenomenon is starting in the point of minimum pressure of the stream, and the cavitation coefficient is ussually expressed as

$$\sigma_{design} = -C_{p\,min} = k_{p\,max}$$

where $C_{p\,min}$ is the minimum pressure coefficient, that is the maximum velocity coefficient ($k_{p\,max}$).

Investigating the incipience of fixed cavitation at profiles, Bailey and Casey [4.4] show that cavitation is actually occurring in the vicinity of the laminar detachment point and not in the minimum pressure point. For defining the cavitation coefficient, they suggest the equation:

$$\sigma_{design} = -C_{p\,det} = \frac{p_\infty - p_{det}}{\frac{\rho}{2}v_\infty^2} \tag{4.8}$$

where $C_{p\,det}$ is the pressure coefficient in the laminar detachment point.

On the other hand

$$\sigma_{ins} = \frac{p_\infty - p_v}{\frac{\rho}{2}v_\infty^2}$$

defines the cavitation coefficient of the installation (the exterior coefficient), since it is determined by the exterior quantities, independent of those characteristic of the stream in the profile region.

It is worth mentioning that the equality of the three coefficients indicates incipient cavitation. Their value identity does not imply physical identity.

Many researchers have extended the definition of these cavitation coefficients — cavitational similarity criteria — by considering also the effects of the gravitational field, the air and gas content of the liquid, and impurities. However, the researches made are of small importance.

The coefficients of dynamic similarity σ_{design} and σ_{inst}, which could be expressed in accordance with Euler, Froude, Reynolds and Weber numbers, are used at present for all turbopumps and hydraulic elements, regardless of the working liquid.

The experimental determination of cavitation coefficients by means of laboratory models, and their application to the real phenomenon or industrial equipment, implies the "scale effect" which should be taken into consideration.

4.1.3 Cavitation Origins

Generally, the incipience pressure of cavitation in industrial liquids, also called the critical pressure, is equal to the vaporization pressure: $p_{cr} = p_v$ at the given working temperature. In the case of homogeneous liquids, the occurrence of cavitational bubbles implies the existence of a tension, high enough to break the liquid. It is known that in such liquids the cohesion forces, the intermolecular forces and the exterior forces are holding the particles together.

The liquid breaking resistance has been measured in static and dynamic conditions by various methods. Worth mentioning in this context are the results of Blake and Temperley [4.8], who give values for water between $13-150$ daN/cm². Davies [4.10] obtained, under dynamic conditions, values between 8 and 14 daN/cm² for the breaking resistance of purified water.

Turbopump producers would be delighted if this were the case in reality, since the pumps could then easily be installed at higher suction heads.

Although the results obtained by various researchers with different methods are rather scattered, they lead to the assumption that various influences affect this parameter: either internal influences, specific to the respective liquid, or external, specific to the separation from the solid surface, which considerably change the value of the breaking resistance or the critical pressure p_{cr}, respectively.

Fischer [4.14] and Frenkel [4.15] developed a theory which assumes some empty spaces in the liquids. They postulate that the liquid state is a pseudo-crystalline one, with a great number of free spaces of radii about 10^{-8} cm. For such dimensions, the calculations show that tensions of $1000-4000$ daN/cm² are required for the occurrence of breakage, and thus of cavitation origins. These values differ considerably from the experimental ones, and thus the generation of cavitation nuclei starting from such empty spaces seems rather doubtful, since their occurrence probability is also negligible.

Impurities in liquids form the so-called "weak points" of low resistance, where the origins of cavitation, that is the first cavitational bubbles, may appear. The impurities influence the occurrence of cavitation only with a considerable modification of the liquid viscosity, density, surface pressure and thermal properties.

Harvey's model [4.18] assumes the existence of gas nuclei in the cracks and grooves of the solid walls. These nuclei become cavitation origins when the pressures fall to the critical level. This model gives a physically consistent explanation for the existence of permanent nuclei in liquids. Harseg obtained sufficient experimental data in good agreement with the suggested model. The latest results mentioned in the literature, both in the field of cavitation and of boiling, supported this model.

In analyzing the conditions of occurrence of cavitation near a plane surface in contact with the liquid and into the liquid mass, Volmer [4.45] showed that the wall nucleation requires a smaller work L_{min} than nucleation in the liquid mass L_{min0}. The reduction of the work necessary for formation of bubbles depends on the solid-liquid contact angle (θ), (see Fig. 4.3, a)

$$L_{min} = L_{min0} \cdot f(\theta), \quad f(\theta) \leqslant 1 \tag{4.9}$$

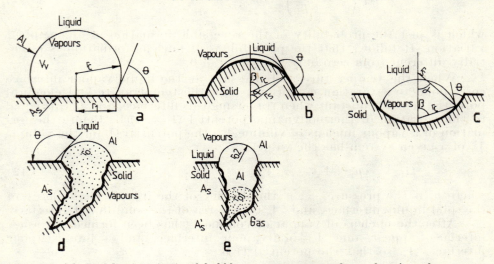

Fig. 4.3. Models of cavitational bubbles occurrence on surfaces of various shapes.

Extending Volmer's theory to curved solid surfaces with hills and pits, Bankoff [4.5] shows that the work necessary for the formation of a nucleus in the case of a hill (Fig. 4.3, b) is higher than in the case of a plane surface, while the occurrence of a nucleus in a spherical pit (Fig. 4.3, c) requires less work than that necessary in the case of a plane surface. Thus, it follows that the microcracks of solid surfaces in contact with the liquid are the most favourable places for bubble nucleation or cavitational origin.

The small overheating values necessary for the occurrence of bubble nucleation is explained by the existence of undissolved air, or gas in the wall microcracks, according to Harvey's model (see Fig. 4.3, d and e).

Consider the system of a liquid in contact with a solid wall, of the form shown in Fig. 4.3, d. The probability of a fluctuation (leading to the formation of a vapour bubble) is:

$$P \sim \exp\left[+\frac{\Delta S_t}{k}\right] \tag{4.10}$$

where ΔS_t is the variation of total entropy of the system due to the fluctuation (k = Boltzmann's constant). But since

$$\Delta S_t = -\frac{L_{min}}{T}$$

where L_{min} is the minimum work, done for the reversible variation of the thermodynamic quantities corresponding to a small part of the system (the rest playing the role of "medium"), and T is the system temperature,

$$P \sim \exp\left(-\frac{L_{min}}{kT}\right) \tag{4.11}$$

which is just the probability of the reversible formation of an incipient variation. It follows that the probability of nucleus formation is essentially different from zero only when $L_{min} \to 0$.

When the temperature and pressure at the liquid-vapour interface correspond to saturation conditions, then the temperature and chemical potential remain constant when traversing this interface, and L_{min} is equal to the variation of thermodynamic potential $\Omega = -pV$. Before the formation of a vapour nucleus of volume V_v, the potential Ω for the volume V_v of a given system has the value

$$\Omega_1 = -p_1 \times V_v + \sigma_{sl} A_s, \tag{4.12}$$

where p_1 is the pressure, σ_{s1} is the density of the free surface energy of the solid-liquid interface, and A_s is the area of the solid-liquid interface.

After the nucleus of vapour of volume V_v has been formed, two new interfaces appear, one A_s solid-vapour interface, and a liquid-vapour interface, A_l, so that the potential Ω is

$$\Omega = -p_v \times V_v + \sigma_{sv} \times A_s + \sigma_{lv} \times A_1 \tag{4.13}$$

The last two equations give the value of the work necessary for the formation of a vapour nucleus

$$L_{min} = (\sigma_{sv} - \sigma_{sl}) A_s + \sigma_{lv} A_1 - V_v \times \Delta p \tag{4.14}$$

In case of equilibrium, if we write $\Delta p = 2\sigma/r_0$, and

$$\sigma_{sv} - \sigma_{s1} = \sigma_{sv} \cos\theta \equiv \sigma_{\cos\theta}.$$

it follows that

$$L_{min} = (A_1 + A_s \cos\theta) - V_v \frac{2\sigma}{r_0} \tag{4.15}$$

Thus, the formation probability of a growing vapour nucleus is

$$P \sim \exp\left\{\frac{1}{kT}\left[\sigma(A_1 + A_s \cos\theta) - V_v \frac{2\sigma}{r_0}\right]\right\} \tag{4.16}$$

When a gas volume V_g is present in the relevant pit before the formation of a vapour nucleus (Fig. 4.3, e) we have the following equation for the minimum work required

$$L_{min} = -\frac{2\sigma}{r_{02}}V_v + \frac{2\sigma}{r_{01}}V_g + \sigma[(A_1 - A_1') + (A_s - A_s')\cos\theta] \tag{4.17}$$

Naturally, equation (4.11) is then changed in accordance with (4.17).
Research at the Laboratory of hydraulic machinery within the Polytechnic Institute of Timișoara (Romania) took into consideration nucleation centres of a more general form than those studied by L. B. Kokarev, of type I and II from Fig. 4.4. Given the condition that

$L_{min} \to 0$ in equation (4.11), it follows that

$$\cos\theta = \frac{1}{A_s}\left(\frac{2V_v}{a} - A_1\right), \quad r_0 = a \tag{4.18}$$

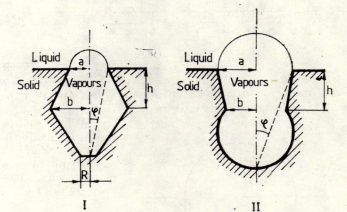

Fig. 4.4. Various forms of nucleation centres.

which expresses the relation between the contact angle θ and the geometry of the nucleation centre. For the centres of type I and II we get the equations

$$\cos\theta = \frac{2}{3}\cdot\frac{h(a-R)(a+b+R)-a^3-a(b^2+b_R)\cot\varphi}{a[(a+b)\sqrt{(a-b)^2+h^2}+(b+R)\sqrt{(R-b)^2+(a\cot\varphi-h)^2}+R^2]} \tag{4.19}$$

$$\cos\theta = \frac{a^3(\cot^3\varphi-2)+a^2h(2-3\cot^2\varphi)+(b^2+h^2)3a\cot\varphi+h(2ab-b^2-h^2)}{3a[(a+b)\sqrt{(a-b)^2+h^2}+b^2+(a\cot\varphi-h)^2]} \tag{4.20}$$

By plotting $\theta = \theta(\varphi)$ for various forms of centres of type I(Fig. 4.5, a) and of type II(Fig. 4.5, b) in the range of the possible values of angles θ ($0 \leqslant \theta \leqslant 180°$) and φ($0 \leqslant \varphi \leqslant 90°$), two ranges become distinct:
(a) the range beneath the $\theta = \theta(\varphi)$ curve, for which $L_{min} > 0$, and
(b) the range above the $\theta = \theta(\varphi)$ curve, where $L_{min} < 0$.

We may then conclude that a growth of range (b) above the $\theta = \theta(\varphi)$ curve, leads to a more rapid actuation of the nucleation centre.

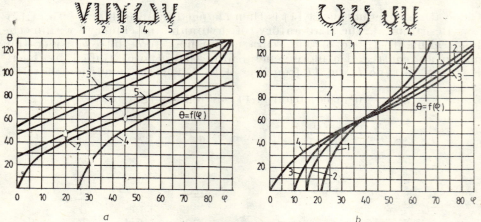

Fig. 4.5. Cavitation origin occurrence domains.

Figure 4.5, a, b shows that the centres of type I and form 4 are the most rapidly actuated ones, while those of type II are actuated only at small angles (those of form I).

When we also consider the gas remaining inside the wall microcracks, the work necessary for the formation of cavitation origins is given by equation (4.17), which for a cylindrical crack (pit) (Fig. 4.6) becomes

$$L'_{min} = 2\pi a^2 \left\{ \cot \varphi [m \sin 2\theta + (1-m)\cos\theta - 1] - \frac{\cos^2 2\theta}{6 \sin^2 \theta} \right\} \quad (4.21)$$

Here, $m = \dfrac{H_0}{a \cot \varphi} < 1$ is a quantity characterizing the amount of the residual gas.

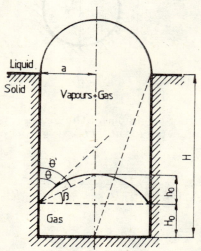

Fig. 4.6. Sketch of a cavitation origin of cylindrical shape filled with gases and vapours.

We see that in the physical range of parameters θ the curve $\theta = f(\varphi)$ is outside, which shows that such a centre is easily actuated in the whole range of parameters θ and φ, if it contains gas.

Thus, the presence of a reduced amount of undissolved gas inside a nucleation centre has a considerable influence on the occurence of cavitation.

4.1.4 Dynamics of Cavitation Bubbles

Besant first mentioned the dynamics of a cavity without gas or vapours in 1859 and Rayleigh gave a good mathematical solution in 1917. In real phenomena, a cavity always contains gas or vapours, or a combination of both. The association of several cavitational bubbles filled with vapours is called *"vaporous cavitation"*, and when the bubbles are filled with gas we speak of *"gaseous cavitation"*.

Rayleigh considers a spherical cavitation bubble in an incompressible liquid of constant pressure $p(\infty)$, with insignificant surface forces and an inner pressure $p_i = 0$, or $p_i =$ constant. Taking R as the bubble radius, with U the speed of collapse or growth of its walls, $p(R)$ the wall pressure and u the liquid velocity at radius r and time t(Fig. 4.7), we may write the potential of liquid motion provided we assume a spherical symmetry and an irrotational radial flow during the bubble collapse from R_0 to R [4.30]

$$\Phi = \frac{R^2}{r} \times \frac{dR}{dt} \qquad (4.22)$$

while from the continuity equation we get

$$\frac{U}{u} = \frac{r^2}{R^2}$$

The kinetic energy of the liquid mass resulting from the work done during the collapse process is

$$dF_c = dm \frac{u^2}{2} = 2\pi u^2 r^2 \, dr \qquad (4.23)$$

or

$$F_c \int_R^\infty 2\pi u^2 r^2 \, dr = 2\pi \rho U^2 R^3 \qquad (4.24)$$

The work done by the liquid during the bubble collapse from R_0 to R is

$$L = \frac{4\pi}{3} p_\infty (R_0^3 - R^3). \qquad (4.25)$$

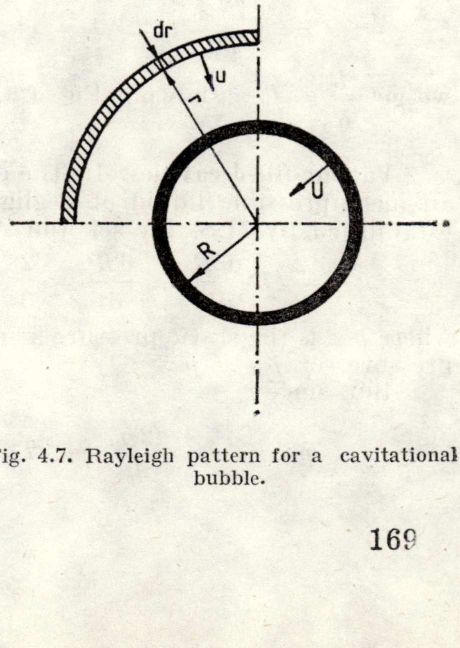

Fig. 4.7. Rayleigh pattern for a cavitational bubble.

By equating the work with the kinetic energy of the liquid mass we get the differential equation of motion of the cavitational bubble according to Anton [4.4]

$$U^2 = \left(\frac{dR}{dt}\right)^2 = \frac{2}{3} p(\infty) \left(\frac{R_0^3}{R^3} - 1\right) \tag{4.26}$$

By the end of the collapse process ($R = 0$), the wall velocity U becomes infinite and thus, theoretically, is changed into an infinite pressure.

The kinetic energy is suddenly changed into elastic pressure energy, leading to very high (theoretically infinite) pressure, or in practice to a pressure of some tens or even hundreds of thousands of atmospheres. Naturally, in the real case the pressure cannot rise to infinity, since a certain amount of gas is always present inside the bubbles and thus, it does not allow for the total bubble collapse ($R = 0$), keeping it to a certain radius R.

Rayleigh also established a solution for the case of a bubble filled with gas of initial pressure p_i and which is isothermally compressed. Thus, we have

$$U^2 = \left(\frac{dR}{dt}\right)^2 = \frac{2p_\infty}{3\rho}\left(\frac{R_0^3}{R^3} - 1\right) - \frac{2p_i}{\rho}\frac{R_0^3}{R^3}\ln\frac{R_0^3}{R^3} \tag{4.27}$$

This equation shows that when p_i is constant the sudden collapse of cavitation bubbles proceeds to a radius $R \neq 0$ where $U = 0$. If $p_i > p_\infty$ the bubble is growing, while if $p_i < p_\infty$ the bubble is collapsing.

The pressure field of the liquid, containing the collapsing bubble with $p_i = 0$, is expressed according to Rayleigh by

$$\frac{p(r)}{p_\infty} - 1 = \frac{R}{3r}\left(\frac{R_0^3}{R^3} - 4\right) - \frac{R^4}{3r^4}\left(\frac{R_0^3}{R^3} - 1\right) \tag{4.28}$$

By introducing the dimensionless time t in equations (4.27) and (4.28)

$$\tau = \frac{\sqrt{\dfrac{p}{\rho}}}{R_0} t \tag{4.29}$$

we get $\dfrac{R}{R_0} = f(\tau)$, shown in Fig. 4.8.

Vapour-filled cavities. In the case of a vapour-filled cavity located in an incompressible liquid of negligible viscosity, the equation of motion established by M.S. Plesset and A. S. Zwick [4.35], is

$$\frac{2}{3} R \frac{d^2R}{dt^2} + \frac{dR^2}{dt} = \frac{2}{3} \rho \left[p_{(R)} - p_{(t)}\right] \tag{4.30}$$

where $p_{(R)}$ is the static pressure at radius R and $p_{(t)}$ is the liquid external pressure, or $p_{(t)} = p_\infty$.

But since

$$p_{(R)} = p_v - \frac{2\sigma}{R} = p_v(T) - \frac{2\sigma}{R} \tag{4.31}$$

then $p_v(T)$ is the equilibrium pressure of vapours at temperature T, and at the wall cavitation bubble, while σ is the water surface tension. The differential equation can be solved when $p_{(t)}$ is known and when we introduce radius R_0, defined by

$$\frac{2\sigma}{R_0} = p_v(T_0) - p_\infty$$

where T_0 is the liquid overheating temperature at a certain distance from the bubble.

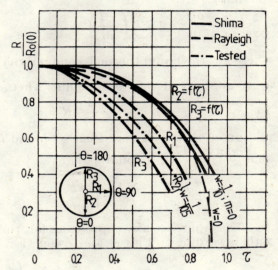

Fig. 4.8. Curves showing the collapse of cavitational bubbles.

Thus, equation (4.30) becomes

$$\frac{R\,d^2R}{dt^2} + \frac{2}{3}\left(\frac{dR}{dt}\right)^2 = \frac{p_v(T) - p_v(T_0) + \dfrac{2\sigma}{R_0}\left(1 - \dfrac{R_0}{R}\right)}{\rho}$$

or

$$\frac{1}{2R^2 \dfrac{d^2R}{dt^2}} \frac{d}{dt}\left[R^3\left(\frac{d^2R}{dt^2}\right)^2\right] = \frac{p_v(T) - p_v(T_0) + \dfrac{2\sigma}{R_0}\left(1 - \dfrac{R_0}{R}\right)}{\rho} \quad (4.32)$$

and by integration we get

$$\left(\frac{d^2R}{dt^2}\right)^2 = \left(\frac{R_0}{R}\right)\left(\frac{d^2R_0}{dt^2}\right)^2 + \frac{4\sigma}{3\rho R_0}\left(1 - \frac{R_0^3}{R^3}\right) - \frac{2\sigma}{\rho R}\left(1 - \frac{R_0^2}{R^2}\right) \quad (4.33)$$

Plesset called this equation 'the Rayleigh solution'.

Plesset admits that the real motion is different from that given by Rayleigh, due to the cooling effect. The heat $\dfrac{dQ}{dt}$ which must be supplied to the bubble per unit time is

$$\frac{dQ}{dt} = \frac{4\pi}{3} L \frac{d}{dt}(R^3 \rho_v) \qquad (4.34)$$

where L is the latent heat of vaporization and ρ_v is the vapour density.

This heat is transferred by conduction from the liquid to the bubble, so that

$$\frac{dQ}{dt} = 4R^2 k \left(\frac{\partial T}{\partial r}\right)_R \qquad (4.35)$$

where k is the liquid thermal conductivity and $\left(\dfrac{\partial T}{\partial r}\right)_R$ is the liquid temperature gradient at the bubble limit. Then

$$\left(\frac{\partial T}{\partial r}\right)_R = \frac{L}{3k} \frac{1}{R^2} \frac{d}{dt}(R^3 \rho_v). \qquad (4.36)$$

and given the small variations of L and k with temperature and the fact that $\dfrac{d\rho_v}{dt} \ll \dfrac{dR^3}{dt}$, it follows that

$$\left(\frac{\partial T}{\partial r}\right)_R = \left(\frac{L\rho_v}{k}\right) \frac{dR}{dt}$$

The approximate expression for the temperature near the bubble wall, according to Plesset and Zwick, is

$$T = T_0 - \left(\frac{D}{\pi}\right)^{1/2} \int_0^t \frac{R^2(x) \left(\dfrac{\partial T}{\partial r}\right)_{r=R(x)}}{\left[\int_x^t R^4(y)\, dy\right]^{1/2}} dx \qquad (4.37)$$

Equations (4.36) and (4.37) define the thermodynamics of the vapour-filled bubble.

The two equations give the general equation of motion if we note that

$$z = \frac{R^3}{R_0^3}; \quad u = \left(\frac{\alpha}{R_0}\right)^4 \times \int_0^t R^4(y)\, dy$$

$$\alpha = \left(\frac{2}{\rho R_0^3}\right)^{1/2}; \quad z' = \frac{dz}{du} \quad \text{and} \quad \mu = \frac{AL\rho_v}{3kR_0\alpha}\left(\frac{D}{\pi\alpha}\right)^{1/2}$$

and if we assume that the vapour pressure could be approximated, for a small overheating above the boiling temperature T_v, at an exterior pressure p_0, by means of a linear function of temperature

$$\frac{p_v(T) - p_0}{\rho} = A(T - T_v)$$

with
$$\frac{dR}{dt} = \left(\frac{\alpha R_0}{3}\right) z^{2/3} \frac{dz}{dt}$$
and thus we get
$$T = T_0 = -\frac{\alpha^2 R^0}{A} \int_0^\mu \frac{p'(v)\,dv}{(k-v)^{1/2}} \tag{4.38}$$

Zwick and Plesset obtained small differences by calculating $R = f(t)$ in accordance with the present theory and with Rayleigh's, maintaining the assumption of spherical cavities up to the final moment of collapse, an assumption which does not correspond to the real phenomenon.

Gas-filled bubbles. The most general case of motion of a cavitational bubble is that in which it contains gas and is to be found in a viscous and compessible liquid. Poritsky [4.36] and Gilmore [4.16] gave satisfactory partial solutions.

Thus Poritsky, dealing with the bubble growth and collapse in a viscous and incompressible liquid of a certain surface tension, established the following equation of motion

$$\frac{R d^2 R}{dt^2} + \frac{3}{2}\left(\frac{dR}{dt}\right)^2 + \frac{4\mu}{\rho R}\frac{dR}{dt} = \frac{p_i(R) - \frac{2\sigma}{R} - p_\infty}{\rho} \tag{4.39}$$

By integrating this equation graphically, he showed that the liquid viscosity leads to a reduction of both growth and collapse speeds of the bubble, while the surface tension reduces the growth speed of the bubble and increases its collapse speed.

Starting from Poritsky's equation, Shu [4.39] proved, by neglecting the surface pressure, that the collapse time is infinite if

$$\mu' = \frac{4\mu}{R_0[(p_\infty - p_i)]^{1/2}} > 0.46$$

which corresponds to a viscosity μ, $-1{,}500$ times higher than that of water.

Gilmore [4.16], accepting Kukwood-Beth's assumption that the disturbances are propagating in the liquid with the sum of the velocity of sound c, in the liquid ($c^2 = dp/d$) and the liquid local velocity u, viz
$$u' = c + u,$$
established the following equation of motion.

$$\frac{RU dU}{dR}\left(1 - \frac{U}{C}\right) + \frac{3}{2}U^2\left(1 - \frac{U}{3C}\right) = H\left(1 + \frac{U}{C}\right) +$$
$$+ \frac{RU}{C}\frac{dH}{dR}\left(1 - \frac{U}{C}\right) \tag{4.40}$$

The capitals here denote the values at the bubble wall, and H is the difference of the liquid enthalpy from the bubble wall to infinity.

Gilmore expressed the liquid pressure $p(R)$ at the bubble wall as

$$p(R) = p_i - \frac{2\sigma}{R} - 4\mu \frac{dR/dt}{R} - \frac{4}{3C^2} \frac{dR}{dt} \frac{dH}{dR} \tag{4.41}$$

In the case in which $|H| \ll C^2$ and $p_i - p_\infty \ll 20{,}000$ daN/cm^2 for water, Gilmore gave the equation

$$U^2 = \frac{2(p_\infty - p_i)}{3\rho_\infty} = \left[\frac{(R_0/R)^3}{\left(1 - \frac{U}{3C}\right)^4} - 1 \right] \tag{4.42}$$

which, for $p_i = 0$ and $(U/C) = 0$, leads to Rayleigh's equation.

Maximum pressures occurring at the collapse of bubbles. When a cavitational bubble collapses, the velocity U of its wall and of the surrounding liquid passes suddenly from a high value to zero, and consequently the liquid kinetic energy is turned into elastic energy of compression, leading to overpressures similar to those known as water hammer.

In the case of water hammer, Jukowski established for the overpressures the relation $\Delta p = \rho_1 C_1 U_1$, where C_1 is the velocity of sound in water.

Parsons and Cook [4.33] were the first to use this relation for calculating the overpressure occurring at the collapse of cavitational bubbles, taking the velocity from Rayleigh's equation. The values he found were too high, and required some corrections. Thus, Engel recommended the relation $\Delta p = \frac{\alpha}{2} \rho_1 C_1 U_1$, where $\alpha \approx 1$ while $\alpha/2$ is due to the assumption of the spherical shape of cavitational bubbles up to the final stage of their collapse. The equations defining the dynamics of gas- or vapour-filled cavitational bubbles, formed in viscous compressible liquids with significant surface tensions, enabled various researchers to calculate more realistically the cavitational overpressures.

By synthesizing some results, Knapp, Daily, and Hammitt [4.25] showed that the calculated pressures transferred to the solid surfaces around the collapse centres of cavitational bubbles are of the order of 1,000 daN/cm^2.

The fact that the alloyed steel used in tools, whose breaking resistance is very high, is destroyed by cavitation makes us accept values even higher than those calculated.

Experimental research for the determination of maximum pressures occurring at the collapse of maximum pressures occurring at the collapse of cavitational bubbles is only beginning, and thus cannot confirm the values calculated.

4.1.5 Cavitational destruction and protection methods

The destructive effects of cavitation have been remarked for a long time at screw ships, hydraulic turbomachinery (turbines and pumps), throttle valves and other hydraulic constructions. More recently, cavitational destruction has been observed at the high temperature pumping installations of nuclear reactors, and at the pumping installations of metallic liquids and in the cooling liquids in combustion engines. Cavitational destruction is noticed ever more frequently in the chemical industry.

Although there is impressive experimental and practical material in this field, disputes are going on among experts regarding a theory underlying the destructive mechanism of cavitation. Some give pride of place to the mechanical theory, while others favour thermogalvanic, electrochemical or chemical causes.

Mechanical theory. This theory attributes cavitational destruction of various materials to the mechanical effect of shocks or microjets.

a. **Destruction by shocks.** The most accepted assumption is that cavitational destruction is the outcome of shocks produced by the waves and overpressures radiating towards the solid walls from the centres of cavitational bubbles as they collapse.

The pattern of pressure waves causing the cavitational destruction is shown in Fig. 4.9, a for a bubble collapsing at the wall, and in Fig. 4.9, b for a bubble collapsing near the wall.

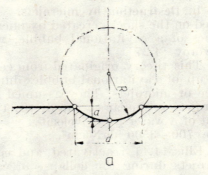

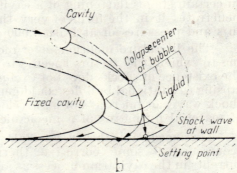

Fig. 4.9. Model of cavitational bubble collapse by means of pressure waves:

a — at the wall; b — near the wall.

This pattern is in good agreement with the observation that the number of destructive shocks is much smaller than the number of individual collapsing bubbles. A qualitative estimation is obtained if we assume that the work done by a collapsing cavity on a solid surface for the production of a permanent detectable deformation is the same as that done during a hardness test.

The centre of the sphere of the test indentation can be considered as the point from which the hydrodynamically concentrated collapsing energy of a bubble is transmitted to the solid wall. The calculated values for the two energies starting from this model lead to a 30 : 1 dispersion, which shows that the model has more a qualitative than a quantitative significance.

Figure 4.9, *b* shows schematically the motion of a cavitational bubble towards its collapse point, and the propagation of pressure waves towards the solid wall. The collapse point will be the nearer to the wall, the greater the cavity dimension at the beginning of the collapse process, and the overpressure and destruction energy will also be higher.

Thus, the pressures occurring at the collapse of cavitational bubbles may or may not reach destructive values, since they depend on the cavitation type, the bubble dimensions, the gas content, the distance of the collapse centre from the wall, the pressure gradient, the collapse speed, the stream velocity, the physical nature of the liquid, the kind of material exposed to the shock, etc.

b. **Destruction by microjets.** The model of cavitational destruction based on the shocks created by microjets of liquid, formed in the collapse stages of a cavitational bubble, was first suggested by Kornfeld and Suworov [4.27].

This theory originated from certain photographs showing the breaking up of cavitational bubbles during the collapse period and the occurrence of microjets. They assumed that the destruction caused by these microjets is similar to the destruction of a sample periodically passed in front of a water jet of velocity, $v = 50 - 250$ m/s. The impact pressures were $700-3500$ daN/cm².

Ellis [4.4] reactivated and confirmed the theory of occurrence of microjets during the collapse of cavitational bubbles. He showed that in certain stages flattening of a cavitational bubble occurs and the liquid penetrates it in the form of very thin microjets with velocity $180-200$ m/s and pressures of at least $3,000$ daN/cm².

The literature mentions ever more frequently the hypothesis according to which the factors specific to shocks also correspond to jets, and the existence of shocks does not exclude that of jets. Consequently, the collapsing cavitational bubbles could generate considerable overpressures and shock waves or microjets of high velocity, leading to the rapid destruction of material by means of a mechanical action.

Thermogalvanic theory. Formulated by Kernn, this theory assumes the occurrence of some corrosive thermodynamic and electromechanical currents in the cavitation process as an outcome of high temperature and of some temperature gradients. Temperatures about $10,000$°C are generated by the compression of the gas contained by the collapsing cavitational bubble, while about $500-800$°C is present in the material due to its straining and breaking under the action of shock waves or microjets. The temperature gradient between the heated region and the ambient medium facilitates an electric current in a conductive medium.

The electrolytic corrosion generated by this current is proportional to its magnitude. Foltyn and Nechleba checked this theory by using a counter-current equal to the electrolytic one measured in the cavitational phenomenon, and thus succeeded in considerably reducing the material destruction. It should be mentioned that the reduction of destruction was obtained only in the case of smooth surfaces, while for rough ones the effect was almost nil.

Luminescent phenomena, similar to those of sonoluminescence, have been observed during the cavitation produced in high temperature regions. It was assumed that this luminescence was caused by recombination of the free ions remaining after thermal dissociation of the molecules on the bubble surface.

Electrochemical theory. Many researchers consider that cavitational destruction is based on the generation of some electric currents between the anodic and cathodic surfaces immersed in electrolytes, and which actually lead to the electrochemical corrosion.

a. **Anodic corrosion** is based on the assumption that the deformed regions of the material, in which its crystalline structure has been distorted, are electrochemically unstable.

The anodic surfaces immersed in an electrolyte will be electrolytically corroded in the vicinity of undistorted metal. This process will last until polarization halts the current flux, or until the anodic surfaces are corroded. The electrolytic corrosion is actuated by cavitation.

b. **Protecting film breakage.** A protecting film is formed in certain materials such as bronze when immersed in water, which can be cathodic relative to the basic material. Such films have been observed on ship propellers made of bronze. It is said that the breakage of this film creates a galvanic pile between metal and film. According to this theory the mechanical action of cavitation leads to a violent electrochemical destruction.

Theory of chemical corrosion. Experimental researches on cavitational destruction lead to the conclusion that chemical corrosion is also present. This conclusion is based on the proper chemical activity in highly corrosive media, or at high temperatures which hasten the chemical process.

The corrosion in an electrolyte is of an electrochemical nature, while the chemical corrosion in non-electrolytes, or dry gases is considered as purely chemical in nature, since it implies heterogeneous reactions of chemical kinetics.

Rhaingans [4.38] compared the corrosive effect of hydrochloric acid and sulphuric acid with that of water on steel and other non-ferrous metals, and concluded that after two hours of exposure, the difference is insignificant.

Conclusions. The numerous theoretical and experimental researches on the mechanism of cavitational destruction have demonstrated that the mechanical effect is always present, and that the other effects — thermodynamic, electrochemical and chemical — are superimposed on it, leading to the intensification and acceleration of cavitational destruction.

4.1.6 The Resistance of Materials to Cavitational Destruction

Of special importance in the construction and operation of turbopumps and hydromechanical equipment is a knowledge of material characteristics during cavitational destruction. These characteristics depend on the physical, mechanical, metallurgical and electrical properties of the material, as well as on the cavitation intensity and stage. Various apparatus and methods have been conceived and constructed as representations of these characteristics.

Representation of cavitational destruction. The representation of cavitational destruction with time in the form $\Delta G = f(t)$ has been generally accepted, irrespective of the used method and device. ΔG is the loss of weight (or ΔV, of volume) which can be measured in certain given conditions with the highest accuracy during a given time interval. Figure 4.10 shows curves obtained for various materials in a Venturi installation, where the water stream velocity was kept at a value $v = 65$ m/s.

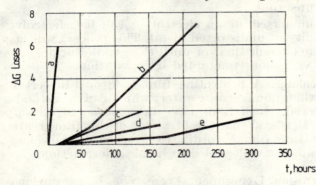

Fig. 4.10. Material loss by cavitational destruction:
a — lead; b — cast iron; c — bronze; d — aluminum; e — steel.

As to the interpretation of cavitational destruction, one of the most important observations made of late is that the destruction rate $\dfrac{\Delta G}{\Delta t}$ depends on the time of exposure to cavitation, t. Figure 4.11 shows such a characteristic of pure iron tested in sodium.

Four zones can be noticed on this curve:
— the incubation zone (*1*), where the weight loss is insignificant;
— the accumulation zone (*2*), where the material absorbs ever more important amounts of energy, leading to considerable weight losses and material ruptures;
— the attenuation zone (*3*), where the destruction rate drops;
— the steady zone (*4*), where a constant destruction rate is being set.
The main cause of a constant destruction rate is supposed to be air accumulation in the deep cracks of the sample attacked.

The investigations made on cavitational destruction of various metals, as well as the analysis of curves $\Delta G = f(t)$ and $\frac{\Delta G}{\Delta t} = f(t)$, have proved that the physicomechanical, chemical, electrical and thermal properties are decisive for the resistance to cavitation. These properties could

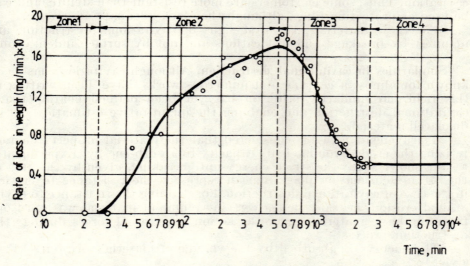

Fig. 4.11. Characteristic zones of cavitational destruction speed.

be defined by the following characteristics: the breaking resistance, elasticity limits, flow limits, strain energy, fatigue limit, hardness, modulus of elasticity, maximum resilience, plasticity, fragility, thermal conductivity, melting point, chemical inertia, oxides adherence to surface, crystalline structure, granulation size, hardening capacity, electrical conductivity, etc.

Depending on the dominant characteristic, the resistance of the material to cavitation could be greater or smaller.

Metallic materilals. Study of these materials revealed that their great hardness, fine and uniform granulation, great strain and fatigue resistance, and significant distortion energy, secure them superior anti-cavitational characteristics.

Special steels have all these qualities and thus are the materials most resistent to cavitational attack. The most resistant material to cavitational destruction is stellite, but since it is too hard and expensive it is less used. Fragile metals, such as cast iron, are rapidly destroyed by cavitation.

Non-metallic materials and polymers. For non-metallic materials, the electrical and chemical phenomena are of secondary importance in cavitational destruction, leaving the mechanical dominant.

The majority of these materials, such as bakelite and glass, are fragile and rapidly destroyed by cavitation. At the other end we find rubbers and elastomers, which can stand 1—10 times elongations, but their moduli of elasticity are low. The ability of these materials to absorb a great part of the deformation energy secures them great resistance to cavitation. Thus, some elastomers are more resistant than stellite and can be used as protecting films, although they lack adequate adherence.

For these materials, the first signs of cavitational destruction are manifest by breakage and dislocations and not by surface indentations.

Similarities in cavitational destruction. Although cavitation has been known for almost a century and numerous works have been written in this field, no similarity criterion has yet been defined characterizing cavitational destruction and enabling the quantitative estimation of the displaced material.

Thiruvengadan [4.43] considered that the material property characterizing the eroded volume is its capacity to absorb energy. Experimental results showed that the strain energy until breaking, given by the area of the strain-deformation diagram, can represent the resistance to erosion in the case of cavitational destruction. For defining the resistance to cavitational erosion, he said that a certain volume of material was torn out from the basic material under the action of external forces during the destruction process.

The energy E_a, absorbed by the volume of material torn out, ΔV, is given by

$$E_a = \Delta V \times S_e \tag{4.43}$$

where S_e is the erosion resistance, which represents the capacity of the material to absorb energy per unit volume under the action of erosion forces.

The absorbed power is

$$P_a = \frac{E_a}{t} = \frac{\Delta V \times S_e}{t} \tag{4.44}$$

The intensity of cavitational destruction, I, can be expressed by the ratio between the power P_a and the eroded surface A_e

$$I = i \times \frac{S_e}{t} \text{ watt/cm}^2 ; \quad I = \frac{P_a}{A_e} = \frac{\Delta V}{A_e} \times \frac{S_e}{t} \tag{4.45}$$

where $\sum_0^N i = i$ is the average erosion depth $i = \frac{\Delta V}{A_e}$

Thiruvengadan calculated the intensity of cavitational destruction I for turbines, pumps and Diesel engines. The values he obtained are presented in Table 4.1.

Although the intensity of cavitational destruction I is defined only in accordance with the specific parameters of material, which indirectly estimate the destruction energies of the collapsing bubbles and microjets, it offers nevertheless a calculation method.

This complex problem demands a solution, mainly for the exploitation of hydraulic machinery, propellers of ships, and hydraulic equipment which operate in cavitation conditions.

An exact calculation is still impossible, but approximation should be made by simultaneously considering the inlet and outlet factors which regulate the cavitation process, including only the dominant ones and neglecting the secondary ones.

TABLE 4.1

Machine type	Intensity of cavitational destruction I, watt/cm^2
Turbines	$(10-5) \times 10^{-1}$
Pumps	10^{-1}
Diesel engines	$(0.5-1) \times 10^{-1}$

4.1.7 Methods of Localizing and Detecting Cavitation

The methods of localizing and detecting cavitation can be classified as follows:

(1) *The method of characteristics*, which detects the starting point of inertial cavitation by the fall in energetic characteristics of the turbopump.

(2) *The method of pressure distributions*, which is based on knowledge of the pressure distributions on a turbopump blade, on the single profile or profile network, on a certain body, measured in water or air, or computed, and which enables determination of the cavitation starting point by means of minimum pressures.

(3) *The acoustical method*, which aims at establishing the noise generated by cavitation in its various development stages.

(4) *The laser method*, which is based on cavitational scattering of the laser rays in a photocell.

(5) *The optical visualization method*, using photography, film and closed-circuit television. This method offers the necessary conditions for a detailed study of cavitation in all its stages. It is applied only to transparent liquids, and cannot be used with non-transparent liquid metals. Since cavitation is an extremely rapid phenomenon, the photographs can be taken only by stages, which would imply a cyclic process, or by means of a high-speed camera, getting 10^6-10^7 images per second, with an exposure time of $10^{-7}-2 \times 10^{-8}$ seconds.

4.1.8 Atenuation of Cavitational Destruction

Attenuation or even prevention of cavitational destruction and the protection of turbopump regions exposed to this process can be achieved as follows:

— By adequate design or redesign of the component elements, to secure operation without cavitation or with an industrially acceptable cavitation. Thus, redesign of the rotor and of its blades is used in many turbopumps where the cavitational attack is strong;

— By the use of highly resistant materials at the parts undergoing cavitational destruction. The blades of turbopump rotors as well as the chambers of axial pumps are made of alloyed steel. Sometimes the walls of these chambers are made of bimetal alloys with metallic or non-metallic protective coatings.

Various metallic or non-metallic protective coatings have been tested with a view to reducing the destructive effect of cavitation on new turbopumps, or after a certain number of operational hours. These materials can be applied by various welding or spraying methods.

Excellent results have been obtained with welded stainless steels or aluminum bronze, despite the intricate and expensive technology.

Sprayed metallic coatings are most tempting, since they are cheap and easy to apply, but the problem of simultaneous melting of the sprayed material and of the layer of basic metal has not yet been solved.

Lichtman predicted a great future for coatings of sprayed elastomers, having in view their resistance to cavitation which is comparable to that of the best metals such as 6B stellite and 71B inconel. Yet the problem of total adherence to the basic material, as well as that of obtaining the prescribed geometry when turbopump blades are concerned, have not been solved so far.

4.2 Equations and Curves Characteristic of Cavitation

4.2.1 Defining Equations of Cavitation Occurrence Conditions

In order to find general equations defining the conditions of cavitation occurrence in a pump, we should write Bernoulli's equations on a streamline between the M_A and M_1 points and then between M_1 and M (Fig. 4.12).

$$\frac{p_s}{\rho g} + \frac{v_A^2}{2g} = \frac{p_1}{\rho g} + \frac{v_1^2}{2g} + z_1 + h_{l1} \quad (4.46)$$

$$\frac{p_1}{\rho g} + z_1 + \frac{w_1^2 - u_1^2}{2g} = \frac{p_M}{\rho g} + z_M + \frac{w_M^2 - u_M^2}{2g} + h_{lM} \quad (4.47)$$

where h_l is the head loss, p_s the pressure in the suction basin, v_1 the stream absolute velocity, and the other quantities are as shown in the figure.

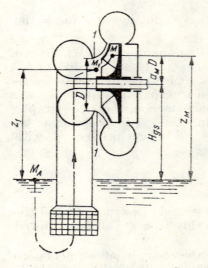

Fig. 4.12. Sketch for setting the general equations for defining cavitation occurrence conditions.

Taking $v_A = 0$, $z_M = h_{gs} + a_M D$, where h_{gs} is the geodesic suction head and D the rotor diameter, we get from equations (4.46) and (4.47) for the pressure p_M at the point M

$$\frac{p_M}{\rho g} = \left(\frac{p_s}{\rho g} - h_{gs} - h_{l1}\right) - \left(\frac{w_M^2 - u_M^2}{2g} - \frac{w_1^2 - u_1^2}{2g} + \frac{w_1^2}{2g} + a_M D + h_{lM}\right)$$

or

$$\frac{p_M - p_{cr}}{\rho g} = \left(\frac{p_s - p_{cr}}{\rho g} - h_{gs} - h_{l1}\right) - \left(\frac{w_M^2 - u_M^2}{2g} - \frac{w_1^2 - u_1^2}{2g} + \frac{v_1^2}{2g} + a_M D + h_{lM}\right) = \Delta h_e - \Delta h_i \qquad (4.48)$$

where p_{cr} denotes the critical pressure at which cavitation occurs and grouping of the terms as Δh_e and Δh_i was made on the basis of which side they belong to, namely, the suction line or the pump.

The form (4.48) of Bernoulli's equation has the advantage of emphasizing the factors determining the pressure difference $\frac{p_M - p_{cr}}{\rho g}$, as well as each of the terms Δh_e (the net positive suction head) and Δh_i (the dynamic pressure loss). Indeed, if Q denotes the flow capacity corresponding to the maximum efficiency and Q_x a certain flow capacity, Anton [4.1] to finds the following equation for Δh_i

$$\Delta h_i = \frac{u_1^2}{2g} [\varepsilon^2 \tan^2 \beta_1 - K_{ux} - K_{p\,max}(\varepsilon^2 \tan^2 \beta_1 - 1) + \varphi_s (1 - \varepsilon)^2 + c\varepsilon^3 + a_M D] \qquad (4.49)$$

where

$$\varepsilon = \frac{Q_x}{Q};$$

c and φ_s are coefficients of the equation

$$h_{l0M} = c\varepsilon^2 + \varphi_s \frac{u_1^2}{2g}(1 - \varepsilon)^2;$$

$$\tan \beta_1 = \frac{c_{m1}}{u_1 - c_{u1}} \simeq \frac{c_{m1}}{u_1}$$

with $c_{u1} \simeq 0$; $k_u = \frac{u_M^2}{u_1^2} - 1$; $k_{p\,max} = \frac{w_{max}^2}{w_1^2} - 1$. For flow capacities of $Q_{1x} \neq Q_1$ we have $\beta_{1x} \neq \beta_1$, where β_1 is the angle made by the blade with the u axis (Fig. 4.13), and additional head losses may appear.

Since h_{l1} may also be expressed as a function of, ε it follows that $\Delta h_e = f_1(\varepsilon)$ and $\Delta h_i = f_2(\varepsilon)$. The $\Delta h_e = f_1(\varepsilon)$ curves are called cavitation inner characteristics. The main problem is the formulation of these characteristic curves. The Δh_e curves are easy to construct, since h_{l1} may easily be expressed as a function of ε. The Δh_i curves can only be experimentally constructed.

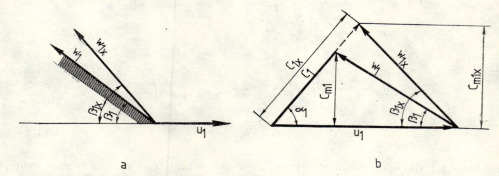

Fig. 4.13. Speed triangle for a radial-flow pump at $Q_{1x} \neq Q_1$.

By superposing the Δh_e and Δh_i curves (Fig. 4.14), we can define the range within which the pump is normally operating ($\Delta h_e > \Delta h_i$), distinguishing it from that in which the pump is working under cavitational conditions ($\Delta h_e < \Delta h_i$).

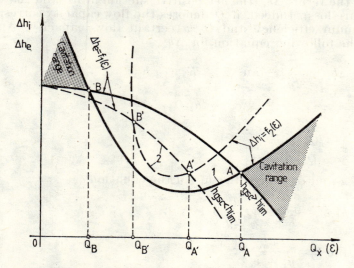

Fig. 4.14. Inside and outside cavitation characteristic curves.

The technical literature [4.40] mentions a simplified form of Bernoulli's equation, written for the section at the free surface of the suction basin and the inlet section into the rotor

$$\frac{p_M - p_{cr}}{\rho g} = \frac{p_s - p_{cr}}{\rho g} - h_{gs} - h_{l1} - \frac{\alpha_1 c_1^2}{2g} - \frac{\lambda w_1^2}{2g} \qquad (4.50)$$

where c_1 is the absolute inlet mean velocity into the rotor, α_1 the Coriolis coefficient, $\lambda \frac{w_1}{2g}$ the pressure drop due to the relative velocity, and λ the dynamic depressure coefficient, (experimentally determined).

We notice that equation (4.50) differs from (4.48) by the structure of the Δh_i expression. The term $\lambda \frac{w_1^2}{2g}$ is hiding the influence of all factors emphasized in equation (4.49).

Experiments have revealed that the coefficient λ has the form shown in Fig. 4.15.

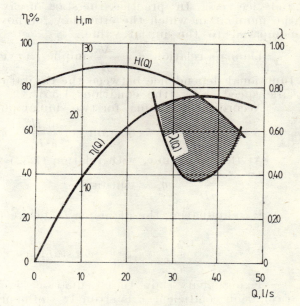

Fig. 4.15. Variation of coefficient λ

Thoma [4.40] argued the hypothesis that the total dynamic depressure Δh_i could be written as a fraction of the total head

$$\frac{c_1^2}{2g} + \lambda \frac{w_1^2}{2g} = \Delta h_i = \tau H \qquad (4.51)$$

where H is the pump total head and σ a coefficient experimentally determined and called the *cavitation coefficient*.

In critical operation conditions, when $p_M = p_{cr}$, equations (4.48) and (4.50) give

$$\sigma_{cr} = \frac{\frac{p_s - p_{cr}}{\rho g} - h_{gs} - h_{l1}}{H} \qquad (4.52)$$

The reason for introducing the cavitation coefficient σ is the hypothesis that for similar pumps the stream velocity (absolute, relative, peripheral) is proportional to $\sqrt{H}$, and for the same pump at different flow capacities, the velocity is also proportional to $\sqrt{H}$, so that for $p_M \geqslant p_{cr}$, $\sigma = \sigma_{cr} = $ constant.

The maximum geodesic suction head is obtained as

$$h_{gs} = \frac{p_s - p_{cr}}{\rho g} - h_{l1} - \sigma H \tag{4.53}$$

For safety reasons, the critical pressure p_{cr} should be equal to the liquid vaporization pressure (itself a function of temperature).

The determination of quantity h_{gs} implies knowledge of coefficient σ and vice-versa. In practice, the start of cavitation is usually defined as the moment in which the efficiency, or power curve, falls by $2-3\%$ as compared to the initial value.

Thoma's relation $\sigma = \dfrac{\Delta h_i}{H}$ implies a very simple structure of the functional dependence between the cavitation coefficient σ and the specific speed, n'_q — [the equation (1.55)].

According to Thoma, for two similar pumps we may write

$$\Delta h_i = \frac{u_1^2}{2g} = \text{constant, or } \frac{\sigma_g H}{n^2 D_1^2} = h_{sc} = \text{constant.} \tag{4.54}$$

At the same time, with relation (1.52) we get

$$\frac{Q}{nD_1^3} = q_s = \text{constant.}$$

By eliminating the diameter D_1 between equations (4.54) and (1.52) we see

$$\sigma = \frac{h_{sc}}{q_s^{2/3}} \times (n'_q)^{4/3} = \text{constant } (n'_q)^{4/3} \tag{4.55}$$

Experiments show that equation (4.55) is very approximate. The cavitation coefficient σ is strongly influenced by the amount of air (or other gas) present in water, by water temperature, and by other factors. Even a comparison of equations (4.49) and (4.54) for the same quantity of contained air shows that Δh_i is a complicated function of the quantities it depends on (e.g. the Re number). However, equations of type (4.55) are often used in practice at the suggestion of various investigators:

1. A. Stepanoff: $\quad \sigma = 2.20 \times 10^{-4}\, n_s^{4/3}$
2. Escher-Wyss: $\quad \sigma = 2.16 \times 10^{-4}\, n_s^{4/3}$
3. C. Pfleiderer: $\quad \sigma = 2.41 \times 10^{-4}\, n_s^{4/3}$
4. I. Anton: $\quad \sigma = 2.29 \times 10^{-4}\, n_s^{4/3}$
5. G. Proskura: $\quad \sigma = 2.20 \times 10^{-4}\, n_s^{4/3}$
6. D. Pavel: $\quad \sigma = 0.001 n_s (1.1 + 0.001\, n_s) + 0.02$
7. Standards of Hydraulic Institute (for axial-flow pumps): $\quad \sigma = 2.05 \times 10^{-4}\, n_s^{4/3}$ \hfill (4.56)

The different values of constant σ are due to the different experimental conditions. For instance, Krisam [4.40] emphasized the difference between the net positive suction head Δh_i, calculated with σ = constant, and that one exprimentally determined for a pump with n_s = constant, but variable speed n (Fig. 4.16).

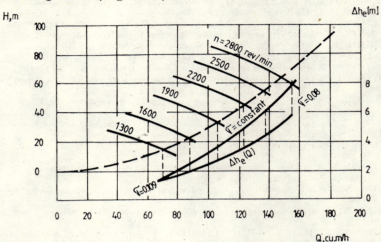

Fig. 4.16. Influence of pump rotation speed, n, on the values of coefficient σ

Rutschi [4.40] stressed the influence of hydraulic efficiency η_h on σ, showing that since η_h is an increasing function of Re, σ is a decreasing function of η_h. Experiments have been conducted showing that the value of cavitation coefficient σ decreases with the increase of pumped liquid temperature.

The cavitation coefficient σ also descreases with increase of number of blades. On the other hand, too many blades decrease the flow section, which leads to a rise of the relative and absolute velocity, and thus to an increased possibility of cavitation. Too many blades also lead to a rise of head losses. Therefore, the optimum number of blades is between 5 and 11 for radial-flow pumps, and between 3 and 6 for axial-flow pumps.

REFERENCES

4.1. ANTON, I., *Curbele caracteristice de cavitație la mașinile hidraulice*. Papers of the Conference on hydraulic machines, September (1964).
4.2. ANTON, I., VEKAS, L., *Studiu asupra activării centrelor de nucleație de diferite microgeometrii și procesul de cavitație*. St. Cerc. Mec. Apl., Tom. 31, No. 3 (1972).
4.3. ANTON, I., VEKAS, L., *A study on the cavitation bubble formation process*. Rev. Roum. Schi. Tech. Mec. Appl., No. 18/6 (1973).
4.4. ANTON, I., *Turbine hidraulice*. Facla Publishing House, Timișoara, 1980.
4.5. BANKOFF, S. G., *Ebullition from solid surfaces in the absence of a free-existing gaseous phase*. Trans. ASME, No. 79 (1957).
4.6. BENJAMIN, T. B., ELLIS, A. T., *The collapse of cavitation bubbles and the pressures thereby produced against solid boundaries*. Phil. Trans. Royal Soc. London, A. Vol. 260 (1966).
4.7. BLAHA, J., *Sangfährigkeit von Propellerpumpen*. Maschinenmarkt, No. 97 (1969).
4.8. BLAKE, F.G., *The tensile strength of liquids*. Harward Acoustics Res. Lab. Tm. 9, June 1949
4.9. CHIVERS. T. C., *Cavitation research on a centrifugal pump*. ASME Paper, No. 27 (1969)
4.10. DAVIES, R. M., *The tensile strength of liquids under dynamic stressing*. Symp., On Cavitation in Hydrodynamics, Proc. NPL (1955).

4.11. DUC, J., *Fenomeni di cavitazione nelle tubazioni di pompaggio.* Acqua Industriale, No.3 (1961)
4.12. DUMOV, V. I., PESKIN, M. A., *Issledovanie kavitaţii v kolese ţentrobejnovo nasosa.* Teploenregetika, No. 12 (1959).
4.13. EISENBERG, P. A., *Mechanics of cavitation, Handbook of fluid dynamics.* Book Co., McGraw-Hill, Book Co., New York, 1961.
4.14. FISCHER, J. C., *The fracture of liquids.* Jr. Appl. Phys. No. 19 (1948).
4.15. FRENKEL, J., *Kinetic theory of liquids.* Calderon Press, Oxford, 1946.
4.16. GILMORE, F. R., *The growth and collapse of a spherical bubble in a viscous compressible liquid.* Calif. Inst. of Tech. Hydrodyn. Lab. Rep. No. 4 (1952).,
4.17. GUITON, P., *The cavitation in pumps.* La Houille Blanche, No. 5—6 (1962).
4.18. HARSEG, E. N., *Removal of gas nuclei from liquids and surfaces.* Jr. Am. Chem. Soc., No. 67 (1945)
4.19. HICKLING, R., PLESSET, M. S., *Collapse and rebound of a spherical bubble in water.* Physics of fluids, No. 7 (1974).
4.20. HLADIS, V. I., *La hauteur d'aspiration des pompes centrifuges et axiales.* L'industrie lourde Tchechoslovaque, No. 3, (1960).
4.21. JACOBS, R. B., *Prediction of cavitation symptomes.* Jr. of Research of the N.B. S., No.3 1961
4.22. JALOVOY, N. S., *Isledovanie vsasivaiuşcih patrubkov energheticeskih nasosov.* Energomaşinostroenie, No. 5 1969.
4.23. KARELIN, V. Ia., *Kavitaţionie iavlenya v ţentrobejnih i osevih nasosah.* Maşinostroenia, Moscow, 1978.
4.24. KARELIN, V. Ia., *Nasosi i nasosnîe stanţii dlea vodosnabjenia i oroşenia.* Stroizdat, Moscow 1966
4.25. KNAPP, R. T., DAILY, J. W., HAMMITT, F. G., *Cavitation.* McGraw-Hill Book Co, New-York, 1970.
4.26. KNAPP, R. T., HOLLANDER, A., *Laboratory investigations of the mechanism of cavitation* Trans. ASME, No. 70 (1948).
4.27. KORNFELD, M., SUWOROV, J. L., *On the destructive action of cavitation.* Jr. Appl. Phys, No. 15 (1944).
4.28. KOVATS, A., DESMUR, G., *Pompes, ventilateurs, compresores.* Paris. Dunod, 1965.
4.29. KRIVCHENKO, G. I., *Nososi i gidroturbini.* Energya. Moscow, 1970.
4.30. LAMB, H., *Hydrodynamics.* Dover Publ., New York, 1945.
4.31. MINAMI. S., KAWAGUCHI, K., HOMMA, I., *Experimental study on cavitation in centrifugal pump impellers.* Bull. ISME, No. 9 (1960).
4.32. NOSKIEVIC, I., *Kavitace.* Academia Praha, 1969.
4.33. PARSONS, C. A., COOK, S, S., *Investigations into the causes of corrosion or erosion of propellers.* Trans. Inst. Nav. Arch.. No. 61 (1919).
4.34. PERNIC, A. D., *Problemi kavitaţii.* Sudostroenie, Leningrad, 1966.
4.35. PLESSET, M. S., ZWICK, S. A., *The growth of vapor bubbles in superheated liquids.* California Inst. Techn. Rep., No. 26—6 (1953).
4.36. PORITSKY, H., *The collapse or growth of spherical bubble or cavity in a viscous fluid.* Proc. first U.S. Nat. Congr. Appl. Mech. ASME (1952).
4.37. RAO-GOVINDA, N. S., *Cavitation—its inception and damage.* Irrigation and power, No.1 (1961)
4.38. RHAINGANS, W. J., *Accelerated cavitation research.* Trans. ASME. Vol. 72—5 (1952)
4.39. SHU, S. S., *Note on the collapse of a spherical cavity in a viscous incompressible fluid.* Proc. first U.S. Nat. Congr. Appl. Mech. ASME (1952).
4.40. STEPANOFF, A., *Pompes centrifuges et helices*, Dunod. Paris, 1961.
4.41. STOIANOVICI, S., *La hauteur d'aspiration des pompes centrifuges.* La technique de l'eau et de l'assainiscment, No. 3 (1969).
4.42. THIRUVENGADAN, A., WARING, S., *Mechanical properties of metals and their cavitation-damage resistance.* Journal of Ship Research, No. 1 (1966).
4.43. THIRUVENGADAN, A *The concept of erosion strength.* ASTM. (1967).
4.44. TOLITZMAN. V, F. *Uvelicenie dopustimoi visoti vsasivania nasosov.* Vodosnabjenie i sanitarnaia tehnika, No. 1 (1961).
4.45. VOLMER, M. *Kinetik der Phasenbildung.* Leipzig, 1939.
4.46. WINTERNITZ, F. A. L. *Cavitation in turbomachines.* Water Power, No. 9—11 (1 957)
4.47. ZWICK, S. A., PLESSET, M. S. *On the dynamics of small vapor bubbles in liquids.* Jn. Math. and Phys., Vol. 33 (1955).

5
Heads, Head Losses and System Head Curves in Flow through Pumping System

5.1 Elementary Flow Theory for the Study of Pipes

5.1.1 Flow Conditions

Studies carried out on liquid displacement inside pipes show characteristic flow patterns in certain given conditions and, as a consequence, the flow can take the following forms:

Laminar flow. At low velocities, coloured particles suspended in a liquid flowing through a given pipe show that the flow takes place in a regular manner (Fig. 5.1, *a*). Over a straight section of the pipe, particle paths are rectilinear and parallel to the pipe centre-line, and the current generated stream-lines seem to slide one along the other while keeping all the time their individual character.

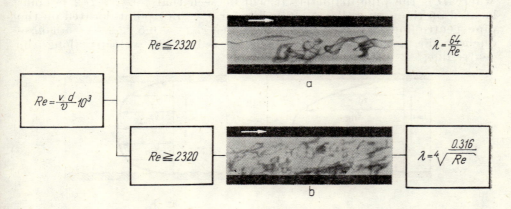

Fig. 5.1. Aspects of liquid flow through a straight pipe of circular cross-section:
a — laminar flow; *b* — turbulent flow.

Transient flow. When, within the same pipe as above, the velocity of flow is higher, the rectilinear path of the stream-lines is replaced by a sinusoidal form which is variable in time, and the flow conditions become unstable.

Heads, Head Losses and System Head Curves

Turbulent flow. Provided the pipe is the same, the flow character is completely changed when the fluid velocity reaches a high value. Any rectilinear path and any regular system disappear (Fig. 5.1, *b*), the particles follow winding unpredictable paths that are continuously crossing each other and changing. As a general rule, these conditions are common in curent practice.

5.1.2 Flow Velocity

Even in the same pipe, the flow velocity v is not uniform at all points of the same section. Indeed, friction means braking action, the effects of which grow continuously from the centre-line to the peripheral walls; hence the velocity will be at maximum in the centre and will decrease along the pipe walls. The velocity profile will vary as a function of the flow pattern (Fig. 5.2.).

Experimental work carried out on laminar flow has shown that the velocity vector reaches its maximum value along the pipe centre-line, falling gradually as it approaches the pipe walls. The velocity distribution curve within the longitudinal section is a parabola (Fig. 5.2, *a*). Velocity v at any point at a distance y from the pipe centre-line is

$$v = 2v_{mean}\left[1 - \left(2\frac{y}{D}\right)^2\right] \tag{5.1}$$

where D is the pipe diameter. It will be seen that for $y = D/2$ (a point located on the pipe wall), $v = 0$, while for $y = 0$ (a point located on the pipe centre-line), $v_{max} = 2v_{mean}$ and hence $v_{mean} = 0.5\, v_{max}$. As a consequence, the mean velocity of the fluid is half the velocity along the centre-line.

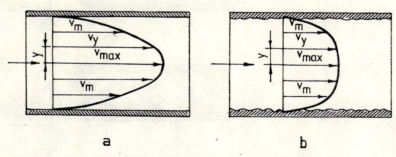

Fig. 5.2. Profiles of liquid flow velocity through a straight pipe:
a — laminar flow; *b* — turbulent flow.

With turbulent flows, the velocity at each point is no longer constant as in laminar flows, but rather varies both in direction and value. The velocity distribution curve will be a flattened parabola (Fig. 5.2, *b*): indeed, within a great part of the section the velocities take values close to the maximum, but within the fluid layer next to the pipe walls the velocity drops more or less linearly down to the limit value $v = 0$ which applies for points located on the walls.

In any point at a distance y from the centre-line, the approximate velocity is

$$v = v_{max} \left(1 - 2\frac{y}{D}\right)^{1/7} \tag{5.2}$$

The vector of the average flow velocity is parallel to the pipe centre-line, while the v_{mean}/v_{max} ratio is no longer constant as with the laminar flow, but has a variable value in the range 0.80—0.86.

5.1.3 Flow Stabilization Length

At the inlet of a laminar flow pipe, the flow conditions are not perfectly laminar; the flow becomes laminar only after a certain pipe length l is traversed, the approximate value of which is

$$l = 0.028 \text{ Re } D \, [\text{m}] \tag{5.3}$$

For $\text{Re} = 2000$, it follows $l = 60 \, D$.

In turbulent flow, the conditions are perfectly set only after the flow has travelled a certain length of pipe:

$$l = 10^5 \frac{D}{\text{Re}} \tag{5.4}$$

Thus, for $\text{Re} = 10^4$ it follows $l = 10 \, D$
$\text{Re} = 10^5$ it follows $l = D$
$\text{Re} = 10^6$ it follows $l = 0.1 \, D$

5.1.4 Viscosity

Viscosity is that property that makes liquids incapable (due to their forces of molecular cohesion) of instantaneous changes in their particle shape and array. In the case of moving liquids, this property is called dynamic viscosity, μ, and is manifest by the resistance offered by one portion of the liquid to the parallel displacement of an adjacent fluid portion. This resistance is due to the internal friction generated by the relative motion of neighbouring fluid layers.

The inverse dynamic viscosity $1/\mu$, is called *fluidity*.

The ratio of the dynamic viscosity μ to the specific mass of a fluid ρ is called the kinematic viscosity ν and is given by

$$\nu = \frac{\mu}{\rho} \, [\text{m}^2/\text{s}] \tag{5.5}$$

which varies as a function of temperature (Fig. 5.3).

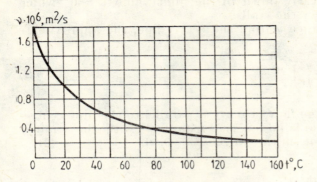

Fig. 5.3. Kinematic viscosity ν of water as a function of its temperature, $t°$.

5.1.5 Roughness

The state of the pipe inner surface is important in flow phenomena. This state is characterized by the *wall roughness*, which is that quality of pipe walls conditioned by the average height of surface roughness, by the variation of the real heights around the average value, and by the shape and pattern of irregularities. In practice, a simplified concept is used which takes into account only the height of irregularities: this concept is called the *absolute roughness*. It is known that roughness affects the fluid flow, the effect being greater in pipes of smaller diameter. Therefore, if it is desired to establish the effect roughness and pipe diameter have on the liquid flow these two factors should be simultaneously taken into account. The quantity used in practice is the ε/d ratio (that is the ratio of average roughness value to the pipe diameter) which is called the *relative roughness*.

The choice of the pipe absolute roughness determines the pressure losses due to friction.

In pipes carrying water, roughness is related both to the wall condition and its evolution with time, and to the physico-chemical characteristics of the liquid. Two situations are to be distinguished.

Non-corrodible, no-settlement pipes. This situation could be met in the case of low-suspension waters carried through pipes of plastic materials, cement asbestos, centrifugally cast concrete, and all other non-corrodible materials, or else materials with smooth inner coatings of high quality.

In such cases, the absolute roughness used in practice is $\varepsilon = 0.1$ mm (due to minimal degradations that cannot be avoided in time) although for new pipes, the theoretically acceptable value is $\varepsilon = 0.03$ mm.

Non-corrodible pipes, with settlements. When such pipes carry waters that are comparatively adverse, corrosive, calciferous, or heavily loaded with suspensions, an average roughness coefficient of $\varepsilon = 2$ mm is taken.

For raw, non-chlorinated, slightly adverse and slightly calciferous waters, the value chosen is $\varepsilon = 1$ mm.

For raw waters almost free of deposits, filtered, which are neither chemically adverse nor calciferous, nor treated with algicide, the value used is $\varepsilon = 0.5$ mm.

The values adopted in practice for the water wall absolute roughness (in average duty conditions) are shown in Table 5.1. for several common pipe materials.

TABLE 5.1. **Absolute roughness established for commercial pipes made of different materials and under various conditions**

Name of material	State of material	Absolute roughness
New cast iron	Asphalted Unasphalted Cemented	0.10—0.15 0.25—0.50 0.025
Worn-out iron	Uniform corrosion tracks Slight up to strong tracks Cleaned after a certain time of use	1.00—1.10 1.50—3.00 1.50
New steel	Varied	0.02—0.50
New steel	Zinc coated Asphalted Cemented Electro-plated	0.10—0.15 0.05 0.025 0.01
Worn-out steel	With uniform corrosion tracks Slight tracks Medium tracks Strong tracks	0.15 0.15—0.40 1.50 2.00—4.00
New asbestos cement	Varied	0.03—0.10
New concrete	Smooth Semi-smooth Rough	0.30—0.80 1.00—2.00 2.00—3.00
Worn-out concrete	After a certain time of use with water	0.20—0.30
New centrifuged concrete	Smooth Rough	0.20—0.50 1.00—2.00
New plastic	Varied	0.03—0.10
Glass, aluminum, brass, lead, or copper (new)	Varied	0.01
Glass, aluminum, brass, lead or copper (worn-out)	Varied	0.03

5.1.6 Reynold's Number

The conditions of liquid flow are determined by the Reynold's number, which is a dimensionless quantity

$$\text{Re} = \frac{vD}{\nu} \tag{5.6}$$

where v is the liquid flow velocity, in m/s;
ν is the liquid kinematic viscosity, in sq.m/s;
D is the pipe diameter, in m.

Reynold's number is a ratio between the liquid inertial and viscosity forces. At low flow velocities the inertial forces are small, the viscosity-forces predominate, and hence a laminar flow sets in, while at high velocities the inertial forces predominate and the flow becomes turbulent.

The values of Reynold's number were experimentally determined for the three flow conditions, i.e.:
— laminar flow: $\text{Re} \leqslant 2000$
— transient flow (critical zone): $2000 < \text{Re} < 4000$
— turbulent flow: $\text{Re} \geqslant 4000$

The changeover from one flow patern to the other was found by Schiller to occur at $\text{Re} = 2{,}320$, which is called the *critical state*. The expression for the critical velocity whereby the laminar flow is changed is:

$$v_{cr} = 2{,}320 \, \frac{\nu}{D}.$$

The above values hold true for the conditions that prevailed when the tests were made, viz straight pipe lengths with circular scross-section and smooth walls.

5.2 Calculation of Linear Head Losses

5.2.1 The Friction Coefficient

For straight pipes of circular cross-section the head losses through friction are obtained from the general equation (5.7) suggested by Darcy and Weisbach

$$h_l = \lambda \frac{l}{D} \frac{v^2}{2g} \tag{5.7}$$

where h_l is the head loss produced through friction, in m;
l — the pipe length, in m;
λ — a dimensionless friction coefficient;
D — the pipe diameter, in m;
v — the liquid mean velocity, in m/s;
g — the gravitational acceleration, in m/s².

Since the head losses are proportional to the pipe length, it is more convenient to use the value of the head loss per unit length (i.e. the value of the hidraulic gradient).

$$J = \frac{h_l}{l} = \lambda \frac{l}{D} \frac{v^2}{2g} \qquad (5.8)$$

where J is the unit head loss, in meters per one meter of pipe length.

Substituting in equation (5.7) the ratio $4Q/\pi d^2$ for velocity v, it results in

$$h_l = \frac{8\lambda l}{\pi^2 g d^5} Q^2 \qquad (5.9)$$

Now, by noting

$$R = \frac{8\lambda l}{\pi^2 g d^2} \equiv S(Q) \quad \text{and} \quad S = \frac{8\lambda}{\pi^2 g d^2} = S(Q)$$

we get

$$h_l = RQ^2 \quad \text{or} \quad h_l = SlQ^2$$

where R is the pipe resistance, in s^2/m^5, while S is the pipe specific resistance, in s^2/m^6.

Because S values are a function of the diameter, they can be calculated for a variety of standard diameters of mass-produced pipes (Fig. 5.4).

Now, by taking S-values from Fig. 5.4 and multiplying by the pipe length and by the square of the flow rate, we find the head loss for the given system.

Substituting for velocity v in equation (5.8) the ratio $4Q/\pi d^2$, we find

$$J = \frac{h_l}{l} = \frac{8\lambda}{\pi^2 g d^5} Q^2 \qquad (5.10)$$

which we can infer that the head loss due to friction is directly proportional to the squared flow rate and varies inversely with the fifth power of the diameter. Actually, this is not always true, because the friction coefficient is itself a function of Q and D and, consequently, modifies the exponents of equation (5.10).

The above general expressions — in any form they are put, show that the calculation of frictional head loss is reduced to the calculation of the friction coefficient λ, which depends on the Reynold's number, Re, and on the pattern of the flow.

Friction coefficient in laminar flow. For Re < 2320, i.e. the conditions of laminar flow, the friction coefficient is found by means of the Hagen-Poiseulle equation

$$\lambda = \frac{64}{\text{Re}} \qquad (5.11)$$

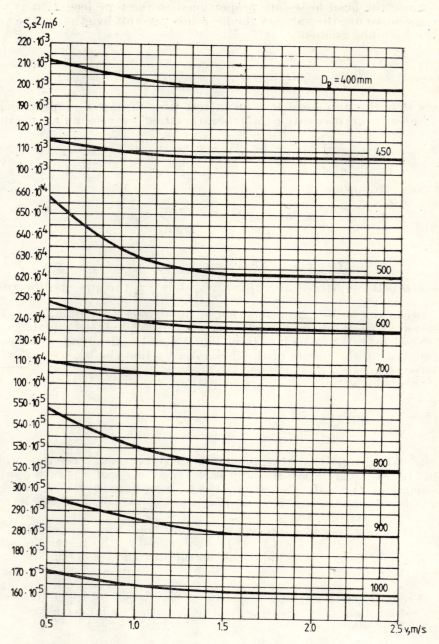

Fig. 5.4. Specific resistance S of pipes with respect to liquid flow velocity v.

which shows that for laminar flow the friction coefficient is independent of the wall roughness. This phenomenon is explained by Prandtl, who assumed the existence of a boundary layer. Since the thickness of this layer varies inversely with Re, it follows that for the low Re values characteristic of laminar flow, the boundary layer is thick. Actually, this thickness is much greater than the surface irregularities which, being swamped in the boundary layer, cannot affect the main stream which therefore flows independently of the roughness.

Thus in the case of laminar flow, according to equation (5.11), the head losses are caused only by viscosity.

Friction coefficient in turbulent flow. For $Re > 2{,}320$, i.e. the conditions of turbulent flow, the friction coefficient depends on roughness and the phenomena are much more complex than in the case of laminar flow. The friction coefficient is more dependent on roughness as the Re values rise, therefore, the thinner the boundary layer, the more the surface irregularities penetrate the main stream.

Three characteristic regions are to be distinguished in turbulent flow:

(1) *The region of hydraulically smooth flow.* For small roughness and Re values, it may happen that the surface irregularities or asperities are flooded in the boundary layers and, as a consequence, the friction coefficient is higher than in the case of laminar flow because the very nature of the flow is different. The energy necessary to maintain the flow is spent both in overcoming the viscosity forces and in mixing the foreign particles into the liquid mass (resulting from the continuous exchange of particles between the various liquid layers, proper to the turbulent flow). In this region, since roughness does not affect the friction losses, the flow can be considered as motion through a smooth pipe and thus is called *hydraulically smooth flow*. The relevant friction coefficient is determined by means of the equation suggested by Prandtl and Kármán:

$$\lambda = \frac{1}{\left(2 \log \dfrac{Re \sqrt{\lambda}}{2 \cdot 51}\right)^2} \tag{5.12}$$

Altschoul [5.3] has shown the limits of application of this relation as to critical Re values, thus

$$Re < Re_{cr\,I} = \frac{23}{\varepsilon/D} = \frac{200\,d}{\varepsilon \sqrt{\lambda}} \tag{5.13}$$

but Re values should not fall below 2,320.

However, equation (5.12) was not corroborated by experimental results, which yielded lower values for the coefficient λ. Therefore it is common practice to make use of the Blasius equation

$$\lambda = \frac{0.316}{\sqrt[4]{Re}} = \sqrt[4]{100\,Re} \tag{5.14}$$

(2) *The region of hydraulically rough flow*. For very high Re values and significant roughness, the boundary layer becomes very thin and hence the roughness predominantly determines the frictional losses. Indeed under such conditions the friction coefficient does not actually depend on Reynolds number, but only on the relative roughness (ε/d) of the pipe. In this region the flow is called *hydraulically rough flow* and the friction coefficient is found by means of the equation suggested by Prandtl and Kármán

$$\lambda = \frac{1}{2 \log \left(\dfrac{\varepsilon}{3.71\, D}\right)^2} \tag{5.15}$$

The application limits of this equation were given by Altschoul with reference to a critical Re value

$$\mathrm{Re} > \mathrm{Re}_{cr\, II} = \frac{560}{\varepsilon/D} \tag{5.16}$$

Experimental results do not contradict equation (5.15).

(3) *The region of intermediate flow*. Between the two extreme regions there is a transitional one, where the friction coefficient depends both on the Reynolds number, and on the relative roughness ε/D; in fact, this region covers almost the entire range met in practice and is characteristic of the roughness of industrially-made pipes.

Within this region, the friction coefficient is found by means of Colebrook's equation (found in 1939 through interpolation) [5.7] and recently checked experimentally by White

$$\lambda = \frac{1}{\left[2 \log \left(\dfrac{\varepsilon}{3.7 d} + \dfrac{2.5}{\mathrm{Re} \sqrt{\lambda}}\right)\right]} \tag{5.17}$$

The application limits of this formula are

$$\mathrm{Re}_{cr\, I} < \mathrm{Re} < \mathrm{Re}_{cr\, II}$$

Now, supposing that we are confronted with the two limit cases, i.e. hydraulically smooth pipe ($\varepsilon/d = 0$) and hydraulically rough pipe ($\mathrm{Re} = \infty$), we find that equation (5.17) reduces to equations (5.12) and (5.15). This means that equation (5.17) holds true over the whole range of turbulent flow and consequently it is considered as the only one established on really scientific grounds, and the one that best translates the real phenomena taking place in turbulent flow. On the other hand, due to its implicit form, direct use of this equation is a very cumbersome task.

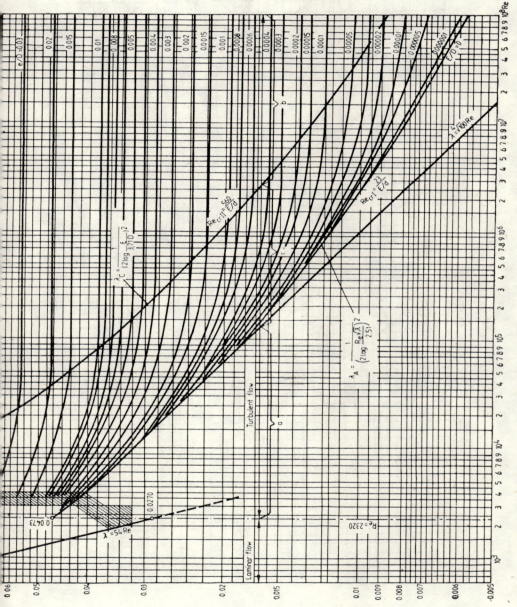

Fig. 5.5. Friction coefficient λ with respect to Reynolds number Re, for various relative roughnesses ε/d of pipe walls.

Heads, Head Losses and System Head Curves

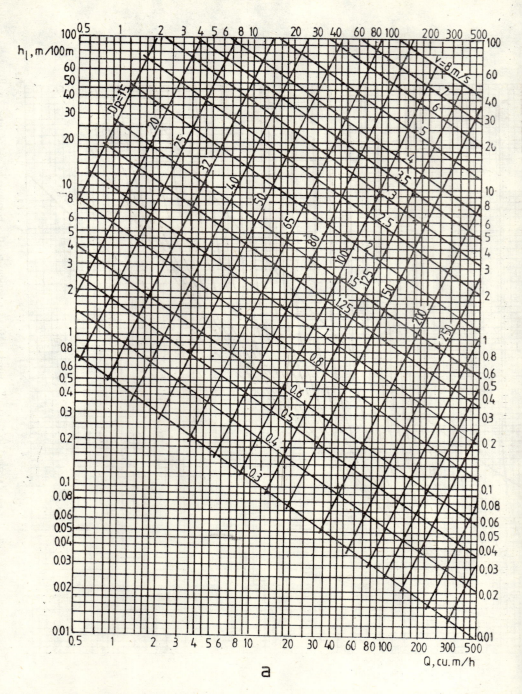

Fig. 5.6. Frictional head loss h_l, in m per 100 m of pipe length, for pipes having absolute wall kinematic viscosity $v = 1.236$ c St (clean water at 12°C). For other

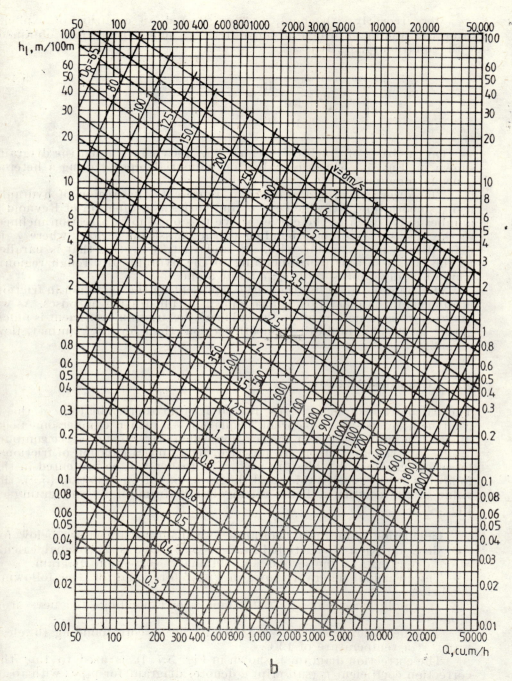

roughness $\varepsilon = 0{,}1$ mm (new cast iron pipes, bituminized inside) and for liquids having values of ε, we apply the correction coefficient, f_d, obtained from Fig. 5.7.

More recently, Altschoul, [5.3] suggested an equation of explicit form, the results of which do not differ very much from those obtained with equation (5.16). It is

$$\lambda = \frac{1}{\left(1.8 \log \dfrac{\text{Re}}{\dfrac{\text{Re}}{10}\dfrac{\varepsilon}{d} + 7}\right)^2} \tag{5.18}$$

On the basis of Colebrook's equation, Mody constructed the diagram shown in Fig. 5.5 which is valid for industrial pipes having a heterogeneous roughness.

The lower curve on the diagram represents the region of hydraulically smooth flow, for which $\varepsilon = 0$, and varies as a function of Reynold's number only. Each of the curves above this one shows a section inclined to the abscissa axis (corresponding to the transition region) where λ is defined by a pair of values (Re, ε/d) and another section that is parallel to the abscissa axis (corresponding to the hydraulically rough region), where λ is determined by the ε/d value alone.

On the same diagram, the curve representing the variation in friction coefficient for laminar flow is also shown for illustrating purposes. As we can see, for the same Re value, the value of the friction coefficient is much grater in the hydraulically smooth region than in the laminar flow region.

5.2.2 Calculation Methods

The calculation of head losses due to friction (determination of the λ coefficient) by means of Colebrook's equation is a rather cumbersome task, due to the implicit form of the relation which leads to successive computations (by trial and error). In practice, for rapid calculation of frictional losses, diagrams and tables are used. The reader will be presented in the following with several tables and diagrams based on equation (5.9), the value of the coefficient being taken directly from Mody's diagram (see Fig. 5.5).

Calculation by means of diagrams. Two diagrams are shown below for the determination of frictional head losses h_l. One is a principal calculation diagram, and the second is an auxiliary correction diagram.

The principal diagram shown in Fig. 5.6 is valid for the following conditions:
— absolute roughness $\varepsilon = 0.1$ mm, corresponding to new iron pipes;
— kinematic viscosity $\varepsilon = 1.2366 \times 10^{-6}$ m^2/s, corresponding to clear water at a temperature of 12°C.

The correction diagram is shown in Fig. 5.7. It is used to find the correction coefficient f_d (subscript d denotes diagram) for pipes with absolute roughness $\varepsilon \neq 1$ (see Table 5.1). The corrected head loss, h_{lc}, is found by multiplying the frictional head loss $h_{l(k=0.1)}$, obtained from Fig. 5.6, by the correction coefficient f_d obtained from Fig. 5.7, thus

$$h_{lc} = h_{l(k=0.1)} \times f_d \qquad (5.18, a)$$

Is found in Fig. 5.6. ─────────┘
Is found in Fig. 5.7. ──────────────┘

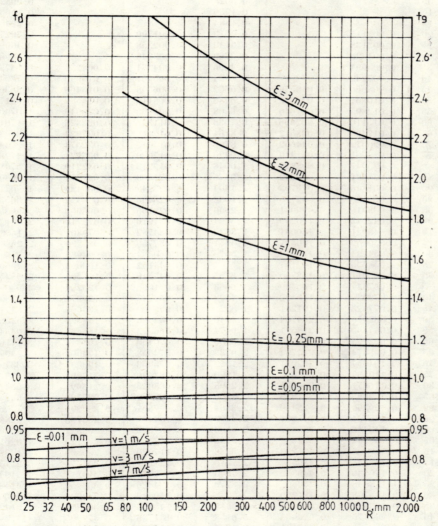

Fig. 5.7. Correction coefficient, f_d, for absolute roughness of pipe walls $\varepsilon \neq 0.1$ mm.

Calculation by means of tables. The table 5.2 has been prepared to assist numerical determination of the unit frictional loss, J.

TABLE 5.2 Giving the unit head loss J, in meters per one linear meter of pipe as function of diameter and the mean velocity:
For new pipes ($\varepsilon = 0.1$ mm) and for existing pipes ($\varepsilon = 2$ mm)

Mean velocity	Pipe diameter 0.040 m			Pipe diameter 0.050 m		
	Pipe section 0.0012566 m²			Pipe section 0.0019635 m²		
	Head loss m/1 m of pipe length		Flow dm³/s	Head loss m/1 m of pipe length		Flow dm³/s
	New pipes	Existing pipes		New pipes	Existing pipes	
0.01			0.0125			0.0196
0.05			0.0628			0.0982
0.10			0.1256			0.1963
0.15			0.1884			0.2945
0.20	0.002115		0.2513	0.001590		0.3927
0.25	0.003138		0.3140	0.002358		0.4909
0.30	0.004329		0.3769	0.003248		0.5890
0.35	0.005694		0.4396	0.004281	0.008237	0.6872
0.40	0.007242		0.5024	0.005451	0.010690	0.7854
0.45	0.008966	0.018576	0.5652	0.006708	0.013468	0.8836
0.50	0.010861	0.022868	0.6280	0.008115	0.016587	0.9817
0.55	0.012895	0.027640	0.6908	0.009668	0.020046	1.0790
0.60	0.015116	0.032856	0.7538	0.011340	0.023826	1.1781
0.65	0.017493	0.038512	0.8164	0.013118	0.027924	1.2763
0.70	0.020072	0.044652	0.8792	0.015013	0.032374	1.3744
0.75	0.022793	0.051212	0.9420	0.017030	0.037128	1.4726
0.80	0.025647	0.058227	1.0048	0.019213	0.042210	1.5708
0.85	0.028681	0.065742	1.0676	0.021509	0.047658	1.6690
0.90	0.031845	0.073703	1.1304	0.023948	0.053429	1.7671
0.95	0.035190	0.082110	1.1932	0.026496	0.059524	1.8653
1.00	0.038546	0.090981	1.2566	0.029155	0.065955	1.9635
1.05	0.042143	0.100299	1.3194	0.031916	0.072710	2.0617
1.10	0.046021	0.110081	1.3822	0.034782	0.079801	2.1598
1.15	0.050052	0.120327	1.4451	0.037750	0.087229	2.2580
1.20	0.054224	0.131019	1.5079	0.040884	0.094980	2.3562
1.25	0.058535	0.142157	1.5707	0.044152	0.103054	2.4555
1.30	0.063011	0.153760	1.6353	0.047549	0.111465	2.5525
1.35	0.067647	0.165809	1.6963	0.051090	0.120200	2.6507
1.40	0.072428	0.178322	1.7592	0.054745	0.129271	2.7489
1.45	0.077423	0.191281	1.8220	0.058509	0.138665	2.8471
1.50	0.082570	0.204704	1.8846	0.062386	0.148396	2.9452
1.55	0.087865	0.218591	1.9474	0.066373	0.158463	3.0434
1.60	0.093293	0.232907	2.0105	0.070459	0.168841	3.1416
1.65	0.098874	0.247704	2.0733	0.074658	0.179568	3.2397
1.70	0.104657	0.262931	2.1362	0.078953	0.190606	3.3379
1.75	0.110597	0.278639	2.1990	0.083420	0.210993	3.4361
1.80	0.116671	0.294775	2.2608	0.088020	0.213691	3.5343
1.85	0.122893	0.311375	2.3236	0.092732	0.225725	3.6324
1.90	0.129260	0.328440	2.3864	0.097557	0.238096	3.7306
1.95	0.135764	0.345951	2.4499	0.102487	0.250790	3.8288
2.00	0.142410	0.363926	2.5132	0.107526	0.263821	3.9270
2.05	0.149244	0.382347	2.5760	0.112669	0.277175	4.0251
2.10	0.156222	0.401232	2.6388	0.117920	0.290865	4.1233
2.15	0.163337	0.420564	2.7016	0.123271	0.304879	3.2215
2.20	0.170586	0.440342	2.7645	0.128712	0.319217	4.3197
2.25	1.178048	0.460601	2.8273	0.134336	0.333904	4.4179
2.30	0.185708	0.481290	2.8888	0.140046	0.348901	4.5160
2.35	0.193518	0.502442	2.9516	0.145863	0.364235	4.6142
2.40	0.201476	0.524058	3.0158	0.151786	0.379905	4.7124
2.45	0.209576	0.546121	3.0786	0.157780	0.395899	4.8106
2.50	0.217815	0.568630	3.1412	0.164058	0.412217	4.9087
3.00	0.307923	0.818833	3.7698	0.233035	0.593597	5.8905
3.50	0.414432	1.114518	4.3981	0.312190	0.807948	6.8723
4.00	0.536204	1.455703	5.0264	0.404498	1.055283	7.8540

TABLE 5.2 (*continued*)

Mean velocity	Pipe diameter **0.060** m			Pipe diameter **0.080** m		
	Pipe section 0.00282 m²			Pipe section 0.00502 m²		
	Head loss m/1 m of pipe length		Flow dm³/s	Head loss m/1 m of pipe length		Flow dm³/s
	New pipes	Existing pipes		New pipes	Existing pipes	
0.01			0.0283			0.0503
0.05			0.1414			0.2513
0.10			0.2827	0.000256		0.5027
0.15	0.000753		0.4241	0.000520		0.7540
0.20	0.001249		0.5655	0.000863	0.001438	1.0053
0.25	0.001856		0.7069	0.001280	0.002213	1.2566
0.30	0.002557	0.004713	0.8482	0.001775	0.003154	1.5080
0.35	0.003364	0.006354	0.9896	0.002336	0.004263	1.7593
0.40	0.004277	0.008262	1.1310	0.002994	0.005539	2.0106
0.45	0.005289	0.010406	1.2723	0.003702	0.006969	2.2620
0.50	0.006412	0.012803	1.4137	0.004467	0.008568	2.5133
0.55	0.007633	0.015466	1.5551	0.005339	0.110347	2.7646
0.60	0.008961	0.018374	1.6965	0.006274	0.010290	3.0159
0.65	0.010388	0.021530	1.8378	0.007280	0.014396	3.2672
0.70	0.011907	0.024955	1.9792	0.008353	0.016680	3.5186
0.75	0.013523	0.028612	2.1206	0.009450	0.019120	3.7699
0.80	0.015223	0.032522	2.2619	0.010646	0.021733	4.0212
0.85	0.017034	0.036682	2.4033	0.011910	0.024515	4.2726
0.90	0.018959	0.041084	2.5447	0.013249	0.027458	4.5239
0.95	0.020968	0.045771	2.6861	0.014651	0.030590	4.7752
1.00	0.023064	0.050715	2.8274	0.016119	0.033895	5.0266
1.05	0.025257	0.055909	2.9688	0.017644	0.037367	5.2779
1.10	0.027556	0.061361	3.1102	0.019241	0.410011	5.5192
1.15	0.029941	0.067073	3.2516	0.020906	0.044828	5.7805
1.20	0.032418	0.073033	3.3929	0.22635	0.048811	6.0319
1.25	0.034975	0.079242	3.5343	0.024420	0.052961	6.2832
1.30	0.037615	0.085709	3.6757	0.026273	0.057283	6.5345
1.35	0.040392	0.092426	3.8170	0.028181	0.061772	6.7858
1.40	0.043257	0.099401	3.9584	0.030145	0.066434	7.0372
1.45	0.046204	0.106624	4.0988	0.032175	0.071261	7.2885
1.50	0.049255	0.114106	4.2412	0.034261	0.076262	7.5398
1.55	0.052392	0.121848	4.3825	0.036478	0.081436	7.7911
1.60	0.055606	0.129828	4.5239	0.038753	0.086769	8.0425
1.65	0.058908	0.138076	4.6653	0.041093	0.092283	8.2937
1.70	0.062308	0.146564	4.8066	0.043490	0.097955	8.5451
1.75	0.065796	0.155320	4.9480	0.045952	0.103807	8.7965
1.80	0.069359	0.164314	5.0894	0.048489	0.109818	9.0478
1.85	0.073003	0.173568	5.2368	0.051089	0.116003	9.2991
1.90	0.076759	0.183080	5.3721	0.053751	0.122360	9.5505
1.95	0.080625	0.192841	5.5135	0.056472	0.128884	9.8018
2.00	0.084576	0.202861	5.6549	0.059253	0.135580	10.0531
2.05	0.088607	0.213129	5.7963	0.062118	0.142443	10.3044
2.10	0.092722	0.223646	5.9376	0.065046	0.149479	10.5558
2.15	0.096914	0.234432	6.0790	0.068032	0.156680	10.8071
2.20	0.101266	0.245457	6.2204	0.071078	0.164049	11.0584
2.25	0.105710	0.256749	6.3617	0.074187	0.171597	11.3097
2.30	0.110234	0.268282	6.5031	0.077350	0.179304	11.5610
2.35	0.114844	0.280072	6.6445	0.080574	0.187184	11.8124
2.40	0.119540	0.292122	6.7859	0.083857	0.195238	12.0637
2.45	0.124318	0.304420	6.9272	0.087196	0.203457	12.3150
2.50	0.129176	0.316967	7.0686	0.090591	0.211842	12.5664
3.00	0.183110	0.456436	8.4823	0.128731	0.305056	15.0795
3.50	0.246110	0.621258	9.8960	0.172875	0.415213	17.5928
4.00	0.318732	0.811442	11.3098	0.224268	0.542321	20.1060

TABLE 5.2 (continued)

Mean velocity	Pipe diameter 0.100 m			Pipe diameter 0.125 m		
	Pipe section 0.00785 m²		Flow dm³/s	Pipe section 0.01227 m²		Flow dm³/s
	Head loss m/1 m of pipe length			Head loss m/1 m of pipe length		
	New pipes	Existing pipes		New pipes	Existing pipes	
0.01			0.0785			0.1227
0.05			0.3927			0.6136
0.10	0.000191		0.7854	0.000144	0.000207	1.2272
0.15	0.000388	0.000604	1.1781	0.000291	0.000449	1.8408
0.20	0.000643	0.001054	1.5708	0.000486	0.000783	2.4544
0.25	0.000956	0.001622	1.9635	0.000726	0.001204	3.0680
0.30	0.001335	0.002312	2.3562	0.001009	0.001712	3.6816
0.35	0.001763	0.003120	2.7489	0.001330	0.002311	4.2952
0.40	0.002248	0.004060	3.1416	0.001701	0.003004	4.9088
0.45	0.002786	0.005111	3.5343	0.002104	0.003785	5.5224
0.50	0.003370	0.006281	3.9270	0.002548	9.004656	6.1360
0.55	0.004009	0.007584	4.3197	0.003037	0.005618	6.7496
0.60	0.004707	0.009006	4.7124	0.003560	0.006568	7.3632
0.65	0.005447	0.10543	5.1051	0.004120	0.007804	7.9768
0.70	0.006245	0.012215	5.4978	0.004726	0.009037	8.5904
0.75	0.007090	0.14000	5.8905	0.005369	0.010356	9.2040
0.80	0.007985	0.015911	6.1830	0.006059	0.011769	9.8176
0.85	0.008931	0.017951	6.6759	0.006765	0.013279	10.4312
0.90	0.009930	0.020108	7.0686	0.007531	0.014878	11.0448
0.95	0.010980	0.022402	7.4613	0.008332	0.016567	11.6584
1.00	0.012080	0.024822	7.8540	0.009166	0.018349	12.2720
1.05	0.013233	0.027365	8.2467	0.010047	0.020228	12.8866
1.10	0.014431	0.030033	8.8394	0.010962	0.022201	13.4992
1.15	0.015673	0.032839	9.0321	0.011913	0.024268	14.1128
1.20	0.016855	0.035756	9.4248	0.012091	0.026424	14.7264
1.25	0.018301	0.038785	9.8175	0.013921	0.028670	15.8400
1.30	0.019692	0.041950	10.2102	0.014988	0.031010	15.9536
1.35	0.021142	0.045237	10.6029	0.016089	0.033440	16.5672
1.40	0.022637	0.048651	10.9956	0.017231	0.035964	17.1808
1.45	0.024197	0.052197	11.3883	0.018406	0.038578	17.7944
1.50	0.025803	0.055849	11.7810	0.019615	0.041285	18.4080
1.55	0.027456	0.059638	12.1737	0.020857	0.044036	19.0216
1.60	0.029149	0.063544	12.5664	0.022140	0.046973	19.6352
1.65	0.30890	0.067581	12.9591	0.023458	0.049957	20.2448
1.70	0.032671	0.071735	13.3518	0.024805	0.053208	20.8624
1.75	0.034514	0.076021	13.7445	0.026200	0.056196	21.4760
1.80	0.036397	0.080423	14.1372	0.027625	0.059450	22.0896
1.85	0.038324	0.084952	14.5293	0.029097	0.062798	22.7032
1.90	0.040296	0.089608	14.9226	0.030588	0.066240	23.3168
1.95	0.042347	0.094385	15.3153	0.032126	0.069772	23.9304
2.00	0.044446	0.099290	15.7081	0.033714	0.073397	24.5440
2.05	0.046589	0.104315	16.1007	0.035335	0.077112	25.1576
2.10	0.048777	0.109468	16.4934	0.036990	0.080921	25.7712
2.15	0.051010	0.114742	16.8861	0.038678	0.034820	26.3848
2.20	0.053285	0.120138	17.2788	0.040437	0.088808	26.9984
2.25	0.055608	0.125665	17.6715	0.042236	0.092894	27.6120
2.30	0.057970	0.131310	18.0642	0.044068	0.097057	28.2256
2.35	0.060377	0.137081	18.4569	0.045960	0.101333	28.8392
2.40	0.062828	0.142978	18.8496	0.047890	0.105692	29.4538
2.45	0.065320	0.148998	19.2423	0.049858	0.110142	30.0664
2.50	0.065853	0.155139	19.6350	0.051862	0.114682	30.6800
3.00	0.096333	0.223402	23.5620	0.073580	0.165143	36.8160
3.50	0.129559	0.304073	27.4890	0.098802	0.224777	42.9520
4.00	0.167589	0.397152	31.4160	0.128004	0.293587	49.0880

TABLE 5.2. (*continued*)

Mean velocity	Pipe diameter **0.150** m			Pipe diameter **0.200** m		
	Pipe section 0.01767 m²			Pipe section 0.03141 m²		
	Head loss m/1 m of pipe length		Flow dm³/s	Head loss m/1 m of pipe length		Flow dm³/s
	New pipes	Existing pipes		New pipes	Existing pipes	
0.01			0.1767			0.3142
0.05	0.000034		0.8836	0.000024	0.000030	1.5708
0.10	0.000114	0.000163	1.7671	0.000079	0.000110	3.1416
0.15	0.000232	0.000352	2.6507	0.000162	0.000238	4.7424
0.20	0.000387	0.000612	3.5343	0.000270	0.000413	6.2832
0.25	0.000578	0.000941	4.4179	0.000400	0.000636	7.8540
0.30	0.000801	0.001336	5.3014	0.000557	0.000903	9.4248
0.35	0.001059	0.001810	6.1850	0.000736	0.001217	10.9956
0.40	0.001351	0.002347	7.0686	0.000940	0.001581	12.5664
0.45	0.001674	0.002948	7.9522	0.001169	0.001989	14.1372
0.50	0.002031	0.003622	8.8357	0.001421	0.002445	15.7080
0.55	0.002421	0.004374	9.7193	0.001692	0.002945	17.2788
0.60	0.002842	0.005187	10.6029	0.001986	0.003491	18.8496
0.65	0.003293	0.006070	11.4865	0.002298	0.004080	20.4204
0.70	0.003777	0.007028	12.3700	0.002642	0.004734	21.9912
0.75	0.004289	0.008054	13.2536	0.002996	0.005433	23.5620
0.80	0.004834	0.009155	14.1372	0.003376	0.006181	25.1328
0.85	0.005411	0.010329	15.0208	0.003784	0.006979	26.7036
0.90	0.006017	0.011572	15.9043	0.004212	0.007824	28.2744
0.95	0.006652	0.012883	16.7879	0.004658	0.008717	29.8452
1.00	0.007316	0.014268	17.6715	0.005122	0.009659	31.4160
1.05	0.008009	0.015722	18.5550	0.005619	0.101648	32.9868
1.10	0.008732	0.017247	19.4386	0.006139	0.011686	34.5576
1.15	0.009487	0.018852	20.3222	0.006680	0.012774	36.1284
1.20	0.010271	0.020527	21.2058	0.007241	0.013909	37.6992
1.25	0.011086	0.022273	22.0893	0.007821	0.015092	39.2700
1.30	0.011933	0.024091	22.9729	0.008424	0.016324	40.8408
1.35	0.012813	0.025978	23.8565	0.009047	0.017603	42.4116
1.40	0.013726	0.027939	24.7401	0.009695	0.018931	43.9824
1.45	0.014667	0.029970	25.6237	0.010362	0.020307	45.5532
1.50	0.015642	0.032072	26.5072	0.011049	0.021737	47.1240
1.55	0.016646	0.034248	27.3908	0.011756	0.023206	48.6948
1.60	0.017684	0.036491	28.2744	0.012480	0.024726	50.2656
1.65	0.018752	0.038809	29.1580	0.013232	0.026297	51.8364
1.70	0.019846	0.041195	30.0415	0.011400	0.027913	53.4072
1.75	0.020970	0.043656	30.9251	0.014790	0.029581	54.9780
1.80	0.022129	0.046184	31.8087	0.015597	0.031294	56.5488
1.85	0.023317	0.048785	32.6922	0.016424	0.033056	58.1196
1.90	0.024533	0.051459	33.5758	0.017268	0.034868	59.6904
1.95	0.025777	0.054202	34.4594	0.018141	0.036727	61.2612
2.00	0.027062	0.057018	35.3430	0.019032	0.038635	62.8320
2.05	0.028374	0.059905	36.2265	0.019942	0.040591	64.4028
2.10	0.029716	0.062863	37.1101	0.020882	0.042596	65.9736
2.15	0.031085	0.065892	37.9937	0.021841	0.044548	67.5444
2.20	0.032497	0.068991	38.8772	0.022831	0.046748	69.1152
2.25	0.033941	0.072165	39.7608	0.023843	0.048899	70.6860
2.30	0.035411	0.075406	40.6444	0.024873	0.051095	72.2658
2.35	0.036911	0.078720	41.5279	0.025924	0.053340	73.8276
2.40	0.038441	0.082107	42.4115	0.026981	0.055635	75.3984
2.45	0.039998	0.085564	43.2951	0.028071	0.057978	76.9692
2.50	0.041583	0.089090	44.1787	0.029180	0.060367	78.5400
3.00	0.059023	0.128291	53.0145	0.041400	0.086929	94.2480
3.50	0.079296	0.174618	61.7503	0.055757	0.118320	109.956
4.00	0.102483	0.228073	70.6860	0.072051	0.154541	125.664

TABLE 5.2 (*continued*)

Mean velocity	Pipe diameter 0.250 m			Pipe diameter 0.300 m		
	Pipe section 0.04908 m²			Pipe section 0.07068 m²		
	Head loss m/1 m of pipe length		Flow dm³/s	Head loss m/1 m of pipe length		Flow dm³/s
	New pipes	Existing pipes		New pipes	Existing pipes	
0.01			0.4909			0.7069
0.05	0.000017	0.000022	2.4544	0.000014	0.000018	3.5343
0.10	0.000060	0.000081	4.9087	0.000048	0.000064	7.0686
0.15	0.000122	0.000175	7.3631	0.000097	0.000139	10.6029
0.20	0.000204	0.000305	9.8175	0.000163	0.000241	14.1372
0.25	0.000303	0.000469	12.2719	0.000244	0.000370	17.6715
0.30	0.000424	0.000668	14.7262	0.000339	0.000527	21.2058
0.35	0.000563	0.000902	17.1806	0.000450	0.000711	24.7401
0.40	0.000720	0.001173	19.6350	0.000574	0.000925	28.2744
0.45	0.000890	0.001477	22.0894	0.000712	0.001164	31.8087
0.50	0.001080	0.001815	24.5437	0.000864	0.001431	35.3430
0.55	0.001286	0.002188	26.9981	0.001031	0.001725	38.8773
0.60	0.001512	0.002594	29.4525	0.001215	0.002046	42.4116
0.65	0.001753	0.003034	31.9069	0.001411	0.002393	45.9459
0.70	0.002013	0.003511	34.3612	0.001622	0.002769	49.4802
0.75	0.002294	0.004024	36.8156	0.001845	0.003170	53.0145
0.80	0.002586	0.004573	39.2700	0.002079	0.003603	56.5488
0.85	0.002896	0.005159	41.7244	0.002326	0.004064	60.0831
0.90	0.003226	0.005781	44.1787	0.002588	0.004556	63.6174
0.95	0.003571	0.006440	46.6331	0.002866	0.005076	67.1517
1.00	0.003935	0.007136	49.0875	0.003157	0.005634	70.6860
1.05	0.004315	0.007867	51.5418	0.003461	0.006200	74.2203
1.10	0.004712	0.008634	53.9962	0.003878	0.006804	77.7546
1.15	0.005123	0.009437	56.4506	0.004110	0.007438	81.2889
1.20	0.005555	0.010276	58.9050	0.004453	0.008099	84.8232
1.25	0.006002	0.011150	61.3593	0.004808	0.008787	88.3575
1.30	0.006464	0.012060	63.8137	0.005177	0.009504	91.8918
1.35	0.006944	0.013005	66.2681	0.005561	0.010249	95.4261
1.40	0.007441	0.013986	68.7225	0.005957	0.011011	98.9604
1.45	0.007956	0.015002	71.1769	0.006365	0.011823	102.4947
0.50	0.008486	0.016055	73.6312	0.006785	0.012653	106.0290
1.55	0.009033	0.017144	76.0856	0.007217	0.013511	109.5633
1.60	0.009593	0.018267	78.5400	0.007659	0.014397	113.0976
1.65	0.010169	0.019428	80.9944	0.008123	0.015311	116.6319
1.70	0.010759	0.020622	83.4487	0.008602	0.016252	120.1662
1.75	0.011364	0.021854	85.9031	0.009090	0.017223	123.7005
1.80	0.011989	0.023120	88.3575	0.009595	0.018221	127.2348
1.85	0.012629	0.024422	90.8118	0.010106	0.019247	130.7691
1.90	0.013285	0.025760	93.2662	0.010635	0.020302	134.3034
1.95	0.013954	0.027133	95.7206	0.011170	0.021384	137.8377
2.00	0.014639	0.028543	98.1750	0.011723	0.022495	141.3720
2.05	0.015345	0.029988	100.6293	0.012288	0.023633	144.9063
2.10	0.016067	0.031469	103.0837	0.012865	0.024801	148.4406
2.15	0.016804	0.032985	105.5381	0.013461	0.025996	151.9749
2.20	0.017564	0.034537	107.9924	0.014070	0.027218	155.5092
2.25	0.018341	0.036126	110.4468	0.014691	0.028470	159.0435
2.30	0.019133	0.037748	112.9012	0.015324	0.029749	162.5778
2.35	0.019940	0.039407	115.3555	0.015969	0.031057	166.1121
2.40	0.020763	0.041102	117.8099	0.016627	0.032393	169.6464
2.45	0.021600	0.042833	120.2643	0.017296	0.033756	173.1807
2.50	0.022465	0.044598	122.7187	0.017988	0.035148	176.7150
3.00	0.031873	0.064222	147.2625	0.025490	0.050613	212.058
3.50	0.042907	0.087413	171.8063	0.034341	0.068890	247.401
4.00	0.055455	0.114173	196.3500	0.044527	0.089979	282.744

TABLE 5.2 *(continued)*

Mean velocity	Pipe diameter **0.350 m**			Pipe diameter **0.400 m**		
	Pipe section 0.09681 m²			Pipe section 0.12566 m²		
	Head loss m/1 m of pipe length		Flow dm³/s	Head loss m/1 m of pipe length		Flow dm³/s
	New pipes	Existing pipes		New pipes	Existing pipes	
0.01			0.0621			1.2566
0.05	0.000011	0.000014	4.8106	0.000010	0.000012	6.2832
0.10	0.000039	0.000052	9.6211	0.000033	0.000044	12.5664
0.15	0.000081	0.000112	14.4317	0.000068	0.000094	18.8496
0.20	0.000135	0.000195	19.2432	0.000115	0.000164	25.1328
0.25	0.000203	0.000298	24.0529	0.000172	0.000253	31.4160
0.30	0.000282	0.000425	28.8634	0.000239	0.000360	37.6992
0.35	0.000374	0.000574	33.6740	0.000317	0.000485	43.9824
0.40	0.000477	0.000747	38.4846	0.000406	0.000631	50.2656
0.45	0.000594	0.000941	43.2952	0.000506	0.000795	56.5488
0.50	0.000721	0.001157	48.1057	0.000615	0.000978	62.8320
0.55	0.000860	0.001396	52.9163	0.000732	0.001180	69.1152
0.60	0.001009	0.001657	57.7269	0.000858	0.001400	75.3984
0.65	0.001172	0.001942	62.5375	0.000996	0.001640	81.6816
0.70	0.001348	0.002252	67.3480	0.001146	0.001899	87.9648
0.75	0.001533	0.002584	72.1586	0.001305	0.002177	94.2480
0.80	0.001730	0.002940	76.9692	0.001472	0.002473	100.5312
0.85	0.001936	0.003320	81.7798	0.001648	0.002790	106.8144
0.90	0.002153	0.003722	86.5903	0.001832	0.003128	113.0976
0.95	0.002383	0.004147	81.4009	0.002026	0.003485	119.3808
1.00	0.002626	0.004595	96.2115	0.002233	0.003861	125.6640
1.05	0.002878	0.005065	101.0221	0.002447	0.004257	131.9472
1.10	0.003142	0.005559	105.8326	0.002672	0.004672	138.2304
1.15	0.003417	0.006077	110.6432	0.002905	0.005106	144.5136
1.20	0.003701	0.006616	115.4538	0.003147	0.005560	150.7968
1.25	0.003998	0.007179	120.2644	0.003399	0.006033	157.0800
1.30	0.004304	0.007765	125.0749	0.003659	0.006525	163.3632
1.35	0.004623	0.008373	129.8855	0.003929	0.007037	169.6464
1.40	0.004952	0.009005	134.6961	0.004208	0.007567	175.9296
1.45	0.005291	0.009660	139.5067	0.004498	0.008117	182.2128
1.50	0.005642	0.010338	144.3172	0.004796	0.008687	188.4960
1.55	0.006004	0.011039	149.1278	0.005107	0.009276	194.7792
1.60	0.006375	0.011762	153.9384	0.005425	0.009884	201.0624
1.65	0.006760	0.012509	158.7490	0.005752	0.010512	207.3456
1.70	0.007155	0.013278	163.5595	0.006087	0.011158	213.6288
1.75	0.007560	0.014071	168.3701	0.006431	0.011825	219.9120
1.80	0.007979	0.014886	173.1807	0.006783	0.012509	226.1952
1.85	0.008403	0.015725	177.9913	0.007143	0.013214	232.4784
1.90	0.008852	0.016586	182.8018	0.007516	0.013938	238.7616
1.95	0.009286	0.017470	178.6124	0.007988	0.014181	245.0448
2.00	0.009745	0.018379	192.4230	0.008288	0.015444	251.3280
2.05	0.010214	0.019309	197.2336	0.008686	0.016226	257.6112
2.10	0.010693	0.020262	202.0441	0.009092	0.017027	263.8944
2.15	0.011188	0.021239	206.8547	0.009513	0.017848	269.1776
2.20	0.011693	0.022237	211.6653	0.009942	0.018687	276.4608
2.25	0.012209	0.023261	216.4759	0.010380	0.019547	282.7440
2.30	0.012734	0.024305	221.2864	0.010826	0.020425	289.0272
2.35	0.013270	0.025373	226.0970	0.011280	0.021322	295.3104
2.40	0.013816	0.026465	230.9076	0.011744	0.022240	301.5936
2.45	0.014371	0.027579	235.7182	0.012215	0.023176	307.8768
2.50	0.014945	0.028716	240.5287	0.012695	0.024131	314.1600
3.00	0.012167	0.041351	288.6345	0.017971	0.034749	376.992
3.50	0.028543	0.056283	336.7403	0.024273	0.047297	439.824
4.00	0.036908	0.073513	384.8460	0.031296	0.061276	502.656

TABLE 5.2 (*continued*)

Mean velocity	Pipe diameter 0.450 m			Pipe diameter 0.500 m		
	Pipe section 0.15904 m²			Pipe section 0.19635 m²		
	Head loss m/1 m of pipe length		Flow dm³/s	Head loss m/1 m of pipe length		Flow dm³/s
	New pipes	Existing pipes		New pipes	Existing pipes	
0.01			1.5904			1.9635
0.05	0.000008	0.000010	7.9522	0.000007	0.000009	5.8175
0.10	0.000029	0.000037	15.9043	0.000025	0.000033	19.6350
0.15	0.000059	0.000081	23.8565	0.000052	0.000070	29.4525
0.20	0.000099	0.000141	31.8087	0.000088	0.000123	39.2700
0.25	0.000149	0.000217	39.7609	0.000131	0.000189	49.0875
0.30	0.000207	0.000309	47.7130	0.000182	0.000270	58.9056
0.35	0.000275	0.000418	55.6652	0.000242	0.000365	68.7225
0.40	0.000352	0.000543	63.6174	0.000310	0.000474	78.5400
0.45	0.000438	0.000684	71.5696	0.000386	0.000597	88.3575
0.50	0.000533	0.000841	79.5217	0.000469	0.000735	98.1750
0.55	0.000636	0.001016	87.4739	0.000560	0.000887	107.9925
0.60	0.000746	0.001205	95.4261	0.000658	0.001053	117.8100
0.65	0.000865	0.001412	103.3783	0.000763	0.001233	127.6275
0.70	0.000994	0.001634	111.3304	0.000875	0.001427	137.4450
0.75	0.001131	0.001872	119.2826	0.000995	0.001635	147.2625
0.80	0.001276	0.002127	127.2348	0.001123	0.001856	157.0800
0.85	0.001429	0.002399	135.1870	0.001258	0.002093	166.8975
0.90	0.001589	0.002688	143.1391	0.001400	0.002343	176.7150
0.95	0.001757	0.002991	151.0913	0.001548	0.002606	186.5325
1.00	0.001936	0.003313	159.0435	0.001704	0.002885	196.3500
1.05	0.002122	0.003652	166.9957	0.001869	0.003180	206.1675
1.10	0.002316	0.004008	174.9478	0.002040	0.003491	215.9850
1.15	0.002520	0.004382	182.9000	0.002219	0.003815	225.8025
1.20	0.002730	0.004771	190.8522	0.002405	0.004154	235.6200
1.25	0.002948	0.005177	198.8044	0.002596	0.004508	245.4375
1.30	0.003174	0.005599	206.7565	0.002794	0.004876	255.2550
1.35	0.003408	0.006038	214.7087	0.003000	0.005258	265.0725
1.40	0.003650	0.006494	222.6609	0.003213	0.005654	274.8900
1.45	0.003901	0.006965	230.6131	0.003436	0.006065	284.7075
1.50	0.004162	0.007454	238.5652	0.003665	0.006491	294.5250
1.55	0.004430	0.007960	246.5174	0.003902	0.006931	304.3425
1.60	0.004706	0.008481	254.4696	0.004144	0.007385	314.1600
1.65	0.004990	0.009020	262.4218	0.004393	0.007854	323.9775
1.70	0.005280	0.009574	270.3739	0.004649	0.008337	333.7950
1.75	0.005578	0.010147	278.3261	0.004911	0.008835	343.6125
1.80	0.005883	0.010734	286.2783	0.005179	0.009347	353.4300
1.85	0.006194	0.011338	294.2305	0.005456	0.009873	363.2475
1.90	0.006518	0.011960	302.1826	0.005741	0.010414	373.0650
1.95	0.006848	0.012598	310.1348	0.006031	0.010970	382.8825
2.00	0.007186	0.013252	318.0870	0.006328	0.011540	392.7000
2.05	0.007530	0.013923	326.0392	0.006632	0.012124	402.5175
2.10	0.007887	0.014611	333.9913	0.006946	0.012723	412.3350
2.15	0.008252	0.015315	341.9435	0.007266	0.013336	422.1525
2.20	0.008623	0.016035	349.8957	0.007593	0.013963	431.7900
2.25	0.009003	0.016773	357.8479	0.007927	0.014605	441.7875
2.30	0.009389	0.017526	365.8000	0.008267	0.015261	451.6050
2.35	0.009783	0.018296	373.7522	0.008613	0.015932	461.4220
2.40	0.010184	0.019083	381.5044	0.008966	0.016517	471.2400
2.45	0.010593	0.019887	389.6566	0.009315	0.017317	481.0575
2.50	0.011008	0.020706	397.6087	0.009697	0.018030	490.8750
3.00	0.015607	0.029817	477.1305	0.013762	0.025964	589.0500
3.50	0.021053	0.040585	556.6523	0.018519	0.035340	687.2250
4.00	0.027034	0.053009	636.1740	0.023976	0.046158	785.4000

TABLE 5.2 (continued)

Mean velocity	Pipe diameter 0.600 m Pipe section 0.28274 m²			Pipe diameter 0.700 m Pipe section 0.38464 m²		
	Head loss m/1 m of pipe length		Flow dm³/s	Head loss m/1 m of pipe length		Flow dm³/s
	New pipes	Existing pipes		New pipes	Existing pipes	
0.01			2.8274			3.8484
0.05	0.000006	0.000007	14.1372	0.000005	0.000006	19.2423
0.10	0.000020	0.000026	28.2744	0.000017	0.000022	38.4846
0.15	0.000041	0.000056	42.4116	0.000034	0.000047	57.7269
0.20	0.000068	0.000095	56.5488	0.000057	0.000080	76.9692
0.25	0.000105	0.000149	70.6860	0.000087	0.000123	96.2115
0.30	0.000146	0.000212	84.8232	0.000121	0.000175	115.4538
0.35	0.000193	0.000287	98.9604	0.000160	0.000236	134.6961
0.40	0.000247	0.000372	113.0976	0.000205	0.000308	153.9384
0.45	0.000307	0.000469	127.2348	0.000255	0.000387	173.1807
0.50	0.000372	0.000577	141.3720	0.000309	0.000473	192.4230
0.55	0.000443	0.000697	155.5092	0.000368	0.000571	211.6653
0.60	0.000521	0.000827	169.6464	0.000433	0.000679	230.9076
0.65	0.000605	0.000969	183.7836	0.000502	0.000795	250.1499
0.70	0.000695	0.001122	197.9208	0.000576	0.000921	269.3922
0.75	0.000790	0.001287	212.0580	0.000655	0.001057	288.6345
0.80	0.000890	0.001463	226.1952	0.000738	0.001202	307.8768
0.85	0.000996	0.001651	240.3324	0.000826	0.001358	327.1191
0.90	0.001107	0.001849	254.4696	0.000917	0.001521	346.3614
0.95	0.001221	0.002059	268.6068	0.001015	0.001681	365.6037
1.00	0.001341	0.002279	282.7440	0.001117	0.001830	384.8460
1.05	0.001472	0.002513	296.8812	0.001224	0.002068	404.0883
1.10	0.001609	0.002758	311.0184	0.001338	0.002272	423.3306
1.15	0.001750	0.003014	325.1556	0.001454	0.002482	442.5729
1.20	0.001897	0.003282	339.2928	0.001562	0.002701	461.8152
1.25	0.002049	0.003561	353.4300	0.001688	0.002934	481.0575
1.30	0.002208	0.003852	367.5672	0.001817	0.003175	500.2998
1.35	0.002372	0.004154	381.7044	0.001946	0.003420	519.4521
1.40	0.002541	0.004467	395.8416	0.002084	0.003680	538.7844
1.45	0.002715	0.004792	409.9788	0.002225	0.003950	558.0267
1.50	0.002896	0.005128	424.1160	0.002376	0.003223	577.2690
1.55	0.003082	0.005476	438.2532	0.002528	0.004512	596.5113
1.60	0.003273	0.005835	452.3904	0.002681	0.004834	615.7536
1.65	0.003469	0.006205	466.5276	0.002843	0.005115	634.9959
1.70	0.003673	0.006578	480.6648	0.003012	0.005437	654.2382
1.75	0.003879	0.006980	494.8020	0.003181	0.005750	673.4805
1.80	0.004090	0.007384	508.9392	0.003356	0.006079	692.7228
1.85	0.004309	0.007800	523.0764	0.003530	0.006424	711.9651
1.90	0.004533	0.008228	537.2136	0.003714	0.006775	731.2074
1.95	0.004761	0.008666	551.3508	0.003901	0.007146	750.4497
2.00	0.004995	0.009117	565.4882	0.004088	0.007508	769.6920
2.05	0.005234	0.009587	579.6250	0.004286	0.007895	788.9343
2.10	0.005477	0.010051	593.7624	0.004484	0.008277	808.1666
2.15	0.005729	0.010536	607.8996	0.004686	0.008673	827.4189
2.20	0.005986	0.011031	622.0368	0.004889	0.009089	846.6612
2.25	0.006249	0.011539	636.1740	0.005103	0.009502	865.9035
2.30	0.006516	0.012057	650.3112	0.005322	0.009944	885.1458
2.35	0.006788	0.012587	664.4484	0.005547	0.010451	904.3881
2.40	0.007066	0.013128	678.5856	0.005773	0.010806	923.6304
2.45	0.007353	0.013681	692.7228	0.006010	0.011268	942.8727
2.50	0.007645	0.014245	706.8600	0.006248	0.011733	962.1150
3.00	0.010841	0.020513	848.232	0.008867	0.016901	1 154.5380
3.50	0.014610	0.027920	989.604	0.011925	0.022997	1 346.9610
4.00	0.018893	0.036476	1 130.967	0.015436	0.030410	1 539.3840

TABLE 5.2 *(continued)*

Mean velocity	Pipe diameter **0.800** m			Pipe diameter **0.900** m		
	Pipe section 0.50265 m²			Pipe section 0.63617 m²		
	Head loss m/1 m of pipe length		Flow dm³/s	Head loss m/1 m of pipe length		Flow dm³/s
	New pipes	Existing pipes		New pipes	Existing pipes	
0.01			5.0205			6.3617
0.05	0.000004	0.000005	25.1328	0.000004	0.000005	31.8087
0.10	0.000014	0.000018	50.2656	0.000012	0.000015	63.6174
0.15	0.000029	0.000039	75.3984	0.000025	0.000034	95.4261
0.20	0.000049	0.000067	100.5312	0.000043	0.000058	127.2348
0.25	0.000074	0.000103	125.6640	0.000064	0.000087	159.0435
0.30	0.000103	0.000147	150.7968	0.000089	0.000124	190.8522
0.35	0.000137	0.000198	175.9296	0.000167	0.000167	222.6609
0.40	0.000174	0.000258	201.0624	0.000150	0.000218	254.4696
0.45	0.000216	0.000324	226.1952	0.000186	0.000274	286.2783
0.50	0.000262	0.000398	251.3280	0.000225	0.000336	318.0870
0.55	0.000312	0.000481	275.4608	0.000268	0.000406	349.8957
0.60	0.000367	0.000572	301.5936	0.000316	0.000483	381.7044
0.65	0.000425	0.000670	326.7264	0.000367	0.000565	413.5131
0.70	0.000489	0.000776	351.8592	0.000421	0.000654	445.3218
0.75	0.000557	0.000890	376.9920	0.000479	0.000749	477.1305
0.80	0.000628	0.001012	402.1248	0.000540	0.000852	508.9392
0.85	0.000703	0.001142	427.2576	0.000605	0.000961	540.7479
0.90	0.000781	0.001279	452.3904	0.000671	0.001077	572.5566
0.95	0.000864	0.001425	477.5232	0.000743	0.001199	604.3653
1.00	0.000952	0.001579	502.6560	0.000817	0.001327	636.1740
1.05	0.001044	0.001741	527.7888	0.000896	0.001461	667.9827
1.10	0.001139	0.001910	552.9216	0.000980	0.001606	699.7914
1.15	0.001239	0.002088	578.0544	0.001065	0.001752	731.6001
1.20	0.001341	0.002274	603.1872	0.001144	0.001910	763.4088
1.25	0.001448	0.002467	628.3200	0.001237	0.002073	795.2175
1.30	0.001559	0.002668	653.4528	0.001332	0.002241	827.0262
1.35	0.001673	0.002877	678.5856	0.001428	0.002420	858.8349
1.40	0.001791	0.003095	703.7184	0.001529	0.002604	890.6436
1.45	0.001914	0.003319	728.8512	0.001632	0.002787	922.4523
1.50	0.002041	0.003552	753.9840	0.001741	0.002983	954.2610
1.55	0.002174	0.003793	779.1168	0.001857	0.003186	986.0697
1.60	0.002309	0.004042	804.2496	0.001968	0.003398	1 017.8784
1.65	0.002449	0.004298	829.3824	0.002086	0.003610	1 049.6871
1.70	0.002593	0.004563	854.5152	0.002208	0.003837	1 081.4958
1.75	0.002740	0.004835	879.6480	0.002337	0.004061	1 113.3045
1.80	0.002890	0.005115	904.7808	0.002461	0.004299	1 145.1132
1.85	0.003044	0.005403	926.9136	0.002594	0.004538	1 176.9219
1.90	0.003202	0.005699	955.0464	0.002726	0.004792	1 208.7306
1.95	0.003363	0.006003	980.1792	0.002862	0.005044	1 240.5393
2.00	0.003530	0.006315	1 005.3120	0.003001	0.005307	1 272.3480
2.05	0.003700	0.006635	1 030.4448	0.003144	0.005578	1 304.1567
2.10	0.003875	0.006963	1 055.5776	0.003296	0.005850	1 335.9654
2.15	0.004052	0.007298	1 080.7104	0.003445	0.006136	1 367.7741
2.20	0.004234	0.007641	1 105.8432	0.003598	0.006424	1 399.5828
2.25	0.004419	0.007993	1 130.9760	0.003757	0.006712	1 421.3915
2.30	0.004611	0.008352	1 156.1088	0.003915	0.007025	1 463.2002
2.35	0.004806	0.008719	1 181.2416	0.004074	0.007319	1 495.0089
2.40	0.005006	0.009094	1 206.3744	0.004240	0.007641	1 526.8176
2.45	0.005209	0.009477	1 231.5072	0.004419	0.007960	1 558.6263
2.50	0.005416	0.009867	1 256.6400	0.004590	0.008288	1 590.4435
3.00	0.007695	0.014209	1 507.968	0.006518	0.011923	1 908.5220
3.50	0.010357	0.019340	1 759.296	0.008782	0.016246	2 226.6090
4.00	0.013405	0.025261	2 010.624	0.011367	0.021204	2 544.6960

TABLE 5.2 (*continued*)

Mean velocity	Pipe diameter 1.000 m			Pipe diameter 1.250 m		
	Pipe section 0.78539 m²			Pipe section 1.22719 m²		
	Head loss m/1 m of pipe length		Flow dm³/s	Head loss m/1 m of pipe length		Flow dm³/s
	New pipes	Existing pipes		New pipes	Existing pipes	
0.01			7.8539			12.2715
0.05	0.000003	0.000004	39.2694	0.000002	0.000003	61.3575
0.10	0.000010	0.000013	78.5389	0.000008	0.000010	122.7150
0.15	0.000022	0.000029	117.8083	0.000017	0.000022	184.0725
0.20	0.000037	0.000051	157.0778	0.000028	0.000038	245.4300
0.25	0.000056	0.000078	196.3472	0.000043	0.000059	306.7875
0.30	0.000078	0.000111	235.6167	0.000060	0.000084	368.1450
0.35	0.000103	0.000150	274.8861	0.000079	0.000113	429.5025
0.40	0.000132	0.000195	314.1556	0.000101	0.000147	490.8600
0.45	0.000164	0.000246	353.4250	0.000125	0.000185	552.2175
0.50	0.000200	0.000308	392.6945	0.000152	0.000227	613.5755
0.55	0.000239	0.000365	431.9639	0.000182	0.000274	674.9325
0.60	0.000280	0.000433	471.2334	0.000213	0.000326	736.2900
0.65	0.000325	0.000507	510.5028	0.000248	0.000382	797.6475
0.70	0.000372	0.000587	549.7723	0.000285	0.000443	859.0050
0.75	0.000423	0.000673	589.0417	0.000324	0.000509	920.3625
0.80	0.000478	0.000765	628.3112	0.000366	0.000579	981.7200
0.85	0.000536	0.000863	667.5806	0.000409	0.000653	1 043.0775
0.90	0.000596	0.000966	706.8501	0.000456	0.000732	1 104.4350
0.95	0.000660	0.001076	746.1195	0.000505	0.000815	1 165.7925
1.00	0.000726	0.001193	785.3980	0.000556	0.000903	1 227.1500
1.05	0.000795	0.001315	824.6484	0.000609	0.000995	1 288.5075
1.10	0.000868	0.001443	863.9279	0.000665	0.001092	1 349.8650
1.15	0.000944	0.001577	903.1973	0.000723	0.001193	1 411.2215
1.20	0.001024	0.001718	942.4668	0.000783	0.001299	1 472.5800
1.25	0.001106	0.001864	971.7362	0.000846	0.001409	1 533.9375
1.30	0.001191	0.002016	1 021.0057	0.000911	0.001524	1 595.2950
1.35	0.001280	0.002174	1 050.2751	0.000979	0.001644	1 656.6525
1.40	0.001372	0.002338	1 099.5446	0.001049	0.001767	1 718.3010
1.45	0.001486	0.002508	1 138.8140	0.001121	0.001895	1 779.7267
1.50	0.001563	0.002684	1 178.0835	0.001196	0.002028	1 840.0825
1.55	0.001663	0.002866	1 217.3529	0.001274	0.002166	1 902.0005
1.60	0.001767	0.003053	1 256.6224	0.001353	0.002307	1 963.4405
1.65	0.001873	0.003247	1 295.8918	0.001434	0.002454	2 024.7975
1.70	0.001983	0.003447	1 335.1613	0.001518	0.002604	2 086.1550
1.75	0.002096	0.003653	1 374.4307	0.001603	0.002760	2 147.5125
1.80	0.002213	0.003864	1 413.7002	0.001691	0.002920	2 208.8700
1.85	0.002332	0.004082	1 452.9696	0.001782	0.003084	2 270.2275
1.90	0.002455	0.004306	1 492.2381	0.001875	0.003253	2 331.5850
1.95	0.002580	0.004535	1 531.5075	0.001971	0.003427	2 392.9425
2.00	0.002708	0.004771	1 570.7880	0.002068	0.003605	2 454.3000
2.05	0.002838	0.005012	1 610.0474	0.002168	0.003787	2 515.6575
2.10	0.002972	0.005260	1 649.3169	0.002269	0.003964	2 577.0150
2.15	0.003108	0.005513	1 688.5863	0.002375	0.004166	2 638.3275
2.20	0.003246	0.005773	1 727.8558	0.002483	0.004361	2 699.7300
2.25	0.003388	0.006038	1 767.1252	0.002593	0.004562	2 761.0875
2.30	0.003532	0.006309	1 806.3947	0.002705	0.004767	2 822.4450
2.35	0.003679	0.006587	1 845.6641	0.002819	0.004976	2 883.8025
2.40	0.003831	0.006870	1 884.9336	0.002936	0.005191	2 945.1600
2.45	0.003985	0.007159	1 924.2030	0.003055	0.005409	3 006.5175
2.50	0.004141	0.007454	1 963.4725	0.003178	0.005632	3 067.8750
3.00	0.005895	0.010734	2 356.194	0.004510	0.008110	3 681.5700
3.50	0.007930	0.014610	2 748.893	0.006084	0.011039	4 295.1650
4.00	0.010259	0.019083	3 141.592	0.007875	0.014418	4 908.7600

TABLE 5.2 *(continued)*

Mean velocity	Pipe diameter 1.500 m			Pipe diameter 1.750 m		
	Pipe section 1.76715 m²		Flow dm³/s	Pipe section 2.40528 m²		Flow dm³/s
	Head loss m/1 m of pipe length			Head loss m/1 m of pipe length		
	New pipes	Existing pipes		New pipes	Existing pipes	
0.01			17.671			24.053
0.05	0.000002	0.000002	88.355	0.000002	0.000002	120.264
0.10	0.000006	0.000008	176.710	0.000005	0.000007	240.528
0.15	0.000013	0.000018	265.065	0.000011	0.000014	360.792
0.20	0.000023	0.000030	353.420	0.000019	0.000025	481.056
0.25	0.000034	0.000047	441.775	0.000028	0.000038	601.320
0.30	0.000048	0.000067	530.130	0.000040	0.000055	721.584
0.35	0.000063	0.000090	618.485	0.000053	0.000074	841.848
0.40	0.000081	0.000117	706.840	0.000068	0.000096	962.112
0.45	0.000101	0.000148	795.195	0.000084	0.000121	1 082.376
0.50	0.000122	0.000182	883.550	0.000102	0.000149	1 202.641
0.55	0.000146	0.000219	971.905	0.000122	0.000181	1 322.905
0.60	0.000172	0.000260	1 060.260	0.000144	0.000215	1 443.169
0.65	0.000200	0.000305	1 148.615	0.000167	0.000251	1 563.433
0.70	0.000230	0.000353	1 236.970	0.000191	0.000291	1 683.697
0.75	0.000261	0.000405	1 325.325	0.000217	0.000334	1 803.961
0.80	0.000295	0.000461	1 413.680	0.000246	0.000380	1 924.225
0.85	0.000330	0.000521	1 502.035	0.000276	0.000429	2 044.489
0.90	0.000368	0.000584	1 590.390	0.000306	0.000481	2 164.753
0.95	0.000406	0.000651	1 678.745	0.000339	0.000536	2 285.107
1.00	0.000447	0.000721	1 767.100	0.000373	0.000594	2 405.281
1.05	0.000490	0.000795	1 855.455	0.000409	0.000654	2 525.545
1.10	0.000535	0.000872	1 943.810	0.000447	0.000718	2 645.809
1.15	0.000582	0.000953	2 032.175	0.000486	0.000785	2 766.073
1.20	0.000631	0.001038	2 120.520	0.000527	0.000854	2 886.337
1.25	0.000682	0.001126	2 208.875	0.000570	0.000927	3 006.601
1.30	0.000735	0.001218	2 297.230	0.000614	0.001003	3 126.865
1.35	0.000789	0.001314	2 385.585	0.000659	0.001081	3 247.129
1.40	0.000845	0.001412	2 473.940	0.000706	0.001163	3 367.393
1.45	0.000903	0.001515	2 562.295	0.000754	0.001247	3 487.657
1.50	0.000963	0.001621	2 650.650	0.000805	0.001335	3 607.922
1.55	0.001025	0.001731	2 739.005	0.000857	0.001425	3 728.186
1.60	0.001089	0.001844	2 827.360	0.000911	0.001519	3 848.500
1.65	0.001155	0.001961	2 915.715	0.000966	0.001615	3 968.714
1.70	0.001223	0.002082	3 004.070	0.001023	0.001715	4 088.978
1.75	0.001292	0.002206	3 092.425	0.001080	0.001817	4 209.242
1.80	0.001363	0.002334	3 180.780	0.001140	0.001922	4 329.506
1.85	0.001436	0.002466	3 269.135	0.001201	0.002030	4 449.770
1.90	0.001512	0.002601	3 357.490	0.001264	0.002142	4 570.034
1.95	0.001589	0.002739	3 445.845	0.001329	0.002256	4 690.298
2.00	0.001669	0.002882	3 534.200	0.001396	0.002373	4 810.562
2.05	0.001749	0.003027	3 622.555	0.001463	0.002493	4 930.826
2.10	0.001833	0.003177	3 710.910	0.001531	0.002617	5 051.090
2.15	0.001916	0.003330	3 799.265	0.001602	0.002742	5 171.354
2.20	0.002003	0.003487	3 887.620	0.001675	0.002872	5 291.618
2.25	0.002092	0.003647	3 975.975	0.001749	0.003004	5 411.882
2.30	0.002182	0.003811	4 064.330	0.001824	0.003138	5 532.146
2.35	0.002274	0.003978	4 152.685	0.001901	0.003277	5 652.410
2.40	0.002378	0.004149	4 241.040	0.001980	0.003417	5 772.674
2.45	0.002462	0.004324	4 329.395	0.002058	0.003561	5 892.938
2.50	0.002557	0.004502	4 417.750	0.002139	0.003708	6 013.203
3.00	0.003633	0.006483	5 301.45	0.003035	0.005340	7 215.843
3.50	0.004891	0.008825	6 185.025	0.004092	0.007268	8 418.484
4.00	0.006350	0.011526	7 068.400	0.005303	0.009493	9 621.124

TABLE 5.2 (continued)

Mean velocity	Pipe diameter 2.000 m			Pipe diameter 2.500 m		
	Pipe section 3.14159 m²			Pipe section 4.90873 m²		
	Head loss m/1 m of pipe length		Flow dm³/s	Head loss m/1 m of pipe lenght		Flow dm³/s
	New pipes	Existing pipes		New pipes	Existing pipes	
0.01			31.416			49.087
0.05	0.000001	0.000002	157.080	0.000001	0.000001	245.437
0.10	0.000005	0.000006	314.159	0.000003	0.000004	490.874
0.15	0.000009	0.000012	471.239	0.000007	0.000009	736.311
0.20	0.000016	0.000021	628.318	0.000012	0.000016	981.748
0.25	0.000024	0.000032	785.398	0.000018	0.000025	1 227.185
0.30	0.000034	0.000046	942.478	0.000026	0.000035	1 472.621
0.35	0.000045	0.000062	1 099.557	0.000035	0.000048	1 718.058
0.40	0.000058	0.000081	1 256.637	0.000044	0.000062	1 963.495
0.45	0.000072	0.000102	1 413.716	0.000055	0.000078	1 208.932
0.50	0.000087	0.000126	1 570.796	0.000067	0.000096	2 454.369
0.55	0.000104	0.000152	1 727.876	0.000080	0.000116	2 699.806
0.60	0.000122	0.000181	1 884.955	0.000094	0.000138	2 945.243
0.65	0.000142	0.000212	2 042.035	0.000109	0.000161	3 190.680
0.70	0.000163	0.000246	2 199.114	0.000125	0.000187	3 436.117
0.75	0.000186	0.000282	2 356.194	0.000143	0.000214	3 681.554
0.80	0.000210	0.000321	2 513.274	0.000161	0.000244	3 926.990
0.85	0.000235	0.000363	2 670.353	0.000181	0.000275	4 172.427
0.90	0.000261	0.000406	2 827.433	0.000202	0.000308	4 417.864
0.95	0.000289	0.000452	2 984.512	0.000223	0.000343	4 663.301
1.00	0.000319	0.000501	3 141.592	0.000246	0.000380	4 908.738
1.05	0.000349	0.000552	3 298.672	0.000270	0.000419	5 154.175
1.10	0.000381	0.000605	3 455.571	0.000295	0.000459	5 399.612
1.15	0.000415	0.000662	3 612.831	0.000321	0.000502	5 645.049
1.20	0.000450	0.000720	3 769.910	0.000348	0.000546	5 890.486
1.25	0.000487	0.000782	3 926.990	0.000376	0.000593	6 135.923
1.30	0.000524	0.000845	4 084.070	0.000405	0.000641	6 381.359
1.35	0.000563	0.000912	4 241.149	0.000435	0.000691	6 626.796
1.40	0.000603	0.000981	4 398.229	0.000467	0.000743	6 872.233
1.45	0.000645	0.001052	4 555.308	0.000498	0.000797	7 117.670
1.50	0.000688	0.001126	4 712.388	0.000531	0.000853	7 363.107
1.55	0.000733	0.001202	4 869.468	0.000566	0.000911	7 608.544
1.60	0.000779	0.001281	5 026.547	0.000601	0.000971	7 853.981
1.65	0.000826	0.001362	5 183.627	0.000638	0.001032	8 099.418
1.70	0.000874	0.001446	5 340.706	0.000675	0.001096	8 344.855
1.75	0.000923	0.001532	5 497.786	0.000714	0.001161	8 590.292
1.80	0.000974	0.001621	5 654.866	0.000753	0.001229	8 835.728
1.85	0.001027	0.001712	5 811.945	0.000794	0.001298	9 081.165
1.90	0.001080	0.001806	5 959.025	0.000836	0.001369	9 326.602
1.95	0.001136	0.001902	6 126.104	0.000878	0.001442	9 572.039
2.00	0.001193	0.002001	6 283.184	0.000922	0.001517	9 817.476
2.05	0.001250	0.002102	6 440.264	0.000966	0.001594	10 062.913
2.10	0.001308	0.002206	6 597.343	0.001011	0.001672	10 308.350
2.15	0.001369	0.002313	6 754.423	0.001057	0.001753	10 553.787
2.20	0.001431	0.002421	6 911.502	0.001105	0.001835	10 799.224
2.25	0.001494	0.002533	7 068.582	0.001154	0.001920	11 044.661
2.30	0.001559	0.002647	7 225.662	0.001204	0.002006	11 290.097
2.35	0.001624	0.002763	7 382.741	0.001254	0.002094	11 535.534
2.40	0.001691	0.002882	7 539.821	0.001307	0.002184	11 780.971
2.45	0.001759	0.003003	7 696.900	0.001361	0.002276	12 026.408
2.50	0.001829	0.003127	7 853.980	0.001416	0.002370	12 271.845
3.00	0.002592	0.004503	9 424.776	0.002015	0.003413	14 726.214
3.50	0.003497	0.006128	10 995.572	0.002712	0.004645	17 180.583
4.00	0.004526	0.008004	12 566.368	0.003517	0.006068	19 634.952

Table 5.2 is the proper calculation table and is constructed for the following conditions:

(a) *absolute roughness*, $\varepsilon = 0.1$ mm, corresponding to smooth or new pipes;

(b) *absolute roughness* $\varepsilon = 2$ mm, corresponding to rough or old pipes.

5.3 Calculation of Local Head Losses

5.3.1 **Local Resistance Coefficient**

Any deformation of liquid stream direction or section leads to a local head loss whose value depends both on the nature and geometry of this deformation and on the velocity field of the streaming liquid entering the fitting. The velocity field is determined by the constructional characteristics of the fittings the liquid has traversed before entering the one where the local head loss occurred. As a consequence, the head loss of two or more series-connected fittings is always higher than the sum of the head losses of these fittings taken separately.

In a similar way to the frictional losses, the local head losses are acting as resistances opposing the liquid motion, and are proportional to the kinetic head. Therefore, the local head loss is expressed in fractions of kinetic head on the basis of the following formula

$$h_{loc} = \xi \frac{v^2}{2g} \tag{5.19}$$

where h_{loc} = local head loss, in m;
$\quad v$ = mean liquid velocity, in m/s;
$\quad \xi$ = local, dimensionless resistance coefficient depending on the geometrical characteristics of the fitting.

This coefficient could be established theoretically in certain cases, especially for fittings showing a cross-sectional change of the liquid stream. In most cases ξ is determined experimentally [5.17; 5.18; 5.28; 5.29]. The mean values of the coefficient are given below for various fittings, frequently met with pumping pipelines.

Direction-changing fittings. Curved and elbow pieces change the flow direction within pipes. The characteristic geometrical quantities of these pieces are the angle of flow direction change α and the curve radius R, usually expressed by the dimensionless quantity R/D. Generally, the curved fittings are those with $R/D > 0.5$ and the elbow ones those with $R/D < 0.5$.

(1) *Bent curve* (Fig. 5.8)

$$\xi_8 = \xi_{loc\,8} + \xi_{lin\,8}$$

$$\xi_{loc\,8} = A_8 B_8; \quad \xi_{lin\,8} = 0.17 \, \lambda \frac{R}{D}$$

A_8 is determined from Fig. 5.8,b in accordance with α, and B_8 in accordance with R/D from Fig. 5.8,c for $r/D = 0.5 - 1.5$, and from Fig. 5.8,d for $R/D > 1.5$.

(2) *Welded curve* (Fig. 5.9)

$$\xi_9 = \xi_{loc\,9} + \xi_{lin\,9}$$

$\zeta_{loc\,9}$ is determined from Fig. 5.9,b in accordance with l/D or R/D; $\xi_{lin\,9} = 3\lambda \frac{l}{D}$.

(3) *Compound elbow* (Fig. 5.10)

$$\xi_{10} = \xi_{loc\,10} + \xi_{lin\,10}$$

ξ_{loc} is determined from Fig. 5.10,b as depending on l/D or R/D; $\xi_{lin\,10} = \lambda l/D$.

Section-changing fittings are produced as divergent or convergent cones:

(1) *Divergent cone* (Fig. 5.11)

$$\xi_{11} = \xi_{loc\,11} + \xi_{lin\,11}$$

$\xi_{loc\,11}$ is determined from Fig. 5.11,b as depending on α and A_i/A_0; $\xi_{lin\,11}$ is determined from Fig. 5.11,c as depending on α and A_i/A_o.

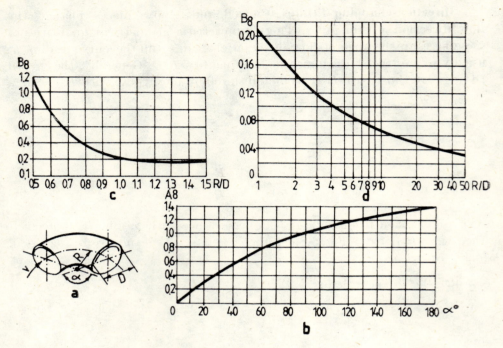

Fig. 5.8. Bent curve:

a — illustrating sketch; b, c, d — graph for the computation of coefficients ξ_{loc_8} (A_8 and B_8).

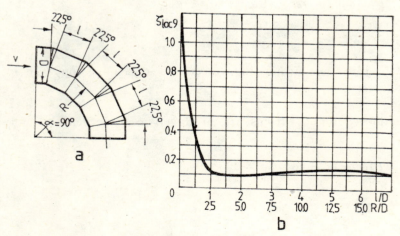

Fig. 5.9. Welded curve:

a — illustrating sketch; b — graph for the computation of coefficient ξ_{loc_9}.

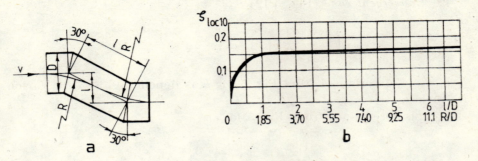

Fig. 5.10 Compound elbow piece:
a — illustrating sketch; b — graph for the computation of coefficient $\xi_{loc\,10}$.

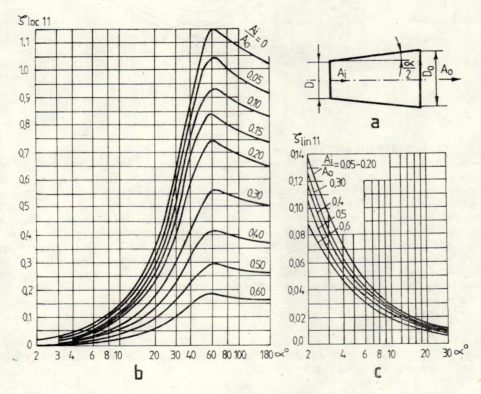

Fig. 5.11. Divergent cone:
a — illustrating sketch; b and c — graphs for the computation of coefficient $\xi_{loc\,11}$.

(2) *Convergent cone* (Fig. 5.12)

$$\xi_{12} = \xi_{loc\,12}\left(1 - \frac{A_o}{A_i}\right) + \xi_{lin\,12}$$

$\xi_{loc\,12}$ is determined from Fig. 5.11, *b* as depending on α for various l/D_0 ratios;

$\xi_{lin\,12}$ is determined from Fig. 5.11, *c*.

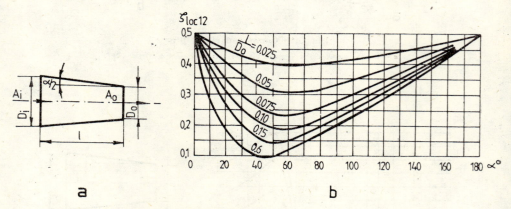

Fig. 5.12. Convergent cone:
a — illustrating sketch; *b* — graph for the computation of coefficient $\xi_{loc\,12}$.

Branched fittings. The head loss occurring at the joint between a branched piece and the main pipe is caused by the shock-like contact of the two streaming liquids. The local liquid resistance in the main pipe is a consequence of the deformation the lateral stream imposes on the main streaming lines. The head loss observed in the branch fitting is due both to the energy the lateral stream has spent in displacing a fraction of the main stream at the joining point, and to the eddies generated by the direction change of the lateral stream. Generally, the head loss is greater in the branched fitting than in the main stream. Sometimes, due to the ejection effect of one stream on the other, no head loss is manifest and, on the contrary, one stream may gain additional energy from its entrainment. In the conventional notation, a minus sign denotes the ξ coefficients, representing cases in which the branch fitting or the main pipe show an energy gain.

In order to obtain the real resistance of a hydraulic circuit, the energy gains, resulting from the products of these coefficients (conventionally negative) with the kinetic head of the respective pipe, should be subtracted from the losses generated by other resistances.

The branched fittings used on pumping pipes have the following shapes:

1. 45° Cylindrical branch   (streams joining), (Fig. 5.13)
2. 90° Cylindrical branch   (streams joining), (Fig. 5.14)

Calculation of Local Head Losses

Fig. 5.13. Cylindrical branch at 45° (streams joining):
a — illustrating sketch; *b* — graphs for the computation of coefficient $\xi_{loc_{13}}$.

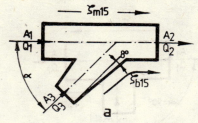
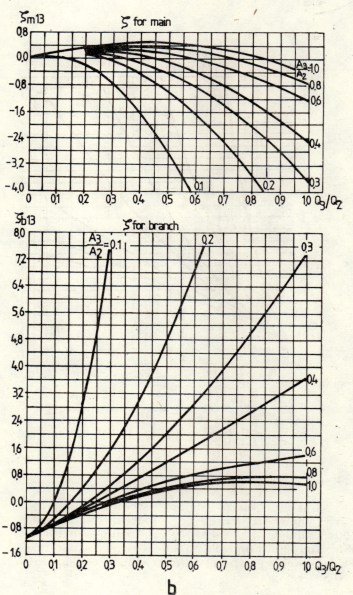

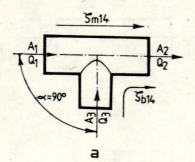
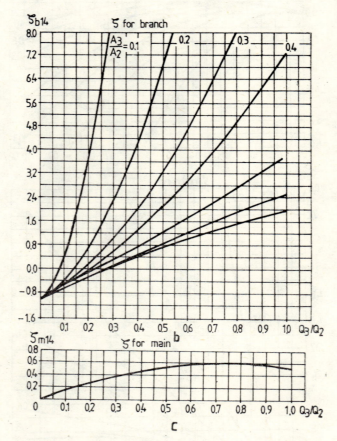

Fig. 5.14. Cylindrical branch at 90° (streams joining):
a — illustrating sketch; b, c — graphs for the computation of coefficient $\xi_{loc\,14}$.

3. 45° and 90° Conical branch (streams joining), (Fig. 5.15)
4. 45° Cylindrical branch (streams separating), (Fig. 5.16)
5. 90° Cylindrical branch (streams separating), (Fig. 5.17)
6. 45° and 90° Conical branch (streams separating), (Fig. 5.18)
7. 45° Conical Y-pipe (streams joining or separating), (Fig. 5.19)

Calculation of Local Head Losses

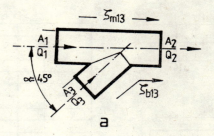

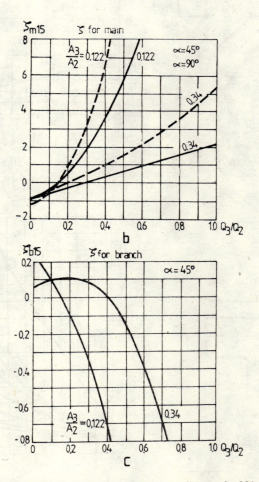

Fig. 5.15. Conical branch at 45° and 90° streams joining):

a — illustrating sketch; b, c — graphs for the computation of coefficient $\xi_{loc_{15}}$ (ξ_m and ξ_b).

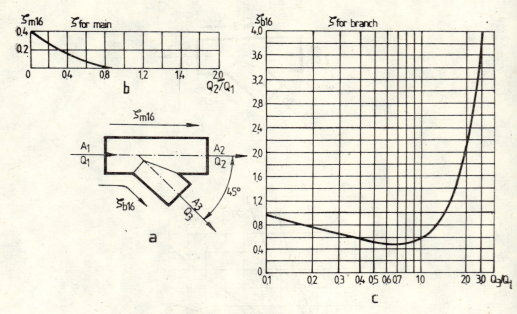

Fig. 5.16. Cylindrical branch at 45° (stream separation):
a — illustrating sketch; b, c — graphs for the computation of coefficient $\xi_{loc_{16}}$ (ξ_m and ξ_b).

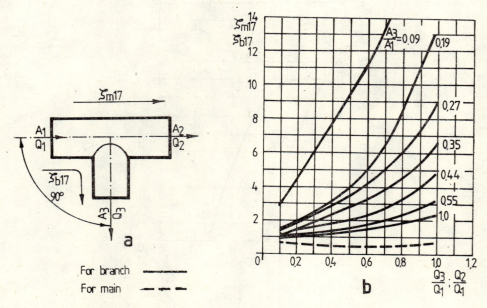

Fig. 5.17. Cylindrical branch at 90° (stream separation):
a — illustrating sketch; b — graph for the computation of coefficient $\xi_{loc_{17}}$ (ξ_m and ξ_b).

Calculation of Linear Head Losses

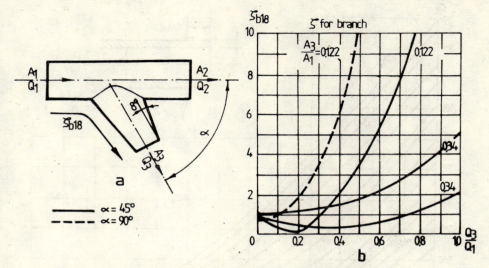

Fig. 5.18. Conical branch at 45° and 90° (stream separation):
a — illustrating sketch; b — graph for the computation of coefficient ξ_{b18}.

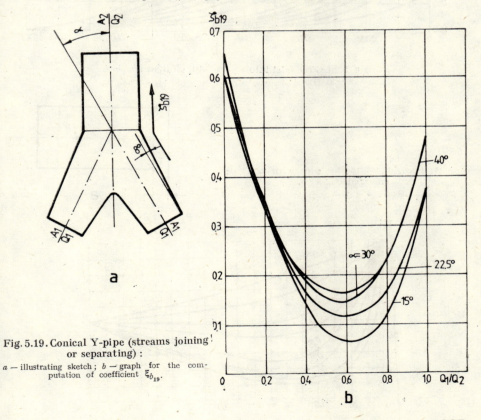

Fig. 5.19. Conical Y-pipe (streams joining or separating):
a — illustrating sketch; b — graph for the computation of coefficient ξ_{b19}.

225

15 — c. 9

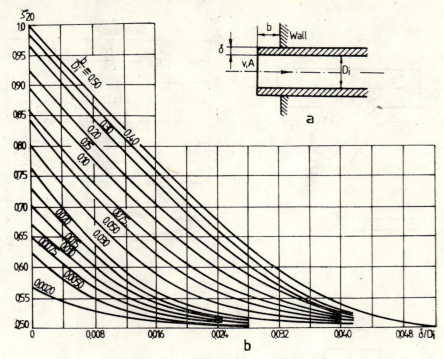

Fig. 5.20. Free inlet into a pipe with straight edges:
a — illustrating sketch; b — graph for the computation of coefficient ξ_{20}.

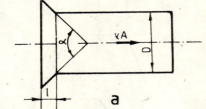

Fig. 5.21. Free inlet into the pipe with conically enlarged edges:
a — illustrating sketch; b — graph for the computation of coefficient ξ_{21}.

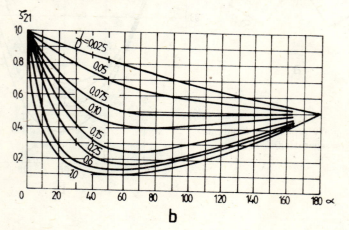

Calculation of Local Head Losses

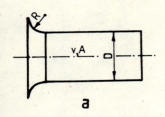

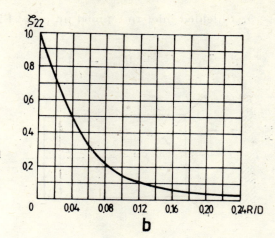

Fig. 5.22. Free inlet into the round-edged pipe:
a — illustrating sketch; *b* — graph for the computation of coefficient ξ_{22}.

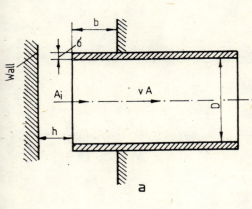

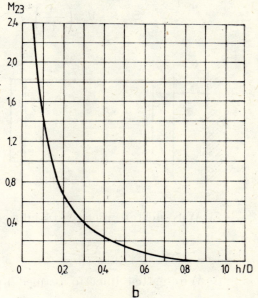

Fig. 5.23. Screened inlet with straight edges:
a — illustrating sketch; *b* — computational graph.

Pipe inlets may have the following shapes:
1. Free inlet to straight-edged pipe   (Fig. 5.20)
2. Free inlet to reamed up pipe   (Fig. 5.21)
3. Free inlet to round-edged pipe   (Fig. 5.22)
4. Shielded inlet to straight-edged pipe (Fig. 5.23)

$$\xi_{23} = \xi_{20} + M_{23}$$

227

5. Shielded inlet to reamed up pipe (Fig. 5.24)

$$\xi_{24} = \xi_{21} + M_{23}\left(\frac{A}{A_i}\right)^2$$

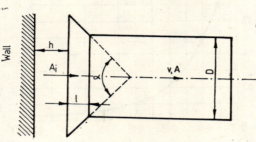

Fig. 5.24. Screened inlet into a pipe with conically enlarged edges.

6. Shielded inlet to round-edged pipe (Fig. 5.25)

$$\xi_{25} = \xi_{22} + M_{23}\left(\frac{A}{A_i}\right)^2$$

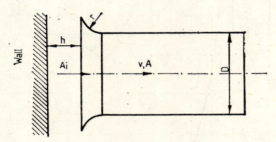

Fig. 5.25. Screened inlet into a round-edged pipe.

7. Wire-netted inlet to straight-edged pipe (Fig. 5.26)

$$\xi_{26} = \xi_{20} + N_{26}$$

8. Wire-netted inlet to reamed up pipe (Fig. 5.27)

$$\xi_{27} = \xi_{21} + N_{26}\left(\frac{A}{A_0}\right)^2$$

9. Wire-netted inlet to round-edged pipe (Fig. 5.28)

$$\xi_{28} = \zeta_{22} + N_{26}\left(\frac{A}{A_0}\right)^2$$

Calculation of Local Head Losses

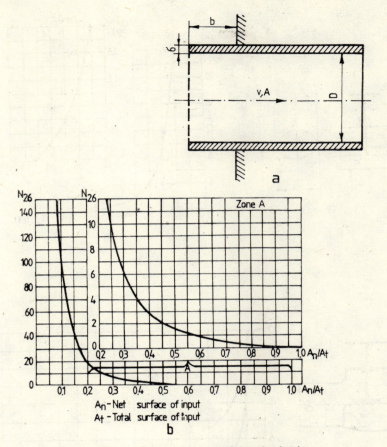

Fig. 5.26. Strainer into a straight-edged pipe:
a — illustrating sketch; *b* — computational graph.

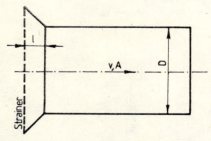

Fig. 5.27. Strainer into a pipe with conically enlarged edges.

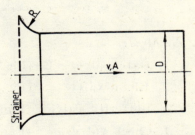

Fig. 5.28. Strainer into a round-edged pipe.

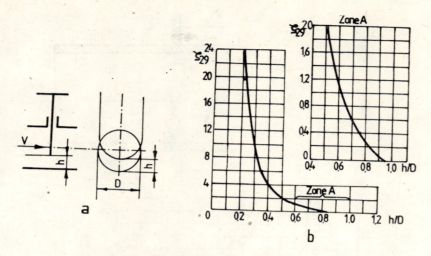

Fig. 5.29. Wedge gate valve:
a — illustrating sketch; *b* — graph for the computation of coefficient ξ_{29}.

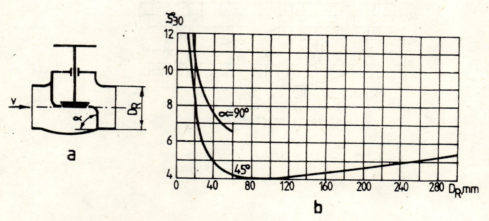

Fig. 5.30. Globe valve (screw-down stop valve):
a — illustrating sketch; *b* — graph for the computation of coefficient ξ_{30}.

Valves:

1. Gate valve   (Fig. 5.29)
2. Globe valve   (Fig. 5.30)
3. Butterfly valve   (Fig. 5.31)
4. Check valve   (Fig. 5.32)
5. Globe check valve (Fig. 5.33)

Calculation of Local Head Losses

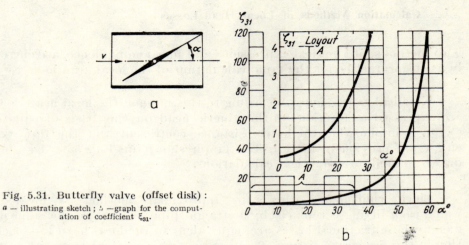

Fig. 5.31. Butterfly valve (offset disk):
a — illustrating sketch; b — graph for the computation of coefficient ξ_{31}.

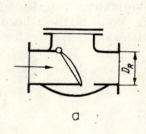

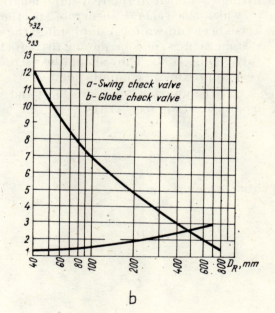

Fig. 5.32. Swing check valve:
a — illustrating sketch; b — graph for the computation of coefficient ξ_{32} and ξ_{33}.

Fig. 5.33. Globe check valve (piston-lift type).

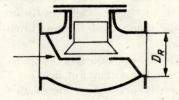

5.3.2 Calculation Methods of Local Head Losses

Two methods, of kinetical expressions and linear equivalence, are currently used in practice for the rapid calculation of the local head loss.

Kinetical expressions. According to this method the local head losses are expressed as fractions of the kinetic head on the basis of equation (5.19). It implies that the local resistance coefficient and the flow velocity v are previously known. The graph shown in Fig. 5.34 has been drawn to facilitate practical calculations.

Linear equivalence. This is an approximate method by means of which each fitting is replaced by a straight pipe segment of the same diameter whose frictional losses are equivalent to the local head loss occurring in the respective fitting. The corresponding equivalent length of the fittings is added to the real pipe length to give a total imaginary length to which the general theory of frictional losses is applied. Two graphs have been drawn for establishing the equivalent lengths of pumping system fittings, one for joining fittings (Fig. 5.35) and the other for shut-off valves (Fig. 5.36). They should be used, with caution, only for rapid and very approximate computation.

By comparing equations (5.7) and (5.19) and denoting the equivalent length by l, we may write

$$\xi \frac{v^2}{2g} = \lambda \frac{v^2}{2g} \cdot \frac{l}{D}$$

from which we get

$$\xi = \lambda \frac{l}{D} \tag{5.20}$$

or

$$l = \frac{\xi}{\lambda} D \tag{5.20,a}$$

The analysis of equation (5.20,a) shows the equivalent length l to depend on the friction coefficient and hence on Re number. This leads to the conclusion that the local head loss does not depend on the flow velocity alone.

The latest investigations in this field seem to confirm this assumption, although the variation law for the local resistance coefficient has not been sufficiently proved to be identical with that of the friction coefficient, λ.

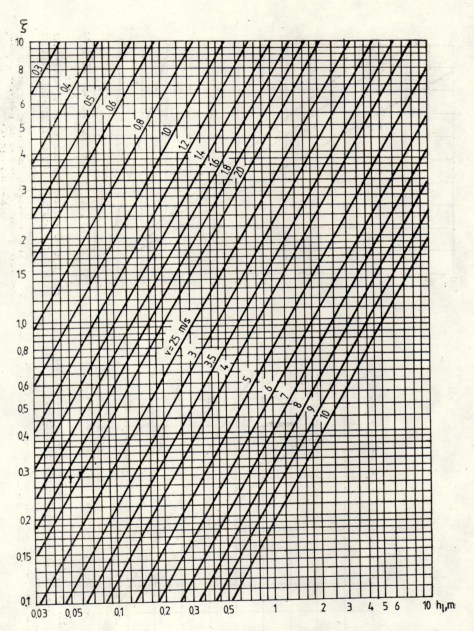

Fig. 5.34. Local head loss, $h_{1\,loc}$ with respect to the local resistance coefficient ξ.

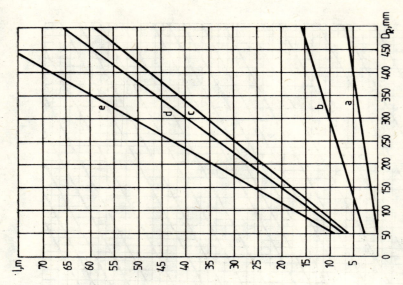

Fig. 5.35. Graph for the conversion of local head losses into length of straight pipeline. Joining fittings:

a — cone (convergent or divergent) with 1 : 15 slope; b — cone with 1 : 10 slope; c — branch at 45°; d — 90° curve; e — sudden passage with $D_1/D_2 = 3/4$; f — double curve at 180°.

Fig. 5.36. Graph for the conversion of local head losses into length of straight pipeline. Shut-off valves:

a — gate valve; b — swing-type check valve; c — piston-lift type check valve (or sucker); d — screw-down stop globe valve; e — screw-down stop angle valve.

5.4 Pumping Lines Calculation

5.4.1 Calculation of the System Pumping Head

A pumping installation is meant to transport liquid from a lower energy level to a higher one. Transport is achieved with the help of two distinct lines: the suction line and the discharge line. The energy required for the liquid transfer from the lower to the higher energetical head is provided by turbopumps. This energy is represented by the hydrodynamic load H_t and is called the *total pumping head of the system*.

Components of the pumping head. The total head of the system H_t has two distinct components: the static head H_{st} and dynamic head H_{dy}.

Static head. This stands for the algebraic sum of the pressure heads and the geodesic head considered between a point in the cross-sectional outlet plane of the system and another point in the cross-sectional inlet plane, where the pressures exerted on the liquid are p_0 and p_i respectively. The pumping head can be expressed as

$$H_{st} = \frac{p_0 - p_i}{\rho g} + H_g \qquad (5.21)$$

The first term of equation (5.21) stands for the pumping pressure (or piezometric) head and is equal to the difference between the outlet piezometric head $p_0/\rho g$ and the inlet piezometric head $p_i/\rho g$, while the second term represents the geodesic head H_g equal to the difference between the discharge piezometric level of the liquid z_0 and the suction piezometric level of the liquid z_i.

The pumping geodesic head consists of two distinct parts, as seen in Fig. 5.37: the suction geodesic head H_{gsc} (equal to the altitude difference between the turbopump axis and the suction altitude z_i) and the discharge geodesic head H_{gds} (equal to the altitude difference between the discharge head z_0 and the turbopump axis).

Dynamic head is the algebraic sum of the kinetic heads and the hydraulic losses between a point located in the cross-sectional outlet plane of the system and another point in the cross-sectional inlet plane of the system, where the liquid velocities are v_0 and v_i, respectively. The dynamic pumping head can be expressed as

$$H_{dy} = \frac{v_0^2 - v_i^2}{2g} + h \qquad (5.22)$$

The first term of equation (5.22) represents the kinetic head and is equal to the difference between the outlet kinetic head of the system, $v^2/2g$, and the inlet kinetic head $v^2/2g$. The second term represents the head loss along the whole system, and is equal to the sum of the suction

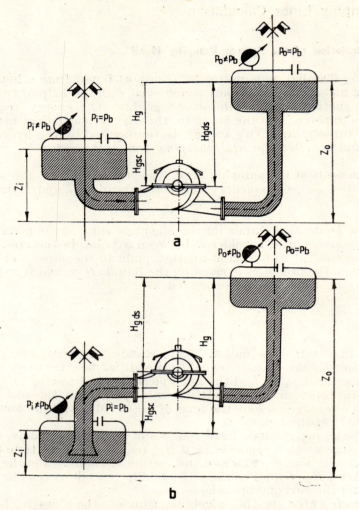

Fig. 5.37. Illustration for the geodesic head of a pumping system:
a — for a pump with negative suction head; b — for a pump with positive suction head.

hydraulic losses h_{sc} (the inlet losses included) and the discharge hydraulic losses h_{as} (the outlet losses included).

The pumping kinetic head shows a small value in the case of systems with a high pressure head (drinking water pumping systems) and a high value in the case of systems with a small pumping head (land water pumping systems). Therefore the kinetic head may be neglected in the first case but should be considered in the second, since it considerably influences the final value of the total pumping head.

Calculation of suction head. The suction head H_{sc} of the system depends on the position of the turbopump axis relative to the suction piezometric level, being negative when the axis is below the suction level and called in this case the *manometric suction head,* or positive when the axis is above the suction piezometric level, and then the called *vacuumetric suction head.*

Manometric suction head. When Bernoulli's equation is applied between sections $i - i$ and $a - a$ of a suction line (coupled to a turbopump) having its centre-line below the suction piezometric level (Fig. 5.38), we get:

$$\frac{p_i}{\rho g} + H_{gsc} + \frac{v_i^2}{2g} = \frac{p_{sc}}{\rho g} + \frac{v_{sc}^2}{2g} + \Sigma h_{sc}(i \ldots a)$$

which, after replacing the sum $\left(\dfrac{p_{sc}}{\rho g} + \dfrac{v_{sc}^2}{2g}\right)$ by H_{sc}, leads to

$$H_{sc} = H_{gsc} + \frac{p_i}{\rho g} + \frac{v_i^2}{2g} - \Sigma h_{sc}(i \ldots a) \tag{5.23}$$

This equation represents the case when the turbopump is sucking from a closed tank, exerting thereby a different pressure on the liquid surface from the atmospheric one. When the turbopump is sucking from

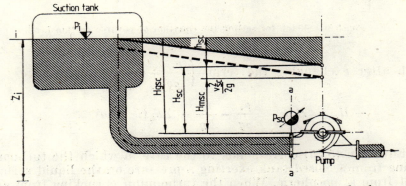

Fig. 5.38. Sketch for setting the equation of negative suction head.

an open tank and the pressure at the liquid surface is atmospheric, that is $p_i = 0$, equation (5.23) becomes

$$H_{sc} = H_{gsc} + \frac{v_{sc}^2}{2g} - \Sigma h_{sc}(i \ldots a) \tag{5.24}$$

In case of already existing pumping systems, the manometric suction head can be established by detection methods. Thus, if we are fitting a manometer connected for the turbopump axis on the turbopump suction stub, we could read directly the manometric suction head (without the kinetic suction head in the manometer coupling point).

Vacuumetric suction head. By applying Bernoulli's equation between the $i-i$ and $a-a$ sections of a suction pipe of a turbopump having its axis above the suction level (Fig. 5.39) we get

$$\frac{p_i}{\rho g} + \frac{v_i^2}{2g} = H_{gsc} + \frac{p_{sc}}{\rho g} + \frac{v_{sc}^2}{2g} + \Sigma h_{sc}(i \ldots a)$$

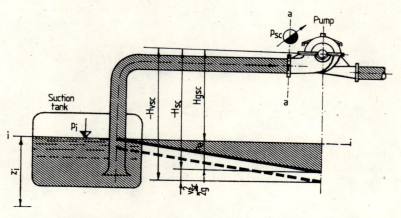

Fig. 5.39. Sketch for setting the equation of positive suction head.

which after exchanging the terms $\left(\dfrac{p_{sc}}{\rho g} + \dfrac{v_{sc}^2}{2g}\right)$ with H_{sc} gives

$$-H_{sc} = H_{gsc} - \frac{p_i}{\rho g} - \frac{v_i^2}{2g} + \Sigma h_{sc}(i \ldots a) \qquad (5.25)$$

Equation (5.25) corresponds to the case in which the turbopump is sucking from a closed tank exerting a pressure on the liquid surface different from atmospheric. When the turbopump is sucking from an open tank and the pressure exerted on the liquid surface is atmospheric, that is $p_i = 0$, equation (5.25) becomes

$$-H_{sc} = H_{gsc} - \frac{v_i^2}{2g} + \Sigma h_{sc}(i \ldots a) \qquad (5.26)$$

The vacuumetric suction head could also be established by detection methods in the case of pumping lines already built. Thus, it could be directly read (without the kinetic suction head in the coupling point of the vacuumeter) by mounting a vacuumeter corrected for the turbopump axis position on the suction stub of the turbopump.

Geodesic suction head is determined in accordance with the position of the minimum suction level and depends on the mounting characteristics of the suction line. Figure 5.40 shows the geodesic suction head for the cases of suction line assembling currently met.

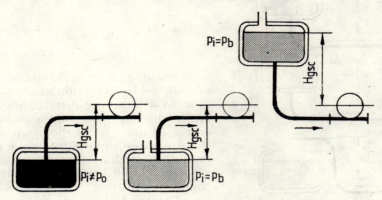

Fig. 5.40. Illustration of geodesic suction head.

Calculation of the discharge head. By applying Bernoulli's equation between the $r-r$ and $o-o$ sections of a discharge line (Fig. 5.41) we get

$$\frac{p_{ds}}{\rho g} + \frac{v_{ds}^2}{2g} = H_{gds} + \frac{p_0}{\rho g} + \frac{v_0^2}{2g} + \Sigma h_{ds}(r \ldots o)$$

which after substituting H_{ds} for the left-hand side gives

$$H_{ds} = H_{gds} + \frac{p_0}{\rho g} + \frac{v_0^2}{2g} + \Sigma h_{ds}(r \ldots o) \tag{5.27}$$

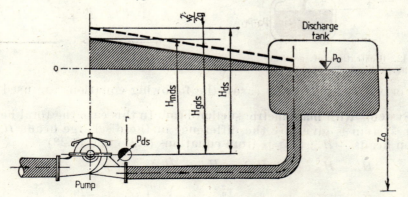

Fig. 5.41. Sketch for setting the equation of discharge head.

This equation applies when the turbopump is discharging into a closed tank and the pressure exerted on the discharge liquid surface is not atmospheric. When the turbopump discharges into an open tank where the pressure exerted on the liquid surface is atmospheric, that is $p_0=0$, equation (5.27) becomes

$$H_{ds} = H_{gds} + \frac{v_{ds}^2}{2g} + \Sigma h_{ds}(r \ldots o)$$

(5.28)

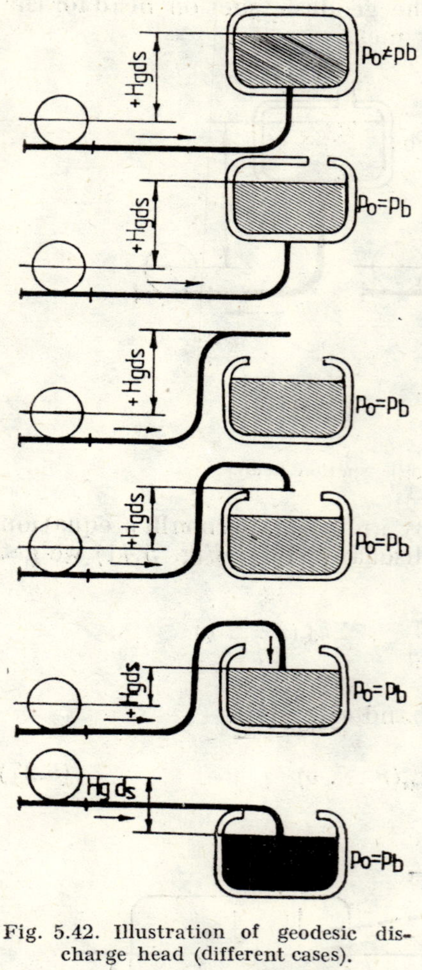

The discharge level could also be established by detection methods in the case of a pumping installation already built. Thus by mounting a manometer, corrected for the turbopump axis, on the discharge nipple, we could directly read the discharge head (without the kinetic suction head in the manometer coupling point).

The geodesic discharge head is determined in accordance with the position of the maximum discharge level, and depends on the mounting characteristics of the discharge line. Thus, Fig. 5.42 shows the geodesic discharge heads for the current cases of discharge line mounting.

Calculation of the total pumping head. The total pumping head is generally calculated with equation

$$H_t = H_{st} + H_{dy}$$

$$H_t = \underbrace{H_g + \frac{p_0 - p_i}{\rho g}}_{\text{Static head}} + \underbrace{\frac{v_0^2 - v_i^2}{2g}}_{\text{Dynamic head}} + h.$$

(5.29)

Fig. 5.42. Illustration of geodesic discharge head (different cases).

For various particular cases, the following equations are used:

Systems with manometric suction head. In this case the total head H_t of the system is given by the difference of the discharge heads H_{ds} and suction heads $+H_{sc}$, that is from equations (5.27) and (5.23)

$$\overset{*}{H}_t = H_{ds} - (+H_{sc}) = H_{ds} - H_{sc})$$

$$H_t = \left(H_{gds} + \frac{p_0}{\rho g} + \frac{v_0^2}{2g} + h_{ds}\right) - \left(H_{gsc} + \frac{p_i}{\rho g} + \frac{v_i^2}{2g} - h_{sc}\right)$$

which gives after solving

$$H_t = H_{gds} - H_{gsc} + \frac{p_0 - p_i}{\rho g} + \frac{v_0^2 - v_i^2}{2g} + h_{ds} + h_{sc} \qquad (5.30)$$

System with vacuumetric suction head. In this case, the total head H_t of the system is given by the sum of the discharge heads H_{ds} and suction heads $-H_{sc}$, that is from equations (5.27) and (5.25)

$$H_t = H_{ds} - (-H_{sc}) = H_{ds} - H_{sc}$$

$$H_t = \left(H_{gds} + \frac{p_0}{\rho g} + \frac{v_0^2}{2g} + h_{ds} \right) + \left(H_{gsc} - \frac{p_i}{\rho g} - \frac{v_i^2}{2g} + h_{sc} \right)$$

which gives after solving

$$H_t = H_{gds} + H_{gsc} + \frac{p_0 - p_i}{\rho g} + \frac{v_0^2 - v_i^2}{2g} + h_{ds} - h_{sc} \qquad (5.31)$$

For a pumping installation already built, the total pumping head could be established by detection methods. Thus, by measuring pressures p_{ds} and p_{sc} and velocities v_{ds} and v_{sc} at the outlet from the discharge nipple and the inlet of the suction stub respectively (see Figs. 5.38, 5.39 and 5.41), we could determine the total head of the system by means of the following equation

$$H_t = 10 \left(\frac{p_{ds} - p_{sc}}{\rho g} \right) + \frac{v_{ds}^2 - v_{sc}^2}{2g} + y \qquad (5.32)$$

where y is the distance between the pressure gauge centre-lines.

5.4.2 Calculation of Economical Diameter for Discharge Pipe

The economical diameter of a discharge line is represented by the minimum value of the sum of the expenses required for the pipe mounting and the cost of the total energy consumed per year.

In 1860, Bresse [5.14] elaborated the following formula for the economical diameter of the discharge pipe

$$d_{ec} = 1.5 \sqrt{Q_{ds}} \qquad (5.33)$$

where d_{ec} is the economic diameter of the discharge pipe, in m;
  Q_{ds} — the flow of the discharge pipe, in m³/s.

This formula leads to an optimum velocity of 0.566 m/s, which is rather small in current economic conditions.

Vibert [5.14] made a thorough study of this matter, considering both the output coefficient and the economic data. He established in 1948 two practical formulae of the form:

$$\text{— for } u = 1.000 : D_{ec} = 1.547 \left(\frac{e}{p}\right)^{0.154} Q_{ds}^{0.46} \quad (5.34)$$

$$\text{— for } u = 0.416 : D_{ec} = 1.350 \left(\frac{e}{p}\right)^{0.154} Q_{ds}^{0.46} \quad (5.35)$$

where u is the pipe output coefficient; $u = 1$ for 24 hours per day use; $u = 0.416$ for 10 hours per day (i.e. 10 out of 24 hours);

D_{ec} — discharge pipe economical diameter, in m;

e/p — ratio between the cost of 1 kWh electric energy and that of 1 kg of iron-made pipe;

Q_{ds} — discharge line flow-rate, in cm/s.

These formulae lead to an optimum velocity between 1.00 and 1.20 m/s which is satisfactory for current economic conditions.

Table 5.3 has been drawn up for rendering practical computations more simple. It includes values of discharge pipe economical diameter d_{ec} depending on the discharged quantity Q_{ds} for $u = 1$ and $u = 0.416$, as well as for several values of the e/p ratio.

TABLE 5.3 **Economic diameter D_{ec} in accordance with the discharge flow Q_{ds} for several values of the output coefficient u and the e/p ratio**

e/p	D_{ec}, mm	
	$u = 1.000$	$u = 0.416$
1/2	$1.39\, Q^{0.46}$	$1.21\, Q^{0.46}$
1/3	$1.30\, Q^{0.46}$	$1.13\, Q^{0.46}$
1/4	$1.24\, Q^{0.46}$	$1.08\, Q^{0.46}$
1/5	$1.21\, Q^{0.46}$	$1.05\, Q^{0.46}$
1/6	$1.17\, Q^{0.46}$	$1.02\, Q^{0.46}$
1/8	$1.12\, Q^{0.46}$	$0.87\, Q^{0.46}$
1/10	$1.08\, Q^{0.46}$	$0.94\, Q^{0.46}$
1/12	$1.05\, Q^{0.46}$	$0.92\, Q^{0.46}$

For special works, computation of the economic diameter of the discharge line can be done by use of the equation

$$C = \frac{1}{t} C_i l + \frac{\rho g Q_{ds} H_{ds}}{102} T C_e \quad (5.36)$$

where Q_{ds} — output of discharge pipe, in m³/s;

H_{ds} — discharge pipe pumping head, in m;

C — yearly cost of pumped water, in dollars/year;

C_i — installation cost of the discharge pipe, in dollars/m;
C_e — cost of the electric energy consumed for pumping, in dollars/kWh;
t — investment amortization time, in years;
T — yearly average time of water pumping, in hours/year;
l — discharge line length, in m
η — turbopump efficiency, in %

Example 5.2 For a sample calculation of economic diameter of a discharge pipe, consider a cast iron pipe of length $l = 10$ km, discharge flow-rate $Q_{ds} = 250$ l/s, discharge geodesic head $H_{gds} = 50$ m and investment pay-off time $t = 20$ years. Given the turbopump efficiency $= 0.75 \eta$ and the cost of electric energy $C_e = 0.30$ dolars/kwh, we require the economic diameter of the discharge line.

Solution. For simplification purposes the economic diameter will be calculated only for 1 m of the pipe and the energy cost only for the energy needed to overcome the hydraulic resistances, since a unit value does not charge the value of economic diameter.

The economic diameter stands for the minimum yearly expenses and is calculated as follows

$$C = \frac{C_{u.p}}{t} + E_p C_p$$

where C = yearly expenses, in dollars/m/year;
$C_{u.p}$ = pipe unitary cost, in dollars/m;
E_p = yearly consumption of electrical energy for pumping.

The yearly electrical energy consumption is

$$E_p = 8\,760\ [\text{hours/year}] \times \frac{Q_{ds} J}{102}\ [\text{kwh}] = 28\,600\ [\text{kwh}]$$

where J is the pipe pitch corresponding to the discharge flow Q_{ds}.

The rest of the calculations are shown in Table 5.4. The economic diameter d_{ec} corresponds to the smallest value of the sum C, that is, $C = 63.50$.

TABLE 5.4 **Calculation of economical diameter of the discharge pipe (considered in Example 5.2)**

D_n mm	v m/s	J	E_p kWh	$E_p C_e$ $/year	C_c $/m	C_c/t $/year	C $/year	Remarks
500	1.30	0.00400	115.00	34.50	796	39.80	74.30	
600	0.90	0.00150	43.00	12.90	1012	50.60	63.50	d_{ec}
700	0.66	0.00067	19.30	5.80	1264	63.20	69.00	
800	0.50	0.00032	9.20	2.80	1520	76.00	78.80	

REFERENCES

5.1. ALTSCHUL, A. D., *Gidravliceskie poteri na trenie v truboprovodah*. Gosenergoizdat Moscow 1863.
5.2. ALTSCHUL, A. D., *Gidravlicheskie soprotivlenia*, Nedra, Moscow. 1970.
5.3. ALTSCHUL, A. D., *Lois essentielles du mouvement turbulent dans les conduites techniques*, Technique sanitaire, No. 6 (1957).

References

5.4. ANDREESCU-CALE, I. *Manual de hidraulică*. Didactic and Pedagogic Publishing House, Bucharest, 1951.
5.5. BENEDICT, P., CARLUCCI, A., *Handbook of specific losses in flow systems*. Plenum Press Data Division, New York, 1970.
5.6. CAUVIN, A., DIDIER, G., *Distribution d'eau dans les agglomerations*. Eyrolles, Paris, 1960.
5.7. COLEBROCK, F., *Turbulent flow in pipes with particular reference to the transition region, between the smooth and rough pipe laws*, Journ.of the Inst. of Civil Enginners, No. 4 (1939).
5.8. DANGHERTY, L., RANZINI, B., *Fluid mechanics with engineering applications*. McGraw-Hill Book Co. New York, 1965.
5.9. DEGRÉMONT, S. G. E. A., *Mémento technique de l'eau*. Stabiliments Degrémont, Paris 1972.
5.10. DUPONT, A., *Hydraulique urbaine* (tome II). Eyrolles, Paris, 1969.
5.11. EGOROV, A. I., *Gidravliceskii rascet truboceatih sistem dlea raspredelenia vodî v vodoprovodnîh ocistnih soorujniah*, Stroizdat, Moscow, 1960.
5.12. ESCANDE, L., *Hydraulique Générale*. Private, Paris, 1948.
5.13. FILONENKO, G. K., *Gidravlicheskoe soprativlenie truboprovodov*. Teploenergetica, No. 4 (1954)
5.14. GOMELA, C., GUERRÉE, H., *La distribution d'eau dans les agglomérations urbaines et rurales*, Eyrolles, Paris, 1970.
5.15. HALBERG, G. M. B. H., *Hydraulische Grundlagen für den Entwurf von Kreiselpumpenanlagen*. Halberg Maschinenbau GmbH and Co., Ludwigshafen, 1972.
5.16. IDELCIK, I. E., *Mémento des pertes de charge*. Eyrolles, Paris, 1969.
5.17. IDELCIK, I. E., *Spravocinik po gidravliceskim soprotivleniam*. Maşinostroenie, Moscow, 1975.
5.18. IONEL, I. I., *Instalaţii de pompare reglabile* Technical Publishing House, Bucharest, 1976.
5.19. KARASSIK, I., J. CARTER, R., *Centrifugal Pumps*, McGraw-Hill Book Co., New York,
5.20. KING, W., *Handbook of hydraulics*. McGraw-Hill Book Co., New York, 1963.
5.21. KOCH, P., *L'alimentation en eau des agglomérations*, Dunod, Paris, 1969.
5.22. LENCASTRE, A., *Manual d'hydraulique générale*, Eyrolles, Paris, 1964.
5.23. LEVIN, L., *Difficultés de calcul des pertes de charge linéaires dans les conduites forcées*. La Houille Blanche, No. 1 (1966).
5.24. MARECHEL, H., *Pertes de charge continues en conduite forcée de section circulaire*. Annales des travaux publics de Belgique, No. 6 (1955).
5.25. McCLAIN, C., *Fluid flow in pipes*. Industrial Press Co., New York, 1963.
5.26. MOODY, L. F., *Friction factor for pipe flow*. Trans. of the ASME, No. 11 (1944).
5.27. MORRIS, M., *Applied hydraulics in engineering*. Rolland Press Co., New York, 1963.
5.28. MÜLLER, W., *Drukverlust in Rohrleitungen*. Energietechnik, No. 7 (1953).
5.29. MÜLLER, W., STRATMAN, H., *Pertes de charge dans les embrachements et collecturs*, Revue technique Sultzer, No. 4 (1971).
5.30. ONIGA, M., *Prepararea apei calde în ansamblurile de locuinţe*. CSCAS—IPCT, Bucharest, 1968.
5.31. RICHTER, H., *Rohrhidraulik*. Springer, Berlin, 1962.
5.32. ROSE, H., *Engineering hydraulics*. John Willey and Sons, 1950.
5.33. VALEMBOIS, J., *Mémento d'hydraulique pratique*. Eyrolles, Paris, 1958.
5.34. WECMAN, A., *Hidraulik*, Bauwesen, Berlin, 1966.

6
Water Hammer Theory and Computation Mechanics

The unsteady motion of liquids is rather frequent in the operation of pumping systems. It occurs whenever the motion condition is changed, that is whenever the limit conditions of the flow are changed by turbo-pumps stopping or starting, valves opening or closing, pipes breaking, power failure at the pump motors, etc. The unsteady condition may introduce significant stresses. The liquid motion through discharge lines of the pumping system may generate overpressures several times higher than the working pressure, as well as significant underpressures, leading to the destruction of components of the installation.

The present chapter presents the calculation of water hammer as a consequence of the unsteady motion of liquids in pumping systems, for use in the establishment and dimensioning of the protection devices required for limitation of the adverse effects of water hammer when the strains introduced exceed those allowed in normal operating conditions. This chapter also includes the basic principles of the water hammer computation technique, the characteristic features of water hammer development as a phenomenon, and approximate and accurate calculation methods, with special stress on their range of validity.

6.1 Basic Principles of Calculation Technique

6.1.1 Water Hammer Phenomenon

The unsteady motion occurring in a hydraulic installation in consequence of a sudden or relatively rapid change of operation conditions, characterized by a significant pressure variation in a short lapse of time, is called *water hammer*. When the velocity varies, the mass of the fluid in motion generates some forces manifest in the form of pressure variation. Due to water compressibility and pipe elasticity, these pressure variations propagate along the pipe as waves with a velocity equal to that of sound, being partially or totally reflected at pipe ends, tanks, or cross-sectional changing branches, and coming back to their starting point for another reflection. Due to these waves, the pressures in various points of the pipe are changed from their initial values determined by the steady flow condition, and vary in time from maximal to minimal values until finally damped by the inner frictions, leading to another steady flow condition imposed by the new conditions of the system elements.

The unsteady motions have an oscillatory character. An unsteady motion occurs in a hydraulic system as a consequence of modification of limit conditions in one or more points of the system. A disturbance of limit conditions leads to local changes of flow rate and pressure being propagated by waves around the hydraulic installation. A disturbance generates two associated waves: a pressure wave, p, and a flow rate wave, q, directly connected to each other. These two elementary waves, the pressure wave and the flow rate wave, form the physical wave or sonic wave [1] carrying the flow rate and pressure changes around the system. Positive and negative waves, with opposite propagation directions, usually occur in a pumping system due to reflection and refraction, or due to their generation in several points. Establishing a common positive direction for abscissae, velocities and flow rates, the waves propagated in the positive direction are called *direct waves*, and those propagated in the negative direction, *backward waves*. Their character is different because the direct wave propagates with $v_0 + a$ velocity, and the backward waves with $v_0 - a$ velocity.

6.1.2 Water Hammer General Equations

Compressibility is one liquid property participating in the development of unsteady motion. Thus, the equations best describing the unsteady motion are those including the compressibility. By neglecting energy dissipation and adopting the perfect fluid model, these equations (identical with those of a vibrating string) are as follows.

When the hydraulic losses are neglected

$$\frac{\partial v}{\partial t} + \frac{\partial (gH)}{\partial S} = 0 \qquad (6.1)$$

$$\frac{\partial (gH)}{\partial t} + a^2 \frac{\partial v}{\partial S} = 0 \qquad (6.2)$$

When the hydraulic losses are considered

$$\frac{\partial v}{\partial t} + \frac{\partial (gH)}{\partial S} = \pm gJ \qquad (6.3)$$

$$\frac{\partial (gH)}{\partial t} + a^2 \frac{\partial v}{\partial S} = 0 \qquad (6.4)$$

where J is the hydraulic gradient.

The general solutions of equations (6.1) and (6.2) are

$$\frac{p}{\varrho g} - \frac{p_0}{\varrho g} = H - H_0 = F(S - at) + f(S + at) \qquad (6.5)$$

$$\frac{a}{g}(v_0 - v) = m(Q_0 - Q) = F(S - at) - f(S + at) \qquad (6.6)$$

---

[1] The name used by Gogu Constantinescu in his work "Theory of Sonics"

where p is pressure;
- p_0 — initial pressure;
- H — pressure head;
- H_0 — pumping head for initial steady pumping conditions;
- v — liquid velocity;
- v_0 — liquid velocity under initial steady conditions;
- Q — liquid flow rate;
- Q_0 — initial flow through the pump;
- F — direct wave of pressure head (propagating in the positive direction of space S);
- f — backward (reflected) wave of pressure head (propagating in the negative direction of space S);
- S — space at current cross-section;
- t — time;
- m — angle coefficient ($m = a/g\omega$);
- ω — pipe cross-section;
- ρ — specific density;
- g — gravity acceleration.

The integration of equations (6.1), (6.2), (6.3) and (6.4) was based on the assumption of a constant wave speed, a, so that they are valid for a pipe segment with unique characteristics, i.e. a pipeline with constant diameter and thickness.

The calculations of pressure heads $H = H(S, t)$ and of flow rates $Q = Q(S, t)$ imply knowledge of functions F and f, which depend on the boundary conditions. Various methods have been devised for the calculation of pressure heads H and of flow rates Q (see Section 6.1.5).

6.1.3 Basic Notations and General Relationships

Water hammer is a phenomenon of an oscillatory character. A change of boundary conditions generates local pressure changes, which are gradually transmitted, due to liquid and pipe material elasticity, forming plane flow rate and pressure waves. By a wave, we mean here a carrier of disturbances of flow rate, or of pressure difference (rise or decrease) which are being propagated. One of the main features of the flow rate and pressure waves is that they are associated, that is they are formed and propagated together, with a determinate relation between them.

Wave speed. The velocity with which the associated waves are propagated relative to the fluid at rest is called the *wave speed*, or velocity of the pressure wave, and is equal to the velocity of sound. This velocity is a function of liquid and pipe characteristics. The speed wave (sometimes called celerity) is determined by means of different formulae in accordance with the homogeneous or non-homogeneous material of the pipe. The following formulae have been simplified for the case of water.

a. *Pipe made of homogeneous materials.* According to Mostkov [6.14] and Streeter [6.22], wave speed is calculated as follows

$$a = \frac{1\,425}{\sqrt{1 + \frac{d_i}{e} \cdot \frac{\varepsilon}{E} k}} \qquad (6.7)$$

where d_i is the pipe inner diameter, in mm;
- e — pipe wall thickness, in mm;
- ε — liquid modulus of elasticity, in da N/cm² :
  for water $\varepsilon = 2.1 \times 9.8 \times 10^3$ daN/cm²;
- E — pipe modulus of elasticity, varying with pipe material:
  - for steel $\quad\quad\quad\quad E = 2.1 \times 9.8 \times 10^2$ daN/cm²;
  - for cast iron $\quad\quad E = 1 \times 9.8 \times 10^5$ daN/cm²;
  - for concrete $\quad\quad E = 2 \times 9.8 \times 10^4$ daN/cm²;
  - for cement asbestos $E = 2 \times 9.8 \times 10^4$ daN/cm²;
- k — a coefficient referring to pipe wall thickness and to its conditions:
  - for pipes with thin walls ($D/e > 25$) and with longitudinal displacements, $k = 1$;
  - for pipes with thin walls ($D/e > 25$) and without longitudinal displacements: either $k = 1 - \mu^2$, or $k = 1$ and wall thickness is equal to e/φ_0;
  - for pipes with thick walls ($D/e < 25$) and with longitudinal displacements: either $k = 2e/D \times (1 + \mu) + D/(D + e)$, or $k = 1$ and wall thickness is equal to e/φ_1;
  - for pipes with thick walls ($D/e < 25$) and without longitudinal displacements: either $k = 2e/D(1 + \mu) + D(1 - \mu^2)/(D + e)$, or $k = 1$ and wall thickness is equal to $e/\varphi_0\varphi_1$.

The φ_0 and φ_1 coefficients are given in Table 6.1 and 6.2 for some values of μ and $2e/D$:

TABLE 6.1 Coefficient φ_0 with respect to the Poisson coefficient for various $2\,e/D$ ratios

μ	2 e/D						
	0.00	0.025	0.05	0.10	0.25	0.50	0.75
0.300	0.910	0.913	0.916	0.921	0.935	—	—
0.250	0.937	0.939	0.942	0.944	0.954	0.965	0.969
0.200	0.960	0.961	0.963	0.965	0.970	0.976	0.987
0.167 (concrete)	0.972	0.974	0.976	0.977	0.980	0.984	0.987

b. *Pipe made of non-homogeneous materials* (reinforced concrete). According to Kiselev, wave speed is calculated as follows

$$a = \frac{1\,425}{\sqrt{1 + \frac{d_i}{e(1 + 9.5\alpha)} \cdot \frac{\varepsilon}{E_b}}} \qquad (6.8)$$

where a is the speed or celerity of the pressure wave, in m/s;
- d_i — pipe inner diameter, in mm;
- e — thickness of pipe wall, in mm;

α — circular reinforcement percentage: $\alpha = 0.015 - 0.050$;
ε — liquid modulus of elasticity, in daN/cm² : for water $\varepsilon = 2.1 \times \times 9.8 \times 10^3$ daN/cm²;
E_b — concrete modulus of elasticity, varying with concrete brand: average value $E_b = 2 \times 9.8 \times 10^4$ daN/cm².

TABLE 6.2 **Coefficient φ_1 with respect to the Poisson coefficient for various $2e/D$ ratios**

μ	2 e/D						
	0.00	0.025	0.05	0.10	0.25	0.50	0.75
0.300	1.000	1.020	1.040	1.082	1.213	—	—
0.250	1.000	1.019	1.037	1.077	1.200	1.425	1.664
0.200	1.000	1.018	1.035	1.077	1.188	1.400	1.627
0.167 (concrete)	1.000	1.017	1.033	1.069	1.181	1.383	1.602

Wave drag. The equation correlating the two associated waves is known as Jukowski's equation

$$p = \rho a v_0 \tag{6.9}$$

It is used in practice in the forms

$$p = \pm zq \tag{6.10}$$

$$H = \pm mq \tag{6.11}$$

where $z = \dfrac{\rho a}{\omega}$ and $m = \dfrac{a}{g\omega}$ are called *wave drag*, corresponding to electromagnetic wave propagation.

To include the two possible travel directions of a wave in our computations, it is convenient to choose arbitrarily a common positive direction for space, velocity and flow rate. We define as direct waves those propagating in the positive direction and as backward waves, those propagating in the negative direction. Depending on the wave kind — direct or backward — the plus or minus sign is valid in equations (6.10) and (6.11).

Water hammer stages. Water hammer is developed in two stages, primary and secondary. *Water hammer primary stage* is the phenomenon taking place during the time in which the relevant pipe section is only under the influence of primary waves, generated by the change of boundary conditions. *Water hammer secondary stage* is the phenomenon taking place during the time in which the relevant pipe section is under the influence of secondary waves superposed on that of primary waves. The primary stage ends when the secondary one reaches the pipe section.

The round-trip or go-return wave propagation time in a pipe length l, equal to the period of primary wave development, is called the *reflection time t* and is given by the equation

$$t = 2\tau = \frac{2l}{a} \tag{6.12}$$

where $\tau = l/a$ is called the *time of travel*.

Wave reflection and refraction. When a wave propagating in a pipe with a drag z_1 reaches a point where the pipe characteristics suddenly change, the wave drag z_2 causes reflection and refraction phenomena (Fig. 6.1).

Denoting the incident wave by φ_1, the reflected wave by φ_2 and the refracted one by φ_3, the following coefficients are defined:

The reflection coefficient of the flow rate wave

$$l_q = \frac{q\varphi_2}{q\varphi_1} = \frac{z_1 - z_2}{z_1 + z_2} \tag{6.13}$$

The reflection coefficient of the pressure wave

$$l_p = \frac{p\varphi_2}{p\varphi_1} = \frac{z_1 - z_2}{z_1 - z_2} \tag{6.14}$$

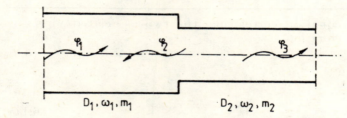

Fig. 6.1. Illustration for wave reflection and refraction.

The refraction coefficient of the flow rate wave

$$r_q = \frac{q\varphi_3}{q\varphi_1} = \frac{2z_1}{z_1 + z_2} \tag{6.15}$$

The refraction coefficient of the pressure wave

$$r_p = \frac{p\varphi_3}{p\varphi_1} = \frac{2z_2}{z_1 + z_2} \tag{6.16}$$

When the incident wave reaches a big size tank ($z_2 = 0$), the above coefficients become

$$l_q = +1 \; ; \; l_p = -k \; ; \; r_q = 2 \; ; \; r_p = 0$$

When the incident wave reaches a closed end of the pipe ($z_2 = \infty$), the above coefficients become

$$l_q = -1 \; ; \; l_p = +1 \; ; \; r_q = 0 \; ; \; r_p = 2$$

Each of the above cases represent frequent conditions at pipe ends, and we are interested mainly in the values of reflection coefficients. It is worth noting that the pressure waves are reflected with the same sign in a closed pipeline and with a changed sign (for instance, overpressure becomes underpressure) in extremely large tanks. This brief approach to reflection and refraction coefficients indicates how hydraulic accumulators for attenuation of water hammer effects function, since they are nothing else than larger or smaller tanks; a wave reaching the accumulator is changed into its exact counterpart (for instance, the pressure rise is turned into a pressure drop) thus reducing the adverse effects of the incident wave.

Boundary conditions. The hydraulic elements participating in the formation, propagation, modification or reflection of waves may be expressed in the $Q-H$ co-ordinate system by characteristic curves. These curves may change with time representing a family of characteristic curves called *the boundary conditions of water hammer equations*.

The turbopump characteristic curves in a pumping installation are represented by a universal characteristic, referring both to the pumping head and to the load torque as functions of flow rate and velocity. In certain cases of water hammer calculation we also need the turbopump universal characteristic, both for positive and negative values of the flow rate. When the turbopump characteristics are not known the graphs shown in Figs. 6.15 to 6.20 should be used.

The characteristic curves of the pipes in a pumping system are represented by second-degree parabolae.

The shut-off valves of a pipe have the characteristic curves represented by a family of second-degree parabolae.

The check valves show, in the flow direction, a characteristic curve represented by a second-order parabola determined by the local hydraulic loss, and in the reverse direction of flow, the characteristic curve represented by a line with equation $Q = 0$.

The characteristic curve of the constant-level tank is represented by a line having the equation $H = H_{st}$.

Pipes with multiple characteristics. A pipe made of several sections of different materials and with different cross-sections and wall thicknesses is called a *pipe with a multiple characteristic*.

In this case several average values should be taken for the general equations of water hammer:

For mean wave speed

$$a_m = \frac{1}{\sum \dfrac{l_i}{a_i}} \tag{6.17}$$

For reflection average time

$$t_m = 2 \sum \frac{l_i}{a_i} \tag{6.18}$$

For average cross-section

$$\omega_m = \frac{\sum \omega_i l_i}{l} \tag{6.19}$$

For average velocity

$$v_m = \frac{Ql}{\sum \omega_i l_i} \tag{6.20}$$

For average wave drag

$$m_m = \frac{\sum a_i l_i}{g \sum a_i l_i} \tag{6.21}$$

where i stands for the respective pipeline section.

Effective stroke time. Each type of valve has its own operating characteristic $Q = f(t)$ and thus affects the flow rate in a different way during the closing and opening periods. The waves resulting from the closing and opening of the valves depend on the dq/dt ratio, and not on the total variation of flow rate during the total closing/opening time.

The time interval during which the valve is kept closed for producing a uniform change rate of the flow, dq/dt, is called *the effective closing time*, t_e.

The tangent to the maximum slope of the flow curve as a function of stem travel (assumed to be uniform with respect to time) gives the effective time (Fig. 6.2).

The effective time can be defined in terms of stem total travel time as follows: $t_e = f(t_t)$, in which f is the ratio of total stroke resulting from the tangent to the maximum slope of the flow calibration line, and t_e and t_t are the effective time and total time, respectively.

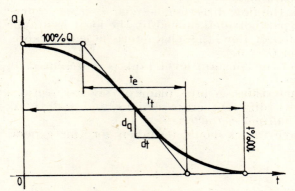

Fig. 6.2. Graphical computation of the effective closing time.

The effective time, t_e, is smaller than the total closing time, t_t, so that $t_e = (0.40 - 0.65)t_t$, and it varies with the type of throttling valve used (Fig. 6.3) [6.7].

The graph shown in Fig. 6.4 was drawn for settling the rise of pressure head ΔH in accordance with the effective closing time t_e, for the case of pipelines with gravity flow and assumig absence of friction in them.

The majority of pipelines with gravity-flow have the constant $\rho \leqslant 5$ and therefore discussion will be confined to pipelines with constant $\rho = 5$ In such cases, if $t_e = \mu$, $\Delta H = \Delta H_{max}$; if $t_e = 2\mu$, $\Delta H = 0.65 \Delta H_{max}$; if $t_e = 3\mu$, $\Delta H = 0.40 \Delta H_{max}$; if $t_e = 4\mu$, $\Delta H = 0.25 \Delta H_{max}$; and if $t_e = 5 \mu$, $\Delta H = 0.15 \Delta H_{max}$. That is the pressure head ΔH is substantially reduced when the effective closing time $t_e = 5\mu$. When the effective time is increased beyond this value, the reduction of pressure head ΔH is rather insignificant, since the $\Delta H(t_e)$ curves start flattening (see Fig. 6.4).

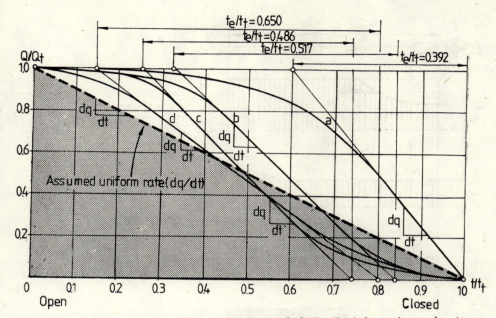

Fig. 6.3. The effective time t_e (with respect to the total closing time) for various valve types:
a — gate valve; b — globe valve, with no wee ports; c — cone valve; d — globe valve with wee ports.

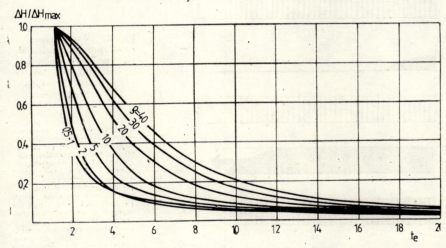

Fig. 6.4. Head rise ΔH (with respect to ΔH_{max}) variation with effective time t_e, for valve closing on pipelines with different ρ constants.

6.1.4 Water Hammer Development

The pressure oscillations generated by the sudden stopping of a turbo-pump develop as shown in a simplified form, in Fig. 6.5. In this figure we

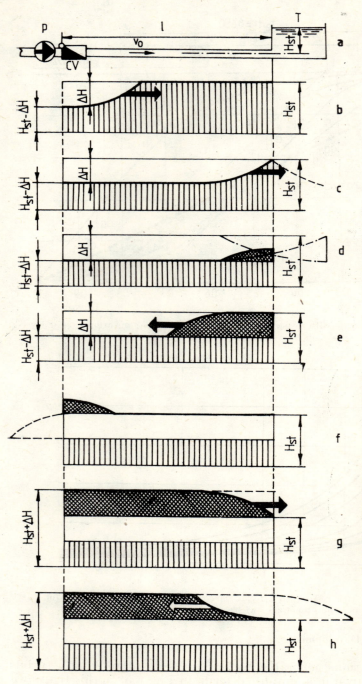

Fig. 6.5. Water hammer development phases in the case of a pipeline with pumped flow (P — pump; T — tank; CV — check valve):

a — the pump operates normally (in continuous regime); b — the pump is suddenly stopped and, consequently, an underpressure wave ($-\Delta H$) is propagating towards the tank; c — arrival of underpressure wave ($-\Delta H$) at the tank; d — the underpressure wave ($-\Delta H$) is reflected from the tank as an overpressure wave ($+\Delta H$) and both waves are superposed in the discharge line; e — the overpressure wave ($+\Delta H$) is propagating towards the pump; f — finding the check valve closed, the overpressure wave ($+\Delta H$) is reflected and superposed over the static head, H_{st}; g — the overpressure wave propagates towards the tank; h — reaching the tank, the overpressure wave is reflected as an underpressure wave ($-\Delta H$) and propagates back to the pump; then the cycle starts again from position b.

assume that the turbopump is coupled to a frictionless pipeline of length l, which is feeding a reservoir and overcomes only the static head H_{st}.

Figure 6.5 shows the following stages :

(1) When the turbopump is stopped, the liquid velocity drops in a very short time from the initial value, v_0, to zero. This change generates an underpressure wave, $-\Delta H$, which is propagated along the pipeline with wave speed a and, as a consequence, the pressure head falls from its initial value H_{st} (Fig. 6.5, b) to the value ($H_{st} - \Delta H$), starting from the turbopump and going to the tank (Fig. 6.5, c).

(2) At time $t = \mu/2$, the pressure wave reaches the level of the liquid in the tank, where it is reflected backwards towards the turbopump as an underpressure wave (Fig. 6.5, d). A pressure head, equal to the static head H_{st} (Fig. 6.5, e), dominates behind this wave.

(3) At time $t = \mu$, the underpressure wave $-\Delta H$ reaches the check valve (in the closed position), where it is reflected forward, towards the tank, as an overpressure wave $+\Delta H$ (that is, without sign change) (Fig. 6.5, f and g). The pressure ($H_{st} + \Delta H$) dominates behind this wave.

(4) At time $t = 1.5\ \mu$, the overpressure wave again reaches the liquid level and is reflected backwards, towards the turbopump (Fig. 6.5, h). The static head H_{st} dominates behind the propagating wave.

(5) When the wave reaches the check valve, the above cycle resumes, except for the wave which is propagating towards the tank with a reduced amplitude (Fig. 6.5, b).

6.1.5 Classification of Computation Methods

The methods used in water hammer computation are classified in the following categories (in their rising order of complexity) :

Approximate computation methods
(a) Approximate methods, based on formulae obtained theoretically or experimentally.
(b) Approximate methods, based on graphs obtained by processing the results of accurate methods.

Accurate computation methods
(a) Accurate methods, based on results obtained by analytical solutions of water hammer equations.
(b) Accurate methods, based on direct integration of water hammer differential equations.

We now present these methods separately.

6.2 Approximate Computation Methods, Based on Formulae

For rapid estimation of the possible minimum and maximum pressure heads generated by water hammer in simple particular cases, several authors [6.18, 6.21] elaborated approximate methods, based on the use of formulae. These formulae, called ready formulae, lead to results which are not accurate enough, but which can be used in preliminary computations, or when the resultant values fall within the design pressures of the pipelines. The formulae used differ in accordance with the nature of the liquid flow in the pipelines. Thus, we present first the formulae used for the computation of gravity-flow pipelines, and then those for the computation of pump discharge lines.

6.2.1 Computation of Gravity-flow Pipelines

The formulae used. The usual computation formulae in the case of gravity-flow pipelines are:
(1) *Michaud's formula*

$$\Delta H = \frac{2lv_0}{gt_c} \qquad (6.22)$$

where ΔH is the rise of pressure head over the static head, in m;
$\quad l$ — pipeline length, in m;
$\quad v_0$ — flow velocity for initial steady conditions, in m/s;
$\quad g$ — acceleration of gravity, in m/s²;
$\quad t_c$ — valve closing time, in s.

The pressure head ΔH stands for the maximum pressure head at the valve (at total closing of the valve) when the closing time is longer than the reflection time, $t_c > 2l/a$, and assuming that velocity varies linearly with time.

The velocity variation law is generally unknown. It depends both on the valve closing law, on valve characteristics, and on the pipe characteristics, and in most cases the velocity variation is non-linear. No rules can be established for finding cases to which the formula can be applied, and therefore it should only be used with caution.

(2) *Waren's formula* (applied to the same conditions as Michaud's)

$$\Delta H = \frac{lv_0}{g\left(t_i - \dfrac{l}{a}\right)} \qquad (6.23)$$

(3) *Johnson's formula*

$$\Delta H = \frac{lv_0}{2g^2 H_0 t_i^2} \left[lv_0 + \sqrt{4g^2 H_0^2 t_i^2 + l^2 v_0^2} \right] \qquad (6.24)$$

This is applied in the same conditions as the formula elaborated by Michaud.

(4) *De Sparre's formula*

$$\Delta H = \frac{2lv_0}{gt_i} \times \frac{1}{2\left(1 - \dfrac{lv_0}{2gt_i H_0}\right)} \qquad (6.25)$$

Applied to the same conditions as Michaud's formula.

Comparison of formulae. In order to compare the approximate formulae with one of the accurate methods (the method of wave characteristics, see Section 6.4.2), they have been applied to the case of gravity-flow pipelines with the following characteristics: pipe length $l = 250$ m, initial velocity $v_0 = 3.58$ m/s, closing time $t_c = 2.1$ s, static head $H_{st} = 50.30$ m, wave speed $a = 981$ m/s and reflection time $t = \mu = 0.509$ s.

We have used for comparison both the absolute value of pressure head rise ΔH, and its relative value ΔH^* (the ratio between the rise of pressure head obtained with the approximate formula and the same rise obtained with the accurate method).

With the accurate method we found $\Delta H = 67$ m and $\Delta H^* = 1$, and with the approximate formulae we obtained the following results:

(1) *Michaud's formula*
$\Delta H = 87.00$ m; $\Delta H^* = 1.30$

(2) *Waren's formula*
$\Delta H = 49.50$ m; $\Delta H^* = 0.74$

(3) *Johnson's formula*
$\Delta H = 65.80$ m; $\Delta H^* = 0.98$

(4) *De Sparre's formula*
$\Delta H = 76.50$ m; $\Delta H^* = 1.14$

Note that the formula given by Jonson is the most realistic.

6.2.2 Computation of Pumping Flow-pipelines

Presentation of formulae used. The formulae used in the case of pumping flow-pipelines are:

a. *Jukowski's formula*

$$\Delta H = \pm \frac{a \Delta v}{g} \tag{6.26}$$

where Δv is the velocity variation, in m/s;
a — wave speed (velocity of pressure wave), in m/s;
g — gravity acceleration, in m/s².

This formula gives the exact value of the pressure wave generated by the closing or opening of a pipeline closing device (shut-off valve or check valve). When the pipeline valve produces a total closing or opening, then $v = v_0$ (steady state velocity). The formula can be used only in this latter case. The valve handling time t_h should be shorter than the reflection time (round-trip wave travel time)

$$t_h < \frac{2l}{a} \tag{6.27}$$

where l denotes the distance from the pipeline closure valve to the first section where an important reflection occurs (reservoir, branch, etc.).

Formula (6.26) can be applied to the pumping flow pipelines only when the flow rate cessation is produced by the total closing of a shut-off valve or check valve, in the conditions mentioned above. The formula is not applicable when the flow rate ceases because the turbopump stops. Jukowski's formula is used for the computation of both the minimum and maximum pressure heads, according to the location of the pipeline closure device and to the flow direction.

b. *Smirnov's formula*

$$\Delta H = H_{st} + H_v + \frac{a}{g} \frac{v_1}{\sqrt{1 + \frac{h_{ds}}{H_0 + H_v}\left(\frac{v_1}{v_0}\right)^2}} \tag{6.28}$$

where H_{st} — is the static head of the system, in m;
$\quad H_v$ — the pressure head corresponding to vacuum generation: $(H_v \approx 8\text{ m})$, in m;
$\quad a\ $ — wave speed (velocity of pressure wave), in m/s;
$\quad g\ $ — gravity acceleration, in m/s²;
$\quad h_{ds}$ — discharge hydraulic losses, in m;
$\quad v_0$ — velocity in conduit for initial steady conditions, in m/s;

$$v_1 = v_0 - \frac{g}{a}(H_t + H_v), \text{ in } m/s;$$

H_t — total pumping head of the system, in m.

This formula also considers friction in the pipeline and may be applied to the situation in which the flow rate ceases due to sudden stopping of the turbopumps and vacuum formation in the pipe. It is valid for pumping installations having long pipelines and small or medium flow rates (e.g. pumping installations used for water supply).

c. *Stoianovici's formula*

$$\Delta H = H\left[\frac{\sqrt{1 + 4(\alpha + \psi)} - 1}{\varphi} - (\alpha - \psi)\right] \tag{6.29}$$

where ΔH is the possible maximum rise of pressure head beyond the static head in m

$$\left.\begin{array}{l} H = \dfrac{av_0}{g}; \qquad \varphi = \dfrac{h_{ds}}{H} \\[6pt] \psi = \dfrac{H_0 + H_v}{H}; \quad \alpha = \dfrac{\sqrt{1 + 4\varphi(1 - \psi)} - 1}{2\varphi} \end{array}\right\} \tag{6.30}$$

This can be applied to the same conditions and assumptions as Smirnov's formula, since the frictional forces are distributed along the pipeline.

When we modify the computation assumptions, the values of coefficients change as follows:

— when the pipe frictional forces are to be found near the reservoir (which stands ready to accept the flow through the turbopump, thus reducing the vacuum cavity formed)

$$\alpha = 1 - (\varphi + \psi), \text{ that is } \alpha + \psi = 1 - \varphi \qquad (6.31)$$

— when vacuum formation is prevented by the cavities being filled with liquid from the suction tank, thus by-passing the turbopumps

$$\alpha = 0 \qquad (6.32)$$

$$\psi = \frac{H_{st}}{H} \qquad (6.33)$$

— when vacuum formation is hindered by the high static head, that is $(H_{st} + H_v)/H \geqslant 1$

$$\alpha = 0 \qquad (6.34)$$

$$\psi = 1 \qquad (6.35)$$

Comparison of the formulae presented. For comparison purposes, the approximate (ready) formulae and the accurate method of wave characteristics (see Section 6.4.2 for Bergeron's solutions) have been applied to the case of a pumping system with a continuous upward-leg discharge line. It has the following characteristics: pipe length $l = 9,000$ m, pipe diameter $D_r = 600$ mm, pipe cross-section $\omega = 0.282$ m², static head $H_{st} = 63$ m, initial flow through the pump $Q_0 = 1,500$ m³/h, velocity in conduit for initial steady conditions $v_0 = 1.48$ m/s, hydraulic loss on the discharge pipe $h_{ds} = 30$ m, system pumping head $H_t = 93$ m, pressure head corresponding to vacuum formation $H_v = 10$ m, wave speed or velocity of pressure wave, $a = 1,120$ m/s.

We needed for comparison both the absolute value of pressure head rise ΔH, and the relative value of this rise ΔH^* (the ratio between the maximum possible rise of pressure head obtained with the approximate formula, and the rise of the same quantity obtained with the accurate method).

The accurate method gave $\Delta H = 100$ m and $\Delta H^* = 1$, while the approximate formulae lead to the following results:

(1) Jukowski's formula

$\Delta H = 169.00$ m; $\Delta H^* = 1.69$

(2) Smirnov's formula

$\Delta H = 137.00$ m; $\Delta H^* = 1.37$

(3) Stoianovici's formula

$\Delta H = 117.50$ m; $\Delta H^* = 1.17$.

When the friction forces are transferred from conduit to the discharge tank, that is when the coefficients are changed in accordance with equation (6.32).

$$\Delta H = 108.50 \text{ m}; \Delta H^* = 1{,}085$$

We notice that Stoianovici's formula is the most realistic.

6.3 Approximate Computation Methods, Based on Graphs

For rapid estimation of the possible maximum and minimum pressure heads, the approximate computation methods used in the analysis of water hammer in pump discharge lines have been elaborated [6.8, 6.9, 6.20]. They are based on the use of graphs or tables, and are applicable to simple situations and specific conditions. Although leading to approximate results, these methods are rapid enough, and eliminate the use of more laborious accurate methods. They can be used for prediction of the minimum and maximum transient pressure heads with about 5% computational accuracy as to the initial pumping head. In most cases this accuracy is satisfactory. When higher accuracy is needed, the exact methods (numerical, graphical or those using computer techniques) should be used. In the following, we have differentiated the methods on the basis of whether the turbopumps of the respective pumping installations were fitted or not with check valves on their discharge sections, since the presence or absence of these check valves leads to different behaviour.

6.3.1 Computation Methods Used for Pumping Installations Having Check Valve at the Pump

We present the methods [6.20] used for rapid computation of the possible rise of pressure heads (overpressures) in the case of pumping installations with turbopumps having check valves on their discharge sections. These methods use graphs or tables for the computation of pipe overpressures ΔH, thus allowing estimation of possible maximum pressure heads along the whole discharge line. These graphs (or tables) are valid under the conditions: (1) when the breaking of a water column occurs at the turbopump, or in the characteristic point (the highest point in the water hammer direct stage), and (2) when the liquid stops its flow through the line suddenly and simultaneously with the turbopump stopping (the flywheel effect of rotating parts of the motor, pump and entrained water is neglected, and heavy check valves are supposed to be fitted on the discharge pipes of the turbopumps). These conditions are satisfied by pumping installations with long discharge lines, medium pumping heads and pumping aggregates with small flywheel effect (the pumping installations used for water supply). The computation methods mentioned above can be used both with unprotected pumping installations, where we assumed rapid-closing check valves leading to sudden stopping of the water column, and with protected pumping installations, using liquid introduction into the discharge pipeline during the underpressure stage for preventing vacuum formation, and thus reducing the overpressures. We give more details of these methods below.

Determination of maximum pressure heads. The vH diagram is drawn in Fig. 6.6 by means of the wave characteristics method (see Section 6.4.2, Bergeron's solutions). For replacing the characteristic line with the head losses we shall use the characteristic curve, as done by Gandenberger [6.5] and Stoianovici [6.20]. It has been postulated that the flow is suddenly stopped at the same time as the turbopumps stop, thus generating a vacuum in the underpressure stage and breaking of the liquid column. A velocity equal, but of reverse sense, to the breaking velocity of the liquid column was supposed to be obtained in the overpressure stage at flow reversal.

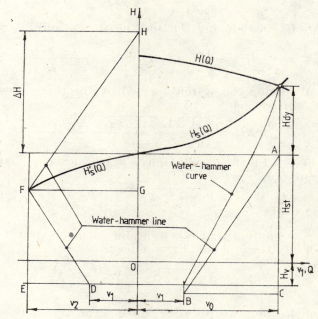

Fig. 6.6. Water hammer computational diagram for the case in which vacuum is created at the pump.

We have used the following notation in Fig. 6.6 :
H_{dy} — dynamic pumping head, in m ;
H_{st} — static pumping head, in m ;
H_v — pressure head corresponding to vacuum formation ($H_v \approx 8$ m), in m ;
v_0 — velocity in pipeline for initial steady conditions, in m/s ;
v_1 — velocity of the liquid near the tank determining the maximum overpressure, in m/s ;

The triangle ABC gives the equation

$$\frac{H_{st} + H_v + H_{dy}\left(\dfrac{v_1}{v_2}\right)^2}{v_0 - v_1} = \frac{a}{g} \qquad (6.36)$$

where g is the acceleration of gravity.

The triangle DEF gives the equation

$$\frac{H_{st} + H_v + H_{dy}\left(\dfrac{v_2}{v_0}\right)^2}{v_2 - v_1} = \frac{a}{g} \qquad (6.37)$$

The triangle FGH gives the equation

$$H = \frac{a}{g} v_2^2 - H_{dy}\left(\frac{v_2}{v_0}\right)^2 \qquad (6.38)$$

By introducing the additional relations:

$$H = \frac{av_0}{g}\,;\quad \varphi = \frac{h_{ds}}{H}\,;\quad \psi = \frac{H_{st} + H_v}{H}\,;\quad \alpha = \frac{v_1}{v_0}$$

and solving the system formed by the three equations (6.36), (6.37) and (6.38), we get

$$\Delta H = H\left[\frac{\sqrt{1 + 4\varphi(\alpha + \psi)} - 1}{\varphi} - (\alpha + \varphi)\right] \qquad (6.39)$$

where ΔH is the maximum rise of pressure head beyond the static pumping head and

$$\alpha = \frac{\sqrt{1 + 4\varphi(1 - \psi)} - 1}{2\varphi} \qquad (6.40)$$

Assuming that vacuum formation is prevented by water penetration into the pipeline, Fig. 6.6 leads to

$$H_v = 0\,;\quad v_1 = 0\,;\quad \psi = \frac{H_{st}}{H}\,;\quad \alpha = 0 \qquad (6.41)$$

In the case of high static heads, that is when $H_{st} + H_v \geqslant H$, we get

$$\psi = 1\,;\quad \alpha = 0 \qquad (6.42)$$

To make the computations easier, two tables and two graphs have been drawn, based on the above equations. Thus, Table 6.3, based on equations (6.39) and (6.40), has been constructed and its values then used for the graph shown in Fig. 6.7, valid for unprotected pumping installations. Table 6.4 and the graph shown in Fig. 6.8 were based on equation (6.39) with $\alpha = 0$, being valid for the protected pumping installations using water admission into the pipeline during the underpressure stage. These

TABLE 6.3 Values of parameter ξ for unprotected pumping installations

φ \ ψ	0.000	0.100	0.200	0.300	0.400	0.500	0.600	0.700	0.800	0.900	1.000
0.00	1.000	1.000	1.000	1.000	1.000	1.000	1.000	1.000	1.000	1.000	1.000
0.01	0.971	0.973	0.974	0.976	0.977	0.978	0.979	0.980	0.980	0.980	0.980
0.05	0.871	0.878	0.884	0.890	0.895	0.899	0.902	0.905	0.907	0.908	0.909
0.10	0.773	0.784	0.794	0.802	0.810	0.816	0.822	0.826	0.830	0.831	0.832
0.20	0.633	0.646	0.658	0.669	0.679	0.687	0.694	0.700	0.705	0.707	0.708
0.30	0.536	0.549	0.561	0.572	0.581	0.590	0.597	0.603	0.607	0.610	0.611
0.40	0.463	0.476	0.487	0.496	0.505	0.513	0.519	0.524	0.528	0.530	0.531
0.50	0.407	0.418	0.428	0.436	0.443	0.449	0.454	0.458	0.461	0.463	0.464
0.60	0.363	0.372	0.380	0.386	0.392	0.396	0.400	0.403	0.405	0.406	0.407
0.70	0.326	0.333	0.334	0.344	0.348	0.352	0.353	0.355	0.356	0.356	0.356
0.80	0.295	0.301	0.305	0.308	0.311	0.312	0.312	0.312	0.312	0.312	0.312
0.90	0.268	0.273	0.276	0.277	0.278	0.277	0.276	0.275	0.274	0.272	0.272
1.00	0.245	0.248	0.250	0.250	0.249	0.247	0.244	0.242	0.240	0.237	0.236

graphs facilitate the computation of maximum pressure head (overpressure) ΔH, that is of the $\xi = \Delta H/H$ ratio starting from coefficients ψ and φ. The demonstration mentioned above is valid for the case in which vacuum is created only near the turbopump.

Making use of the diagram shown in Fig. 6.9, drawn in the same way as that shown in Fig. 6.6, we can establish the conditions in which

TABLE 6.4 Values of parameter ξ for protected pumping installations

φ \ ψ	0.000	0.100	0.200	0.300	0.400	0.500	0.600	0.700	0.800	0.900	1.000
0.00	0.000	0.100	0.200	0.300	0.400	0.500	0.600	0.700	0.800	0.900	1.000
0.01	0.000	0.100	0.199	0.298	0.397	0.495	0.593	0.690	0.787	0.884	0.980
0.05	0.000	0.099	0.196	0.291	0.385	0.476	0.566	0.654	0.741	0.826	0.909
0.10	0.000	0.098	0.192	0.283	0.370	0.454	0.536	0.614	0.689	0.762	0.832
0.20	0.000	0.096	0.185	0.268	0.345	0.416	0.483	0.545	0.603	0.657	0.708
0.30	0.000	0.094	0.179	0.254	0.322	0.383	0.438	0.488	0.533	0.574	0.611
0.40	0.000	0.093	0.172	0.241	0.302	0.354	0.400	0.440	0.475	0.505	0.531
0.50	0.000	0.091	0.166	0.230	0.283	0.328	0.366	0.398	0.425	0.447	0.464
0.60	0.000	0.089	0.161	0.219	0.267	0.305	0.337	0.362	0.381	0.396	0.407
0.70	0.000	0.088	0.156	0.209	0.251	0.285	0.310	0.329	0.343	0.352	0.356
0.80	0.000	0.086	0.151	0.200	0.237	0.266	0.286	0.300	0.308	0.312	0.312
0.90	0.000	0.085	0.146	0.191	0.224	0.248	0.264	0.274	0.278	0.277	0.272
1.00	0.000	0.083	0.142	0.183	0.212	0.232	0.244	0.249	0.249	0.245	0.263

the graph of Fig. 6.7 remains valid both for the case of vacuum formation near the turbopump, and for vacuum formation in the highest point. Two cases can be distinguished: in the first the maximum pressure heads are generated before vacuum obstruction in the highest point, and in the second, they are produced after.

In the first case, as we can see from Fig. 6.9, the maximum pressure head is determined either by the $\overline{IJ}$ line or by the $\overline{MN}$ line, and depends

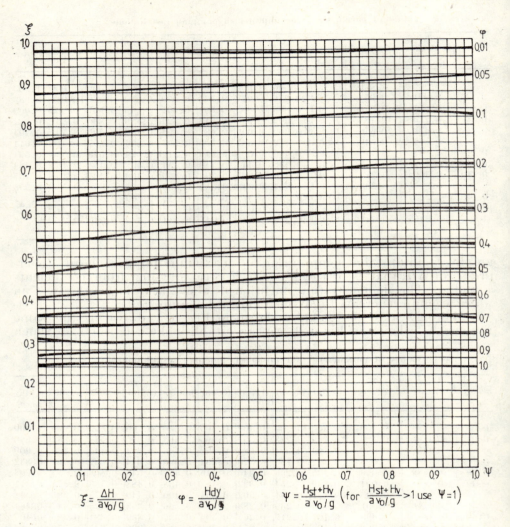

Fig. 6.7. Maximum head ξ at the pump, in accordance with coefficients φ, and ψ for unprotected pumping installations.

on the highest point characteristic. In this case it is the curve *I* that is convenient.

In the second case, as we can also see in Fig. 6.9, the maximum pressure head is determined either by the $\overline{KL}$ line, or by the $\overline{PO}$ line, and is again dependent on the highest point characteristic. In this case, the most convenient is curve *II*.

Note that the graph shown in Fig. 6.7 gives an exact pressure head only when the points *I*, *K*, *M*, and *P* coincide with the point *G* on fig. 6.9.

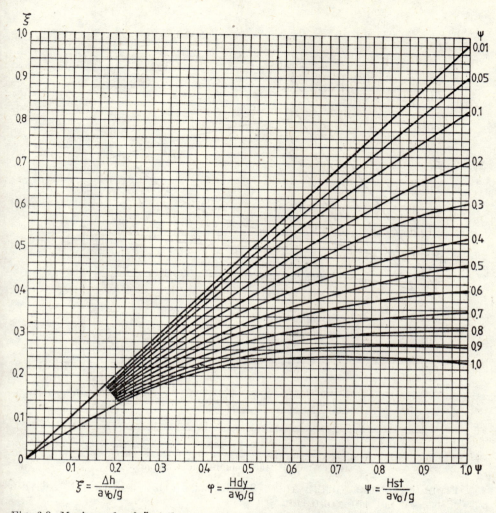

Fig. 6.8. Maximum head ξ at the pump, in accordance with coefficients φ and ψ, for protected pumping installations (by liquid admission into the discharge pipe).

The $\overline{FR}$ line, corresponding to the tank characteristic, has the value

$$\overline{FR} = H_{dy}\left(\frac{v_2}{v_0}\right)^2 \qquad (6.43)$$

while the $\overline{FR}$ line corresponding to the highest point characteristic has the value

$$\overline{FR} = (1-y)H_{st} + H_v + xH_{dy}\left(\frac{v_2}{v_0}\right)^2 \qquad (6.44)$$

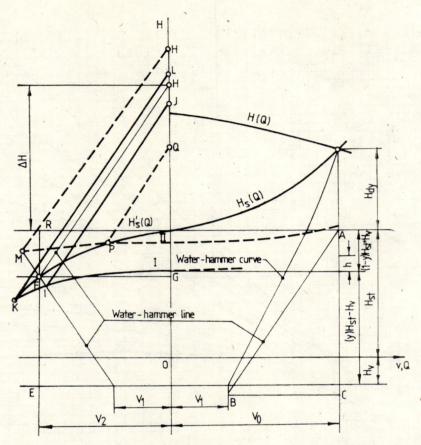

Fig. 6.9. Water hammer computational diagram for the case in which vacuum is created at the summit (high point).

From equations (6.42) and (6.43), with the additional relation $\beta = v^2/v_0$, we get the condition

$$\frac{(1-x)H_{dy}}{(1-y)H_{st} + H_v} = \frac{1}{\beta^2} \tag{6.45}$$

which appears also as condition (6.55), and leads to condition (6.56). Also

$$\beta = \frac{1 - \sqrt{1 - 4\xi\varphi}}{2\varphi} \tag{6.46}$$

which is derived from equation (6.38).

Estimation of protection tank volume. For preventing vacuum formation, the volume of the protection tank should be large enough to obstruct air admission into the pipeline, thus modifying the whole phenomenon. The tank location depends on the longitudinal profile of the discharge line. We meet cases in which the tank is placed near the suction basin, by-passing the turbopump, as well as situations in which the tank is placed in the highest point of the pipeline.

(a) *Tank located near the turbopump*. Making use of the characteristics method and neglecting friction, the tank volume can be calculated by means of the vH diagram shown in Fig. 6.10.

The tank volume is computed from the equation

$$V_P = \mu \omega V_m r \qquad (6.47)$$

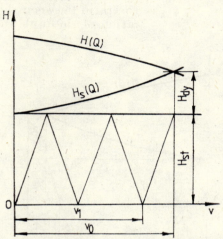

Fig. 6.10. Diagram for the computation of a protection tank located near the pump.

where V_P is the volume of the tank located near the pumping station, in m³;
ω — pipe cross-section, in m²;
μ — round-trip wave travel time (reflection time) equal to $2l/a$, in s;
v_m — mean velocity, equal to $v_0/2$, in m/s;
r — number of reflections.

The number of reflections can be obtained from the equation

$$r = 1 + \frac{v_1}{2(v_0 - v_1)} = 1 + \frac{v_0 - \dfrac{H_{st}}{a/g}}{2\dfrac{H_{st}}{a/g}} = \frac{1}{2}\left(1 + \frac{1}{\psi}\right), \qquad (6.48)$$

since

$$\frac{H_{st}}{v_0 - v_1} = \frac{a}{g}$$

Starting from equations (6.47) and (6.48) and introducing the initial flow Q_0 through the pump, we get

$$V_{sp} = \frac{Q_0}{2}\frac{1}{a}\left(1 + \frac{1}{\varphi}\right) \qquad (6.49)$$

(b) *Tank located near the highest point*. Again by using the method of characteristics (Bergeron's solution), the tank volume is computed by means of the vH diagram shown in Fig. 6.11, assuming simplifying hypothesis (leading to a re — cover volume), such as:
— the flow rate of the first pipeline section is zero;
— no vacuum is created at the pumping section since the distance is small enough for the inertia of pumping aggregates to prevent its formation;
— friction is neglected.

By similar methods to equation (6.49), the volume of the tank located at the line summit has the value:

$$V_L = \frac{Q_0}{2} \frac{1(1-x)}{a} \left[1 + \frac{1}{(1-y)\psi} \right] \qquad (6.50)$$

where x is the ratio between the length of the first section of the pipe and the total length of the pipe;

y is the ratio between the static head corresponding to the location of the tank and the total static head.

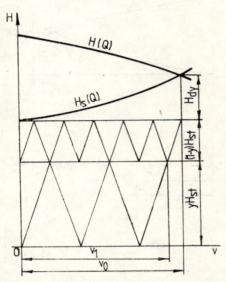

Fig. 6.11. Diagram for the computation of a protection tank located at the line summit (high point).

Validity range of computation method. The analyzed method is applicable to pumping installations with one or more turbopumps and a discharge line with a unique characteristic. The pipeline is coupled to turbopumps be means of check valves at one end, and coupled at the other end either to a storage tank (transmission stations) or to a distribution system (distribution stations).

Since the discharge pipeline follows the land irregularities, it may have various high points influencing the development of water hammer. Thus, the pumping installation may have a discharge line following a horizontal route which becomes ascending near the tank (Fig. 6.12). In this case, vacuum may be created only at the turbopump outlet during the underpressure stage, as a consequence of turbopumps stopping. The pumping system may also have an ascending discharge line at the turbopump outlet, forming one or more high points (Fig. 6.13). In this case, vacuum may be created both at turbopumps and at the high points.

The notations used in Fig. 6.12 and Fig. 6.13 are:

l = pipeline length, in m;
H_{st} = static pumping head, in m;
x — ratio between the pipeline length up to the high point and the total pipeline length;
y — ratio between the static head of the high point and the static head of the discharge tank.

We can therefore differentiate two kinds of pumping installations, (a) without high points and (b) with high points.

(a) *Pumping installations without high points*. In these installations, vacuum can be created only at the turbopump, depending on the magni-

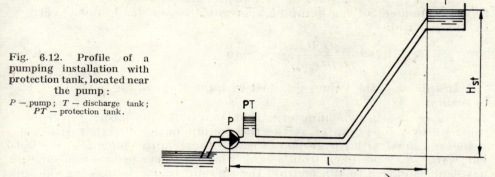

Fig. 6.12. Profile of a pumping installation with protection tank, located near the pump:
P — pump; T — discharge tank; PT — protection tank.

tude of the flywheel effect of the rotating parts of the motor, pump, and entrained water. Here too, we may distinguish two cases:

(a.1) *Vacuum is not created at the pump*. This situation occurs when the flywheel moment of the pumping aggregates is high, and can be verified with the equation

$$K = 1{,}789 \frac{\rho Q_0 H_0}{GD^2 \eta_0 n_0^2} \frac{2lx}{a} < 10 \qquad (6.51)$$

where ρ is the liquid specific density, in kg/m³;
- Q_0 — the initial flow through the pump, in m³/s;
- H_0 — pumping head for initial steady pumping conditions, in m;
- GD^2 — the flywheel effect of the rotating parts of motor, pump, and entrained water, in kg/m²;
- η_0 — initial efficiency (pump efficiency at rated speed and head) in %;
- n_0 — initial rotation speed (rated pump speed), in rev/min;
- l — pipeline length, in m;
- a — wave speed (velocity of pressure wave), in m/s.

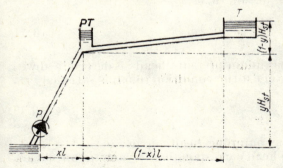

Fig. 6.13. Profile of a pumping installation with protection tank located at the line summit:
P — pump; T — discharge tank; PT — protection tank.

In this case, the values obtained with the graph of Fig. 6.8 are higher than the real ones, but the exact values can be obtained by making use of Paramakian's graphs [6.15].

(a.2) *Vacuum is created at the pump.* This situation occurs when the flywheel moment of the pumping aggregates is too small, and is verified by the equation

$$k = 1{,}789 \frac{\rho Q_0 H_0}{GD^2 \eta_0 n_0^2} \frac{2lx}{a} \geqslant 10 \tag{6.52}$$

In this case, the values obtained with the graph of Fig. 6.6 correspond to reality.

For preventing vacuum formation, that is, for diminishing overpressures, we may inject water at the turbopump outlet. In that case, the values obtained with the graph of Fig. 6.8 are accurate enough. The injection water may be taken from a tank whose size has been established by means of equation (6.49), or from the suction basin, by-passing the turbopump by means of a by-pass pipe.

(b) *Pumping installation with high points.* In these pumping installations, vacuum can be created both at the turbopump and at the high points, depending on the magnitude of the flywheel moment of the pumping aggregates and on the position of the high points. Two distinct cases arise.

(b.1) *Vacuum is not created at the turpopump and the high point.* This situation occurs when the flywheel moment of the pumping aggregates is high enough and is verified by equation (6.52) for the high point and by equation (6.53) for the turbopump

$$K = 1{,}789 \frac{\rho Q_0 H_0}{GD^2 \eta_0 n_0^2} \frac{2lx}{a} < 10 \tag{6.53}$$

In this situation, the overpressures can be determined by means of the graph shown in Fig. 6.7. A water tank, sized by means of equation (6.50), is placed in the high point for diminishing the overpressures. Thus, the overpressures can be determined by the graph of Fig. 6.8.

(b.2) *Vacuum is created at the turbopump and the high point.* This situation occurs when the flywheel moment of the pumping aggregates is too small, and is verified by equation (6.54)

$$K = 1{,}789 \frac{\rho Q_0 H_0}{GD^2 \eta_0 n_0^2} \frac{2lx}{a} \geqslant 10 \tag{6.54}$$

The value of the possible maximum pressure heads is determined with the graph of Fig. 6.7 and it is exact if the condition (6.55) is satisfied

$$\frac{(1-x)H_{dy}}{(1-y)H_{st} + H_v} = \frac{1}{\beta^2} \tag{6.55}$$

where H_{dy} — the dynamic pumping head, in m;
$H_v \approx 8$m — the pressure head corresponding to vacuum formation;
β — a coefficient:

$$\beta = \frac{1 - \sqrt{1 - 4\xi\varphi}}{2\varphi}. \tag{6.56}$$

With condition (6.57)
$$\frac{(1-x)H_{dy}}{(1-y)H_{st}+H_v} \neq \frac{1}{\beta^2} \qquad (6.57)$$
the real value of the maximum pressure head is higher than that computed with the graph of Fig. 6.7 and exact computation requires the use of the wave characteristics method (see Bergeron's solutions in Section 6.4.2).

When the pumping installation is protected by means of a tank sized in accordance with equation (6.60), the maximum pressure heads are substantially diminished to a value determined by means of the graph of Fig. 6.8.

Applicability field. Table 6.3 and the graph in Fig. 6.7. can be used in the following situations:
— pipeline without high point, when $K \geq (\varphi + \psi)$.
— pipeline with unprotected high point, when $K < (\varphi + \psi)$;
— pipeline with unprotected high point, when
$K \geq (\varphi + \psi)$ and $x \times K < (\varphi + \psi)$;
— pipeline with unprotected high point, when $x \times K \geq (\varphi + \psi)$ and
$$\frac{(1-x)H_{dy}}{(1-y)H_{st}+H_v} = \frac{1}{\beta^2}$$

Table 6.4 and the graph in Fig. 6.8 can be used in the following cases:
— pipeline without high point, when $K > (\varphi + \psi)$ (in this case an exact computation is required when the resulting pressures are too high);
— pipeline without high point, when $K \geq (\varphi + \psi)$ and water is injected near the pump (*by-pass* from suction tank);
— pipeline with high point, when $x \times K < (\varphi + \psi)$ and water is injected through the high point;
— pipeline with high point, when $x \times K < (\varphi + \varphi)$ and water is injected both near the pump and through the high point.

Example 6.1. **Calculation of the possible maximum pressure heads for pumps with check valves.** Let us consider a pumping installation made of several turbopumps fitted with check valves and a steel common discharge header. Assume the turbopump characteristics are known: flywheel moment $GD^2 = 60$ daN/m², initial rotation speed (rated pump speed) $n_0 = 1,450$ rev/min, initial efficiency (pump efficiency at rated speed and head), $\eta_0 = 86\%$ also the system characteristics: initial flow through the pump, $Q_0 = 0.90$ m³/s; pumping head for initial steady pumping conditions, $H_0 = 68$ m; pipeline length, $l = 14,800$ m; nominal diameter of the discharge header, $D_r = 1,000$ mm, pipe wall thickness, $e = 8$ mm; length, $xl = 5,000$ m; height, $yl = 40$ m; static head, $H_{st} = 50$ m; pressure head corresponding to vacuum formation, $H_v = 8$ m; and the maximum pressure head allowable, $H_{max\ a} = 100$ m. We desire to find the maximum pressures obtained, by means of graphs shown in Figs. 6.7 and 6.8.

Solution
(1) *Wave speed is computed by equation* (6.8)
$$a = \frac{1,425}{\sqrt{1+\frac{1000}{8} \times \frac{1}{100}}} = 950 \text{ m/s}$$

(2) *Additional data*

$$A_L = 0.875 \times 1^2 = 0.785 \text{ m}^2; \quad v_0 = \frac{0.9}{0.785} = 1.13 \text{ m/s}$$

$$x = \frac{5{,}000}{14{,}000} = 0.338; \quad y = \frac{40}{50} = 0.8$$

$$K = 1{,}789 \frac{1{,}000 \times 0.3 + \times 68}{60 \times 0.86 \times 1{,}450^2} \cdot \frac{2 \times 14{,}800}{950} = 10.5,$$

$$xK = 0.338 \times 10.5 = 3.55; \quad av_0/g = 950 \times 1.13/9.81 = 109$$

$$\psi = \frac{50 + 8}{109} = 0.595; \quad \varphi = \frac{18}{109} = 0.165.$$

From Fig. 6.7 for $\varphi = 0.595$ and $\varphi = 0.165$ we get $\xi = 0.739$. Thus, it follows that

$$\Delta H = \ = 109 \times 0.739 = 81 \text{ m}$$

and hence the maximum pressure head on the pipeline at the turbopump is

$$H = H_{st} + \Delta H = 50 + 81 = 131 \text{ m} > 100 \text{ m}$$

Since $\varphi + \psi = 0.595 + 0.165 = 0.760$, it follows that $K > (\varphi + \psi)$ and $xK > (\varphi + \psi)$, and thus equations (6.30) and (6.31) should be checked

$$\frac{(1 - 0.338)19}{(1 - 0.8)50 + 8} = 0.662$$

$$\beta = \frac{1 - \sqrt{1 - 4 \times 0.739 \times 0.165}}{2 \times 0.165} = 0.865$$

$$\frac{1}{\beta^2} = \frac{1}{0.865^2} = 1.33 \neq 0.662.$$

It follows that the real maximum pressure head is higher than the computed one, and thus the pumping installation should be protected. To this end, two open tanks should be located at the pump and in the high point.

From Fig. 6.8 for:

$$\psi = 50/109 = 0.46 \text{ and } \varphi = 18/109 = 0.165 \text{ we get } \xi = 0.398.$$

It follows that

$$\Delta H = 0.398 \cdot 109 = 43 \text{ m}$$

and hence the maximum pressure head on the pipeline at the turbopump is

$$H = H_{st} + \Delta H = 50 + 43 = 93 \text{ m} < 100 \text{ m}.$$

Since $Kx > (\varphi + \psi)$, water injection is needed both at the turbopump and in the high point.

The volume of water to be injected in the pipeline is determined as follows:

(1) *At the turbopump with equation* (6.24)

$$V_P = \frac{0.9}{2} \times \frac{14{,}800 \times 0.338}{950}\left(1 + \frac{1}{0.8 \times 0.595}\right) = 7.35 \text{ m}^3,$$

(2) *In the high point, with equation* (6.25)

$$V_L = \frac{0.9}{2} \times \frac{14{,}800(1 - 0.338)}{950}\left(1 + \frac{1}{(1 - 0.8)\times 0.595}\right) = 48.50 \text{ m}^3$$

The cross-section of the branch pipe, used for the main pipe coupling to the tank, is

$$\omega_b = \frac{Q}{4} = \frac{0.9}{4} = 0.225 \text{ m}^2$$

which corresponds to a pipe having nominal diameter $D_r = 600$ mm.

6.3.2 Computation Methods Used for Pumping Installations without Check Valve at the Pump

These computation methods [6.8 and 6.9] are used for rapid estimation of the possible maximum and minimum pressure heads of installations without check valves at the turbopump. The methods are valid both for unprotected pumping installations where a reverse flow through the pumps is assumed, and for the protected ones where either limitation of the reverse rotation of the pumps or its prevention is assumed. The reverse flow through the pumps can be allowed in the case of pumping installations for land irrigation, caracterized by small pumping heads and short pumping pipelines, while the prevention or limitation of the reverse rotation may be met with pumping installations for water supply which have medium pumping heads and long pipelines.

Since in pumping installations without check valves, the turbopumps pass through all operating conditions (operating as pump, brake, or turbine), the following quantities should be known: the pumping head and the turbopump load torques, the maximum and minimum pressure heads, and the flow rate reversing times. For rapid estimation of these quantities we give below a series of graphs, valid for turbopumps with rapidity $n_s \leqslant 200$. The computation errors resulting from use of these graphs are $\pm 5\%$ of the initial pumping head.

The following notations were used for graph preparation:

- Q — turbopump flow rate;
- Q_R — rated flow through the pump (for maximum efficiency at the operating speed);
- Q^* — Q/Q_R;
- H — turbopump pumping head;
- H_R — rated pumping head;
- H^* — H/H_R;
- T — turbopump load torque;
- T_R — turbopump load torque for maximum efficiency at the operating speed (corresponding to Q_R and H_R);
- T^* — T/T_R;
- n — turbopump speed;
- n_R — rated pump speed;
- n^* — n/n_R;

$H_{min.t}$ — minimum head at turbopump during transient;
$H^*_{min.t} = H_{min.p}/H_R$;
$H_{min.p}$ — minimum transient head at mid-length of the discharge pipeline;
$H^*_{min.p} = H_{min.p}/H_R$;
$H_{max.t}$ — maximum transient head at turbopump;
$H^*_{max.t} = H_{max.t}/H_R$;
$H_{max.p}$ — maximum transient head at mid-length of the discharge pipeline;
$H^*_{max.p} = H_{max.p}/H_R$;
H_{st} — static head;
h_l — hydraulic loss (friction head);
$h_l^* = h_l/H_R$;
ρ — av_R/gH_R, the pipeline constant;
τ — $= 1/(k2l/a)$, a coefficient;
k — $= (1{,}789 \, \rho Q_R H_R)/(GD^2 \eta_R n_R^2)$, turbopump constant;
η_R — rated efficiency (maximum);
v_R — fluid velocity in the discharge line for rated pump discharge (for maximum efficiency at the operating speed);
GD^2 — moment of inertia of the rotating parts of motor, turbopump, and entrained fluid.

Determination of possible flow and rotation conditions of turbopumps. In order to perform a water hammer analysis of a pumping system it is

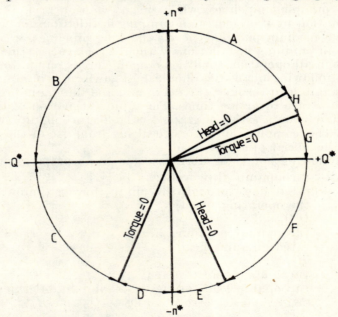

Fig.6.14. Polar-type diagram showing possible flow and rotation conditions for turbopumps:

A — normal pump; B — energy dissipation (braking); C — normal turbine; D — energy dissipation (braking); E — reverse pump; F — energy dissipation (braking); G — reverse turbine; H — energy dissipation (braking).

necessary to know the complete (universal) characteristics, i.e. the head-discharge and torque-discharge diagrams for both the natural and the reverse rotations and flows through the pump. Stepanoff [6.19] examined the complete pump characteristics and gave several examples. The various possible combinations of head, discharge, speed and torque of turbopumps are shown in Fig. 6.14 in the speed-discharge plane ($n-Q$), using the polar-type plot, contrived by Kármán and reported by Knapp [6.11]. The only regions of Fig. 6.14 which are of interest in water hammer analysis are those of energy dissipation (braking conditions) in the first, second and third quadrants, of normal pump operation (pumping condition), and of normal turbine operation (turbine-like condition), denoted by H, B, D, A, and C, respectively, in Fig. 6.14. When the turbopump complete (universal) characteristics are not available, the graphs shown below may be used. They give the pumping head, H, and the load torque, T, in accordance with the flow rate through the turbopump, Q, and the rotation speed, n. The graphs have been drawn on the basis of Knapp's measurements [6.11] for each characteristic operating condition as follows:

a. *Graphs for the determination of pumping conditions.* These conditions occur when the turbopump has a normally decelerated rotation and the liquid in the pipeline has a normal motion towards the tank.

The graphs shown in Figs. 6.15 and 6.16 help the determination of pumping heads and load torques, respectively, for the pumping conditions in region A (see Fig. 6.14).

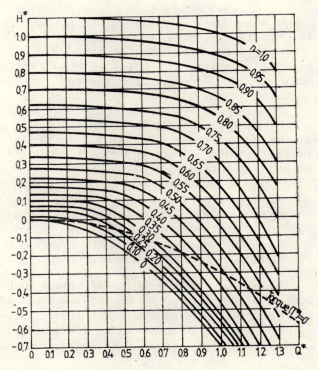

Fig. 6.15. Head-discharge (H^*-Q^*) diagram for the regions of normal pump (region A) and energy dissipation (region H.)

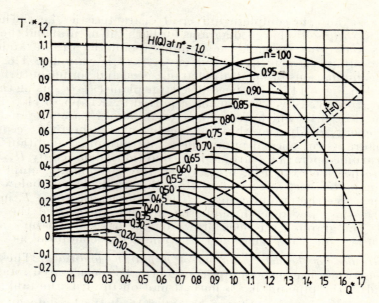

Fig. 6.16. Torque-discharge (T^*-Q^*) diagram for the regions of normal pump (region A) and energy dissipation (region H).

b. *Graphs for the determination of braking conditions*. These conditions occur when the turbopump has a normally decelerated rotation, while the liquid in the pipeline shows a reverse motion towards the turbopump under the influence of the static head, H_{st}.

The graphs shown in Figs. 6.17 and 6.18 assist in finding the pumping heads and the load torques, respectively, for the braking conditions, present in region B (see Fig. 6.14).

The pumping heads and load torques, respectively, for the braking conditions presented in region H (see Fig. 6.14) can be determined by means of the graphs shown in Figs. 6.15 and 6.16, while in the case of braking conditions present in region D (see Fig. 6.14), we may use the graphs shown in Figs. 6.19 and 6.20.

c. *Graphs for the determination of turbine-like conditions*. These conditions occur when a turbpump has a reversley accelerated rotation speed, acting as a turbine, while the liquid in the pipeline is flowing reversely towards the turbopump.

The graphs shown in Fig. 6.19 and 6.20 help to establish the pumping heads and load torques for the conditions present in region C (see Fig. 6.14).

Graphs for the calculation of unprotected pumping installations. We present several graphs used for the determination of the possible minimum and maximum pressure heads, and of the times of flow reversal. They are valid only when the pumping installation is unprotected (except for the discharge tank at the end of the discharge pipeline).

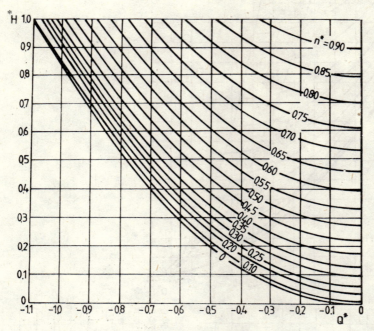

Fig. 6.17. Head-discharge (H^*-Q^*) diagram for the regions of energy dissipation (region B).

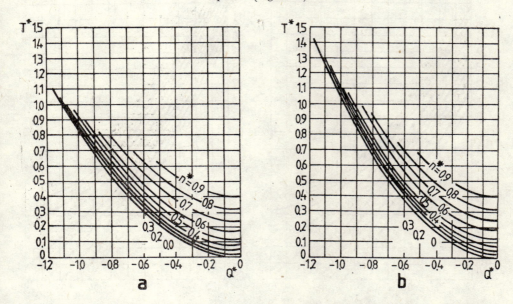

Fig. 6.18. Torque-discharge (T^*-Q^*) diagram for the region of energy dissipation (region B):

a, b — pumps with rated efficiency $\eta_R = 0.8$ and $\eta_R = 0.9$, respectively.

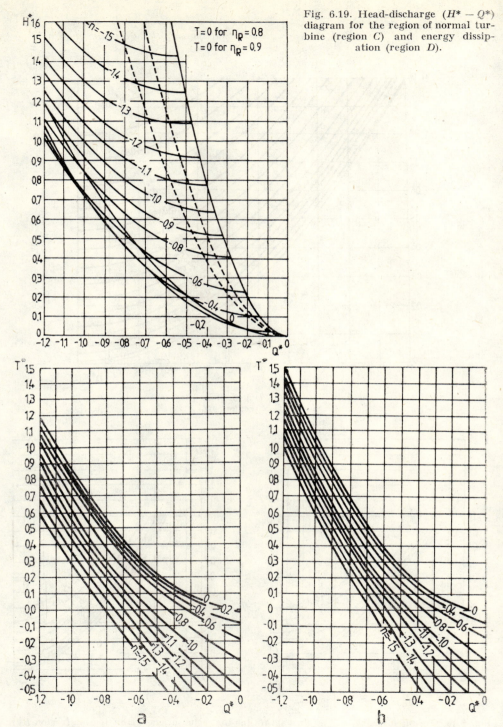

Fig. 6.19. Head-discharge ($H^* - Q^*$) diagram for the region of normal turbine (region C) and energy dissipation (region D).

Fig. 6.20. Torque-discharge ($T^* - Q^*$) for the region of normal turbine (region C) and energy dissipation (region D):

a, b — pumps with rated efficiency $\eta_R = 0.8$ and $\eta_R = 0.9$, respectively.

Approximate Computation Methods Based on Graphs

a. *Graphs for the calculation of minimum pressure heads.* The graphs shown in Fig. 6.21, *a* and *b* are used for the determination of maximum possible pressure heads at the turbopump and at the mid-length of the discharge line, for an unprotected pumping installation. The times at which the minimum pressure heads occur are represented in Fig. 6.21 by different lines:

— solid lines denote the minimum pressure heads produced when the flow rate $Q = 0$;

— dashed lines denote the pressure heads occurring at the reverse flow rate.

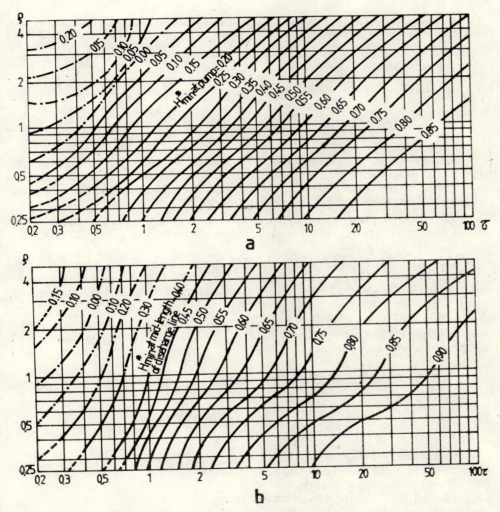

Fig. 6.21. Minimum head (H^*_{min}) at the pump (chart *a*) and at mid-length of discharge line (chart *b*) after power failure.

The graphs shown in Fig. 6.21, *a* and *b* have been drawn on the assumption that the discharge losses are zero. In most cases, however, we should consider these losses. The Fig. 6.21 graph can be used in such cases too, but only if the minimum pressure heads are referred to the static head H_{st} and not to the initial pumping head H_R, and thus, $H^*_{min\,t} = H_{min\,t}/H_{st}$, $H^*_{min\,p} = H_{min\,p}/H_R$, respectively.

In this case, the static pumping head H_{st} can be established by

$$H_{st} = (1 - h_l)H_R \tag{6.58}$$

These approximations are valid only if $H_{min} \geqslant 0$.

Example 6.2. **Computation of minimum pressure heads for a pumping installation with no check valves at the turbopump.** Assuming as known the coefficient $\rho = 2$, the coefficient $\tau = 8$, and the head loss on the discharge pipe $h_{ds} = 0.2$, we want to find the minimum possible pressure heads at the turbopump and at mid-length of the discharge line for the case in which the pumping installation is unprotected.

Solution. From Fig. 6.21, *a*, *b*, for $\rho = 2$ and $\tau = 8$, we get $H^*_{min.t} = 0.45$ and $H^*_{min.p} = 0.70$. Thus

$$H_{min.t} = H^*_{min.t}H_{st} = H^*_{min.t}(1 - 0.2)H_R = 0.45 \times 0.8 H_R = 0.36 H_R$$

$$H_{min.p} = H^*_{min.p}H_{st} = H^*_{min.p}(1 - 0.2)H_R = 0.70 \times 0.8 H_R = 0.56 H_R.$$

b. *Graphs for the calculation of maximum possible pressure head rise.* The graphs shown in Fig. 6.22 and 6.23 help to determine the maximum possible pressure heads at the turbopump and at mid-length of the discharge line for an unprotected pumping installation.

The dashed curves are extrapolations of the numerical data which could not be accurately measured in the region where H_{max} is insignificant for the values of ρ and τ.

The effect of liquid friction in the pipeline on the maximum pressure heads is obvious. If the discharge losses $h_l > 0.2 H_R$, the maximum pressure head, that rises at the turbopump and at mid-length of discharge line, does not exceed the initial pumping head.

When the hydraulic discharge losses $h_l \leqslant 0.2 H_R$, the maximum pressure heads at the turbopump and at mid-length of discharge line can be determined by the equations

$$H_{max.t} = \left[1 + \frac{(0.2 - h_l^*)(H^*_{max.t} - 1)}{0.2}\right] H_R \tag{6.59}$$

$$H_{max.p} = \left[1 + \frac{(0.2 - h_l^*)(H^*_{max.p} - 1)}{0.2}\right]\left(H_R - \frac{1}{2} h_l\right) \tag{6.60}$$

c. *Graphs for the calculation of times of flow reversal.* The graphs shown in Fig. 6.24 help to determine the flow-rate reversing time intervals (measured since power failure at the pump motor and expressed in multiples of the round-trip wave travel time, $\mu = 2\,l/a$, at the turbopump) for an unprotected pumping installation. This graph, unlike others, also considers the hydraulic discharge losses, h_l.

Approximate Computation Methods Based on Graphs

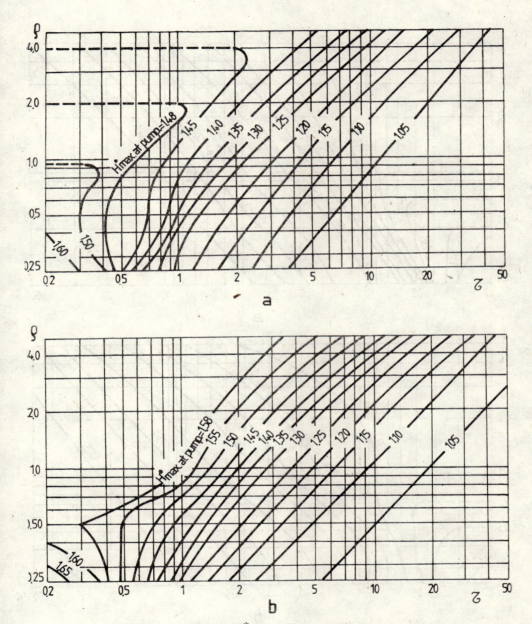

Fig. 6.22. Maximum head (H_{max}^*) at the pump after power failure:
a, b — pumps with rated efficiency, $\eta_R = 0.8$ and $\eta_R = 0.9$, respectively.

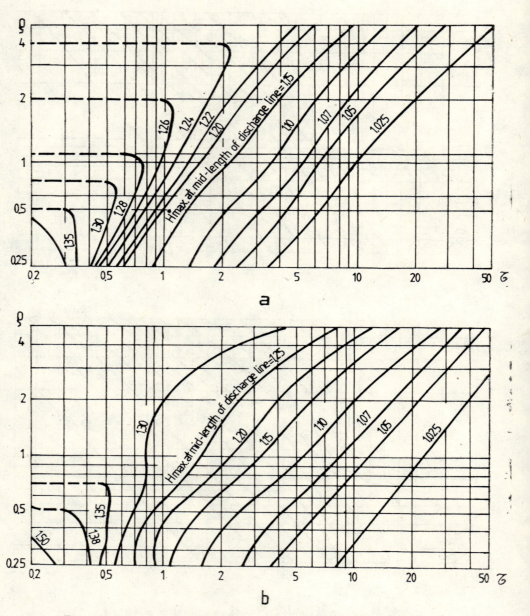

Fig. 6.23. Maximum head during transient at mid-length of discharge line:
a, b — pumps with rated efficiency, $\eta_R = 0.8$ and $\eta_R = 0.9$, respectively.

The time of flow reversal gives indications on the optimum law and optimum closing time of a slow-operating check valve (closing according to a certain programme), located in the discharge hydraulic circuit of each turbopump. This item of information is frequently required when programming valve closure for water hammer control.

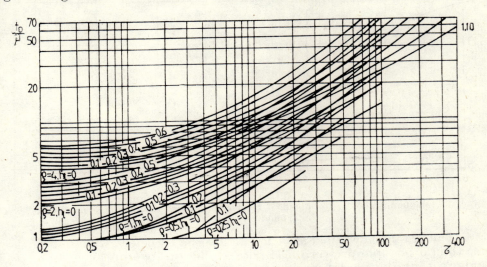

Fig. 6.24. Time of flow reversal t_0, in multiples of $\mu = 2l/a$, from power failure to flow reversal at the pump.

Graphs for the calculation of protected installations. For diminishing the maximum pressure head accompanying the reverse transient flow through turbopumps, adequate measures of prevention and limitation of the reverse-direction rotation should be taken, making use of various protecting devices.

Two graphs are shown below which facilitate estimation of the maximum possible pressure head rise after the use of such devices.

The graph shown in Fig. 6.25 is used for estimation of the maximum pressure heads at turbopumps protected by ratchet mechanisms. Unlike the other graphs, the maximum pressure head is here referred to the static head H_{st} and not to the rated pumping head H_R, so that in this case $H^*_{max.t} = H_{max.t}/H_{st}$. Analysis of this graph reveals that the protection achieved by means of a ratchet mechanism is efficient only when $\rho < 0.5$ and $\tau < 1$.

These are characteristic conditions for pumping installations having turbopumps with small moment of inertia (GD^2) and a pumping system with high head.

The effect of the pipeline liquid on the maximum pressure head rise is important in preventing reverse rotation. Thus, investigations carried out so far [6.9] reveal that when the hydraulic losses in the discharge pipeline $h_l > 0.1\, H_R$, the maximum pressure head rise does not exceed the static head H_{st} of the system, even if $\rho = 0.25$.

283

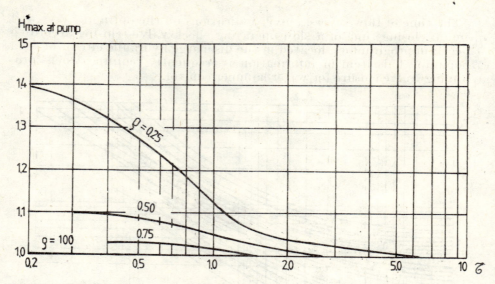

Fig. 6.25. Maximum head at the pump, if reverse rotation of the pump is prevented by a ratchet mechanism.

The graph shown in Fig. 6.25 can also be used for estimation of the maximum possible pressure head rise at the turbopump for a protected pumping installation by means of a two-stage operating check valve (the first stage corresponds to a partial but rapid closing of the reversed flow through the turbopump, while the second stage corresponds to a total but slow closing action).

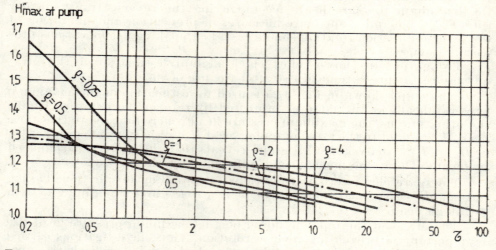

Fig. 6.26. Maximum head at the pump for the case in which the protection check valve has the following two operating states: an optimum opening when the flow $Q = 0$, and a linear closing after reverse flow during a time interval $t = 5\mu$.

Analysis of this graph shows that the protection achieved with a two-stage check valve is efficient when > 0.5 and $\rho > 1$. These conditions are characteristic of a pumping installation having turbopumps with medium moment of inertia (GD^2) and medium pumping head.

The graph shown in Fig. 6.26 has been drawn up on the assumption that the discharge loss is zero. In most cases we should, however, consider these losses, but the maximum pressure head should be referred to the static head H_{st} and not to the rated pumping head H_R, and thus
$$H^*_{max.t} = H_{max.t}/H_{st} \text{ instead of } H^*_{max.t} = H_{max.t}/H_R$$
This graph is valid for the case in which the protection check valve has the following operating stages:
(a) *an optimum opening when the flow rate* $Q = 0$;
(b) *a linear closing after flow reversal in a time interval of* $t = 5\mu$.

6.4 Exact Computation Methods, Based on Analytical Solution of Water Hammer Equations

Several authors [6.2], [6.5], [6.17] have worked out computation methods for accurate determination of the possible minimum and maximum pressure head rises after a water hammer occurrence in general cases, as well as for the analysis of water hammer in pump discharge lines with pumps driven by electric motor. These methods are based on the analytical solution of water hammer equations. The most important ones are the method of gradual elimination of water hammer functions (Allievi method) and the method of solution characteristics of water hammer equations (Schnyder-Bergeron method). We can use these methods when the results, obtained with the approximate (ready) methods, lead to values above the acceptable strain values.

6.4.1 The Method of Gradual Elimination of Water Hammer Functions

This method was worked out by Allievi [6.1] in 1903 as a numerical method. It is based on gradual elimination of functions F and f from equations (6.5) and (6.6), referred to the times $t_1, t_2, \ldots, t_{i-1}, t_i$. Between two consecutive times t_i has the form $t_i = t_{i-1} + \mu$, where $\mu = 2\,l/a$ stands for the reflection time (round-trip wave travel time) in a pipeline of length l.

Allievi proved that functions F and f are interdependent, and thus the determination of function F alone is sufficient. It is also sufficient to determine function F in one point alone (at the discharge valve, for $x = 0$) at various times, which establishes the development of the phenomenon in other points of the pipeline by simple computations. This method can be used for water hammer study during the closing stroke of a valve, and even when this closing stroke is over.

Allievi's method is rather tedious, since it does not allow for incipient approach of water hammer, and does not reflect the influence of the closing stroke on the phenomena taking place in the pipeline.

6.4.2 The Method of Characteristics in Solving Water Hammer Equations

The method of characteristics was elaborated by Bergeron as a graphical method in 1953, and systematically presented in 1960 [6.2]. This improved graphical method is based on interpretation of functions F and f as direct and reflected waves, represented by lines on a pressure-flow rate diagram which are the characteristic curves $H(Q)$ of the two disturbing elements at the two ends of the pipeline. These characteristics are succesively drawn for times delayed by $2l/a$. The method includes consideration of complicated pipeline configurations, head losses, and non-linear variation of cross-section values.

The case where head losses are not considered. The characteristic lines. Let us consider a certain part between sections A and B of a pipeline of length l, where the cross-section ω and the wall thickness e are constant, and where the liquid flows with a mean velocity v_0 (Fig. 6.27). Let us take Q_i as the flow rate and H_i as the pressure head in this section, at time t_i when, for certain reasons, the boundary conditions have not been disturbed.

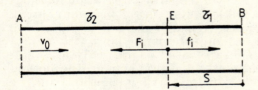

Fig. 6.27. Illustration for the deduction of the equations of water hammer characteristic lines.

The quantities Q_i and H_i are formed due to the intersection of two waves in section E: the first wave F_i travelling in a reverse sense to the flow velocity, that is from E to A, and the second wave f_i moving in the sense of the flow velocity, that is from E to B (see Fig. 6.27).

The equations expressing the relation between the flow rate Q, the pressure head H and functions F and f in any section of the pipeline, including section E, are (6.5) and (6.6) which are given below in simplified form

$$H - H_0 = F + f \tag{6.61}$$

$$m(Q_0 - Q) = F - f \tag{6.62}$$

These two equations lead to the determination of quantities H_i and Q_i in section E at time t_i, given the functions F and f. Since these functions are varying with time, an observer, fixed in section E, can by no means establish a relation between Q and H.

When we assume that the observer is moving with the wave speed along the pipeline in the reverse sense of the flow velocity v_0, that is following wave F, then for him the function F would have a constant value F_i, equal to that corresponding to section E. It follows that the observer, moving with wave F, sees only the quantities H, Q and f as variables.

When the observer is moving from E to A with a wave speed a, wave F_i would seem constant to him, and thus equations, (6.61) and (6.62) become

$$H - H_0 = F_i + f \tag{6.63}$$

$$m(Q_0 - Q) = F_i - f. \tag{6.64}$$

By eliminating function f from these equations, we obtain

$$H - H_0 + m(Q_0 - Q) = 2F_i = \text{constant} \tag{6.65}$$

This equation is valid for any section of the pipeline for an observer moving with the wave F, and is thus valid for section E. Hence

$$H_i - H_0 + m(Q_0 - Q_i) = 2F_i = \text{constant} \tag{6.66}$$

Subtracting equation (6.66) from (6.65), we get

$$H - H_i = m(Q - Q_i) \tag{6.67}$$

This equation represents *a characteristic line PT* (Fig. 6.28) in the $Q - H$ plane for an observer moving with wave F. This line passes, at time t_i, through point E_i having co-ordinates Q_i, H_i, and has the angular coefficient $+ m = + \dfrac{a}{g\omega}$

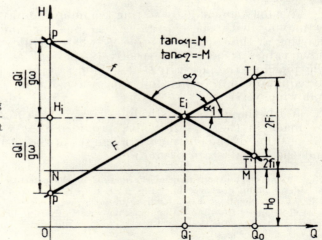

Fig. 6.28. Diagram showing the water hammer characteristic lines (head losses are not taken into account).

If we assume an observer moving along the pipeline with a speed a, in the sense of the flow velocity, that is following wave f, then for him function f would have a constant value f_i, equal to that corresponding to section E. It follows that the observer moving with wave f will consider as variables only the quantities H, Q and F.

When the observer is moving from E to B with a speed a, wave f_i appears to him as constant and, as a consequence, equations (6.61) and (6.62) become

$$H - H_0 = F + f_i \tag{6.68}$$

$$m(Q_0 - Q) = F - f_i \tag{6.69}$$

By eliminating function F from these two equations, we get

$$H - H_0 - m(Q_0 - Q) = 2f_i = \text{constant} \tag{6.70}$$

This equation is valid for any section of the pipeline for an observer moving with wave F and in particular for section E. Then we have

$$H_i - H_0 - m(Q_0 - Q_i) = 2f_i = \text{constant} \tag{6.71}$$

Subtracting equation (6.71) from (6.70) we get

$$H - H_i = -m(Q - Q_i) \tag{6.72}$$

which represents a characteristic line PT' (see Fig. 6.28) in the $Q - H$ plane, for an observer moving with wave f. This line passes, at time t_i through point E_i with the co-ordinates Q_i, H_i, and has the angular coefficient $-m = -\dfrac{a}{g\omega}$.

The lines F and f are symmetric about the horizontal line passing through point E_i.

We may conclude by stating the following physical law. If an observer is moving with speed a along the pipeline and follows wave F or f, starting at time t_i from section E where the flow rate and pressure head have values Q_i and H_i respectively, he will see that, in any section of the pipeline, the flow rate Q and the pressure head H are linearly related by equation (6.67) when following wave F, and by equation (6.72) when following wave f.

With the characteristic lines (6.67) and (6.72), all computations can be made in the $Q - H$ plane, provided that we know at each time t the position of the characteristic curves of the system at the end sections, and the initial state of motion.

Example 6.3. **Application of the method of characteristics to the pump discharge line.** Consider a pump discharge line (Fig. 6.29) for which we know the imaginary point B_t (Fig. 6.29, b) corresponding to the flow rate Q and pressure head H at time t in section B, which is at a distance l from the turbopump. We wish to find the imaginary point A, corresponding to section A at the turbopump.

Solution. Let us take DE as the turbopump characteristic at time $t + l/c$. The desired point will be somewhere on this curve. We consider an observer starting from B at time t, who follows the wave F (the positive sense of the velocity being from turbopump to the tank). The characteristic line corresponding to this observer passes through the point B_t and has the angular coefficient $\tan \alpha = +m$ (line F). The crossing point of this line with the characteristic curve of the pump gives us the imaginary point $A_{t + \dfrac{l}{c}}$.

Example 6.4. **Application of the Schnyder-Bergeron method to water hammer computation after power failure at the pump motor.** After cutting off the power supply to the pump motor, the turbopump passes through all of the three characteristic conditions of operation, due to its own inertia and to that of the liquid column in the pipeline :

— the pumping condition, similar to the condition before the power supply cut-off but with a slower rotation speed ;

— the braking condition, when the liquid in the pipeline starts a reverse motion towards the turbopump due to the static head H_{st}, while the turbopump, due to its own inertia (flywheel effect) GD^2, goes on rotating in the positive sense ;

— the turbine-like condition when, due to the weight of the column of liquid with head H, the turbopump is reversely rotating in the negative sense and is acting as a turbine.

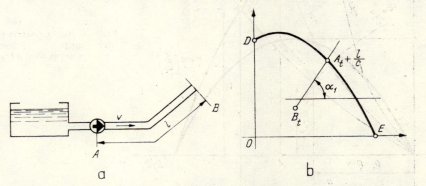

Fig. 6.29. Application of the method of characteristics to the discharge pipe of a pump :
a — installation view; b — determination of imaginary point.

Solution. The first problem to be solved in water hammer computation in discharge lines is the determination of turbopump characteristic curves at various times, which are then crossed by the pipeline characteristic lines. We can do exact computations when we know the turbopump topogram (complete characteristic) in all the three operating conditions, in the case of missing check valves, or in the pumping condition only in the case where a check valve is fitted.

When we do not have this topogram but know the turbopump characteristic curve for fixed speed ($n = n_0$), then the turbopump stopping curve may be approximately drawn, on the basis of the relations $Q/Q_0 = n/n_0$, $H/H_0 = n^2/n_0^2$, and $P/P_0 = n^3/n_0^3$ resulting from the similarity conditions, and that the curves $\eta_0 =$ constant are parabolae passing through the origin of axes Q, H (Fig. 6.30).

For a given value, the position of point Q, H is located on the parabola. To this end, we have only to multiply by α the abscissa ρ of point S, or by α^2 the ordinate x of point S, thus obtaining point S_k. We then pass to another parabola and find the locus of points S_k, which stand for the curve QH for a given α.

When we have only the value $\eta = \eta_0$ corresponding to the operating flow rate, we take $\eta = \eta_0$ as constant throughout the $Q-H$ plane and make use of relations :

$$Q/Q_0 = n/n_0 \; ; \; H/H_0 = n^2/n_0^2 \; ; \; P/P_0 = n^3/n_0^3.$$

After power failure, the motion of the rotating parts of the turbopump is described by the differential equation

$$I \frac{d\omega}{dt} = -T_r = -\frac{P}{\omega_0}$$

After integration, it becomes for the latter case

$$n = \frac{n_0}{1 + n_0 k_2 t}$$

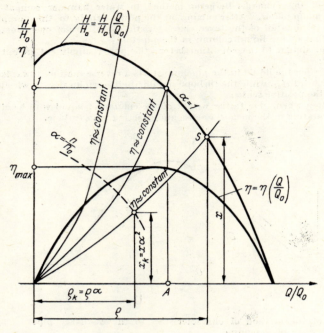

Fig. 6.30. Construction of turbopump characteristic curves at $Q \neq Q_0$

where $k_2 = \dfrac{900\, P_0}{\pi^2 I n_0^3}$;

n_0 = rated pump speed;
P_0 = rated pump power.

When the stopping curve has been drawn (Fig. 6.31, b), we choose the time unit, usually $\mu \leqslant \tau/2$ where $\tau = 2l/a$ is the reflection time (round-trip wave travel time). Taking for t (in the last equation) the values $\mu, 2\mu, 3\mu$ etc. we may determine the corresponding n values and then draw the characteristic curves $H = f(Q)$ for the n values found in this way.

Figure 6.31, c shows the graphical computation of water hammer in the conditions shown in Fig. 6.31, a. The time unit has been taken as $\mu = \tau/2$ and, making use of the pump stopping curve, we may draw the characteristic curves $H - Q$ of the pump corresponding to the times μ, 2μ, 3μ, etc. P denotes the pipeline cross-section at the pump and T the pipeline cross-section at the tank. The imaginary points P_0, T_0, T_1 in the QH plane correspond to the initial values Q_0 and H_0. The imaginary point P_2 is obtained following an observer starting from section T at time 1, and reaching section P at time 2. Thus, point P_2 is located at the intersection of line F that passes through T_1, with the pump characteristic curve corresponding to time 2. The imaginary point T_3 is located at the intersection of line f passing through P_2, with the line $H = H_0$, because if the discharge line were horizontal, the section T would be in T' and the tank would have the head H_0.

At time $t = 8\mu$ the flow rate in section P has an almost zero value ($Q=0$) and the check valve is closing. Due to the check valve, negative flow rate will occur only in the pipeline near this valve. In section P, the characteristic is given by the line $Q = 0$. Thus, the points P_i for $i \geqslant 8$ will be found on the $Q = 0$ characteristic.

Since we did not consider the head losses, the pressure oscillations and the flow rate oscillations are not dying out and the imaginary points T_9, P_{10}, T_{11}, P_{12}, T_{13} ... are located in the corners of a rhomb. The imaginary points P_1, T_2, P_3 can be found in the same way, the imaginary point P_1 being at the intersection of the characteristic line F that passes through T_0, with the $H-Q$ curve corresponding to the speed n_1.

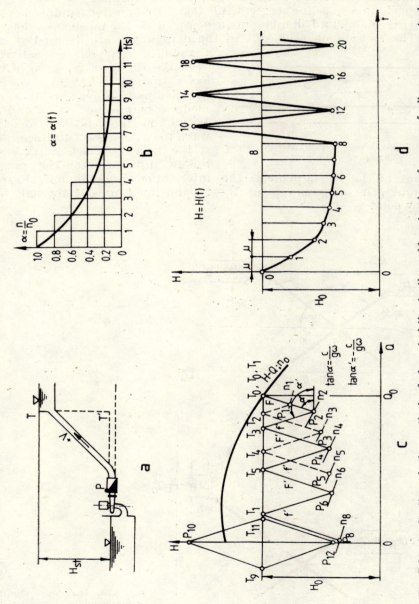

Fig. 6.31. Application of the method of characteristics to the sudden stopping of a pump feeding an open tank (without considering head losses):

a — installation view; b — pump stopping curve; c — water hammer computational diagram; d — diagram of pressure head variation with time.

The case when the head losses are considered. In this case, the linear head losses are replaced by equivalent local head losses by some imaginary diaphragms located in some sections of the pipeline.

Let us take a pipeline segment AB along which water hammer is propagated (Fig. 6.32). Two diaphragms are placed in the pipeline sections A and B. The sections on each side of the diaphragms are denoted A and a, and B and b, respectively. We take the positive sense of velocity from A to B.

Let us consider C as an intermediate part of sections a and B. If t_1 is the time needed for the wave to cover the aC distance, and t_2 the time in which it traverses the BC distance, then the imaginary point in the H–Q plane is the intersection point of line F' passing through the imaginary point $B_i - t_1$, and line f, passing through the imaginary point $a_i - t_2$.

Fig. 6.32. Head losses considered in water hammer study.

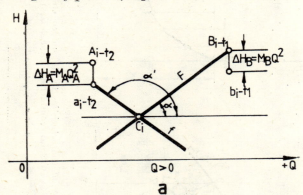

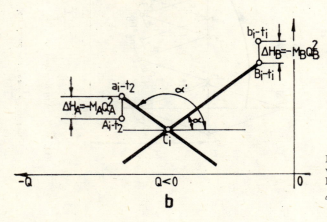

Fig. 6.33. Diagrams showing water hammer characteristic lines (head losses considered):

a — for the case when $Q > 0$; b — for the case in which $Q < 0$.

Since sections A and a are separated by an infinitely small distance, the pressure head wave will be simultaneously in section A and in section a, and the flow rate, associated with this wave, will have the same value in both sections A and a. In these two sections, only the pressure head is different, having the value $\Delta H_A = M_A Q_A^2$, where M_A is the diaphragm resistance modulus, equivalent to the resistance modulus of a pipeline region of length l_A.

It follows that the imaginary point a_{i-t_2} is located on the same vertical line as the imaginary point a_{i-t_2}, at a distance $\Delta H_A = M_A Q_A^2$ above the point $a_i - t_2$ when $Q > 0$ (Fig. 6.33,b). On the other hand, it also follows that the imaginary point $b_i - t_1$ is located on the same vertical line as the imaginary point $B_i - t_i$, at a distance $\Delta H_B = M_B Q_B^2$ beneath the point $B_i - t_i$ if $Q > 0$ (Fig. 6.30,a) and above the point $B_i - t_i$ if $Q < 0$ (Fig. 6.33,b).

Example 6.5. Water hammer computation for a pumping installation, considering the head losses. We resume the application 6.4, considering the head losses on the discharge pipeline and equating them to the local head losses produced on a diaphragm, located at the downstream end of the discharge pipeline (Fig. 6.34).

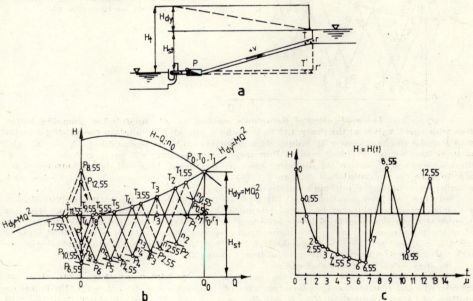

Fig. 6.34. Application of the method of characteristics to the sudden stopping of a pump feeding an open tank (head losses considered).

a — installation view; b — water hammer computational diagram; c — diagram of pressure head variation with time.

The imaginary points T_1 are on the curve of head losses, while the imaginary points r_i are on the $h = H_0$ line. It is worth mentioning that for getting the minimum and maximum pressure heads in section P, we considered the wave starting from point T at time $t = -0.45\mu$ ($R_{-0.45}$) and reaching the point P at time $t = 0.55$ ($P_{+0.55}$).

Solution. The graph of pressure head variation is drawn for section P. Given the pressure head and the flow rates in the points T and P, we can easily establish the pressure head and flow rates in any intermediate point X of the pipeline. For instance, if X is at distance x from P and $(1-x)$ from T, the imaginary point X_i is determined by cutting across the line F, starting from T at the time $i - \left(1 - \dfrac{x}{l}\right)$, with the line f starting from P at the time $i - \dfrac{x}{l}$.

For $\dfrac{x}{l} = 0.6$ in Fig. 6.35 the imaginary points $X_{2,4}$ were determined at the intersection of line F from T_2, with the line f, from $P_{1,8}$, and the imaginary point $X_{3,4}$, at the intersection of the line F from T_4, with the line f from $P_{3,8}$ etc.

When the pump stopping time is very short as related to the reflection time $\tau = 2l/a$, small pressure heads (underpressures) or even cavitation may occur in the pipeline. These phenomena may also occur when the discharge line shows sections with maximum points.

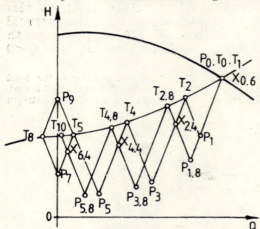

Fig. 6.35. Determination of imaginary points for Example 6.5.

Example 6.6. **The application of Bergeron's method to an unprotected pumping installation with check valves at the pump.** Let us consider a pumping installation comprising a turbopump P and a discharge line with unique characteristic feeding an open tank T. A check valve has been inserted for avoiding reverse flow into the turbopump discharge circuit. Since the hydraulic losses produced on the discharge lines are significant, they will also be considered. We desire to draw the computational diagram vH, making use of the method of characteristics, and assuming an unprotected pumping installation.

Solution. We have denoted by P the pipeline cross-section at the turbopump and by T the pipeline cross-section near the tank. The linear head losses occurring in the discharge line are equivalent to the local head losses produced by a diaphragm, located at the downstream end of the pipe (near the tank). The turbopump stopping phase lasts for a very short lapse of time $\mu \leqslant \tau/2$, so that the first wave starting from the tank finds the pump stopped and the check valve closed, which prevents reverse flow through the pump.

We may meet two particular cases:

(1) *The case in which vacuum is not created at the turbopump* (Fig. 6.36). Here the characteristic line reaches the boundary condition of the check valve, $Q = 0$, before the boundary condition of vacuum formation $H = H_v$. In this case, a primary wave F starts from the tank in the opposite direction of flow velocity (the characteristic line with the line slope $\alpha = +a/g\omega$) at initial time P_0 and reaches the pump at time P_1. Since the discharge valve is closed, a secondary wave f is reflected towards the tank in the direction of flow velocity (the slope line $\alpha' = -a/g\omega$) and reaches the tank at time $T_{1.5}$ for further reflection between the sections P and T. The imaginary points T are located on the characteristic curve of the system. Since we assumed hydraulic losses, the pressure and flow-rate oscillations gradually fade, and hence the imaginary points P_i and T_i progressively approach the point D.

(2) *The case in which vacuum is created at the turbopump* (Fig. 6.37), that is when the characteristic line reaches the vacuum formation boundary condition $H = H_v$ before the boundary condition of the check valve $Q=0$. In this case, the pressure head drop, at time P_1, produced vacuum at the turbopump, i.e. has broken the column of liquid. The water quantity, flowing through the turbopump, represented by the boundary condition $H(Q)$, cannot prevent

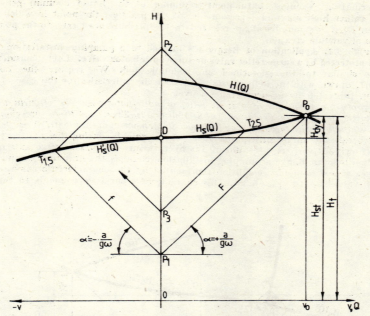

Fig. 6.36. Application of water hammer computational diagram to a pumping installation with check valve at the pump and without protection (the case in which vacuum is not created near the pump).

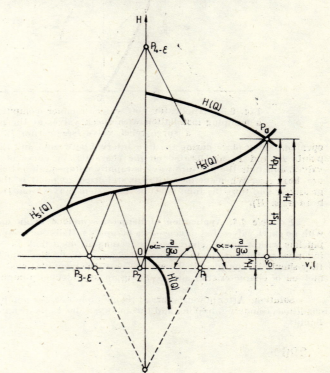

Fig. 6.37. Application of water hammer computational diagram to a pumping installation with check valve but without protection (the case in which vacuum is created at the pump).

vacuum formation. Vacuum confinement (rejoining of the liquid column) generates waves leading to rather high maximum pressure heads. In practice, the point in which the vacuum is confined, $P_{3-\varepsilon}$, is symmetrical (as referred to the pressure head axis) to the point P_1, which is the time of vacuum creation.

Example 6.7. **Application of Bergeron's method to a pumping installation with check valves and protected by a surge relief valve or surge anticipator valve.** Let us consider a pumping installation similar to that described in Example 6.6. We require the computational diagram vH, drawn with the help of the method of characteristics for the case of a pumping installation, protected by a surge relief valve.

The protection by means of a surge relief or anticipator valve is efficient when $\rho < 0.5$ and $\tau < 1$ (the notation is explained in Section 6.3.2). These conditions are characteristic of pumping installations with turbopumps which have small moment of inertia (GD^2) and long pipelines.

In general, the surge anticipator valve opens quickly during the underpressure stage, and closes slowly, at various rates, during the overpressure stage. The surge anticipator valve, fitted to the discharge line of a pumping installation, changes the graphical computation, since it introduces a new boundary condition represented by the characteristic curve of the surge anticipator valve $H_{rv}(Q)$ (Fig. 6.38). This characteristic curve corresponds to the maximum

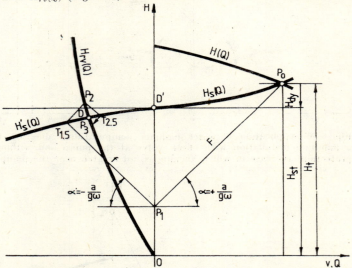

Fig. 6.38. Application of water hammer computational diagram to a pumping installation with check valves at the pump, protected by a surge relief valve (see section 7.3.5).

opening, taking place during the time interval between μ and 1.5 μ. In this case, the imaginary points P_2 and P_3 are located on the characteristic curve of the surge relief or anticipator valves, and thus the pressure wave is rapidly damped, almost completely preventing the pressure head rise. When the liquid flow is stabilized at point D, the surge relief valve closes progressively and thus the final point of the system moves from D to D' (located on the pressure head axis, H).

Example 6.8. **Application of Bergeron's method to an unprotected pumping installation with no check valves.** Let us consider a pumping installation compresing a pump P and a discharge pipeline with a unique characteristic feeding an open tank T. No valve has been fitted, allowing reverse flow on the turbopump discharge circuit. Since the hydraulic losses are significant, they should be considered. We wish to draw the computational diagram vH, making use of the method of characteristics and assuming an unprotected pumping installation.

Solution. After power failure at the pump motor, the pump shows all of the three operating conditions (pumping, braking and turbine-like), due to its own inertia and that of the column of liquid.

Exact Computation Methods Based on Analytical Solution

During the initial pumping condition the turbopump characteristic curve $H(Q)$ and that of the system is $H_s(Q)$, while the initial driving point is P_0. In the final turbine-like condition the turbopump characteristic curve is $H(Q)$, that of the system is $H_s(Q)$ and the final balance point is D (Fig. 6.39).

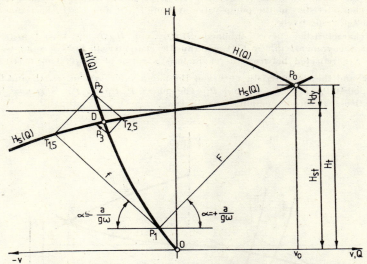

Fig. 6.39. Application of water hammer computational diagram to a pumping installation without check valve at the pump and without protection.

The imaginary points P_i are, at different times, on the characteristic curve $H_s(Q)$.

With no check valve, the boundary condition of this valve $Q = 0$ changes into the boundary conditions of the reverse flow through the pump (since the pump acts as a local hydraulic resistance) represented by the characteristic curve of the pump $H(Q)$, from which the pipeline characteristic lines are reflected. The oscillation damping arises around a negative flow-rate point, which shows that this flow rate is still flowing through the pump until the liquid has left the pipeline, without discharging the tank at the end of the discharge line.

Example 6.9. Application of Bergeron's method to a pumping installation with no check valves and protected by a slow-closing valve. Let us consider a pumping installation similar to that described in example 6.8. We take the turbopump speed in the turbine-like condition as exceeding the allowed critical value, and thus leading to turbopump deterioration. In this case, a slow-closing valve should be fitted on the discharge circuit of each turbopump. We require the computational diagram vH, making use of the method of characteristics and assuming a pumping installation protected by means of a slow-closing valve.

Solution. Figure 6.40 shows the characteristic curve $H(Q)$ of the turbopump under the turbine-like condition, and the characteristic curves of the valve, $H_{cv1}(Q)$ and $H_{cv2}(Q)$, corresponding to the times μ and 2μ, respectiv-

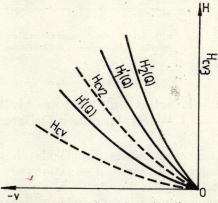

Fig. 6.40. Building of the characteristic curves for the pump-slow-closing valve system (the case of a pumping installation without check valve at the pump but protected by a slow-closing valve).

ely. At time 3μ the valve is completely closed, and thus its curve $H_{cv}(Q)$ is superposed on the pressure head axis.

For creating new boundary conditions, we combine the characteristic curve of the pump $H(Q)$ with the characteristic curves of the valve $H_{cv1}(Q)$ and $H_{cv2}(Q)$, respectively. We thus obtain the characteristics of the pump-valve system, $H_1(Q)$ and $H_2(Q)$, corresponding to the times μ and 2μ, respectively.

The characteristic curves obtained, $H_1(Q)$ and $H_2(Q)$, are then transferred on the computational diagram vH in Fig. 6.41, and thus the characteristic lines F and f can be drawn. These lines are reflected between the characteristic curve of the system $H_s(Q)$ and $H'_s(Q)$, respectively, and the characteristic curves of the pump-valve system, $H'_1(Q)$ and $H'_2(Q)$ in the imaginary points P_0, P_1, $T_{1,5}$, P_2 and P_3. The points $P_0, P_1, P_2, P_3 \ldots D$ give the pressure heads in section P at various computational times.

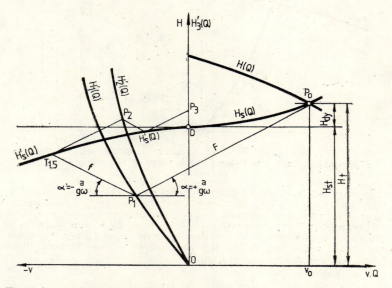

Fig. 6.41. Application of water hammer computational diagram to a pumping installation without check valve at the pump but protected by a slow-closing valve.

6.5 Exact Computation Method, Based on Direct Integration of Water Hammer Differential Equations

The computation methods based on direct integration of water hammer differential equations are (1) the physical (sonic) waves method, and (2) the computational wave method. Both lead to extremely accurate results, but are very tedious. This disadvantage can be removed by using computer techniques with programs allowing for rapid computation of the minimum and maximum pressure heads occurring as a consequence of water hammer.

6.5.1 Physical (sonic) waves method

Given the oscillatory character of water hammer, this method facilitates the computation of flow rate and pressure variations with time by adding together the different waves propagating along the pipeline. Since the waves produced directly by the change of boundary conditions, and others reflected and refracted, are all propagated through the pipeline, the number of waves to be compounded is in practice so large that this method can only be applied in very simple cases.

The physical waves method consists in superposing the effects of various waves passing through a pipeline section

$$H = H_0 + \Sigma h_a + z h_i = H_0 + m g_a - m g_i \tag{6.73}$$

$$Q = Q_0 + g_a + g_i \tag{6.74}$$

where H_0 is the initial pumping head;
Q_0 — the initial flow rate;
h_a, h_i — the direct and indirect pressure waves that passed through the respective section up to the computation time;
q_a, q_i — the direct and indirect flow-rate waves that passed through the respective section up to the computation time.

The physical waves method can be used for simple and rapid computations, as well as for better understanding of the development of the water hammer phenomenon and of the way in which various devices change the character of water hammer.

6.5.2 Computational waves method

This method was worked out by Riemann in 1860 for the general case of gas streams. He established the invariance properties of some expressions including state parameters that do not explicitly contain the physical waves. The method was applied to the particular case of water hammer by Schneyder [6.17] in 1929 and then by Bergeron [6.2] in 1935, the latter also making a general presentation of the method and putting it on a systematic basis. It uses a mathematical representation of the phenomenon of unsteady motion by means of invariant equations estimated at a series of positions, corresponding to the laws of motion of the direct or backward waves. These equations are called *computational waves* or *Riemann invariances*.

The computational waves facilitate writing of systems of equations made up of a number of equations equal to the number of unknowns.

The computational waves method is in fact the method of wave characteristics for partial differential equations of hyperbolic type.

Determination of general equations. Starting from the physical (sonic) waves in equations (6.73) and (6.74), multiplying the first by the wave drag m, and substituting in the second the pressure waves of equation (6.9), we get

$$mQ = mQ_0 + \Sigma mq_a + \Sigma mq_b$$
$$H = H_0 + \Sigma mq_a - \Sigma mq_b$$

By adding and subtracting these equations we obtain

$$H + mQ = (H_0 + mQ_0) + 2\Sigma mq_a \tag{6.75}$$
$$H - mQ = (H_0 - mQ_0) + 2\Sigma mq_b \tag{6.76}$$

The left-hand sides of these two equations do not include waves or wave functions, but the state parameters Q and H and the resistance coefficient m. The right-hand sides include wave elements of only one type, namely direct waves q_a in equation (6.75) and backward waves q_b in equation (6.76). This important result can be summed up as follows:

— the expression $H + mQ$ has a direct wave character, that is, its value $(H + mQ)M, t_0$ computed at a point M (Fig. 6.42, a) at time t_0 remains the same in various sections of the pipeline when they are related by means of the direct waves propagation law

$$\omega = \omega_0 + a(t - t_0)$$

— the expression $H - mQ$ has a backward wave character, that is its value remains invariant in the co-ordinate sections related by the backward waves propagation law

$$\omega = \omega_0 - a(t - t_0)$$

The expressions $H + mQ$ and $H - mQ$ are computational waves or Riemann invariances. Riemann denoted them by R and S, respectively, and we preserve this notation here.

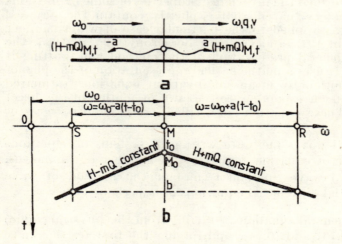

Fig. 6.42. Propagation of the computational waves (Riemann invariances):
a — through the pipe; b — in the physical plane.

Their properties are illustrated in Fig. 6.42. The following forms, equivalent to equations (6.75) and (6.76), may also be written

$$H + mQ = C_1 + F(\omega - at) ; (H + mQ)_{M,t_0} = (H + mQ)_{R,t} \tag{6.77}$$

$$H - mQ = C_2 + f(\omega + at) ; (H - mQ)_{M,t_0} = (H - mQ)_{R,t} \tag{6.78}$$

The relations including $H + mQ$ also demonstrate the above-mentioned property: the value of this expression computed at point M at time t_0 is to be found at times t at the point R of co-ordinates

$$\omega = \omega_0 + a(t - t_0).$$

A similar property is also found for the relation $H - mQ$.

Thus we are not speaking of physical wave propagation (that is, variations of the flow rate and pressure) but of the propagation of some mathematical expressions including the flow rate and pressure as global values.

Since this method makes use of global values of parameters and not of their variations, it is much more practical. In addition, since various boundary conditions are usually expressed as relations between global values of parameters, this method may also be used in cases in which the pipeline shows intricate boundary conditions, such as pumps with programmed closing valves, hydraulic accumulators, etc.

Most advantage of this method may be taken by using the finite differences applied to numerical, graphical or mixed forms. The solution by the method of finite differences consists of replacing the differential equations of motion with finite differences and solving them by specific methods [6.3] and [6.22].

Application of equations to nodes. In order to apply the computational waves method, the pipeline should be divided into sections. The points separating the pipeline sections are called *nodes*. The values of the desired parameters are obtained only at nodes and at times representing multiples of the computation time. Two types of nodes generally arise from division of the pipeline into sections: inner nodes and end nodes.

(a) *Inner nodes*. These nodes, located along the pumping pipeline, may be simple or with diaphragm.

(a.1) *Simple nodes*. These are used when the wave travel time $(2l/a)$ for the whole pipeline is too long and we must divide the pipeline into several sections.

Given the flow rates and pressure heads at time j for all the nodes of the pumping pipe, the conditions at a certain node k, at the time $j + 1$ (where $\Delta t = 1$ is the necessary time for traversing one section) could be computed with equations (6.77) and (6.78) using the notation from Fig. 6.43. Thus, we get for the direct wave

$$H_{k,j+1} + m_k Q_{k,j+1} = H_{k-1,j} + m_k Q_{k-1,j} = R''_{k-1,j} \tag{6.79}$$

and for the backward wave

$$H_{k,j+1} - m_{k+1}Q_{k,j+1} = H_{k+1,j} - m_{k+1}Q_{k+1,j} = S'_{k+1,j} \qquad (6.80)$$

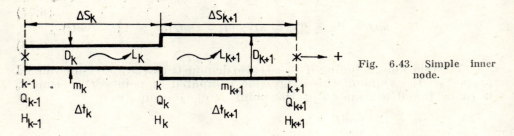

Fig. 6.43. Simple inner node.

In these equations the first index denotes the node, and the second, the time. Two equations with two unknowns: $H_{k,j+1}$ and $Q_{k,j+1}$ result, that is the pressure head and flow rate at the node k at time $j+1$.

(a.2) *Nodes with diaphragm.* These are mostly used for including the effect of hydraulic losses. In this case equations (6.79) and (6.80) are also valid, but the parameters will generally be different before and after the node.

Denoting the parameters before the node by primes, and those after the nodes by double primes, and considering the diaphragm resistance modulus M_k (Fig. 6.44), the two equations become

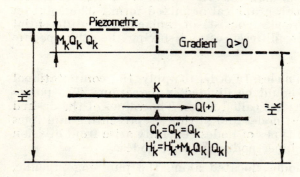

Fig. 6.44. Inner node with diaphragm.

$$Q'_{k,j+1} = Q''_{k,j+1} = Q_{k,j+1} \qquad (6.81)$$

$$H''_{k,j+1} = H'_{k,j+1} - M_k Q_{k,j+1} \times |Q_{k,j+1}| \qquad (6.82)$$

These equations may also be used when the pipeline has a real local resistance, such as rapid or slow-closing valves, check valves, etc. Depending on the case, M_k may be constant or vary in accordance with a known law.

(b) *End nodes.* These nodes are located at the beginning and at the end of the pumping pipeline. Depending on its place we have only one wave equation, for a direct or a backward wave. The remaining equations required for closing the system are obtained from the boundary conditions and vary from case to case.

The following devices may be located in the end nodes of a pumping pipeline:

(b.1) *Turbopump*. Water hammer may occur when the power supply to the turbopumps motor is suddenly cut off.

Assuming that the turbopump is located at the last node, we use the equation given by the direct wave:

$$n_1 = n_0 - \frac{1{,}789 g \Delta t \varrho g}{\pi^2 G D^2 \eta} \left(\frac{Q_0 H_0}{n_0} + \frac{Q_1 H_1}{n_1} \right) \tag{6.83}$$

Assuming the turbopump has pumping head H dependent on the flow rate Q and the speed n, and that its rotating parts have a total inertia moment $J = GD^2/4g$, we may write

$$H_{k,j+1} + m_k Q_{k,j+1} = R''_{k-1,j} \tag{6.84}$$

$$H_{k,j+1} = H_a + H_{j+1} \tag{6.85}$$

$$Q_{k,j+1} = Q_{j+1} \tag{6.86}$$

The quantities H, Q and n are related to each other by the turbopump complete (universal) characteristic (topogram)

$$H_{j+1} = f(Q_{j+1}, n_{j+1}) \tag{6.87}$$

The rotation speed at time $j+1$ can be related to the speed n at time j by means of the turbopump equation of motion for its rotating parts

$$J = \frac{d\omega}{dt} = T_m - T \tag{6.88}$$

where J is $GD^2/4g$ the moment of inertia in $daN \cdot m \cdot s^2$;
Ω — angular speed, in rad/sec;
n — pump speed, in rev/min;
T_m — the motor driving torque, in daN.m;
T — pump load torque, in $daN \cdot m$

Writing this equation in finite differences, we get

$$n_{j-1} = n_j + \frac{60 g \Delta t}{\pi G D^2} [(T_{m,j} + T_{m,j+1}) - (T_j + T_{j+1})] \tag{6.89}$$

The load torque is also a function of the flow rate and speed, so that

$$T_{j+1} = f(Q_{j+1}, n_{j+1}) \tag{6.90}$$

For accurate computations we need the complete (universal) characteristic of the pump, both as regards the pumping head H and the load torque T, as functions of the flow rate and speed. When the pump universal characteristic is not available, we may use Table 6.5.

TABLE 6.5 a **Performance characteristics** $H - Q - n$, predicted for a pump with specific

$\dfrac{n}{n_R}$	−2.0	−1.8	−1.6	−1.4	−1.2	−1.0	−0.8	−0.6	−0.4	−0.2
−2.0	4.50	4.60	4.70	4.75	4.80	4.85	4.95	5.00	5.05	5.10
−1.8	3.40	3.50	3.60	3.65	3.70	3.75	3.85	3.90	3.95	4.00
−1.6	2.20	2.30	2.40	2.50	2.60	2.80	3.00	3.10	3.15	3.20
−1.4	1.12	1.40	1.65	1.88	2.06	2.20	2.35	2.40	2.43	2.45
−1.2	0.32	0.70	1.00	1.25	1.40	1.55	1.65	1.73	1.77	1.78
−1.0	−0.23	0.03	0.35	0.60	0.85	1.00	1.10	1.20	1.24	1.27
−0.8	−0.68	−0.40	−0.17	0.08	0.32	0.50	0.62	0.72	0.78	0.83
−0.6	−1.00	−0.75	−0.50	−0.30	−0.10	0.10	0.23	0.35	0.42	0.47
−0.4	−1.30	−1.15	−0.83	−0.50	−0.35	−0.18	−0.04	0.07	0.15	0.19
−0.2	−1.70	−1.55	−1.25	−0.85	−0.60	−0.40	−0.23	−0.10	−0.01	0.03
0.0	−2.30	−2.00	−1.55	−1.10	−0.80	−0.57	−0.37	−0.20	−0.10	−0.05
0.2	−2.40	−2.30	−1.70	−1.25	−0.90	−0.65	−0.44	−0.27	−0.12	−0.03
0.4	−2.50	−2.40	−1.75	−1.35	−1.05	−0.75	−0.48	−0.27	−0.09	0.04
0.6	−2.55	−2.50	−1.80	−1.45	−1.15	−0.80	−0.50	−0.24	−0.02	0.13
0.8	−2.60	−2.55	−1.90	−1.55	−1.20	−0.85	−0.50	−0.18	0.10	0.30
1.0	−2.90	−2.60	−2.10	−1.70	−1.30	−0.88	−0.47	−0.10	0.20	0.45
1.2	−3.20	−2.90	−2.40	−1.90	−1.40	−0.90	−0.33	0.10	0.50	0.75
1.4	−3.60	−3.20	−2.70	−2.10	−1.45	−0.80	−0.20	0.40	0.75	1.05
1.6	−3.90	−3.60	−3.00	−2.25	−1.40	−0.60	0.10	0.70	1.00	1.35
1.8	−3.95	−3.90	−2.90	−1.90	−1.10	−0.30	0.50	1.00	1.30	1.65
2.0	−4.00	−3.70	−2.50	−1.60	−0.90	−0.20	0.90	1.30	1.50	1.70

TABLE 6.5 b **Performance characteristics** $H - Q - n$, predicted for a pump with specific

$\dfrac{n}{n_R}$	−2.0	−1.8	−0.6	−1.4	−1.2	−1.0	−0.8	−0.6	−0.4	Q/Q_R −0.2
										H/H_R
−2.0	4.20	4.40	4.50	4.55	4.60	4.65	4.70	4.75	4.80	4.85
−1.8	3.00	3.30	3.40	3.50	3.70	4.00	4.20	4.30	4.40	4.45
−1.6	1.70	2.20	2.50	2.80	3.10	3.80	4.00	4.20	4.30	4.40
−1.4	0.50	1.10	1.60	2.00	2.30	2.50	2.60	2.70	2.80	3.00
−1.2	−0.20	0.10	0.65	1.10	1.40	1.70	1.80	1.70	2.00	2.30
−1.0	−1.40	−0.75	−0.12	0.30	0.70	1.00	1.20	1.25	1.30	1.55
−0.8	−2.30	−1.60	−1.00	−0.40	0.05	0.40	0.62	0.80	0.75	0.92
−0.6	−3.20	−2.50	−1.70	−1.10	−0.55	−0.15	0.15	0.35	0.45	0.50
−0.4	−4.00	−3.30	−2.50	−1.75	−1.15	−0.60	−0.25	0.00	0.15	0.20
−0.2	−4.05	−4.00	−3.30	−2.40	−1.70	−1.10	−0.62	−0.30	−0.07	0.03
0.0	−4.10	−4.05	−4.00	−3.00	−2.15	−1.50	−0.95	−0.55	−0.25	−0.07
0.2	−4.15	−4.10	−4.05	−3.30	−2.50	−1.80	−1.20	−0.70	−0.36	−0.12
0.4	−4.20	−4.15	−4.10	−3.60	−2.80	−2.00	−1.35	−0.80	−0.43	−0.24
0.6	−4.25	−4.20	−4.15	−3.80	−3.00	−2.20	−1.45	−0.95	−0.60	−0.35
0.8	−4.30	−4.25	−4.20	−4.15	−3.00	−2.30	−1.70	−1.20	−0.70	−0.62
1.0	−4.35	−4.30	−4.25	−4.20	−3.10	−2.60	−2.00	−1.35	−1.00	−0.90
1.2	−4.40	−4.35	−4.30	−4.25	−3.50	−3.00	−2.20	−1.60	−1.40	−1.30
1.4	−4.45	−4.40	−4.35	−4.30	−4.25	−3.30	−2.50	−2.00	−1.90	−1.75
1.6	−4.50	−4.45	−4.40	−4.35	−4.30	−4.25	−2.80	−2.60	−2.45	−2.25
1.8	−4.55	−4.50	−4.45	−4.40	−4.35	−4.30	−4.25	−3.20	−3.00	−2.80
2.0	−4.60	−4.55	−4.50	−4.45	−4.40	−4.35	−4.30	−4.25	−4.00	−3.40

Exact Computation Methods Based on Direct Integration

speed $n_s = 127$

Q/Q_R										
0.0	0.2	0.4	0.6	0.8	1.0	1.2	1.4	1.6	1.8	2.0
H/H_R										
5.12	5.15	5.18	5.20	5.22	5.25	5.27	5.30	5.35	5.40	5.50
4.02	4.05	4.07	4.10	4.12	4.15	4.20	4.25	4.30	4.35	4.40
3.22	3.25	3.27	3.30	3.40	3.50	3.55	3.60	3.65	3.70	3.75
2.47	2.50	2.53	2.55	2.65	2.90	3.10	3.40	3.45	3.50	3.50
1.80	1.80	1.90	2.00	2.19	2.50	2.70	3.10	3.40	3.45	3.45
1.29	1.30	1.35	1.45	1.65	1.90	2.25	2.70	3.10	3.35	3.40
0.85	0.85	0.92	1.03	1.25	1.55	1.90	2.30	2.70	3.20	3.35
0.47	0.49	0.58	0.72	0.95	1.25	1.60	2.00	2.50	3.00	3.30
0.21	0.24	0.33	0.49	0.72	1.00	1.35	1.75	2.25	2.75	3.25
0.06	0.09	0.18	0.35	0.60	0.85	1.15	1.55	2.00	2.50	3.00
0.00	0.05	0.10	0.27	0.48	0.75	1.00	1.40	1.80	2.30	2.80
0.03	0.06	0.09	0.20	0.40	0.65	0.90	1.32	1.70	2.12	2.65
0.11	0.13	0.15	0.20	0.35	0.53	0.90	1.25	1.65	2.05	2.52
0.22	0.24	0.28	0.33	0.43	0.65	0.95	1.33	1.65	2.05	2.52
0.40	0.43	0.47	0.55	0.65	0.82	1.10	1.40	1.70	2.10	2.55
0.57	0.60	0.65	0.70	0.80	0.95	1.15	1.40	1.75	2.15	2.65
0.90	0.91	0.95	1.02	1.10	1.20	1.40	1.60	1.85	2.20	2.67
1.20	1.28	1.35	1.42	1.47	1.56	1.69	1.81	1.92	2.20	2.70
1.50	1.60	1.75	1.82	1.85	1.87	1.90	1.95	1.98	2.22	2.73
1.75	1.78	1.80	1.87	1.90	1.92	1.95	2.00	2.08	2.25	2.75
1.80	1.82	1.85	1.92	1.95	2.00	2.05	2.10	2.20	2.30	2.80

speed $n_s = 530$

0.0	0.2	0.4	0.6	0.8	1.0	1.2	1.4	1.6	1.8	2.0
4.90	4.95	5.00	5.05	5.05	5.15	5.20	5.25	5.30	5.35	5.40
4.50	4.55	4.60	4.95	5.00	5.10	5.15	5.20	5.25	5.30	5.35
4.45	4.50	4.55	4.60	4.95	5.05	5.10	5.15	5.20	5.25	5.30
3.50	3.70	4.00	4.10	4.90	5.00	5.05	5.10	5.15	5.20	5.25
2.70	3.10	3.40	3.45	4.10	4.90	5.00	5.05	5.10	5.15	5.20
1.90	2.30	2.75	3.10	3.40	4.10	4.90	5.00	5.05	5.10	5.15
1.20	1.50	2.00	2.50	3.00	4.00	4.80	4.90	5.00	5.05	5.10
0.70	0.90	1.30	1.85	2.90	3.90	4.70	4.80	4.90	5.00	5.05
0.30	0.50	0.85	1.60	2.50	3.50	4.50	4.70	4.80	4.90	5.00
0.08	0.22	0.63	1.25	2.00	3.00	4.00	4.20	4.70	4.80	4.90
0.00	0.10	0.37	0.85	1.50	2.30	3.10	3.70	4.20	4.70	4.80
−0.04	0.04	0.10	0.37	0.80	1.50	2.35	3.10	3.70	4.20	4.70
−0.12	0.02	0.07	0.22	0.45	0.80	1.40	2.25	3.00	3.70	4.20
−0.25	−0.10	0.03	0.15	0.37	0.67	1.00	1.50	2.30	3.20	3.20
−0.45	−0.25	−0.02	0.08	0.26	0.55	0.90	1.35	1.80	2.45	2.90
−0.75	−0.45	−0.15	0.03	0.17	0.43	0.75	1.20	1.70	2.25	2.75
−1.00	−0.65	−0.35	−0.04	0.08	0.30	0.60	1.00	1.50	2.05	2.70
−1.35	−1.00	−0.65	−0.20	0.03	0.19	0.45	0.85	1.30	1.80	2.50
−1.80	−1.30	−1.05	−0.45	−0.06	0.08	0.30	0.65	1.15	1.60	2.20
−2.30	−1.70	−1.35	−0.75	−0.27	0.03	0.16	0.50	0.90	1.40	2.00
−2.75	−2.10	−1.80	−1.25	−0.55	−0.10	0.07	0.30	0.70	1.15	1.70

TABLE 6.5 c Performance characteristics $H-Q-n$, predicted for a pump with specific

$\dfrac{n}{n_R}$	−2.0	−1.8	−1.6	−1.4	−1.2	−1.0	−0.8	−0.6	−0.4	−0.2
−2.0	4.00	4.20	5.00	5.90	6.00	6.50	6.70	6.75	6.80	6.85
−1.8	2.50	3.00	3.70	4.60	5.00	5.50	5.70	5.75	5.80	5.85
−1.6	1.00	1.80	2.50	3.30	3.50	4.00	4.50	4.60	4.70	4.75
−1.4	−0.10	0.60	1.30	2.00	2.50	3.00	3.30	3.50	3.55	3.60
−1.2	−1.20	−0.40	0.25	0.65	1.50	2.00	2.30	2.50	2.65	2.90
−1.0	−2.00	−1.25	−0.60	0.00	0.50	1.00	1.40	1.70	1.90	2.35
−0.8	−2.80	−2.00	−1.30	−0.70	−0.20	0.25	0.65	0.95	1.15	1.50
−0.6	−3.40	−2.70	−2.00	−1.30	−0.75	−0.30	0.05	0.35	0.60	0.75
−0.4	−3.80	−3.10	−2.50	−1.80	−1.20	−0.75	−0.30	−0.05	0.15	0.28
−0.2	−3.80	−3.10	−2.50	−1.90	−1.40	−1.00	−0.60	−0.30	−0.08	0.05
0.0	−3.80	−3.10	−2.50	−1.85	−1.40	−1.00	−0.60	−0.35	−0.15	−0.07
0.2	−4.50	−4.00	−3.15	−2.50	−1.85	−1.45	−1.00	−0.70	−0.37	−0.22
0.4	−4.70	−4.20	−3.90	−3.20	−2.70	−2.20	−1.70	−1.25	−0.90	−0.65
0.6	−4.80	−4.30	−4.10	−3.70	−3.50	−3.00	−2.50	−2.00	−1.50	−1.15
0.8	−4.90	−4.40	−4.20	−4.00	−3.80	−3.60	−3.50	−3.00	−2.33	−1.80
1.0	−4.95	−4.90	−4.85	−4.80	−4.70	−4.50	−4.40	−3.80	−3.20	−2.75
1.2	−5.00	−4.95	−4.90	−4.90	−4.80	−4.60	−4.50	−4.40	−4.10	−3.70
1.4	−5.05	−5.00	−4.95	−4.95	−4.85	−4.65	−4.60	−4.50	−4.20	−3.80
1.6	−5.10	−5.05	−5.00	−4.95	−4.90	−4.70	−4.65	−4.55	−4.25	−3.85
1.8	−5.15	−5.10	−5.05	−5.00	−4.95	−4.90	−4.80	−4.70	−4.50	−4.20
2.0	−5.20	−5.15	−5.10	−5.05	−5.00	−5.00	−4.90	−4.80	−4.60	−4.40

TABLE 6.5 d Performance characteristics $T-Q-n$, predicted for a pump with specific

$\dfrac{n}{n_R}$	−2.0	−1.8	−1.6	−1.4	−1.2	−1.0	−0.8	−0.6	−0.4	−0.2
−2.0	−3.80	−3.50	−3.50	−3.50	−2.70	−2.60	−2.45	−2.30	−2.15	−1.80
−1.8	−3.20	−3.00	−3.00	−3.00	−2.40	−2.30	−2.15	−2.00	−1.85	−1.55
−1.6	−2.60	−2.50	−2.50	−2.50	−2.10	−2.00	−1.85	−1.70	−1.55	−1.30
−1.4	−2.00	−2.00	−2.00	−2.00	−1.80	−1.70	−1.55	−1.40	−1.25	−1.05
−1.2	−1.40	−1.50	−1.52	−1.52	−1.48	−1.40	−1.25	−1.10	−0.95	−0.80
−1.0	−0.85	−0.95	−1.00	−1.03	−1.04	−1.00	−0.94	−0.85	−0.75	−0.60
−0.8	−0.32	−0.45	−0.54	−0.60	−0.68	−0.68	−0.66	−0.60	−0.50	−0.40
−0.6	0.10	0.00	−0.13	−0.23	−0.30	−0.37	−0.38	−0.37	−0.34	−0.24
−0.4	0.40	0.33	0.22	0.10	0.00	−0.07	−0.13	−0.17	−0.17	−0.12
−0.2	0.95	0.75	0.58	0.40	0.27	0.15	0.07	0.00	−0.04	−0.05
0.0	1.50	1.25	1.00	0.80	0.60	0.45	0.30	0.17	0.08	0.03
0.2	2.00	1.80	1.60	1.35	1.15	0.85	0.62	0.40	0.25	0.12
0.4	2.80	2.60	2.40	2.00	1.65	1.35	1.05	0.75	0.50	0.28
0.6	3.60	3.40	3.20	2.60	2.25	1.90	1.45	1.10	0.75	0.46
0.8	4.40	4.20	4.00	3.20	2.80	2.50	2.00	1.45	1.10	0.75
1.0	5.20	5.00	4.80	3.80	3.40	3.00	2.50	1.95	1.40	1.00
1.2	6.00	5.80	5.60	4.50	4.00	3.50	3.00	2.35	1.80	1.35
1.4	6.80	6.60	6.40	5.00	4.60	4.00	3.50	2.80	2.25	1.65
1.6	7.60	7.40	7.20	5.60	5.20	4.50	4.00	3.20	2.70	2.00
1.8	8.40	8.20	8.00	6.20	5.80	5.00	4.50	3.60	3.15	2.30
2.0	9.20	9.00	8.80	6.80	6.40	5.50	5.00	4.00	3.60	2.60

Exact Computation Methods Based on Direct Integration

speed $n_s = 950$

Q/Q_R										
0	0.2	0.4	0.6	0.8	1.0	1.2	1.4	1.6	1.8	2.0

H/H_R										
6.90	6.95	7.00	7.05	7.10	7.15	7.20	7.25	7.30	7.35	7.40
5.90	5.95	6.00	6.05	6.10	6.15	6.20	6.25	6.30	6.35	6.40
4.80	4.85	4.90	4.95	5.00	5.20	5.65	5.75	5.85	5.95	6.05
3.65	3.80	3.85	4.10	4.65	5.15	5.60	5.70	5.80	5.90	6.00
3.20	3.40	3.60	4.00	4.60	5.10	5.55	5.65	5.75	5.85	5.95
2.60	2.90	3.30	3.80	4.40	5.00	5.50	5.60	5.70	5.80	5.90
1.80	2.25	2.70	3.20	3.70	4.00	4.30	4.40	4.50	4.60	5.25
1.00	1.40	1.85	2.30	2.70	3.00	3.20	3.40	3.70	4.30	5.20
0.45	0.75	1.00	1.30	1.70	2.20	2.60	3.00	3.50	4.20	5.00
0.12	0.28	0.45	0.70	1.10	1.50	2.00	2.70	3.30	4.00	4.50
0.00	0.07	0.20	0.40	0.70	1.10	1.15	2.15	2.75	3.40	4.00
−0.10	0.00	0.12	0.29	0.52	0.85	1.20	1.65	2.15	2.75	3.85
−0.35	−0.15	0.02	0.16	0.48	0.75	1.15	1.65	2.10	2.60	3.75
−0.75	−0.40	−0.20	0.65	0.35	0.82	1.07	1.45	1.95	2.55	3.50
−1.33	−0.85	−0.50	−0.25	0.09	0.50	0.92	1.40	1.80	2.45	3.25
−2.15	−1.50	−0.95	−0.68	−0.30	0.10	0.65	1.20	1.70	2.30	3.00
−3.00	−2.25	−1.50	−1.20	−0.80	−0.40	0.20	0.82	1.45	2.05	2.75
−3.35	−2.80	−2.25	−1.80	−1.38	−0.90	−0.45	0.30	1.00	1.70	2.50
−3.50	−3.30	−3.00	−2.50	−2.10	−1.60	−1.00	0.50	0.30	1.18	2.00
−4.00	−3.90	−3.70	−3.40	−3.00	−2.40	−1.80	−1.20	−0.50	0.50	1.30
−4.20	−4.10	−4.00	−3.95	−3.90	−3.30	−2.75	−2.00	−1.25	0.50	0.60

speed $n_s = 127$

Q/Q_R										
0.0	0.2	0.4	0.6	0.8	1.0	1.2	1.4	1.6	1.8	2.0

T/T_R										
−1.70	−1.40	−1.20	−1.25	−1.40	−1.80	−1.85	−1.85	−2.25	−2.90	−3.80
−1.45	−1.20	−1.05	−1.10	−1.25	−1.60	−1.70	−1.75	−2.25	−2.90	−3.80
−1.20	−1.00	−0.90	−0.95	−1.10	−1.40	−1.60	−1.75	−2.25	−2.90	−3.80
−0.94	−0.80	−0.75	−0.80	−0.95	−1.20	−1.50	−1.75	−2.25	−2.90	−3.80
−0.68	−0.58	−0.58	−0.65	−0.80	−1.00	−1.40	−1.75	−2.25	−2.90	−3.80
−0.48	−0.40	−0.39	−0.50	−0.70	−1.00	−1.35	−1.75	−2.25	−2.90	−3.80
−0.30	−0.25	−0.27	−0.40	−0.70	−1.00	−1.40	−1.85	−2.35	−2.95	−3.80
−0.17	−0.15	−0.20	−0.37	−0.67	−0.97	−1.40	−1.80	−2.35	−2.90	−3.80
−0.08	−0.09	−0.18	−0.37	−0.65	−0.95	−1.35	−1.75	−2.25	−2.80	−3.85
−0.04	−0.06	−0.16	−0.33	−0.60	−0.90	−1.30	−1.70	−2.10	−2.75	−3.80
0.00	−0.06	−0.12	−0.30	−0.58	−0.90	−1.25	−1.65	−2.05	−2.70	−3.40
0.05	−0.03	−0.09	−0.27	−0.50	−0.80	−1.15	−1.58	−2.00	−2.60	−3.30
0.12	0.02	−0.08	−0.23	−0.42	−0.70	−1.05	−1.45	−1.90	−2.50	−3.10
0.26	0.09	−0.03	−0.17	−0.35	−0.65	−0.95	−1.35	−1.80	−2.35	−2.85
0.40	0.22	0.05	−0.10	−0.30	−0.55	−0.85	−1.25	−1.65	−2.15	−2.65
0.65	0.38	0.18	−0.02	−0.22	−0.46	−0.80	−1.15	−1.55	−2.05	−2.50
0.90	0.60	0.34	0.10	−0.13	−0.35	−0.70	−1.05	−1.45	−1.90	−2.40
1.25	0.87	0.60	0.30	0.00	−0.24	−0.60	−0.95	−1.35	−1.80	−2.30
1.60	1.10	0.80	0.60	0.15	−0.13	−0.50	−0.85	−1.25	−1.70	−2.20
1.90	1.35	1.00	0.90	0.30	0.00	−0.40	−0.75	−1.15	−1.60	−2.10
2.20	1.60	1.20	1.10	0.70	0.15	−0.30	−0.65	−1.05	−1.50	−2.00

TABLE 6.5 e **Performance characteristics** $T - Q - n$, **predicted for a pump with specific**

$\dfrac{n}{n_R}$	Q/Q_R								
	−2.0	−1.8	−1.6	−1.4	−1.2	−1.0	−0.8	−0.6	−0.4	−0.2
	T/T_R								
−2.0	−3.70	−4.00	−4.25	−4.30	−4.30	−4.10	−4.10	−4.30	−4.40	−4.50
−1.8	−3.00	−3.25	−3.45	−3.50	−3.50	−3.30	−3.30	−3.50	−3.60	−3.80
−1.6	−2.25	−2.45	−2.55	−2.60	−2.60	−2.40	−2.25	−2.55	−2.85	−3.20
−1.4	−1.40	−1.65	−1.85	−2.00	−2.00	−2.00	−1.75	−1.85	−2.15	−2.50
−1.2	−0.50	−0.85	−1.15	−1.35	−1.45	−1.50	−1.40	−1.30	−1.50	−1.75
−1.0	−0.40	−0.00	−0.50	−0.75	−0.90	−1.00	−1.02	−0.92	−0.95	−1.20
−0.8	1.25	0.75	0.30	−0.10	−0.37	−0.56	−0.65	−0.65	−0.58	−0.70
−0.6	2.20	1.60	1.00	0.55	0.16	−0.10	−0.27	−0.36	−0.36	−0.38
−0.4	3.20	2.50	1.80	1.20	0.75	0.30	0.08	−0.10	−0.16	−0.16
−0.2	4.00	3.30	2.65	2.00	1.35	0.85	0.45	0.18	0.03	−0.03
0.0	4.25	4.20	3.50	2.75	2.00	1.40	0.90	0.50	0.25	0.07
0.2	4.30	4.25	3.70	3.50	2.80	2.10	1.40	0.90	0.50	0.20
0.4	4.35	4.30	4.10	3.70	3.50	2.75	2.00	1.30	0.75	0.35
0.6	4.40	4.35	4.20	4.10	3.70	3.40	2.50	1.70	1.00	0.70
0.8	4.45	4.40	4.30	4.20	4.10	4.00	3.00	2.00	1.40	1.20
1.0	4.50	4.45	4.35	4.30	4.20	4.10	3.30	2.50	2.00	1.85
1.2	4.55	4.50	4.40	4.35	4.30	4.20	3.50	3.30	3.00	2.75
1.4	4.60	4.55	4.45	4.40	4.35	4.30	3.70	3.60	3.50	3.30
1.6	4.65	4.60	4.50	4.45	4.40	4.35	4.30	4.25	4.20	4.10
1.8	4.70	4.65	4.55	4.50	4.45	4.40	4.35	4.30	4.25	4.20
2.0	4.75	4.70	4.60	4.55	4.50	4.45	4.40	4.35	4.30	4.25

TABLE 6.5 f **Performance characteristics** $T - Q - n$, **predicted for a pump with specific**

$\dfrac{n}{n_R}$	Q/Q_R								
	−2.0	−1.8	−1.6	−1.4	−1.2	−1.0	−0.8	−0.6	−0.4	−0.2
	T/T_R								
−2.0	−3.60	−4.00	−4.05	−4.10	−4.15	−4.20	−4.20	−4.25	−4.35	−4.45
−1.8	−2.60	−3.10	−3.40	−3.70	−3.80	−3.90	−3.95	−4.00	−4.20	−4.30
−1.6	−1.50	−2.10	−2.50	−2.90	−3.10	−3.20	−3.10	−3.30	−3.60	−4.00
−1.4	−0.30	−1.00	−1.50	−2.00	−2.25	−2.30	−2.25	−2.45	−2.75	−3.10
−1.2	0.65	0.00	−0.65	−1.05	−1.45	−1.66	−1.70	−1.70	−1.95	−2.30
−1.0	1.10	0.85	0.20	−0.30	−0.70	−1.00	−1.20	−1.15	−1.25	−1.55
−0.8	2.25	1.60	1.00	0.50	0.00	−0.40	−0.65	−0.75	−0.75	−0.95
−0.6	2.75	2.15	1.50	1.00	0.60	0.20	−0.15	−0.35	−0.42	−0.50
−0.4	3.00	2.45	1.90	1.30	0.95	0.60	0.25	0.00	−0.15	−0.20
−0.2	2.40	2.05	1.70	1.30	1.05	0.75	0.45	0.25	0.05	0.04
0.0	2.50	2.00	1.50	1.15	0.90	0.60	0.40	0.25	0.10	0.05
0.2	3.00	2.75	2.65	2.20	1.70	1.30	1.00	0.65	0.40	0.22
0.4	4.55	4.00	3.75	3.30	2.75	2.20	1.70	1.30	0.90	0.65
0.6	4.60	4.55	4.50	4.00	3.50	3.00	2.50	2.00	1.50	1.20
0.8	4.65	4.60	4.55	4.50	4.45	3.80	3.50	3.00	2.50	1.90
1.0	4.70	4.65	4.60	4.55	4.50	4.45	4.00	3.50	3.20	2.80
1.2	4.75	4.70	4.65	4.60	4.55	4.50	4.45	4.40	3.90	3.80
1.4	4.80	4.75	4.70	4.65	4.60	4.55	4.50	4.45	4.40	3.90
1.6	4.85	4.80	4.75	4.70	4.65	4.60	4.55	4.50	4.45	4.40
1.8	4.90	4.85	4.80	4.75	4.70	4.65	4.60	4.55	4.50	4.45
2.0	4.95	4.90	4.85	4.80	4.75	4.70	4.65	4.60	4.55	4.50

speed $n_s = 530$

−0.0	0.2	0.4	0.6	0.8	1.0	1.2	1.4	1.6	1.8	2.0
−4.60	−4.70	−4.80	−4.85	−4.90	−4.95	−5.00	−5.75	−6.55	−6.95	−7.05
−3.90	−3.95	−4.00	−4.05	−4.10	−4.15	−4.90	−5.70	−6.50	−6.90	−7.00
−3.50	−3.70	−3.80	−3.90	−4.05	−4.10	−4.85	−5.65	−6.45	−6.85	−6.95
−2.80	−3.20	−3.50	−3.80	−4.00	−4.05	−4.80	−5.60	−6.40	−6.80	−6.90
−2.10	−2.50	−2.80	−3.10	−3.25	−3.70	−4.75	−5.50	−6.35	−6.75	−6.85
−1.45	−1.75	−2.10	−2.50	−2.90	−3.65	−4.70	−5.50	−6.30	−6.70	−6.80
−0.90	−1.20	−1.50	−1.90	−2.65	−3.60	−4.70	−5.50	−6.30	−6.70	−6.80
−0.50	−0.70	−1.00	−1.50	−2.40	−3.50	−4.70	−5.50	−6.30	−6.70	−6.80
−0.25	−0.38	−0.68	−1.35	−2.30	−3.30	−4.00	−4.70	−5.50	−6.30	−6.70
−0.07	−0.20	−0.55	−1.15	−1.90	−2.80	−3.60	−4.70	−5.50	−6.30	−6.70
0.00	−0.10	−0.35	−0.75	−1.40	−2.20	−2.90	−3.60	−4.70	−5.50	−6.30
0.06	0.00	−0.15	−0.40	−0.85	−1.50	−2.30	−3.10	−3.90	−4.70	−5.50
0.25	0.10	0.00	−0.20	−0.50	−1.00	−1.70	−2.50	−3.30	−4.00	−4.70
0.50	0.30	0.17	0.00	−0.30	−0.70	−1.25	−2.00	−2.75	−3.50	−4.25
0.87	0.65	0.40	0.22	0.00	−0.38	−0.87	−1.45	−2.20	−3.00	−3.80
1.40	1.10	1.15	0.50	0.30	0.30	0.50	−1.00	−1.70	−2.50	−3.30
2.00	1.65	1.30	0.87	0.65	0.40	0.00	−0.35	−1.20	−2.00	−2.75
2.95	2.75	1.95	1.35	1.05	0.75	0.47	0.00	−0.60	−1.40	−2.20
4.00	3.80	2.90	2.00	1.60	1.25	0.90	0.57	0.00	−0.65	−1.60
4.10	3.85	3.60	2.75	2.20	1.80	1.40	1.00	0.65	0.00	−0.70
4.15	3.95	3.90	3.40	2.90	2.50	2.00	1.60	1.20	0.75	0.00

speed $n_s = 950$

−0.0	−0.2	0.4	0.6	0.8	1.0	1.2	1.4	1.6	1.8	2.0
−4.75	−4.80	−4.90	−4.95	−5.00	−5.05	−5.10	−5.20	−5.25	−5.30	−5.35
−4.70	−4.75	−4.80	−4.85	−4.90	−4.95	−5.05	−5.15	−5.20	−5.25	−5.30
−4.60	−4.65	−4.70	−4.75	−4.80	−4.90	−5.00	−5.10	−5.15	−5.20	−5.25
−3.70	−4.50	−4.55	−4.65	−4.75	−4.85	−4.95	−5.05	−5.10	−5.15	−5.20
−2.80	−3.50	−4.30	−4.60	−4.70	−4.80	−4.90	−5.00	−5.05	−5.10	−5.15
−2.00	−2.50	−3.30	−4.50	−4.60	−4.70	−4.80	−4.90	−5.00	−5.05	−5.10
−1.30	−1.75	−2.50	−3.70	−3.50	−3.60	−3.70	−3.80	−4.10	−4.30	−4.40
−0.70	−1.10	−1.65	−2.00	−2.35	−2.75	−3.00	−3.50	−3.90	−4.00	−4.10
−0.80	−0.62	−0.90	−1.15	−1.45	−1.80	−2.30	−2.90	−3.50	−3.60	−3.70
−0.09	−0.25	−0.37	−0.60	−0.85	−1.25	−1.65	−2.15	−2.65	−3.10	−3.30
0.00	−0.05	−0.10	−0.25	−0.40	−0.65	−0.90	−1.25	−1.70	−2.10	−2.65
0.10	0.00	−0.12	−0.28	−0.43	−0.60	−0.75	−1.00	−1.30	−1.60	−1.90
0.35	0.20	0.00	−0.22	−0.50	−0.80	−1.10	−1.40	−1.70	−2.00	−2.30
0.75	0.50	0.30	0.05	0.30	−0.65	−1.15	−1.60	−2.00	−2.45	−2.90
1.40	0.90	0.65	0.50	0.10	0.30	−0.80	−1.40	−2.00	−2.50	−3.00
2.20	1.60	1.25	1.00	0.65	0.10	−0.40	−1.00	−1.65	−2.35	−3.05
3.00	2.40	1.85	1.55	1.25	0.85	0.20	−0.40	−1.05	−1.85	−2.75
3.80	3.20	2.65	2.50	2.00	1.60	1.00	0.20	−0.40	−1.20	−2.00
3.90	3.80	3.70	3.60	3.00	2.40	1.90	1.25	0.30	−0.30	−1.25
4.40	4.35	4.30	4.25	4.20	3.60	3.00	4.25	1.50	0.30	−0.30
4.45	4.40	4.35	4.30	4.25	4.10	4.00	3.40	2.60	1.60	0.50

This is in fact a series of six tables. The first three give the pumping head H, and the next three the load torque T, as functions of the flow rate Q and the pump speed n. The three pairs of tables are valid for three specific rotation speeds, and interpolations are possible. The tables have been based on the diagrams Knapp, obtained by measurements [6.11].

For less accurate computations, especially in graphical or mixed solutions, we may also use equation

$$T_{j+1} = \frac{\rho Q_{j+1} H_{j+1}}{\frac{\pi n_{j+1}}{30}} \tag{6.91}$$

which gives great errors for speeds close to zero.

When several turbopumps are discharging through the same line, they may be replaced with an equivalent turbopump. When identical turbopumps are coupled in parallel, the equivalent turbopump would have the same pumping head and speed as each turbopump separately, while its flow rate and inertia moment could be found by summing up the flow rates and moments of inertia, respectively, of the turbopumps in the pumping system. When identical turbopumps are coupled in series, the equivalent turbopump would have the same flow rate and rotation speed as each turbopump separately, while the pumping head and moment of inertia are summed.

(b.2) *The tank*. Assuming that the liquid level in the tank is kept constant, the tank will have the equation

$$H_k = H_{st} \tag{6.92}$$

(b.3) *The valve*. Assuming a rapid-closing valve, it has the equation

$$Q_k = 0 \tag{6.93}$$

Solution of the system of equations. This may be done numerically, graphically, or in a mixed way:

(1) *The numerical solution* of equations may be carried out manually, mechanically, or electronically. Computer techniques secure shorter times for water hammer computation, for the most varied and complicated boundary conditions of the pumping pipelines, with many inner nodes and computation time intervals. For instance, an improved electronic computer can compute 1,000 computation time intervals in about three minutes for a pipeline divided into ten sections and having three diaphragms and a turbopump.

(2) *The graphical solution* of the system of equations is possible only in the case of very simple arrangements with simple inner nodes, or with end nodes having simple boundary conditions (such as constant-level tanks, quick-closing valves, slow-closing valves).

The graphical solution is applied to the plane of variables $Q-H$ where the boundary conditions are as follows:

(a) For *the constant-level tank* — a line parallel to the Q axis, at a given constant level;

(b) For *the quick-closing valve* — a line superposed to the H axis, that is $Q = 0$;

(c) For *the slow-closing valve* — at different times the resistance modulus M_k takes different values and forms a family of parabolae, each of them being valid for a certain moment.

As for the wave equations, they can be put in the form

— *direct wave*:

$$H_{k,j+1} - H_{k-1,j} = -m_k(Q_{k,j+1} - Q_{k-1,j}) \qquad (6.94)$$

— *backward wave*

$$H_{k,j+1} - H_{k+1,j} = m_{k+1}(Q_{k,j+1} - Q_{k+1,j}) \qquad (6.95)$$

Thus, the direct wave is represented in the $H-Q$ plane as a line with an angular coefficient $(-m_k)$, while the backward wave is represented as a line with angular coeficient $(+m_{k+1})$.

The situation at an inner node (Fig. 6.45) can be found using the two above wave equations. We represent the points $(k-1, j)$ and $(k+1, j)$ with the co-ordinates $(Q_{k-1,j}, H_{k-1,j})$ and $(Q_{k+1,j}, H_{k+1,j})$ and then draw a line through the point $(k-1, j)$ parallel to the line of the direct wave, and through the point $(k+1, j)$ parallel to the line of the backward wave. The intersection of these two lines gives the desired point $(k, j+1)$ having the co-ordinates $(Q_{k,j+1}, H_{k,j+1})$.

When we have boundary conditions at the end nodes, the solution is found at the crossing point of the direct wave with the backward wave,

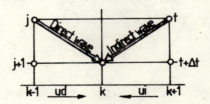

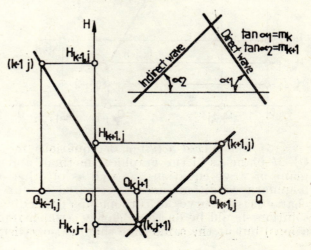

Fig. 6.45. Computational diagram for the graphical solution of a simple inner node.

at the end or at the beginning of the pipeline with the relevant boundary condition, depending on node location.

When an additional condition occurs at an inner node, such as in the case of a diaphragm, then the solution is no longer found at the intersection of the two lines. In this case (Fig. 6.46), we are in fact looking for the situation before the diaphragm (H'_k, Q_k) and after it (H''_k, Q_k). Since the point K' stands for the situation before the diaphragm, the node K is on the line of the direct wave, while the node K is on the line of the backward wave, since point K'' stands for the situation after the diaphragm. Both points (K' and K'') are also on the same line parallel to H (the same flow rate) so the difference $H'_k - H''_k$ is equal to the head loss $M_k Q_k |Q_k|$ in the diaphragm:

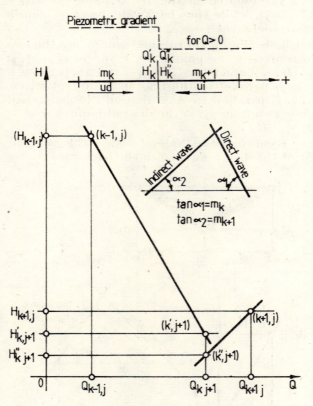

Fig. 6.46. Computational diagram for the graphical solution of an inner node with diaphragm.

(3) *The mixed solution* of equations means the use of the same $Q-H$ plane as in the graphical method, and the representation in this plane of wave equations, as well as of other equations known from the boundary conditions and represented as curves $H = f(Q)$, such as the characteristic curves of turbopumps depending on the speed. The desired solution should be on these curves (which represent a part of the equations) but at the same time should also satisfy the rest of the equations

in the system. Thus, the solution may be found only by successive verifications of the equations which cannot be graphically represented.

In the case of mixed solutions, the more intricate equations, in particular, the implicit ones, may be simplified (changed to explicit) by neglecting certain factors. This obviously leads to other approximations, and hence to results even less close to reality.

Example 6.10. **Water hammer computation by the computational waves method** [6.23]. Let us consider a pumping installation of four turbopumps coupled in parallel, each having a rated flow through the pump $Q_R = 1.25$ cu.m/s, rated pumping head $H_R = 50$ m, rated pump speed $n_R = 570$ rev/min, moment of inertia $GD^2 = 145$ daN.m², pump efficiency at rated speed and head $\eta_R = 0.8$, connected to a steel pipeline having an initial flow through the pump $Q_0 = 5$ cu.m/s, static head $H_{st} = 50$ m, dynamic head $H_{dy} = 3$ m, line length, $l = 500$ m, rated diameter $D_R = 1.50$ m, wall thickness $e = 10$ mm, elasticity modulus, $E = 2.1 \times 10^5 \times 9.81$ daN/cm². We desire the computation diagram and scheme.

Solution. We find the turbopump specific speed n_s from the equation

$$n_s = 3.65 \times 570 \, \frac{\sqrt{1.25}}{50^{3/4}} = 124$$

Then, making use of Tables 6.5 we draw the turbopump complete (universal) characteristic.

The turbopumps of the pumping system have been replaced in computations by an equivalent turbopump having a rated flow through the pump $Q_R = 5$ cu.m/s, rated pumping head $H_R = 50$ m, rated pump speed $n_R = 570$ rev/min, the moment of inertia $GD^2 = 580$ daN.m² and pump efficiency at rated speed $\eta_R = 0.8$.

Given the reduced head loss on the discharge pipeline, it has been neglected.

We find wave speed from equation (6.7)

$$a = \frac{1{,}425}{\sqrt{1 + \frac{1500}{10} \times \frac{1}{100}}} = 900 \text{ m/s}$$

The time necessary to travel the pipeline is found

$$\Delta t_1 = \frac{1}{a} = \frac{500}{900} = 0.555 \text{ s}$$

and the pipeline drag is

$$m = \frac{a}{g\omega} = \frac{900}{10 \times 1.76} = 50 \, \frac{\text{m}}{\text{cu.m/s}}$$

Since the turbopump universal characteristic is given in relative (dimensionless) values, and the solution is graphical, it is more convenient to make the whole computation in relative values, and hence

$$Q^* = Q/Q_R; \; H^* = H/H_R; \; T^* = T/T_R; \; n^* = n/n_R.$$

With these relative values, the equation of motion of the rotating parts of motor, pump, and entrained water, at the cut-off of the motor torque, becomes

$$n^*_{j+1} = n^*_j - K^* - (T^*_{r,j} + T^*_{r,j+1})$$

where

$$K^* = \frac{1{,}800 \times g \times \Delta t \times \rho g Q_R H_R}{\pi^2 \times (GD^2) \times \eta \times n_R^2} = \frac{1800 \times 9.81 \times 0.139 \times 1{,}000 \times 5 \times 50}{(3.14)^2 \times 580 \times 0.8 \times (570)^2} = 0.415$$

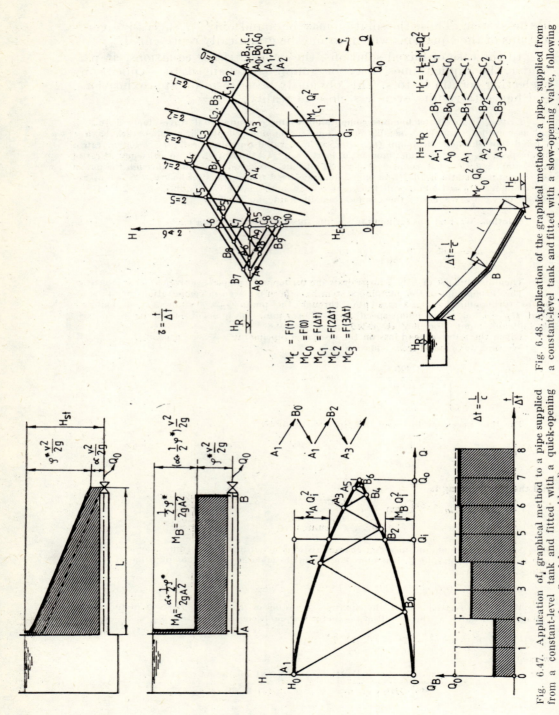

Fig. 6.48. Application of the graphical method to a pipe, supplied from a constant-level tank and fitted with a slow-opening valve, following a prescribed programme (head losses not taken into account).

Fig. 6.47. Application of graphical method to a pipe supplied from a constant-level tank and fitted with a quick-opening valve (head losses considered).

We find the wave drag for the equivalent turbopump

$$m^* = m \frac{Q_R}{H_R} = 50 \times \frac{5}{50} = 5$$

For the turbopump characteristics, the time necessary to travel over the whole discharge line is too long to be taken as the computation time. Therefore, the pipeline had to be divided into four equal sections, starting from turbopump to the tank in the following way: AB, BC, CD, DE. Thus, the computation time is the time necessary to travel one section

$$\Delta t = 0.555/4 = 0.139 \text{ s}.$$

We draw the scheme of computational waves (Fig. 6.49) and the computational diagram (Fig. 6.50).

The computation of turbopump nodes (denoted with A) is done by trials. The condition at the turbopump at a certain time $j+1$ is found on the line representing the backward wave sent from B at a previous time j; we choose a point on this line and find its speed (by interpolation on the solid curves $H-Q-n$). This speed is denoted by n'_{j+1}. The dashed lines, representing the curves $T-Q-n$, assist in finding the load torque denoted with T'_{j+1} corresponding to the selected point (flow rate and speed).

Then we compute the load torque denoted with T'_{j+1} corresponding to the selected point (flow rate and speed), and then $n^{*''}_{j+1}$ (the values n^*_j, T^*_j at previous time are already known):

$n^{*''}_{j+1} = n^*_j - K^*(T^*_j - T^{*'}_{j+1})$. If $n^{*'}_{j+1} = n^{*''}_{j+1}$, then the chosen point is the solution; if not, we have to choose another point.

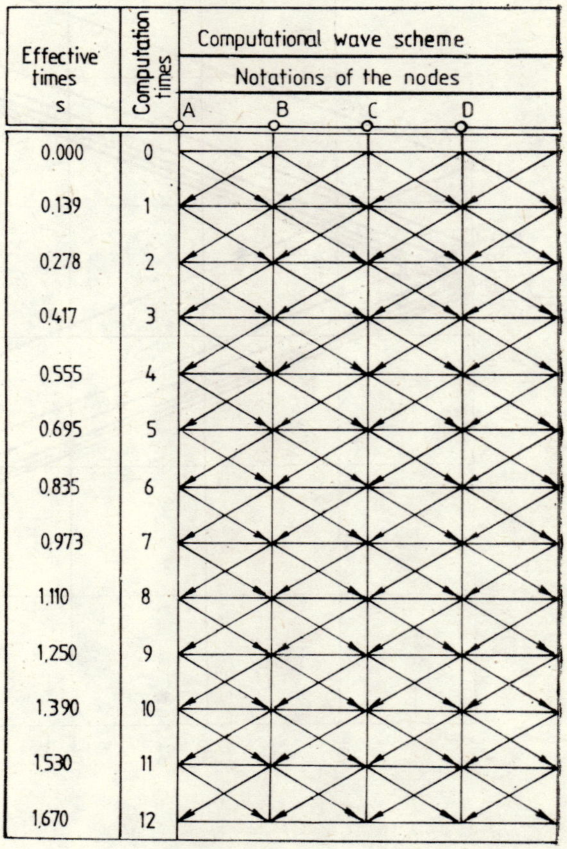

Fig. 6.49. Computational waves for Example 6.10.

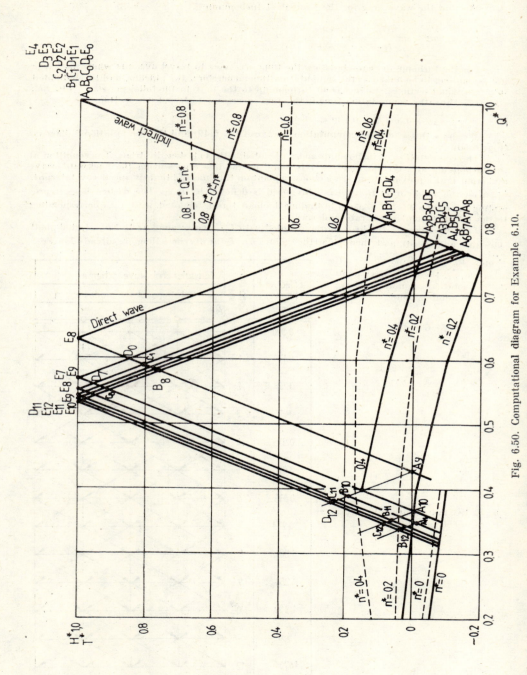

Fig. 6.50. Computational diagram for Example 6.10.

Exact Computation Methods Based on Direct Integration

TABLE 6.6 **Computations for various successive approximations**

Time	Approx.	$n^{*'}_{j+1}$	$T^{*'}_{j+1}$	$T^*_j + T^{*'}_{j+1}$	$K^*(T^*_j + T^{*'}_{j+1})$	$n^{*''}_{j+1}$
0	—	1	1	—	—	—
1	1 2 3	0,6 0,514 0,493 0,490	0,38 0,27 0,24 0,240	1,38 1,27 1,24	0,572 0,528 0,514	0,428 0,472 0,486
2	1 2	0,437 0,377 0,360	0,175 0,11 0,10	0,415 0,35	0,172 0,145	0,318 0,345
3	1 2	0,360 0,300 0,300	0,10 0,04 0,04	0,20 0,14	0,083 0,058	0,277 0,302
4	1 2	0,300 0,275 0,275	0,04 0,01 0,01	0,08 0,05	0,033 0,021	0,267 0,279
5	1 2 3	0,250 0,262 0,266 0,266	−0,01 0 0,005 0,005	0 0,01 0,015	0 0,00415 0,0062	0,275 0,271 0,268
6	1 2	0,266 0,264 0,264	0,005 0,0 0	0,01 0,005	0,004 0,002	0,262 0,264
7	1	0,264 0,264	0 0	0	0	0,264
8	1	0,264 0,264	0 0	0	0	0,264
9	1 2	0,200 0,239 0,239	0,035 0,060 0,060	0,035 0,06	0,145 0,025	0,250 0,239
10	1	0,200 0,200	0,04 0,04	0,1	0,40	0,199
11	1 2	0,200 0,180 0,175	0,04 0,03 0,025	0,08 0,07	0,033 0,029	0,167 0,171
12	1 2	0,175 0,16 0,157	0,030 0,02 0,02	0,055 0,05	0,023 0,021	0,152 0,154

The computations for various successive approximations of the times computed on the computational diagram shown in Fig. 6.49 have summarinzed in Table 6.6.

Use of the computational waves method is rather tedious, but the computer programs secure a rapid and extremely accurate computation of water hammer.

REFERENCES

6.1. ALLIEVI, L., *Teoria generale del moto perturbato dell' acqua nei tubi in pressione*. Ann. Ing. Arch. Italiani, Milano (1903).
6.2. BERGERON, L, *Water hammer in hydraulics and wave surges in electricity*. John Wiley and Sons, New York, 1961.
6.3. CIOC, D., *Hidraulică*. Didactic and Pedagogic Publishing House, Bucharest, 1975.
6.4. DONSKY, B., *Complete pump characteristics and the effects of specific speed on hydraulic transients*. Journal of Basic Engineering, No. 12 (1961).
6.5. GANDENBERGER, W., *Grundlagen der graphischen Ermitlung der Druckschwankungen in Wasserversorgenleitungen*. München, 1950.
6.6. IONEL, I. I., *Instalații de pompare reglabile*. Technical Publishing House, Bucharest, 1976.
6.7. KERR, S. L., *Effect of valve operation on water hammer*. Journal of American Water Works Association, No, 1, (1960).
6.8. KINO, H., *Water hammer control in centrifugal pump systems*. Journal of Hydraulics Division, No. 5 (1968.)
6.9. KINO, H., KENNEDY, F., *Water hammer charts for centrifugal pump systems*. Journal of the Hydraulics Division, No. 5 (1965).
6.10. KITTREDGE, C. P., *Hydraulic transients in centrifugal pump systems*. Transactions ASME, No. 78 (1956).
6.11. KNAPP, R. T., *Complete characteristic of centrifugal pumps and their use in the prediction of transient behaviour*. Transaction ASME, No. 11, (1937).
6.12. KRIVCENKO, G. I., *Ghidromecaniceskie perehodnie professi v ghidroenergheticeskih ustanovkah*, Energhia, Moscow, 1975.
6.13. LESCOVICH, J. E., *The control of water hammer by automatic valves*. Journal of American Water Works Association, No. 5 (1967).
6.14. MOSTKOV, M. A., BASKIROV, A. A., *Rasceti ghidrovliceskovo udara*. Mașinostroenie, Moscow 1968.
6.15. PARMAKIAN, J., *Pressure surge at large pumps installations*. Transaction ASME, No. 8. (1953).
6.16. ROSICH, E., *Le coup de béllier dans les pompages*. La technique de l'eau et de l'assainissément No. 7 (1968).
6.17. SCHNEYDER, O., *Druckstösse in Pumpensteileitungen*. Bauzeitung. No. 22 und 23 (1929).
6.18. SMIRNOV, D. N., *Issledovanie ghidrauliceskovo udara v napornih vodovodah nassosnih stanții*. Mașghiz, Moscow 1959.
6.19. STEPANOFF, A. I., *Pompes centrifuges et helices*. Dunod, Paris, 1961.
6.20. STOIANOVICI, S., *Une méthode pour calcul simplifié du coup de bélier dans les installations pour le pompage d'eau*. La technique de l'eau et de l'assainissément, No. 6 (1969).
6.21. STOIANOVICI, S., *Considerațiuni asupra calculului loviturii de berbec prin metode expeditive*. Hidrotehnica, No. 9 (1964).
6.22. STREETER, L., WYLE, B., *Hydraulic Transients*, New-York, McGraw-Hill Book Co 1967.
6.23. I. C. B., *Îndrumător privind calculul loviturii de berbec si alegerea măsurilor de protecție contra acesteia*. Buletinul Construcțiilor, No. 8 (1975)

7
Automated Control of Pumping Systems by Hydraulic Control Valves

According to their fundamental functions, all control loops can be split into two distinct classes: (1) *corrective controls*, whose aim, as the name implies, is to correct continuously the values of operating parameters (pressure or flow rate); and (2) *protective controls*, whose aim is to protect the pumping system against damage caused by surges or water hammer under certain operating conditions (transient conditions).

Hydraulic controls are known to have several advantages. These include simple design, long service life, and high powers (high forces and speeds). Moreover the pressurized fluid flowing through the system controlled can be used as a control agent (data carrier), which makes hydraulic controls an attractive and, at the same time, economical alternative for the automation of liquid pumping systems.

7.1 Basic Design Analysis of Hydraulic Control Valves Used in Pumping System

7.1.1 The Control Valve as a Component of the Automated System

Let us consider the automated control system shown in Fig. 7.1. The purpose of the system is to keep the output signal x_0 (the actual value of the controlled variable) from the controlled process AP and the feedback signal x_f within the limits set for the point signal (i.e. the desired value) by an input signal x_i. When the output signal changes under the action of disturbance z, the measuring cell, i.e. the transducer T, conveys infor-

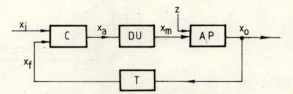

Fig. 7.1. Diagrammatic representation of the components of an automatic control system:

AP — automated process; T — transducer; C — controller; DU — driving unit; x_i — input signal; x_f — feed-back signal; x_0 — output signal; x_a — acting signal; x_m — the variable through which the automated process is controlled; z — disturbance.

mation about this change to the control element C, via the feedback signal x_f. The controller finds the deviation of the controlled variable from the input signal, then generates an actuating signal x_0 to monitor the automated process via the actuator AU, which is mounted at the input

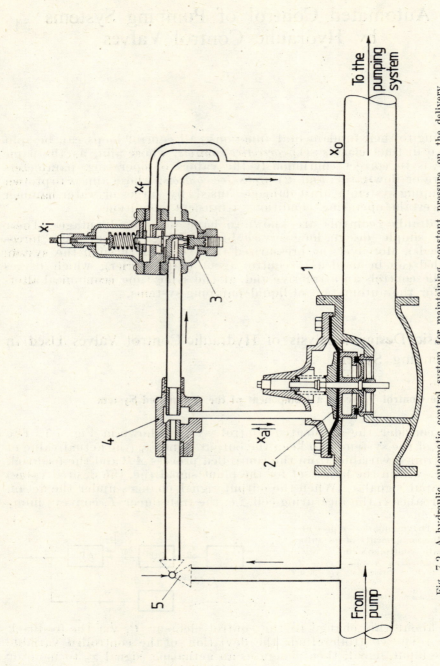

Fig. 7.2. A hydraulic automatic control system for maintaining constant pressure on the delivery side of a pumping loop:

1 — throttle valve; *2* — hydrostatic servomotor; *3* — pressure reducing control; *4* — ejector; *5* — needle valve.

of the controlled process to act upon the controlling variable. As a result of this action, correction x_c is changed so as to offset the effect of the disturbance, and, by the end of the transient period the controlled variable resumes its pre-set value.

An example of an automated pressure control system, taken from actual practice, is illustrated in Fig. 7.2. The pressure is controlled at the turbo-pump outlet, and the operating principle can be readily seen from the block diagram in Fig. 7.1.

In actual practice, automated control systems are often more sophisticated than the one illustrated in these figures. However, from the aspect of the actuator, they all raise the same problems. We shall therefore take the system described above as a standard, and make a few useful observations which are generally valid for other arrangements as well. In the first place, we note that the system includes one actuator through which the controller acts on the process to be monitored (the turbo-pump discharge pipe); the actuator changes the value of the input variable to the automated process as a function of the control signal received. Secondly, we see that the automation device, including controller and actuator, is connected to the process in two places: at the input through the actuator, and at the output via the controller. Both connections confront the designer of the automated system with severe problems. The actuator (the control valve) should be chosen on the basis of laborious calculations which take into account a great many feasibility aspects: energetic, engineering, economic, etc.

Essentially, the whole actuator can be split into two parts (Fig. 7.3): the final control element (the control valve proper), and the hydrostatic driving servomotor.

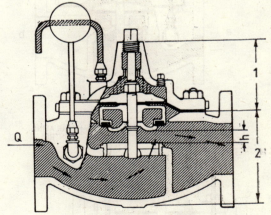

Fig. 7.3. Diagrammatic representation of an acting element:
1 — driving servomotor; *2* — the control member proper; *h* — rod travel; *Q* — flow rate through the valve.

The input variable to the control valve is the hydraulic control signal x_a from the controller, while the output variable can be, say, the fluid flow of the controlled process. Under the action of the hydraulic actuating signal x_a, the stem of the driving servomotor travels over distance h, which is a measure of the input variable to the control valve. By a system of mechanical links and levers, the distance h travelled by the valve stem controls the cross-sectional area of the free fluid passage through the control valve, and hence it controls the fluid flow of the process. Thus the fluid flow through the valve is made a function of the actuating signal x_c.

Now, any control valve included in an automated control system has its own static and dynamic characteristics. These affect both the behaviour of the automated system in steady conditions and the effective performance of the control process. Any user who has to choose a control valve has two alternatives: (a) to go for a valve with linear static and dynamic characteristics, or (b) to select a valve with non-linear static and dynamic characteristics, which instead offers other convenient features. Choice (a) should be made whenever all other components of the automated system in view also have linear characteristics. In such a case, in order to avoid disturbing the linearity of the automated system thus built, a control valve with linear static and dynamic characteristics will be appropiate. The second situation occurs when the other components of the automated system form a non-linear sub-system. Then the control valve should be chosen so as to compensate for the lack of linearity intro-

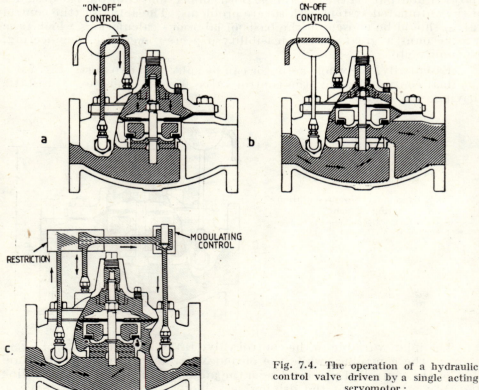

Fig. 7.4. The operation of a hydraulic control valve driven by a single acting servomotor:

a — on-position; b — off-position; c — modulating control.

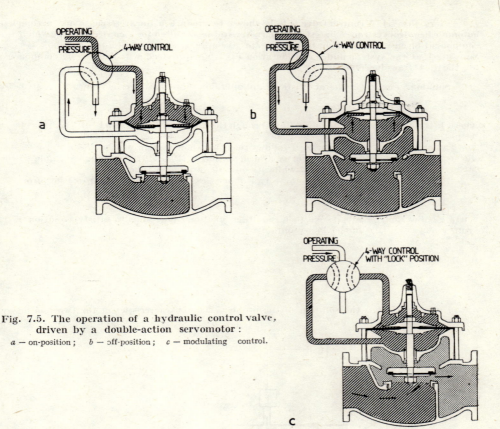

Fig. 7.5. The operation of a hydraulic control valve, driven by a double-action servomotor:
a — on-position; b — off-position; c — modulating control.

duced by the other components, so that the system as a whole will be linear.

The hydrostatic servomotor can be either of two kinds: single-acting (i.e. the pressure of the working liquid is active only on one side of the differential piston), or double-acting (the working liquid is active on both sides of the differential piston).

The operation of a control valve depends on the typical hydrostatic servomotor used as a driving member. If the control valve is driven by a single-acting servomotor, the valve closes (Fig. 7.4, a) when the working liquid flows out of the latter, and any intermediate value between the input and the output pressures of the working liquid will result in a corresponding throttling action (Fig. 7.4, c).

If the control valve is driven by a double-acting servomotor, the valve will close (Fig. 7.5, a) when the working liquid fills the higher chamber while simultaneously flowing out of the lower one. It will open (Fig. 7.5, b) when the working liquid flows out of the higher chamber while simultaneously filling the lower chamber. A throttling action (Fig. 7.5, c) occurs when the working liquid partially fills both chambers.

323

Example 7.1. **A control valve is to be chosen to produce a linear response for a complete automated control system.** What static characteristic is required for a control valve which is a component of an automated control system, to give a linear characteristic for the whole system, knowing that the system under consideration (a) is not provided with linearizing components for the flow transducer.

Solution. The static characteristic of a diaphragm flow transducer is known to be given by

$$x_f = K Q^2 \tag{7.1}$$

where K is a constant, Q is the liquid flow through the pipe, and x_f the signal at the transducer output (the feedback signal).

Since the controller is linear, it follows that the whole system will be linear if the transducer-pipe control-valve sub-system is also made linear (Fig. 7.6); the input variable of the latter is the actuating signal x_a, while its output variable is signal x_f.

We seek to obtain, between these two quantities, a linear relationship of the form

$$x_f = a x_a \tag{7.2}$$

where a is the slope gradient. We then require a function $Q(x_a)$ so as to satisfy equation (7.2) From equations (7.2) and (7.1) we obtain

$$a x_a = K Q^2$$

and hence

$$Q = \frac{1}{\sqrt{K}} \cdot \sqrt{x_a} \tag{7.3}$$

Equation (7.3) is the solution to the problem. The characteristics of the two non-linear components are depicted in Fig. 7.6.

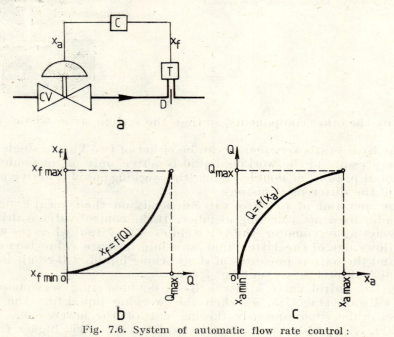

Fig. 7.6. System of automatic flow rate control:
a — principle diagram; *b* — static characteristic of transducer; *c* — static characteristic needed for the control valve to linearize the system; *D* — diaphragm; *T* — transducer; *C* — controller; *CV* — control valve.

7.1.2 Operation Principle of a Hydraulic Control System

As a rule, the control valve in any hydraulic control system will take the place shown in Fig. 7.7, where the hydraulic system includes a pressure supply (the pump), the pipe (pumping system), the control valve and the process to be automatically monitored.

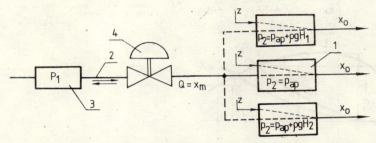

Fig. 7.7. Hydraulic system including supply-pipe-control valve-automated process :
1 — automated process; *2* — pipe; *3* — supply; *4* — control valve; p_1 — supply pressure; p_2 — pressure of the automated process; H_1, H_2 — head differentials between the supply and the automated element; Q — liquid rate of flow.

Under the action of disturbance z, the output (controlled) variable x_0, which is a physical parameter such as pressure, etc., of the controlled process, changes its value.

The control valve acts on the input variable (signal) x_a of the process so as to bring the output variable x_0 within the pre-set limits. The input signal x_e to the automated process is a liquid flow. The automated control process can be set at level H_1, H_2 or O (zero) with respect to the pressure source.

The control action exerted by the control valve on the liquid rate of flow is achieved through changing the hydraulic resistance of the valve (Fig. 7.8). If the supply (consumer) pressure is p_1 while the pressure in the automated process is p_{ap}, then the pressure loss across the hydraulic pipe-control valve system will be

$$\Delta p_s = p_1 - p_2 = p_1 - (p_{AP} \pm \rho g H_{1,2}) \tag{7.4}$$

The pressure loss throughout the system can be expressed as the sum of pressure losses through the pipe Δp_p and across the control valve Δp_v

$$\Delta p_s = \Delta p_p + \Delta p_v \tag{7.5}$$

It is required to keep variable x_e within the pre-set limits, x_{eo}, whatever the value of disturbance z.

Point F on Fig. 7.8 represents the hydraulic system operating at its rated parameters. By changing the rated value of disturbance z_F within the boundary limits z_{min} and z_{max}, the rated operational point F progressively shifts as a consequence of the action of the control valve, until the effect of the disturbance is eventually eliminated.

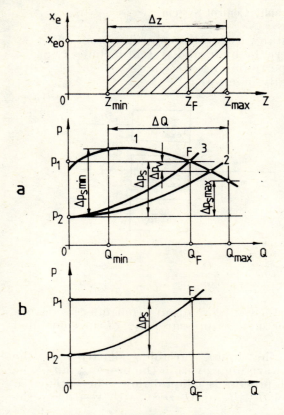

Fig. 7.8. Operating characteristics of hydraulic system including supply pipe-control valve-automated process:

a — when the supply pressure is variable; b — whec the supply pressure is constant; 1 — characteristit of the pressure supply; 2 — pipe characteristic withoun the control valve; 3 — pipe characteristic with the control valve; p_1 — supply pressure; Q_{min}, Q_{max} — minimum and maximum flow rates of the control medium; z_{min}, z_{max} — minimum and maximum values of the disturbance; x_{eo} — set values for the output signal x_e.

In a graphical representation, the pressure losses across the control valve and the pipe, as a function of the flow rate, appear as shown in Fig. 7.8,a if the pressure applied is variable, and as shown in Fig. 7.8, b if it is constant.

Two important conclusions can be reached when operation of the control valve follows the pattern shown in Fig. 7.7. First the values for the maximum and minimum rates of flow, Q_{max} and Q_{min} respectively, required for the control medium, are determined as a function of the minimum and maximum values of the disturbance, z_{min} and z_{max} respectively. The minimum and maximum values of liquid flow through the control valve are of crucial importance for optimal choice of the latter. Second, the pressure loss across the hydraulic system ΔP_s is variable over the same range as that of the flow rate. The same is true for the pressure loss across the pipe ΔP_p and the control valve ΔP_v. Of the three pressure losses mentioned above, ΔP_s can be constant in some conditions, e.g. when the control valve is mounted between two tanks, kept at a constant pressure.

The flow of an ideal liquid through a conduit is described by the continuity and Bernoulli's equations

$$\omega \rho v = C_1 \qquad (7.6)$$

$$\frac{\rho v^2}{2} + p + \rho g H = C_2 \tag{7.7}$$

where ω is cross-sectional area of the conduit;
- v — liquid velocity;
- ρ — liquid density;
- p — liquid pressure;
- H — reference level;
- C_1, C_2 — constants.

In the hydraulic system shown in Fig. 7.7, the rate of flow is changed by the throttling action of the control valve. Figure 7.9 gives a diagrammatical representation of the throttling principle, emphasizing the cross-

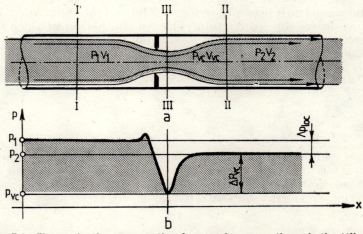

Fig. 7.9. Change in the cross-sectional area of passage through throttling:
a — pipe under throttling proper; *b* — pressure change within the area of the throttled cross-section.

sectional area of free passage of the valve at a certain instant. Beyond section *I*, where the pressure is p_1 and the velocity v_1, the liquid stream is gradually contracted as a result of the throttling action until it reaches the section of maximum contraction *III* (vena contracta, *vc*), where the pressure and velocity reach their maximum values p_{vc} and v_{vc}, respectively. In other words, the pressure energy (associated with the second term in Bernoulli's equation) drops, part of it being converted into kinetic energy, so that the value of the first term in Bernoulli's equation increases. Beyond section *III*, where the cross-sectional area is minimum, the velocity of the liquid stream gradually decreases while its pressure increases to reach, across section *II*, the values v_2 and p_2, respectively. Beyond the throttling point, the pressure of the liquid stream never reaches the original value p_1. This can be explained by the fact that in actual conditions, a liquid flow has to overcome frictional forces as it crosses the throttling section. This is taken into consideration by assigning to Bernoulli's equation a correction factor Δp_{loc}, which stands for the remanent pressure due to friction. Thus, applying Bernoulli's equation for sections *I* and *III*, it results in

$$\frac{\rho v_1^2}{2} + p_1 = \frac{\rho v_{vc}^2}{2} + p_{vc} + \Delta p_{loc} \qquad (7.8)$$

which can be written

$$(p_1 - p_{vc}) - \Delta p_{loc} = \frac{\rho}{2}(v_{vc}^2 - v_1^2) \qquad (7.9)$$

to stress the difference between the two kinds of pressure losses. Term $(p_1 - p_{vc})$ represents the recoverable pressure loss, converted into kinetic energy and again into pressure energy. The second term Δp_{loc} stands for the pressure losses due to friction. The control action of the valve relies exactly on this remanent pressure loss, denoted in Fig. 7.8 and in equation (7.5) by Δp_v.

Applying the continuity equation to the conditions prevailing in section *I* and *III* results in

$$\omega_1 v_1 = \omega_{vc} v_{vc} \qquad (7.10)$$

and

$$\omega_1 v_1 = \mu \, \omega_v v_{vc} \qquad (7.11)$$

respectively, where ω_{vc} denotes the area of the minimum cross-section of the liquid stream across section *III* as a function of the area of the throttled cross-section ω_v, while μ is a contraction coefficient.

Equation (7.11) can be written

$$v_1 = \mu \frac{\omega_v v_{vc}}{\omega_1}$$

which, substituted in equation (7.9), leads to

$$(p_1 - p_{vc}) - \Delta p_{loc} = \frac{\rho}{2}\left(1 - \mu \frac{\omega_v^2}{\omega_1^2}\right) v_{vc}^2$$

and

$$(p_1 - p_{vc}) - \Delta p_{loc} = \alpha_c \frac{\rho v_{vc}^2}{2} \qquad (7.12)$$

where $\alpha_c = 1 - \mu^2 \omega_v^2/\omega_1$.

The remanent pressure loss Δp_{loc} is currently expressed by

$$\Delta p_{loc} = \frac{\rho v_{vc}^2}{2} \xi \qquad (7.13)$$

where ξ is a dimensionless coefficient, called the "local loss coefficient". Remanent pressure losses, of the kind experienced across the throttled section considered, are currently called local pressure losses. Based on relation (7.13), the rate of liquid flow passing through the throttled port under consideration can be readily expressed

$$Q = \frac{1}{\sqrt{\xi}} \omega_v \sqrt{\frac{2}{\rho} \Delta p_{loc}} \tag{7.14}$$

Another factor to be born in mind is the linear pressure loss, caused by frictional forces developed between the liquid stream and the pipe walls. This is currently expressed by the Darcy-Weisbach relation

$$\Delta p_{lin} = \rho \frac{v^2}{2} \cdot \frac{l}{D} \lambda \tag{7.15}$$

where l is the length, D diameter of the pipe, v mean velocity of the liquid stream, ρ liquid density, and λ friction coefficient. Relation (7.15) allows for the rate of liquid flow through the pipe to be expressed as a function of the linear pressure loss. Thus, putting the velocity as expressed in relation (7.15) in an explicit form, and multiplying the result by the pipe cross-sectional area ω_1, results in

$$Q = \sqrt{\frac{D}{\lambda} \omega_1} \sqrt{\frac{2}{\rho} \Delta p_{loc}} \tag{7.16}$$

In the hydraulic system shown in Fig. 7.4, the pipe-control valve sub-system, where the pressure loss Δp_s takes place, includes a number of fittings, straight lengths of conduit, etc., in addition to the control valve. Therefore, taking into account the expressions found for the local and linear pressure losses mentioned above, equation (7.5) can be written in the form

$$\Delta p_s = \sum_{i=1}^{m} \rho \frac{v_i^2}{2} \xi_i + \sum_{i=1}^{m} \rho \frac{v_j^2}{2} \cdot \frac{l_j}{d_j} \lambda_j + \rho \frac{v_v}{2} \xi_v \tag{7.17}$$

where v_v and ξ_v denote the liquid velocity and the coefficient of local resistance across the control valve, respectively.

7.1.3 Static Behaviour of Control Valves

Intrinsic characteristic of control valves. As shown in Section 7.1.1, the input signal (variable) to the control valve is the distance of travel of the throttling member, and the output variable, the rate of liquid flow through the valve. However, since the rate of liquid flow is a function of all hydraulic resistances present in the hydraulic pipe-control valve system, the flow rate is not a unique function of the position of the valve stem. In other words, for the same stem travel, we can expect several flow-rate values, depending on the hydraulic system considered. To assist both the designer and the user in their choice, some intrinsic characteristics, established for certain standard operating conditions, should therefore be specified for the control valve. This leads to the concept of the intrinsic characteristic of the control valve.

From a hydraulic aspect, any control valve can be considered as a local resistance, so the rate of flow through the valve can be expressed by equation (7.14), with the sole difference that the free cross-sectional area ω_v is a variable quantity. However, any change in the free cross-sectional area implicitly brings about a change in the local loss coefficient ξ, in the pressure loss Δp_v [denoted by Δp_{loc} in equation (7.14)] and in the liquid density ρ. For simplifying purposes, variables ξ and ω_v were related thus

$$K_v = \sqrt{2}\, \frac{1}{\sqrt{\xi}}\, \omega_v \tag{7.18}$$

and in terms of K_v relation (7.14) becomes

$$Q = K_v \sqrt{\frac{\Delta p_v}{\rho}} \tag{7.19}$$

Rearranging,

$$K_v = \frac{Q}{\sqrt{\dfrac{\Delta p_v}{\rho}}} \tag{7.20}$$

which affords a definition of the physical meaning of this quantity. Thus, $K_v^{(1)}$ is defined as the flow rate, in m³/h, of a liquid of density $\rho = 1$ kg/dm³, passing through the control valve and suffering thereby a pressure loss $\Delta p_v = 1$ daN/cm². This liquid is water, since its density is 1 kg/dm³ for all practical purposes over the temperature range extending from 5 to 30°C. In the standard conditions already mentioned above, the flow rate value, that is K_v, does not depend on any other hydraulic resistances existing along the conduit, but only on the resistance of the control valve which, in turn, is a function of travel "h" of the valve stem. Consequently, we can write:

$$K_v \equiv K_v(h) \tag{7.21}$$

which is the *intrinsic characteristic* of the control valve, to be specified whenever an order is placed or a delivery is made.

The above is valid for all types of control valves. In what follows, the author deals only with globe valves, the type most frequently used in pumping systems. Along the years, valves having various shapes for the throttling member were tried. This situation led to a wide variety of valve intrinsic characteristics, with little reproducibility or standardization. However, during the last two decades, as a result of advances made in the conception, design and production of automation systems and devices, a certain stability has been achieved in the shapes and types of intrinsic characteristics required for the globe valves. The conclusion was thus reached that, for the successful application of automated control in pumping

---

[1] In the U.S.A., the symbol C_v is used, defined as that waterflow, in U.S. gallons/min, which, passing through the control valve, causes a pressure drop of 1 psi. The conversion ratio between K_v and C_v is $C_v = 1.167\, K_v$; $K_v = 0.856\, C_v$.

systems, only two typical intrinsic characteristics were needed, i.e. linear and logarithmic (Fig. 7.10).

The important points to be specified on the intrinsic characteristic are:

K_{v0} — the value for which the theoretical intrinsic characteristic $K_v(h)$ intersects the K_v axis;

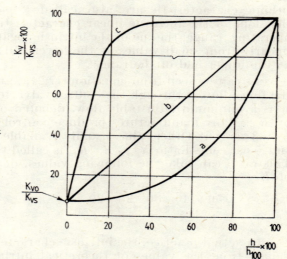

Fig. 7.10. Intrinsic characteristics:
a — logarithmic; b — linear; c — with quick closing (opening) action.

K_{vs} — the K_v value at the rated travel, as designed for a typical, mass-produced valve. This value is to be specified by the constructor; a control valve can have but one K_v, which lies between (1 ± 0.1) times K_{vs};

$K_v 100$ — value of K_v at the rated travel, as determined by measurements; this value should lie within the tolerance range of K_{vs}.

The analytical expression for the linear characteristic is

$$\frac{K_v}{K_{vs}} = \frac{K_{vo}}{K_{vs}} + \left(1 - \frac{K_{vo}}{K_{vs}}\right)\frac{h}{h_{100}} \qquad (7.21,a)$$

and for the logarithmic intrinsic characteristic

$$\frac{K_v}{K_{vs}} = \frac{K_{vo}}{K_{vs}} \exp\left[\ln\left(\frac{K_{vs}}{K_{vo}}\right)\frac{h}{h_{100}}\right] \qquad (7.22)$$

where h_{100} is the rated travel of the control valve stem. Differentiating equation (7.22)

$$\frac{d\left(\dfrac{K_v}{K_{vs}}\right)}{d\left(\dfrac{h}{h_{100}}\right)} = \left(\ln\frac{K_{vs}}{K_{vo}}\right)\frac{K_v}{K_{vs}} \qquad (7.23)$$

which shows that the slope of the intrinsic logarithmic characteristic is proportional, at any point, to the value of K_v in the point considered. In other words, this shows that for any change, dh, in the valve stem travel, there will always be the same change dK_v, as a percentage of K_v in the considered point; in other words, the same ratio dK_v/K_v holds true for both changes above. For this reason, control valves with an intrinsic logarithmic characteristic are also called *"equal percentage"* valves. While the slope of the intrinsic linear characteristic is constant over the whole operating range, the one of the intrinsic logarithmic characteristic varies linearly from small values at the lower end of the range, to high values at the upper end of the range.

As can be seen from equations (7.21a) and (7.22), and from Fig. 7.10, the flow allowed through the control valve, for a travel $h = 0$, is not zero; it is a minimum adjustable flow, denoted on the intrinsic characteristic by K_{vo}. Many manufactures produce control valves with $(K_{vo}/K_{vs}) 100 = 3.33$ per cent. Other values usually available for $(K_{vo}/K_{vs}) 100$ are 2 ; 2.5 ; 4 ; and 5 per cent. Ratio $R_v = K_{vs}/K_{vo}$ is called the *"control ratio"* of the valve. This ratio can take the following values, depending on the above values of K_{vo}

$$\frac{50}{1}\ ;\ \frac{40}{1}\ ;\ \frac{30}{1}\ ;\ \frac{25}{1}\ ;\ \frac{20}{1}$$

In actual practice, the intrinsic characteristics of globe control valves depart from the values of theoretical intrinsic characteristics shown in Fig. 7.10. The most important deviations are recorded near the shut-off point. The permissible departures of actual from theoretical characteristics are set by standard specifications on the basis of experimental tests. Tolerances are established for the slope angle and for the control ratio. Thus, over a range covering at least $h = 10 - 100\%$, the slope of the experimental characteristics cannot exceed that of the theoretical one by more than ± 30 per cent. Similarly the control ratio K_{vs}/K_{vo} cannot be more than 10 per cent below the value of the control ratio specified for the control range claimed for the valve. For illustration purposes, a comparison is made in Fig. 7.11 between theoretical and experimental characteristics of the two basic control modes (linear and logarithmic).

In Fig. 7.11 another specific value of K_v is introduced, namely K_{vv}. This represents the lowest K_v value for which the tolerance allowed for the slope of the intrinsic characteristic is still valid. As can be seen, K_{vv} differs from K_{vo} inasmuch as it corresponds to stem travels different from zero; generally, it will be higher than K_{vo}.

Some values of coefficients K_{vs}, K_{vv} and K_{vo} for Clayton hydraulic control valves, built by Clayton Automatic Valves Co., U.S.A., are reported in Table 7.1.

Example 7.2. **Estimation of K_v**. It is required to find the value of $K_{v\,max}$ for the optimal control valve for the automated control of hydraulic pressure through a given drinking-water pumping system, knowing that

$Q_{v\,max} = 30$ m³/h ; pressure loss across the valve for the maximum flow rate, $\Delta p_{v\,max} = 0.3$ daN/cm² ; $\rho = 1\,000$ kg/m³.

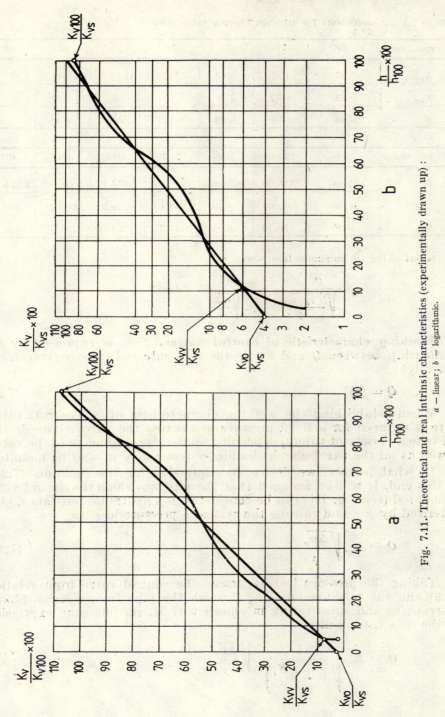

Fig. 7.11. Theoretical and real intrinsic characteristics (experimentally drawn up):
a — linear; *b* — logarithmic.

TABLE 7.1 K_v coefficients for different Clayton valve sizes

Valve size, in mm	20	25	32	40	50	65	80
K_{vs}	5.99	10.27	18.83	22.25	41.08	56.49	84.74
K_{vv}	0.89	1.52	2.80	3.30	6.15	8.40	12.70
K_{vo}	0.30	0.51	0.93	1.10	2.05	2.80	4.20

Valve size, in mm	100	150	200	250	300	350	400
K_{vs}	175.48	414.30	636.86	963.85	1,473.17	1,807.87	2,428.47
K_{vv}	26.40	62.00	95.50	143.43	219.22	269.02	361.37
K_{uo}	8.80	20.70	32.00	47.77	73.02	89.61	120.37

Solution. For the maximum flow rate

$$K_{v\,max} = \frac{Q_{v\,max}}{\sqrt{\frac{\Delta p_{v\,max}}{\rho}}} = \frac{30}{\sqrt{\frac{0.3}{1}}} = 54.50$$

Working characteristic of control valves. This is expressed by the relationship between Q and h (i.e. the flow rate and the stem travel, respectively)

$$Q = f(h) \qquad (7.24)$$

When establishing the working characteristic of a control valve, there is no need for a fixed pressure loss across the valve; indeed, this will take a variety of values, depending on the size and make of the valve, as well as on the particular hydraulic system where it is to be mounted.

In what follows, we derive an expression for the function (7.24). To this end, it will be assumed that the pipe on which the control valve is mounted (see Fig. 7.4) can be considered as a hydraulic resistance, characterized by K_p and causing the system a pressure loss Δp_p

$$Q = K_p \sqrt{\frac{\Delta p_p}{\rho}} \qquad (7.25)$$

Taking the pressure loss Δp_v across the control valve from relation (7.19) and the pressure loss Δp_p through the pipe from relation (7.25), re-arranging and substituting in equation (7.5), the following expression for the flow rate results

$$Q = K_v \frac{1}{\sqrt{1 + \left(\frac{K_v}{K_p}\right)^2}} \sqrt{\frac{\Delta p_s}{\rho}} \qquad (7.26)$$

Now, assuming that the control valve is fully open, that the flow rate is constant, and that the fluid is incompressible so that K_p stays constant, then the expression for the maximum flow will be

$$Q_{100} = K_p \sqrt{\frac{\Delta p_{p\,100}}{\rho}} \qquad (7.27)$$

If expressed for the control valve, the same flow rate can be written as

$$Q_{100} = K_{v100} \sqrt{\frac{\Delta p_{p\,100}}{\rho}} \qquad (7.28)$$

from which the ratio K_{v100}/K_p is obtained, and may be substituted in equation (7.27)

$$Q = K_p \frac{1}{\sqrt{1 + \left(\frac{K_v}{K_{v\,100}}\right)^2 \cdot \frac{\Delta p_{p\,100}}{\Delta p_{v\,100}}}} \sqrt{\frac{\Delta p_s}{\rho}} \qquad (7.29)$$

Now, putting $\Delta p_{p100} = \Delta p_{s100} - \Delta p_{v100}$ for the pressure loss through the pipe with the valve fully open, we can write

$$Q = K_v \frac{1}{\sqrt{1 + \left(\frac{K_v}{K_{v\,100}}\right)^2 \left(\frac{\Delta p_{s\,100}}{\Delta p_{v\,100}} - 1\right)}} \sqrt{\frac{\Delta p_s}{\rho}} \qquad (7.30)$$

Dividing equation (7.30) by (7.28) results in

$$\frac{Q}{Q_{100}} = \frac{1}{\sqrt{1 + \frac{\Delta p_{v\,100}}{\Delta p_{s\,100}}\left[\frac{1}{\left(\frac{K_v}{K_{v\,100}}\right)^2} - 1\right]}} \sqrt{\frac{\Delta p_s}{\Delta p_{s\,100}}} \qquad (7.31)$$

which is valid for incompressbile fluids of low viscosity.

Relation (7.31) is used to draw up the working characteristic Q/Q_{100} as a function of stem travel h/h_{100}, utilising the intrinsic characteristic $K_v/K_{v100} = f(h/h_{100})$ given by equation (7.21) or (7.22). In the expression for Q/Q_{100}, ratio $\Delta p_{v100}/\Delta p_{s100}$ is a parameter characteristic of the hydraulic system where the control valve is mounted, while ratio $\Delta p_s/\Delta p_{ps100}$ can either be constant or vary, as a function of the travel.

The working chara̧ctceristic is expressed above in terms of the flow rate, corresponding to the maximum lift of the stem, Q_{100}. It should however be mentioned that it is not uncommon to see this characteristic expressed in another way, by reference to the maximum flow rate Q_{max}, which applies when the valve is fully open and the pressure loss Δp_{p100} through the pipe is nil. In this situation, the control valve is the sole hydraulic resistance in the system and hence one can write $\Delta p^*_{v\,100} = \Delta p^*_{s\,100}$. In such conditions

$$Q_{max} = K_{v\,100} \sqrt{\frac{\Delta p^*_{v\,100}}{\rho}} \qquad (7.32)$$

and

$$Q_{max} = K_{v\,100} \sqrt{\frac{p^*_{s\,100}}{\rho}} \qquad (7.33)$$

Dividing equation (7.30) by (7.33)

$$\frac{Q}{Q_{max}} = \frac{1}{\sqrt{\dfrac{1}{\left[\dfrac{K_v}{K_{v\,100}}\right]^2} + \dfrac{\Delta p_{s\,100}}{\Delta p_{v\,100}} - 1}} \sqrt{\frac{\Delta p_s}{\Delta p^*_{s\,100}}} \qquad (7.34)$$

A better understanding of the quantities introduced by relations (7.31) and (7.34) is provided by Fig. 7.12. As can be seen in the diagram, there is a clear difference between Q_{max} and Q_{100} on the one hand, and $\Delta p^*_{s\,100}$, $\Delta p_{s\,100}$ and Δp_s on the other.

There are hydraulic systems where the pressure loss across the whole system Δp_s is not a function of the flow rate (see Fig. 7.14), or varies very little, so that the ratio $\Delta p_s/\Delta p_{s\,100}$ in equation (7.31) can be taken roughly as unity. Then

$$\frac{Q}{Q_{100}} = \frac{1}{\sqrt{1 + \dfrac{\Delta p_{v\,100}}{\Delta p_{s\,100}}\left[\dfrac{1}{\left(\dfrac{K_v}{K_{v\,100}}\right)^2} - 1\right]}} \qquad (7.35)$$

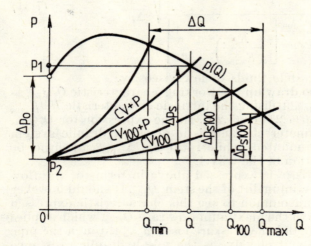

Fig. 7.12. Diagrammatic representation of quantities Q_{max}, Q_{100}, ΔP_s, ΔP_{s100}, ΔP^*_{s100}. Legend for the characteristics of the pumping loop:

CV_{100} — within the hydraulic system, only the control valve exists, in the fully-open position; $CV_{100} + P$ — within the hydraulic system the control valve is fully open and there is also the pipe; $CV + P$ — the hydraulic system including the control valve and the pipe; the former is in an intermediate position.

It follows that if the pressure loss has a constant value over the whole system, the working characteristic (as expressed by the co-ordinates Q/Q_{100} and h/h_{100}) does not depend on the control valve size, but only on the nature of its intrinsic characteristic, i.e.: linear or logarithmic. For various values of the parameter $\Delta p_{v\,100}/\Delta p_{s\,100}$, one can draw a family of working characteristics of the linear control valve (Fig. 7.13, a) and another family for the logarithmic control valve (Fig. 13, b), both characteristics being drawn for the same control ratio.

The procedure is to take, for various values of travel h/h_{100}, the corresponding values for $K_v/K_{v\,100}$, given on the diagrams in Fig. 7.11, a and b, and then introduce these values into relation (7.35). The operating characteristic $Q/Q_{100} = f\,(h/h_{100})$ is thus established by means of the intrinsic characteristic $K_v/K_{100} = f(h/h_{100})$.

From Fig. 7.13, a and b, and equation (7.35), we can see that, if $\Delta p_{s\,100} = \Delta p_{v\,100}$, the working characteristic is coincident with either the linear or the logarithmic characteristic, according to the type of control valve. The two typical intrinsic characteristics are represented in Fig. 13, a and b, for which $\Delta p_{v100}/\Delta p_{s100} = 1$. These situations are met in actual practice when the pipe section carrying the control valve is very short and has few local resistances, so that pressure losses through the pipe are negligible with respect to pressure losses in the control valve, i.e. $\Delta p_p \ll \Delta p_p$

When the pressure losses through the pipe represent a significant proportion of the losses, reported for the whole system (including the pipe and the control valve), as

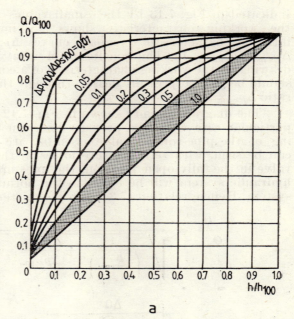

a

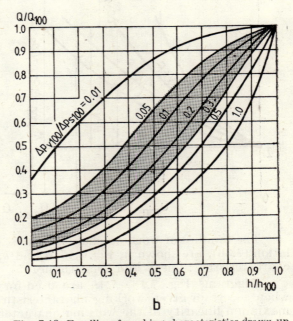

b

Fig. 7.13. Families of working characteristics drawn up for control valves with $K_{vs}/K_{v0} = 25/1$:
a — family of linear characteristics; b — a family of logarithmic characteristics.

indicated in Fig. 7.13 by the small values of the ratio $\Delta p_{v\,100}/\Delta p_{s\,100}$, the working characteristic shows an important deviation from the intrinsic characteristic (compare curves $\Delta p_{v100}/\Delta p_{s100} = 0.01$ with curves $\Delta p_{v\,100}/\Delta p_{s\,100} = 1$). The conclusion of what was said above is that when drawing up the working characteristic of a control valve, due consideration should be paid to pressure losses through the pipe, as expressed by the ratio $\Delta p_{v\,100}/\Delta p_{s\,100}$.

The diagram already shown in Fig. 7.12 reappears in Fig. 7.14 for the particular case when the pressure losses in the hydraulic system consisting of the pipe-control valve assembly are constant. In such a case, as can be seen, with zero pressure losses in the pipe and with the control valve in the fully open position, the pressure loss $\Delta p_{s\,100}$ over the whole hydraulic system will be equal to the normal pressure losses Δp_s in the system. Hence, relation (7.34) can be written

$$\frac{Q}{Q_{max}} = \frac{1}{\sqrt{\frac{1}{\left(\frac{K_v}{K_{v\,100}}\right)^2} + \frac{\Delta p_{s\,100}}{\Delta p_{v\,100}} - 1}} \tag{7.36}$$

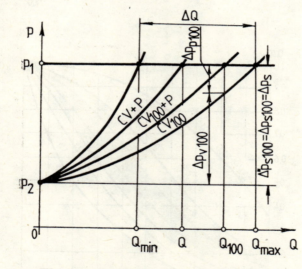

Fig. 7.14. Same as fig. 7.12, for the case when $\Delta P_s =$ constant.

The families of working characteristics, drawn up on the basis of relation (7.36), are shown in Fig. 7.15, a and b, for the two typical intrinsce characteristics: linear and logarithmic.

Diagrams Fig. 7.13—7.15 are used for the choice of control valves when a certain given working characteristic is required. If, for instance, maximum linearity of the working characteristic is required, we may use either a linear or a logarithmic valve, depending on the ratio of valve, pressure losses to system pressure losses. Thus, *for* $\Delta p_{v100}/\Delta p_{s100}$ *ratios from 0.5 to 1, linear valves are most suitable for linearity of the working characteristic*

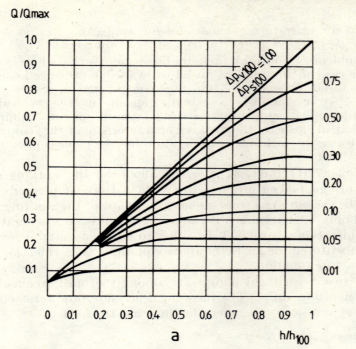
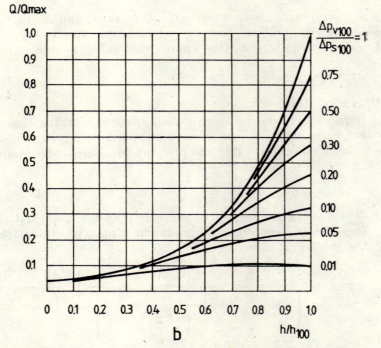

Fig. 7.15. Families of working characteristics at the maximum flow rate for control valves with $K_{vs}/K_{vo} = 50/1$:
a — family of linear characteristics; b — family of logarithmic characteristics.

while logarithmic valves are recommended for $\Delta p_{v100}/\Delta p_{s100}$ *ratios from 0.05 to 0.3, since they achieve better linearity of the characteristic.*

The problem of constant pressure losses across the whole system is a particular case of the problem raised by variable pressure losses across the system, and depends on the pipe-control valve. Since pressure loss-changes, over the whole system, depend on the typical device used to build up the pressure (centrifugal pump, piston pump, etc.), it follows that in this case the working characteristic of the control valve should be drawn up with due consideration for the typical pressure generating unit used.

Equations (7.31) and (7.34), derived above for the working characteristics, are of a general nature, in terms of the pattern of pressure changes Δp_s across the system. As a consequence, the shape of the working characteristics will only be determined if the particular pressure generating unit under consideration is specified. In order to do so, a suitable expression should be substituted for Δp_s in the above equations. As a rule, the expression is a flow rate function.

In practice, centrifugal pumps are among the most frequently used pressure generating units. The theoretical pressure-flow rate characteristic of the typical pressure generating unit has the form

$$p = An^2 + Bn Q - C Q^2 \tag{7.37}$$

where p is the pressure; n — speed; Q — flow rate; and A, B, C, — positive constants. In practice, however, at constant speeds the characteristics of centrifugal pumps may be approximated by the relation

$$p = p_0 - C_1 Q^2 \tag{7.38}$$

where C_1 is a positive constant.

Our purpose is to derive an expression for the working characteristic of a control valve when the pressure is supplied by a centrifugal pump. Bearing in mind equation (7.38), the pressure loss across the whole system (Fig. 7.12) can be expressed as

$$\Delta p = \Delta p_0 - C_1 Q^2 \tag{7.39}$$

Substituting equation (7.39), which is derived for a normal flow rate and for a flow rate Q_{100} corresponding to the fully open position of the throttling device, in (7.31), yields

$$\frac{Q}{Q_{100}} = \frac{1}{\sqrt{1 + \frac{\Delta p_{v\,100}}{\Delta p_{s\,100}}\left[\frac{1}{\left(\frac{K_v}{K_{v\,100}}\right)} - 1\right]}} \sqrt{\frac{\Delta p_0 - C_1 Q^2}{\Delta p_0 - C_1 Q_{100}^2}}$$

$$\tag{7.40}$$

and

$$\frac{Q}{Q_{100}} = \frac{\sqrt{\delta}}{\sqrt{1 + \frac{\Delta p_{v\,100}}{\Delta p_{s\,100}}\left[\frac{1}{\left(\frac{K_v}{K_{v\,100}}\right)} - 1\right](\delta - 1) + 1}} \qquad (7.41)$$

where $\delta = \Delta p_0/C_1 Q_{100}^2$.

Substituting the value of δ for a given pump, we can draw up the family of working characteristics for either the linear or the logarithmic valve. A similar expression for Q/Q_{100} can be derived also for Q/Q_{100}; here however, $\delta = \Delta p_0/C_1 Q_{max}^2$.

Example 7.3 **Mathematical analysis of the working characteristic of a control valve.** It is required to find the working characteristic of a logarithmic valve mounted on the feed-water supply pipe of a steam boiler. Data available: pressure loss across the fully open valve $\Delta p_{v\,100} = 1.3$ daN/cm^2; boiler pressure $p_2 = 48$ daN/cm^2; pressure at the pump outlet in idle-running conditions shut-off valve) $p_1 = 64$ daN/cm^2; constant $C_1 = 6 \times 10^{-3}$; maximal flow rate $Q_{100} = 40$ m^3/h.

Solution. With the above data, we can write

$$\delta = \frac{\Delta p_0}{C_1 Q_{100}^2} = \frac{64 - 48}{6 \times 10^{-3} \times 40^2} = 1.67$$

$$\Delta p_{s100} = \Delta p_0 - C_1 Q_{100}^2 = 16 - 6 \times 10^{-3} \times 40^2 = 6.4 \text{ daN/cm}^2$$

Substituting the above results in equation (7.41) gives

$$\frac{Q}{Q_{100}} = \frac{\sqrt{1.67}}{\sqrt{\left[1 + 0.203\left(\frac{1}{\left(\frac{K_v}{K_{v100}}\right)^2} - 1\right)\right]0.67 + 1}}$$

Function $Q/Q_{100} = f(h/h_{100})$ can be diagrammatically illustrated by using the $K_v/K_{v\,100} = f(h/h_{100})$ diagram shown in Fig. 7.11, *b*. The point diagram obtained is shown in Fig. 7.16. On the same diagram, the dotted line represents, for comparison purposes, the working characteristic of the same valve when the pressure losses across the whole system are constant.

In Section 7.1.3, the variable K_{vv} was defined as that value of K_v for which the slope tolerances for the intrinsic characteristic are still satisfied. For section $K_{vv} - K_{vs}$ on the experimental intrinsic characteristic (Fig. 7.17), the two allowed tolerances are not satisfied any more. To K_{vv} on the intrinsic characteristic it corresponds a certain value of the flow rate on the working characteristic. This value is known as the *"minimum adjustable flow rate"*, and is mathematically derived by substituting K_{vv} in relations (7.31) and (7.35). Looking at Fig. 7.13, one can easily see that, for any value $\Delta p_{v\,100}/\Delta p_{s\,100}$, the value of the minimum adjustable flow rate is higher than K_{vv} on the intrinsic characteristic.

In practice, however, it was found that K_{vv} is generally higher than K_{vo}, as defined by the intersection point of the intrinsic characteristic and the ordinate. This is due to the many difficulties encountered when trying to establish an adequate mathematical model for simultating the behaviour of the shut-off device within the valve closing range, and to the many difficulties encountered in the mechanical operation of the shut-off device.

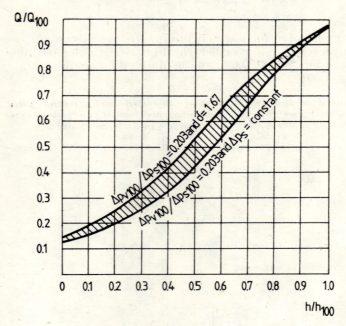

Fig. 7.16. Working characteristic of the control valve illustrated in Fig. 7.3.

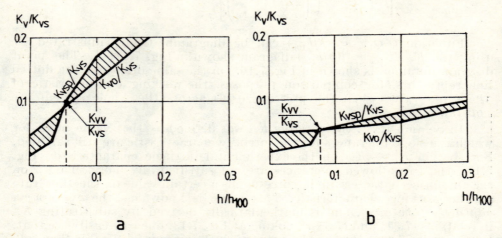

Fig. 7.17. Details of intrinsic characteristics:
a — linear; b — logarithmic;

The fact that K_{vv} is greater than K_{vo} has an immediate and detrimental effect on the control valve performance, which is no longer $K_{v\,100}/K_{vo}$ but $K_{v\,100}/K_{vv}$. Accordingly, the manufacturers are primarily concerned to design and build their control valves so that K_{vv} be as close as possible to K_{vo}.

As a general rule, bearing in mind their operational function, control valves should not provide a perfect shut-off state in the closed position. To achieve this condition, a soft ring (for instance, of polytetrafluorethylene) is inserted on the valve seat.

On the intrinsic characteristic shown in Fig. 7.17, the value of K_v, as experimentally established for $h/h_{100} = 0$, was denoted by $K_{vl}(K_l$ for leak). The presence of K_{vl} brings about a certain value on the working characteristic, this value being called *"leak flow"*; it is established from equation 7.20 where K_{vl} is substituted for K_v

$$Q_l = K_{vl} \sqrt{\frac{\Delta p_v}{\rho}} \qquad (7.42)$$

The *VDI/VDE* German standards specify certain boundary values for K_{vl}, meant to limit the leak flow. Thus, for globe valves, $K_{vl} \leqslant 0.0005\,K_{vs}$. As can be seen both from the stated limit values and Fig. 7.17, the leak flow is much lower than the minimum adjustable flow. However, from the aspect of the control process, these small values cannot possibly have any detrimental effect.

Example 7. 4. **Mathematical analysis of the minimal adjustable flow and of the maximum leak flow of a control valve.** It is required to find the minimum adjustable flow, the control ratio and the maximum leak flow of a linear valve, mounted as part of a hydraulic system with a steady pressure loss across the whole system, Δp_s. Data available: $\Delta p_{v\,100}/\Delta p_{s\,100}=0.6$; $Q_{100} = 130$ m³/h; $K_{vs} = 70$; $K_{vo}=0.04\,K_{vs}$; $K_{vv} = 0.049\,K_{vs}$; $K_{v\,100} = 71$; $\Delta p_v=15$ daN/ m²; and $\rho = 0.870$ kg/m³.

Solution. Using equation (7.35), we get

$$\frac{Q}{Q_{100}} = \frac{1}{\sqrt{1 + 0.6\left[\left(\dfrac{1}{\dfrac{0.049 \times 70}{71}}\right)^2 - 1\right]}} = 0.0621$$

and the minimum adjustable flow rate

$$Q = 0.0621 \times 130 = 8.07 \text{ m}^3/\text{h}$$

The actual control ratio is

$$\frac{K_{v\,100}}{K_{vv}} = \frac{71}{0.049 \times 70} = \frac{20.7}{1}$$

In our case, the control valve will have a working characteristic $\Delta_{v\,100}/\Delta p_{s\,100} = 0.6$ over the range 8.07—130 m³/h; below 8.07 m³/h, the working characteristic is no longer within the permitted tolerance range. Likewise, it will be noticed that the actual control ratio (20.7/1) is smaller than the theoretical ratio (25/1).

Applying relation (7.42), it gives

$$Q_{l\ max} = K_{vl max}\sqrt{\frac{\Delta p_v}{\varrho}} = 0.005 \times 70 \sqrt{\frac{15}{0.87}} = 1.45 \text{ m}^3/\text{h}$$

It will be noticed that the maximum leak is very small with respect to the maximum flow of 130 m³/h. The actual leak flow can take values below 1.45 m³/h, which is closer to the imposed minimizing requirements.

Development of cavitation within control valves. The cavitation phenomenon consists of the development and collapse of vapour and gas bubbles within the flowing liquid that evolve through vaporization, as a result of the local pressure dropping below a critical value. Once built up, these bubbles (also called cavities) are carried by the stream to areas with pressures above the critical level, where the vapour suddenly condenses and causes pressure surges of tens or even hundreds of thousands of atmospheres.

The critical pressure for onset of the cavitation phenomenon is a function of the physical characteristics of the liquid, as well as of the amount, size and distribution within the liquid mass of the gas nuclei. With a good safety margin, the critical pressure is taken as the vaporization pressure at the temperature of the liquid.

As already said, when a fluid passes through a control valve its pressure drops from p_1 to p_2. During the passage, the pressure goes through a minimum value which, generally, is lower than p_2. If the fluid is a liquid, whenever the pressure falls to the value of the vapour tension corresponding to the system temperature, vaporization occurs. In the range where the pressure of the fluid stream is lower than the vapour pressure, the reverse phenomenon of vapour condensation takes place, also known under the designation of *implosion of vapour bubbles* (see Section 4.1.1).

The change in liquid pressure, as it passes through a control valve with and without the vaporization phenomenon, is shown in Fig. 7.18. As can be seen, this phenomenon of cavitation in a liquid flow, associated with vaporization and condensation phenomena, is highly undesirable because it adversely affects the valve components, and in particular the closing member and the seat. Indeed, when in service, control valves, subject to cavitation, have to withstand vibration, erosion (due to pressure surges) and corrosion (due to the chemical action of gases, evolved from the liquid towards the metallic walls of the valve body), and we do not yet know of a single material which could be claimed to be perfectly resistant to cavitation (see Section 4.1.6).

The operation of control valves in cavitation conditions is a difficult problem, from the aspect of both material strength and dimensioning analysis. For, in actual practice, the cavitation phenomenon is not as simple as described above; in fact, it is rather highly complex and not yet fully understood. Research on this topic is currently in progress, the main targets being to get a better insight into the physics of the phenomenon, to devise a mathematical apparatus for design analysis, and to establish adequate measures to eliminate, or at least to reduce, cavitation.

As long ago as 1962, Bauman suggested a *"critical flow rate"* coefficient, C_f, to correct most of the errors in the value of Q when computed from equation (7.19); with this coefficient, the above relation becomes

$$Q = C_f K_v \sqrt{\frac{p_1 - p_{vap}}{\rho}} \ m^3/h \qquad (7.43)$$

where p_{vap} is the liquid vapour pressure at the flow temperature.

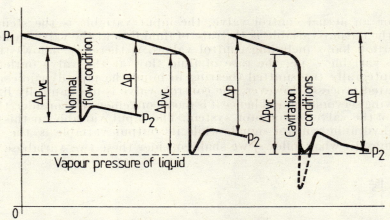

Fig. 7.18. Pressure changes within a control valve in two flow conditions: normal, or with cavitation.

The drawback of this coefficient lies in the fact that it has to be established for each particular valve, like coefficient K_v, which is time consuming and hence costly.

Comparison of equations (7.43) and (7.19) results in

$$p_1 - p_2 = C_f^2 (p_1 - p_2) \qquad (7.44)$$

The value of expression $C_f^2(p_1 - p_{vap})$ is a measure of the maximum differential pressure Δp_{max} for which the flow conditions are still normal. At $\Delta p_{real} > \Delta p_{max}$, cavitation will develop.

If $\Delta p_{real} < \Delta p_{max}$, we shall use p_{real} in calculations, whilst if $\Delta p_{real} > \Delta p_{max}$, the latter value will be used.

Example 7.5. **Mathematical analysis as a check against cavitation.** It is required to choose the most adequate control valve in the following conditions: pressure, upstream, $p_1 = 11.5$ daN/cm²; pressure, down stream, $p_2 = 4.2$ daN/cm²; differential pressure $\Delta p = 7.3$ daN/cm²; temperature, upstream, 80°C; vapour pressure at 80°C, $p_{vap} = 0.5$ daN/cm² (where no cavitation occurs).

Solution. Assuming that $C_f = 0.68$ for the valve likely to be used, then

$$\Delta p_{max} = C_f^2 (p_1 - p_{vap}) = 0.68^2 \ (11.5 - 0.5) = 5.1 \ daN/cm^2$$

However, since $p_1 - p_2 = 7.3$ daN/cm² > 5.1 daN/cm², it follows that with such a valve, cavitation will occur.

Let us then assume that another typical valve will be used, for which $C_f = 0.90$. Then

$$\Delta p_{max} = C_f^2(p_1 - p_{vap}) = 0.90^2 \, (11.5 - 0.5) = 8.9 \text{ daN/cm}^2$$

And since in such a case, $\Delta p_{max} < \Delta p$, it follows that no cavitation will occur.

7.1.4. Dynamic Behaviour of Control Valves

For an actual control valve, the input variable is the stem travel while the output variable is the rate of fluid flow at the valve outlet. With automation loops including control valves, it therefore follows that this output variable — i.e. the rate of fluid flow at the valve outlet — has to be integrally transmitted (bearing in mind the pipe dynamics) to the automated process. However, the control valve is mechanically linked to its driving servomotor and hence it is most convenient to consider the dynamics of the valve-servomotor system. The input variable to this system is the hydraulic control signal, while its output variable is the rate of fluid flow. In what follows, we shall consider these two variables.

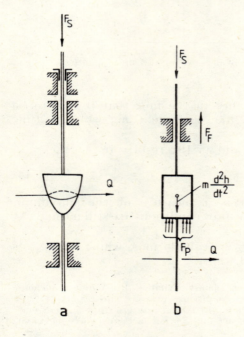

Fig. 7.19. Diagrammatic representation for calculation purposes:
a — moving part of a control valve; *b* — its mechanical equivalent.

The moving section of a globe control valve is diagrammatically shown in Fig. 7.19, *a*; the equivalent mechanical diagram of the same is illustrated, for dynamic conditions, in Fig. 7.19, *b*.

The equation for the equilibrium of forces in a vertical projection leads to

$$m\frac{d^2h}{dt} + \mu\frac{dh}{dt} + F_P = F_S \qquad (7.45)$$

where m is the mass of the movement;
 h — stem travel;
 μ — a coefficient to account for the viscous friction;
 F_P — force resulting from the fluid differential pressure on the shut-off member;
 F_S — driving force developed by the servomotor.

If the valve is driven by a hydraulic servomotor (see Fig. 7.20), then the expression for force F_S will be

$$F_S = AP_m - kh \qquad (7.46)$$

where A is the active area of the diaphragm;
 P_m — pressure on the membrane;
 k — spring constant.

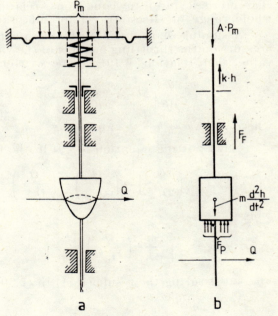

Fig. 7.20. Diagrammatic representation for calculation purposes, when the driving unit is a diaphragm servomotor:
a — moving section of a globe control valve; *b* — its mechanical equivalent.

Substituting equation (7.46) in (7.45), it results

$$m\frac{d^2h}{dt^2} + \mu\frac{dh}{dt} + kh = AP_m - F_P \qquad (7.47)$$

Looking at relation (7.45) above, we see that we need to introduce into it the valve flow rate, which is the true output variable; to this end, we make use of relation (7.31) whereby the relationship between stem travel h and flowrate Q through the intermediary of K_v is expressed

$$\frac{Q}{Q_{100}} = \frac{1}{\sqrt{1 + \frac{\Delta p_{v\,100}}{\Delta p_{s\,100}}\left[\frac{1}{\left(\frac{K_v}{K_{v\,100}}\right)} - 1\right]}} \sqrt{\frac{\Delta p_v}{\Delta p_{s\,100}}} \tag{7.48}$$

Now, equations (7.47) and (7.48) build up between them the mathematical model simulating the dynamical conditions through the control valve.

To obtain a simpler form for $Q/Q_{100} = f(h)$, expression (7.31) can be put in a linear form around the operation point corresponding to $h = h_0$

$$\left(\frac{Q}{Q_{100}}\right) = \left(\frac{Q}{Q_{100}}\right)_{h=h_0} + \left[\frac{d}{dK_v}\left(\frac{Q}{Q_{100}}\right)\frac{dK_v}{dh}\right]\Delta h \tag{7.49}$$

For control valves so chosen as to maximize their linear operation characteristic in static conditions, relation (7.49) can be used over the whole range of steam travel for all practical purposes.

Assuming that force F_p, which arose from the action of the differential pressure on the throttling member of the valve, is ascribed a certain average value, then relation (7.49) can be written

$$\frac{m}{K}\cdot\frac{d^2}{dt^2}(\Delta h) + \frac{\mu}{K}\frac{d}{dt}(\Delta h) + \Delta h = \frac{A}{K}\Delta p_m \tag{7.50}$$

where $\Delta h = h - h_0$ and $\Delta p_m = p_m - p_{mo}$.

Now, putting equation (7.49) in the form

$$\Delta h \left(\frac{Q}{Q_{100}}\right) = \left(\frac{Q}{Q_{100}}\right) - \left(\frac{Q}{Q_{100}}\right)_{h=h_0} = a\,\Delta h \tag{7.51}$$

where

$$a = \left[\frac{d}{dK_v}\left(\frac{Q}{Q_{100}}\right)\frac{dK_v}{dh}\right]_{h=h_0}$$

and substituting the result (7.51) in (7.50), we obtain

$$\frac{m}{k}\frac{d^2}{dt^2}(\Delta q) + \frac{\mu}{k}\frac{d}{dt}(\Delta q) + \Delta q = \frac{Aa}{k}\Delta p_m \tag{7.52}$$

where

$$\delta = \frac{Q}{Q_{100}}.$$

Applying the Laplace operator to the differential equation (7.52), and putting ratio $\Delta q(s)/\Delta p_m(s)$ in an explicit form, it leads to

$$\frac{\Delta q(s)}{\Delta p_m(s)} = \frac{K_r}{T_2 s^2 + T_1 s + 1} \qquad (7.53)$$

where

$$T_2 = m/K, \; T_1 = \mu/K \text{ and } K_r = A\,a/k.$$

Transfer function (7.53) describes, to a fair approximation, the dynamic behaviour of the control valve around the working set point. The difficulty in this case lies in practical determination of time constants T_1 and T_2, as well as of transfer constant K_r. The solution to this problem implies an experimental inquiry into the various types and sizes of control valves, with due consideration of the actual working conditions in view, which is a task costly both in time and material. For illustrative purposes, it may be mentioned that time constant T_1 takes values below 1 to 2 s, and T_2, below 0.2^2s (s = second).

7.1.5. Choice of control valves

This section is concerned with the choice of control valves, in view of the process to be automated, the already existing control equipment, the technical data available in the specifications published by the constructors, etc. Other aspects related to the choice of control valves are also born in mind, such as the intrinsic operating characteristics, etc., which were given detailed consideration in previous chapters. Here we merely summarize what was said before on this topic, and emphasize some aspects shown in practice by the choice of control valves.

Problems that usually confront us in the choice of a control valve are of two kinds: those raised by the original technical data, and those raised by the choice procedure.

Problems raised by the original technical data. The choice of a control valve can only be made when the following basic data are available.

(1) *The path of the pipe onto which the control valve will be mounted* and details of the local resistances existing along this path.

(2) *The liquid to be handled*: its physical and chemical data, with particular emphasis on pressure, temperature, purity (presence or absence of suspensions), etc.

(3) *A summary of the process to be automated in general and an accurate* (as far as possible) statement of its static characteristic. When this is not possible, at least the qualitative form of the static characteristic should be described. These data are needed for determination of the valve working characteristic.

(4) *The values of the minimum, Q_{min}, rated, Q_r, and maximum, Q_{max} flow rates*, as derived from analysis of the unwanted variables interfering with the automated process.

Stages in the selection procedure. Any selection process should comprise the following stages:

(1) *Estimation of pressure losses across the pipe in maximum, rated and minimum* flow conditions, using relation (7.17), Tables 5.2 and diagrams shown in Figs. 5.8—5.33 (see Chapter 5).

(2) *Using data supplied at point number (3) above*, i.e. the value found for the pressure loss through the pipe in maximum flow conditions and the diagrams shown in Fig. 7.13, the most adequate operating characteristic is chosen. This can either be linear, when all other components of the automated control system, taken together, form a linear sub-assembly, or non-linear, when the other components of the system, taken together, show a lack of linearity that has to be compensated, or when the end in view is to obtain an automated non-linear system having certain features. Once the valve working characteristic has been chosen, the pressure losses across the valve are implicitly established through the ratio $\Delta p_{v\,max}/\Delta p_{s\,max}$. At the same time the question of a linear or logarithmic valve must be settled.

(3) *The pressure to be developed at the outlet of the pressure generating unit is found for* Q_{max}

$$p_{1\,max} = p_2 + \Delta p_{s\,max} \pm \varrho g H$$

and an adequate unit is chosen from the specifications supplied by pump manufacturers; in doing so, the static characteristic $p(Q)$ of the pressure generating unit over the whole range of flow rates is also established.

(4) *The pressure loss across the system should be calculated.* For the pipe-control valve, the $\Delta p_{s\,min}$ for Q_{min} is found from the operating characteristic of the pressure generating unit $p(Q)$ (Fig. 7.21): knowing the pressure loss across the pipe and across the pipe-control valve system (for Q_{min}), the pressure losses across the valve are found thus

$$\Delta p_{v\,min} = \Delta p_{s\,min} - \Delta p_{p\,min}$$

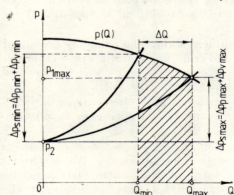

Fig. 7.21. Characteristic of the system. Pipe-control valve-pressure supply:

$\Delta p_{p\,max}$, $\Delta p_{v\,max}$ — pressure losses through the pipe and across the control valve at the maximum flowrate; $\Delta p_{p\,min}$, $\Delta p_{v\,min}$ — pressure losses through the pipe and across the control valve at the minimum flow rate.

(5) *Depending on the physical and chemical properties of the liquid, on pressure losses across the valve and on the work temperature, the type of control valve is chosen.*

(6) *The intrinsic characteristic*, K_v, is found after going through the following stages:

— $K_{v\,max}$ is estimated for the maximum flow rate, and $K_{v\,min}$ for the minimum flow rate;

— K_{vs} is chosen according to the producer specifications, bearing in mind the following considerations:

(a) *Since* $K_{v\,100}$ specified by the valve producer will be somewhere between $0.9\,K_{vs}$ and $1.1\,K_{vs}$, the value chosen for K_{vs} should be greater than $K_{v\,max}$. Adding to these values an additional allowance of $+15$ per cent, to satisfy the conditions imposed by the control process, the following relation is obtained

$$K_{vs} \approx 1.25\,K_{v\,max} \tag{7.54}$$

When the process to be controlled is poorly known, and changes in the flow rate above the value of Q_{max} assumed in the calculations are likely to occur, then the value of K_{vs} can be increased to $1.8\,K_{v\,max}$; if, however, the perturbations likely to develop are small, then one may choose: $K_{vs} \approx 1.25\,K_{v\,max}$.

(b) *For the same considerations as those outlined above in* (a), *but for the minimum flow rate*

$$\frac{K_{v\,min}}{K_{vc}} \geqslant 1.2 \tag{7.55}$$

(7) Check to make sure the condition:

$$\frac{K_{vs}}{K_{v\,min}} < R_c \tag{7.56}$$

is fulfilled, R_c being the control ratio. When this condition is not satisfied, choose another valve, having the same K_{vs}, but a higher R_c, or look for another design solution.

(8) *Find* K_c for the rated flow, Q_r. To this end, proceed as to item 4 above. This value is needed to assess the operating position of the control valve in the normal conditions that will prevail for most of the valve service life.

Example 7.6. **Choice of a control valve.** It is required to choose a control valve for the automated pressure control system of a drinking-water pipe, supplying the distribution network of a high building. The supply pipe hydraulic system, where the control valve is to be mounted, is characterized by the following data. The pressure generating unit is a radial multistage (centrifugal) pump; the pipe is made of steel tube, rated diameter $D_r = 100$ mm; its service parameters are $Q_{max} = 34$ m^3/h, $Q_r = 30$ m^3/h $Q_{min} = 17$ m^3/h; the maximal pressure loss $\Delta p_{p\,max} = 0.6$ daN/cm^2; the rated pressure loss $\Delta p_{pr} = 0.45$ daN/cm^2; the minimum pressure loss $\Delta p_{p\,min} = 0.20$ daN/cm^2; and the static pressure head through the pumping system $p_{st} = 4.5$ daN/cm^2.

Solution

(1) *In this case*, the process to be automatically controlled takes place in the pipe, and the input and output variables are pressure values. The static characteristic of the process to be automatically controlled is linear. Since it is required to achieve, in so far as possible, a linear characteristic for the whole system and since the pressure losses expected are small, we choose a linear control valve for which

$$\Delta p_{v\,max}/\Delta p_{s\,max} = 0.33$$

Thus

$$\frac{\Delta p_{v\,max}}{\Delta p_{v\,max} + \Delta p_{p\,max}} = 0.33$$

and hence

$$\Delta p_{v\,max} = 0.30 \text{ daN/cm}^2$$

Thus, the pressure loss throughout the system

$$\Delta p_{s\,max} = \Delta p_{p\,max} + \Delta p_{v\,max} = 0.60 + 0.30 = 0.90 \text{ daN/cm}^2$$

(2) *Find the pressure required at the outlet* of the pressure generating unit for the maximum flow rate Q_{max}. It is

$$p_{s\,max} = p_2 + \Delta p_{s\,max} = 4.50 + 0.90 = 5.40 \text{ daN/cm}^2$$

On the basis of point $M(Q_{max}, p_{s\,max})$, we choose the radial pump needed. This has, say, the characteristic shown in Fig. 7.22.

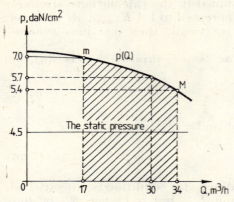

Fig. 7.22. $p(Q)$ characteristic of the radial pump considered in the application illustrated in Fig. 7.6.

(3) *Estimation of values $\Delta p_{p\,min}$ and $\Delta p_{v\,max}$*. From Figs. 7.21 and 7.22 it follows that $\Delta p_{s\,min} = 2.50$ daN/cm^2; since $\Delta p_{p\,min}$ is known to be equal to 0.20 daN/cm^2, the result will be

$$\Delta p_{v\,min} = \Delta p_{s\,min} - \Delta p_{p\,min} = 2.50 - 0.20 = 2.30 \text{ daN/cm}^2$$

(4) *We choose a globe valve* for the rated pressure of $p_r = 10$ daN/cm^2.
(5) *We find the value K_v* of the valve, for the maximum flow rate

$$K_{v\,max} = 34\sqrt{\frac{1}{0.3}} = 62.20$$

and for the minimum flow rate

$$K_{vmin} = 17\sqrt{\frac{1}{2.3}} = 11.20$$

Assuming that $K_{vs} \approx 1.25$ and $K_{v\,max} = 1.25 \times 62.20 = 78$, we look in Table 7.1 and choose a valve for the rated pressure of $p_r = 10$ daN/cm^2 with the following characteristic data: $D_r = 65$ mm; $K_{vs} = 56.50$; $K_{vv} = 8.40$; $K_{vo} = 2.8$; and $R_c = 20$.

(6) *Check to make sure if condition* $K_{v\,min}/K_{vv} \geqslant 1.2$ is met

$$\frac{K_{v\,min}}{K_{vv}} = \frac{11.20}{8.40} = 1.33 > 1.20$$

(7) *Check to make sure the control ratio* is well within the prescribed value

$$\frac{K_{vs}}{K_{v\,min}} = \frac{56.50}{11.20} \approx 5 \ll R_c = 20$$

(8) *Finally estimate* K_{vr} for the rated flow rate $Q_r = 30$ m³/h

$$\Delta p_{vr} = 1.20 - 0.45 = 0.75 \text{ daN/cm}^2$$

$$K_{vr} = 30 \sqrt{\frac{1}{0.75}} = 34.80$$

7.2 Control of System Operation by Corrective Regulating Valves

If the pump operates at a constant speed, it becomes necessary to introduce an artificial source of friction loss into the system, this friction loss being controlled by changes in pressure, flow rate or level. Such an artificial friction loss is provided by a pressure-reducing valve, a flow-rate controller, or a water-level regulator, consisting basically of two elements: (1) a main throttling valve, and (2) a controlling pilot which determines the setting of this valve, depending on fluctuations of pressure, flow rate or level in the system.

7.2.1. Controls to Regulate Pressure

Two categories of controls for maintaining pressure can be distinguished:
 — *pressure-reducing controls*, when maximum pressure values should be limited irrespective of variations in system consumption;
 — *pressure-sustaining controls*, when a minimum value of pressure should be secured in a main line, irrespective of variations in secondary lines.

Pressure-reducing controls. The main variable of such a control is the flow rate, Q, in the pumping system. Thus, they secure a constant pressure at the pump outlet, regardless of variations in the system consumption (Fig. 7.23).

The outlet pressure of a fixed-speed pump is known to be variable and dependent on the flow rate discharged into the system. Thus, for keeping a constant outlet pressure, a pressure-reducing valve should be introduced on the pump discharge line (Fig. 7.24). This valve consists of a main throttling valve and some kind of pilot control system.

The main throttling valve (Fig. 7.25) is a hydraulically-operated diaphragm actuated on a differential piston principle (exposing a larger surface to the closing forces than to the opening ones). The main throttling valve functions are exclusively determined by those of the chosen control pilots.

The pressure-reducing command is automatically given by a normally open pressure-reducing pilot control (Fig. 7.26) that is open as long as it is under pressure. This pilot responds to the action of closing forces resulting from water pressure on the surface of its diaphragm. It tends to close when the water pressure downstream from the main valve rises over a prescribed value, equal to the maximum allowed pressure. The reducing control is sensitive to slight pressure changes and immediately controls the main valve to maintain the desired line pressure.

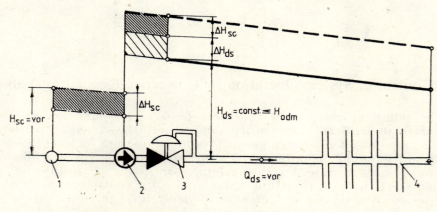

—— The hydraulic gradient with pressure-reducing valve
– – – The hydraulic gradient without pressure-reducing valve

Fig. 7.23. Hydraulic gradient for a pumping plant with pressure reducing valve in discharge line:
1 — town main; *2* — boosting station; *3* — pressure reducing valve; *4* — secondary distribution system.

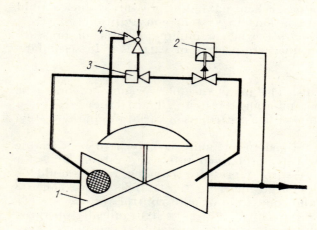

Fig. 7.24. Schematic piping arrangement of the pressure reducing valve:

1 — main valve; *2* — pressure reducing control; *3* — ejector; *4* — flow control, that provides a dampening action to eliminate pulsation of very low flow rates.

The pressure-reducing valve may include a combination of several functions when equipped additionally with proper pilots for the respective purposes. Thus, it could be used in two possible variants:

(1) *With an auxiliary swing check valve*, associated with a hydraulic check valve (Fig. 7.27), to achieve a combination of pressure-reducing

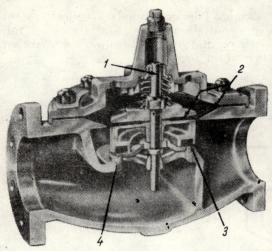

Fig. 7.25. Remote control main (hytrol) valve:
1 — axis; *2* — differential piston; *3* — setting; *4* — seat.

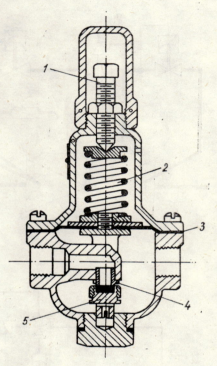

Fig. 7.26. Pressure reducing control:
1 — adjusting screw; *2* — spring; *3* — diaphragm; *4* — seat; *5* — disc retainer.

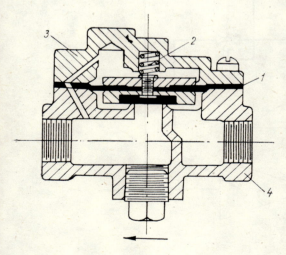

Fig. 7.27. Auxiliary hydraulic check valve:
1 — diaphragm; *2* — spring; *3* — cover; *4* — body.

and check valve functions (Fig. 7.28, *a*). This combined valve has typical applications in pump discharge lines of boosting installations in tall buildings, where it is provided with regulation by flowmeter and direct suction from the variable pressure main line.

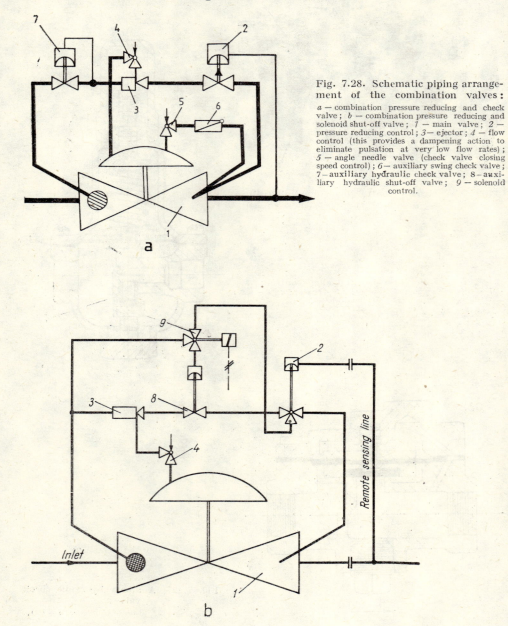

Fig. 7.28. Schematic piping arrangement of the combination valves:

a — combination pressure reducing and check valve; *b* — combination pressure reducing and solenoid shut-off valve; *1* — main valve; *2* — pressure reducing control; *3* — ejector; *4* — flow control (this provides a dampening action to eliminate pulsation at very low flow rates); *5* — angle needle valve (check valve closing speed control); *6* — auxiliary swing check valve; *7* — auxiliary hydraulic check valve; *8* — auxiliary hydraulic shut-off valve; *9* — solenoid control.

(2) *With a solenoid pilot normally open* (that is, open as long as de-energized) to provide a combination of pressure-reducing and solenoid shut-off valve (Fig. 7.28, *b*). This valve has typical applications to reversible pumping installations used for drinking-water supply lines and fire-control lines of tall buildings with two or more pressure zones (see Fig. 14.12).

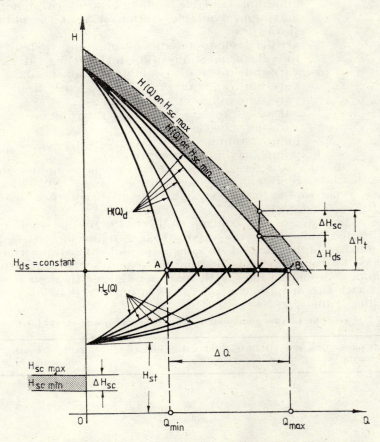

Fig. 7.29. Head-flow rate curves of a pump, provided with a pressure-reducing valve on its discharge line.

All pressure-reducing valves, either simple or combined, secure a a constant outlet pressure, irrespective of successive or simultaneous variations in consumption and inlet pressure.

Figure 7.29 shows the operating diagram of a fixed-speed pump provided with a pressure-reducing valve at its outlet.

The notation of the diagram is the following:

$H(Q)_{sc\,min}$ is pump characteristic corresponding to minimum suction head;

$H(Q)_{sc\,max}$ — pump characteristic corresponding to maximum suction head;
$H(Q)_d$ — decreased pump characteristic due to additional head losses accounted for by control valve;
AB — pump characteristic downstream from the control valve;
$H_{sc\,min}$ — minimum available suction head at the pump inlet;
$H_{sc\,max}$ — maximum available suction head at the pump inlet;
H_{st} — static head;
$H_s(Q)$ — artificial characteristic of pumping system obtained for various positions of main throttling valve;
ΔH_{sc} — variation range of available suction head;
ΔH_{ds} — variation range of available discharge head;
ΔH_t — variation range of total head;
ΔQ — variation range of pump flow rate due to consumption variations in the system;
Q_{min} — pump minimum flow rate;
Q_{max} — pump maximum flow rate.

The artificial characteristic, resulting from the reduction of pump pressure, has been represented by segment AB (Fig. 7.29). This characteristic is a horizontal line (a straight line with constant ordinate) for which $H_{ds} = \text{constant} = H_{max}$.

The extension of the controlled flow rate to the small flow rate zone is limited, since there the valve stroke is reduced, and the pressure coming out can no longer be kept constant; and cavitation, which might occur, has destructive effects on the valve body. For preventing these shortcomings, discharge lines with diameters larger than 80 mm should be fitted with two or more control valves of unequal diameters, mounted in parallel (Table 7.2).

TABLE 7.2 **Sizing the Clayton combination pressure reducing and check valve**

Individual pump flow rate, in dm³/s	Main control valve diameter, in mm	Auxiliary control valve diameter, in mm
3.80	40	—
7.00	50	—
9.50	65	—
12.60	65	40
15.75	80	40
18.90	80	50
22.05	80	65
25.20	80	65
28.35	100	40
31.50	100	50
34.65	100	65
37.80	100	80
63.00	150	40

Remark: One main valve per pump may by used from 40 mm through 65 mm valves. In sizes of 80 mm and larger, a main valve and smaller auxiliary valve is required.

The diameter of a pressure-reducing valve should be chosen after a rigorous hydraulic calculation (see Section 7.1.5), or otherwise the results may be quite unsatisfactory.

The most usual applications of pressure-reducing valves with simple or combined functions are:

(1) *For maintaining constant outlet pressure* of pumps in pressure-boosting plants with automatic control by flowmeter[1], and where pumps are directly fed from a town main line (whose pressure is inherently variable). In that case the pressure-reducing valve shown in Fig. 7.30, *a* has been used, built according to the scheme of Fig. 7.28, *a*.

(2) *For automatic changeover* of a reversible pump from drinking-water to fire-control service. In that case the pressure-reducing valve shown in Fig. 7.30, *b* can be used, built according to the scheme of Fig. 7.28, *b*.

(3) *For pressure control in pressure-reducing stations* of water-supply systems divided into zones (Fig. 7.31), [7.7].

Pressure-sustaining controls. Their variable is the static head, H_{st}, of the system. Thus, they secure the minimum pressure available in the system, irrespective of static head variations.

Fig. 7.30. Combination automatic valves (Clayton Co.):

a — combined pressure-reducing and check valve; *b* — combined pressure-reducing and solenoid shut-off valve.

Let us consider the pumping system as a distribution system fed by pumping. Actually, the modification of static head in a distribution system takes place in one of the following situations (Fig. 7.32)[2]:

(1) *When the same distribution system covers the consumptions* of two zones of which one has higher topographical levels (zone *I*), and the

---

[1] Solution designed by the author as head of project for water-supply system at the "Inter-Continental" Hotel, Bucharest (Romania) [7.4].

[2] Author's original solutions for Bucharest water-supply distribution system.

359

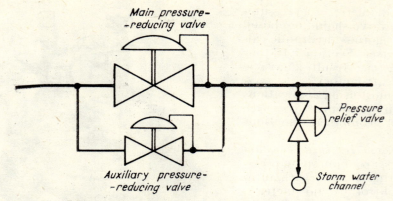

Fig. 7.31. Manifold-type pressure-reducing regulation station.

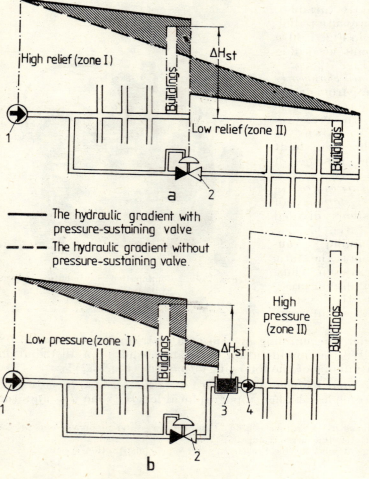

Fig. 7.32. Piezometric profile of a pumping installation, provided with pressure-sustaining valves on its discharge line:
— a — two-zones system of different topographic levels; b — two delivery systems connected by an open tank; 1 — pumping station; 2 — pressure-sustaining valve; 3 — open buffer tank; 4 — boosting station.

other lower (zone *II*) (Fig. 7.32, *a*). This situation might occur in the case of a pump-closed system (without tanks), located on an area made of two distinct topographic zones, although the system has a single pressure zone (with a maximum admitted pressure of 6 or 7 daN/cm^2).

(2) *When two different distribution systems*, one fed by pumping (zone *I*) and the other by boosting (zone *II*), are located at the same topographic level and are connected by means of an open tank, necessary for pump suction in zone *II* (Fig. 7.32, *b*). This situation is similar to the case in which a town main delivery system (zone *I*) is connected by an open tank to a supply system belonging to a residential area or an industrial zone.

These phenomena are better understood by means of the diagram shown in Fig. 7.33. It is the operation diagram of a pumping instal-

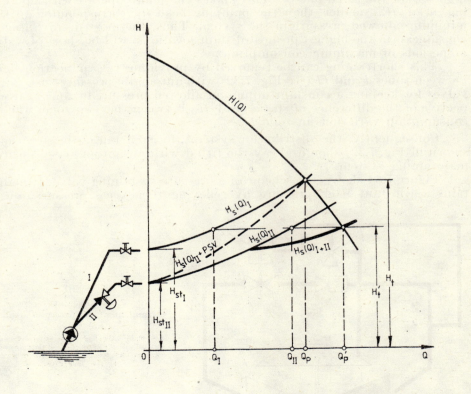

Fig. 7.33. Head flow-rate curves of a pumping installation, provided with a pressure-sustaining valve on its discharge line.

lation, located at the beginning of zones I (see Figs. 7.32, a and b). The diagram makes use of the following notation:

$H(Q)$ is pump characteristic;
$H_s(Q)_I, H_s(Q)_{II}$ — system characteristics;
$H_s(Q)_{I+II}$ — summed up characteristics of the system;
$H_s(Q)_{II} + PSV$ — system corrected characteristic (resulting from additional head loss, introduced in the system by pressure-sustaining valve PSV);
Q_I, Q_{II} — partial flow rates in zones I and II, respectively;
Q_p, Q'_p — corrected pumped and natural flow rates;
H_t, H'_t — corrected and natural pumping heads;
$H_{st\,I}, H_{st\,II}$ — static heads of systems I and II.

Analysis of the operation diagram shown in Fig. 7.33 reveals that, although the consumption of the two topographical zones is equal, the flow rate Q_{II} discharged in zone II is higher than the flow rate Q_I discharged in zone I. By summation, we find a total flow rate Q'_p at the pump higher than the rated one Q_p. This makes the pump operate with a pumping head H'_t, smaller than the pumping head H_t, corresponding to the minimum allowed pressure in the system. Thus, the most disadvantaged consumers (in the highest or most distant locations) will be short of water at periods of maximum consumption.

This shortcoming can be removed by providing the connection lines between zones I and II (see Fig. 7.32) with automatic pressure-sustaining valves for keeping a constant minimum allowed pressure in zone I, irrespective of the difference of static head H_{st} between the two zones and of consumption variations in zone II.

Consequently, the distribution systems, found in one of the situations shown in Fig. 7.32, should always be fitted with an automated pressure-maintaining system.

Figure 7.34 shows the scheme of a pressure-sustaining valve. It maintains a constant back pressure by relieving the excess pressure down-

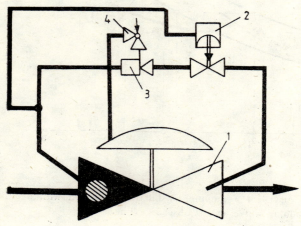

Fig. 7.34. Schematic piping arrangement of pressure-sustaining valve:

1 — main valve; 2 — pressure-sustaining control; 3 — ejector; 4 — flow control (secures slow opening and smooth operation at low flow rates).

stream. When the inlet pressure tends to be higher than the set pressure, the valve adjusts to maintain constant inlet pressure. This valve is controlled by a pressure-relief pilot control normally closing (that is, closing when not under pressure), (Fig. 7.35). The pilot opens under water pressure in the line upstream of the valve, and shuts-off when this pressure drops below the minimum set pressure of the respective pilot. The main valve follows the same operation mode as its pilot.

The pressure-sustaining valve can have other functions too, when equipped with supplementary pilots, suitable for the respective purposes. Thus, the valve can be provided with the following:

(1) *A check pilot*, to provide a combination of pressure-sustaining and check valve (Fig. 7.36, *a*).

(2) *A pressure pilot*, *normally open* (see Fig. 7.26) to provide a combination of pressure-sustaining and pressure-reducing valve (Fig. 7.36, *b*).

Figure 7.37 shows two types of pressure-sustaining valves, one simple and the other combined.

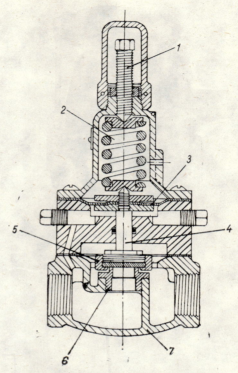

Fig. 7.35. Pressure-sustaining control:
1 — adjusting screw; *2* — spring; *3* — diaphragm; *4* — stem; *5* — disk; *6* — seat; *7* — body.

7.2.2 Controls for Maintaining Flow Rate

Flow rate control can either be based on direct measurement of the flow rate itself, or on the measurement of some other quantity, such as the pressure differential through pumps, etc.

a. *Controls based on flow-rate measurement*. The variable in this case is the dynamic pressure in the pumping line, H_{dy}. The control has the role of keeping a constant flow rate in the line, irrespective of variations in the dynamic pressure.

Let us consider a pumping line made of two members: the main delivery line, *I*, and a plant branch line, *II*, with different diameters and supplied in series by means of pump *1* (Fig. 7.38).

When the excess flow rate of pipeline *II* is not controlled, the piezometric line of the system can drop, so that the most disadvantaged consumers (in the highest and remotest locations) connected to line *I* are short of water during periods of peak demand. The excess flow rate of line *II* is generated by the tank *2*, found at the end of this line. This

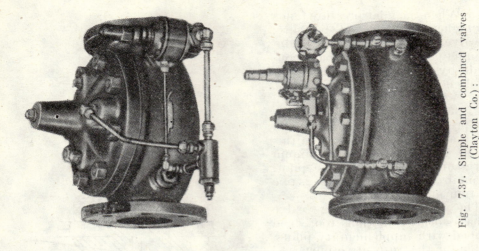

Fig. 7.37. Simple and combined valves (Clayton Co.):
a — pressure-sustaining valve; *b* — combined pressure-reducing and pressure-sustaining valve.

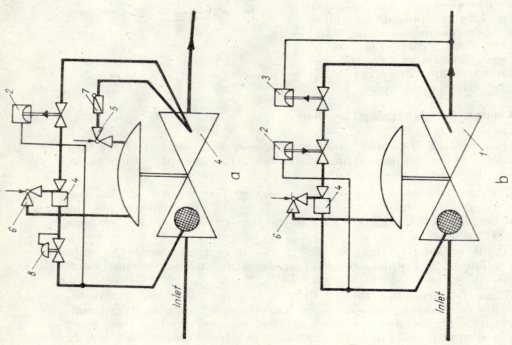

Fig. 7.36. Schematic piping arrangement of combined valves:
a — combined pressure-sustaining and check valve; *b* — combined pressure-sustaining and pressure-reducing valve; *1* — main valve; *2* — pressure-sustaining control; *3* — pressure-reducing control; *4* — ejector; *5* — angle needle valve (check valve closing speed control); *6* — flow control (secures smooth operation at low flow rates); *7* — swing check

shortcoming can be prevented by mounting a flow-rate controller at the junction point between lines *I* and *II*. This flow-rate controller will limit the value of mininum supply flow rate at tank *2*, irrespective of the opening of the float valve mounted on the tank supply pipeline. The limitation of flow rate in line *II* secures the necessary supply pressure to consumers connected to line *II*.

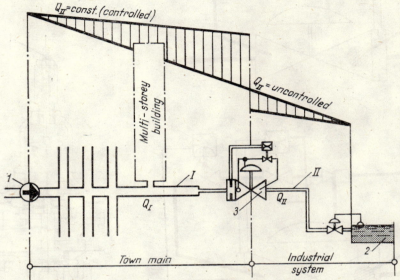

Fig. 7.38. Piezometric branch through a pumping installation, provided with flow-rate controller on the suction line of an industrial plant:
1 — pump; *2* — reservoir; *3* — flow-rate controller; *I* — water main; *II* — industrial branch.

Figure 7.39 shows schematically a flow-rate controller. This maintains constant flow rate very closely, regardless of changing line pressure. This valve is controlled by a differential pressure pilot control which is normally closed (Fig. 7.40). The pilot control is actuated by the differential pressure produced across an orifice plate in the main line. The pilot opens with rising differential pressure created in the orifice plate (which is proportional to the square of flow rate) and shut-off when differential pressure drops below the set value. The main valve follows the same mode of operation as its pilot.

The flow-rate controller can also have a pressure-reducing function when fitted with an additional pressure-reducing pilot control (normally open) (see Fig. 7.26). This produces a combined flow-rate controller and pressure-reducing valve (Fig. 7.39, *b*).

This combined valve can be used on branches of a large transmission main supplying several urban communities. An example is the 80 km transmission main of $D_r = 1500$ mm between Rabat and Casablanca, also supplying on the way Mohamedia, Qued and Casablanca towns in Morocco. In this case the combined valve prevents excessive consumption

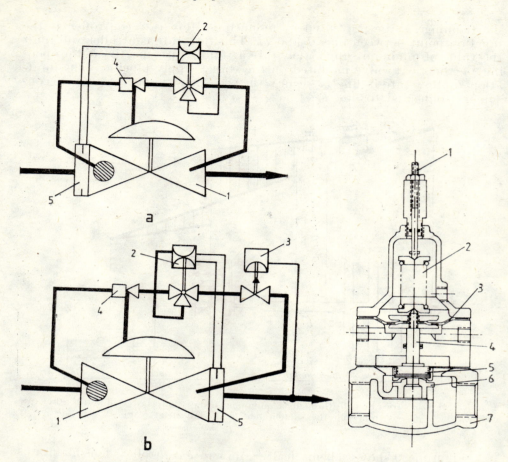

Fig. 7.39. Flow rate controllers:
a — simple flow-rate controller; b — combined flow-rate controller and pressure-reducing valve; 1 — main valve; 2 — differential pilot control; 3 — pressure-reducing control; 4 — ejector; 5 — orifice plate.

Fig. 7.40. Differenţial pilot control:
1 — adjusting stem; 2 — spring; 3 — diaphragm; 4 — stream; 5 — disk retainer; 6 — seat; 7 — body.

on the transmission main and at the same time, keeps a constant delivery pressure in its branches, as long as their flow rate is below a maximum setting.

Figure 7.41 shows the operating diagram of a pumping installation comprising a pump and a pumping system with two serially-jointed pipes, I and II, of different diameters. The latter pipeline discharges into a tank whose level is regulated by a float valve. This valve operates like a variable hydraulic resistance and consequently, the dynamic head within pipeline II will be variable. For preventing any exaggerated pressure drop within pipeline I at filling of the tank, the flow-rate controller should be mounted on pipeline II, limiting the maximum supply flow rate to the respective tank.

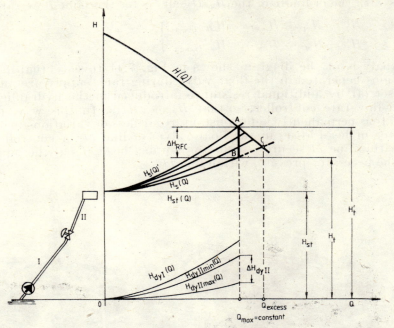

Fig. 7.41. Head flow-rate curves of a pump, provided with a flow-rate controller at the end of the discharge line.

The notation on Fig. 7.41 is the following:

$H(Q)$ — is pump characteristic;
$H_{dy\,I}(Q)$ — pipeline I characteristic;
$H_{dy\,II\,min}(Q)$, $H_{dy\,II\,max}(Q)$ — pipeline II characteristics corresponding to minimum and maximum openings of the float valve orifice;
$H_s(Q), H_s(Q)'$ — system characteristics for $H_{dy\,II\,min}$ and $H_{dy\,II\,max}$, respectively;
H_{st} — static head;
H_t, H_t' — total head for $H_{dy\,II\,min}$ and $H_{dy\,II\,max}$, respectively;
Q_{excess} — pumping installation flow rate without flow-rate control;
Q_{max} — maximum flow rate, limited by a flow-rate controller;
$\Delta H_{dy\,II}$ — variation range of local head loss in float valve.
ΔH_{RFC} — variation range of local head loss in flow-rate controller.

The resulting characteristic curve $H_s(Q)$ of the pumping system has been obtained by summing up the ordinates of $H_{dy\,I}(Q)$ and $H_{dy\,II}(Q)$

curves with the ordinate of the $H_{st}(Q)$ curves for the same flow rate. Thus

$$H_t = H_{st} + H_{dy\,I} + H_{dy\,II\,min}$$
$$H'_t = H_{st} + H_{dy\,II} + H_{dy\,II\,max}$$

Analysis of the diagram shown in Fig. 7.41 reveals that head loss variations generated in the float valve during tank supply are automatically set off by additional resistance introduced in the hydraulic circuit of the flow-rate controller, so that $H_{dy\,II} = H_{RFC}$. In this way, delivery flow rate is permanently kept constant, regardless of variations in dynamic head (or of head losses in the float valve) of line II. A controlled flow rate within line II secures adequate pumping head H'_t at the pump, and thus the necessary pressure within line II.

Fig. 7.42. Flow-rate controllers (Clayton Co.):
a — simple flow-rate controller; b — combined flow-rate controller and pressure-reducing valve.

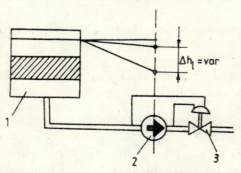

Fig. 7.43. Piezometric profile of a filter plant, coupled to a pump, provided with a differential relief valve on its discharge line:
1 — vacuum filter; 2 — pump; 3 — differential relief valve.

Figure 7.42 shows two types of flow-rate controller, a simple one and a combined one.

b. *Controls based on differential pressure measurement.* This type of regulation has the task of keeping a constant differential pressure through the pump, and thus also a constant total pumping head, regardless of variations in inlet pressure and consumption at the pump outlet. It secures a constant differential pressure across a pump, regardless of fluctuations in suction pressure head or downstream demand.

One of the typical applications of this control is for flow-rate control through a filtered medium (Fig. 7.43). The flow rate of filtered water resulting from vacuum filters should be kept constant, irrespective of head loss variations through the filtering medium. By controlling the filtered

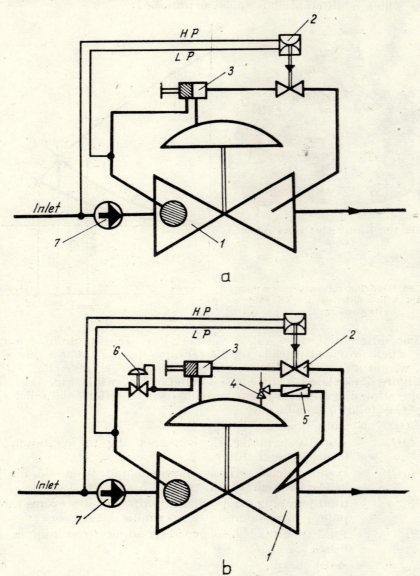

Fig. 7.44. Schematic piping arrangement of differential relief valve:
a — simple valve; *b* — combination valve; *1* — main valve; *2* — differential pilot control; *3* — strainer and needle valve; *4* — angle needle valve (check valve closing speed control); *5* — auxiliary swing chech valve; *6* — auxiliary hydraulic check valve; *7* — pump.

flow rate, we obtain an improvement of vacuum filter efficiency. To this end, a differential relief valve should be mounted on the discharge line of each pump, according to the scheme shown in Fig. 7.43. The valve is actuated by a pilot control system (see Fig. 7.40) sensing at two points across which a differential is to be maintained.

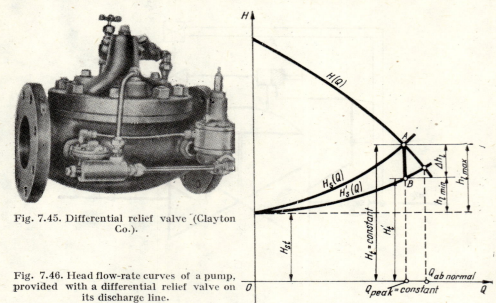

Fig. 7.45. Differential relief valve (Clayton Co.).

Fig. 7.46. Head flow-rate curves of a pump, provided with a differential relief valve on its discharge line.

The differential relief valve can also have the function of a check valve when provided with a supplementary check pilot (see Fig. 7.25). Thus, we get a combined differential relief and check valve (Fig. 7.44, b).

Figure 7.45 shows a differential relief valve, while Fig. 7.46 shows the operation diagram of a pump, provided with a differential relief valve, using the following notation:

- $H(Q)$ is pump characteristic;
- $H_s(Q)$ — system characteristic corresponding to maximum head losses through filter;
- $H_s(Q)'$ — system characteristic corresponding to minimum head losses through filter;
- H_t — total pumping head (equal to differential pressure through pump) corresponding to maximum head losses;
- H_t' — total pumping head corresponding to minimum head losses;
- H_{st} — static head;
- $h_{l\ max}$ — maximum head losses through filtering medium;
- $h_{l\ min}$ — minimum head losses through filtering medium;
- Δh_l — variation range of head losses through filtering medium.

Analysis of the diagram shown in Fig. 7.46 reveals that after washing of filters, when head losses through the filtering medium are minimum, a pumping head drop and a drop of differential pressure through the pump are observed, which lead to a rise of filtered flow rate and adverse consequences on filtering installation efficiency. For preventing this drawback, a combined differential relief and check valve should be mounted on the discharge line of each pump. This valve maintains a constant differential pressure through the pump, that is a constant outlet flow rate, regardless of variations in head losses through the filtering medium caused by the filter becoming progressively clogged with impurities.

7.2.3 Water-level Regulations

The role of these regulations is to keep a constant level within a tank, regardless of successive or simultaneous variations of inlet or outlet flow rate.

a. *Water level regulation in storage reservoirs.* Generally, these reservoirs are underground and the maximum level of stored water coincides approximately with ground level. In such conditions, water-level regulation is done by means of float valves. These valves can be of non-modulating or modulating types.

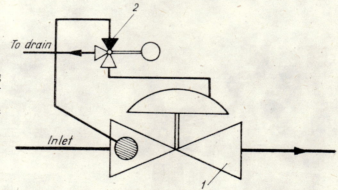

Fig. 7.47. Schematic piping arrangement of non-modulating float valve:
1 — main valve; *2* — remote float control

Non-modulating float valve (Fig. 7.47) is a discontinuous, non-throttling valve. Its role is to keep constant two levels, a maximum and a minimum (Fig. 7.48). Thus, the valve opens entirely (Fig. 7.49 *a*) when the liquid level reached a minimum setting value, and closes completely (Fig. 7.49, *b*) at a maximum level value. The opening and closing values are adjustable. Built to operate only in a completely open position, these valves are meant to be used in heavy duty regimes (for water loaded with suspensions or impurities).

Modulated float valve (Fig. 7.50) has a continuous action, achieving smooth and progressive throttling of the liquid stream. It is controlled by a differential pressure pilot (see Fig. 7.40) receiving signals from a remote float control. This valve transmits level variations to the main valve, which adjusts or modulates to keep a constant liquid level in the reservoir, also off-setting flow rate variations in both delivery and consumption branches. In other words, we may say that this valve balances the inlet flow rate in direct proportion to the outlet flow rate for keeping a constant liquid level within the reservoir (Fig. 7.51).

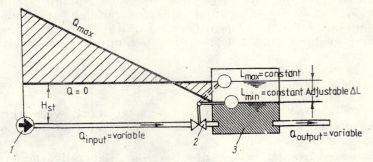

Fig. 7.48. Piezometric profile of a pumping system, provided with a non-modulating float valve on the pump discharge line:
1 — pump; *2* — non-modulating float valve; *3* — storage tank.

Fig. 7.49. Operation cycles of a non-modulating float valve:
a — opening cycle; *b* — closing cycle.

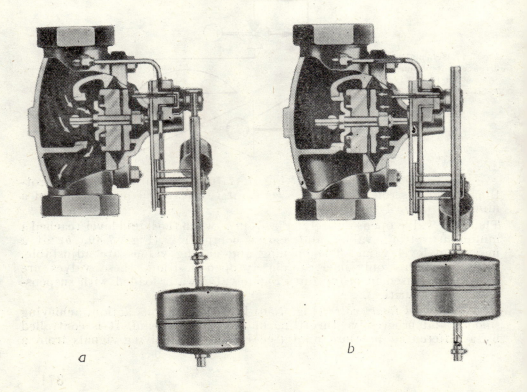

Control of System Operation

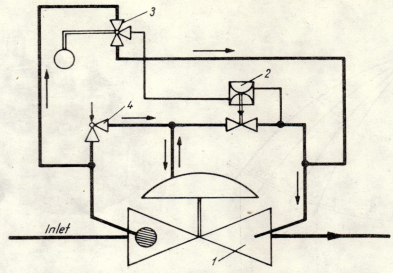

Fig. 7.50. Schematic piping arrangement of modulating float valve :
1 — main valve; *2* — pressure differential control; *3* — remote float control; *4* — angle needle valve.

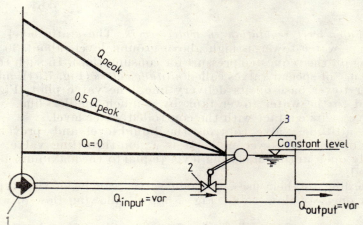

Fig. 7.51. Piezometric profile of a pumping installation, provided with a modulating float valve on the pump discharge line :
1 — pump; *2* — modulating float valve; *3* — storage tank.

This valve can be mounted so that it controls either the inlet flow rate or the outlet flow rate at the reservoir. Thus, a rising level can determine either a valve closing, or a valve opening. In either operational condition, the flow-rate variation range through the valve is relatively large ($\Delta Q = 1:2$) (Fig. 7.52).

Figure 7.53 shows a modulating float valve.

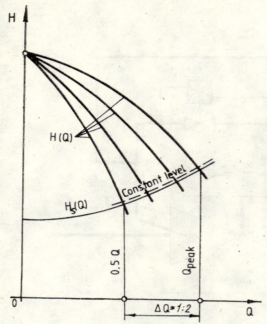

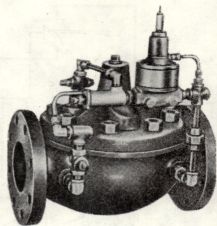

Fig. 7.53. Modulating float valve. (Clayton Co.)

Fig. 7.52. Head flow-rate curves of a pump, provided with a modulating float valve.

b. *Water level regulation in water towers*. The static level of liquid, stored in a water tower, is high above ground level (about 30 to 60 m, depending on the requested pressure for consumption). In such conditions, we make use of special valves called *altitude valves* (Fig. 7.54), mounted at the water tower base on its delivery line. The valve pilot (Fig. 7.55) is connected to the water tower tank by a much smaller line, without a float valve at its contact with the controlled water level.

The altitude valve controls the liquid level and prevents water overflow out of the tower tank. It is a non-throttling valve, and thus stays fully open until the closing level, [equal to the maximum tank level] is reached.

According to their uses, altitude valves can be of two types:
— a one-way flow valve (Fig. 7.56, *a*), allowing flow of water only one way;
— a two-way flow valve (Fig. 7.56, *b*), allowing water flow in both senses.

The two-way flow valve allows both tower supply and water delivery to consumers by means of a single line (Fig. 7.57, *a*). This valve can be used for water towers whose tanks store only compensation water volume for normal use, and not also a fire reserve.

The one-way flow altitude valve secures only tower feeding. Water withdrawal from the tank and its delivery to consumers is done by a check valve, mounted on a by-pass line around the altitude valve, or by a separate delivery line (Fig. 7.57, *b*). This mounting scheme can also

Control of System Operation

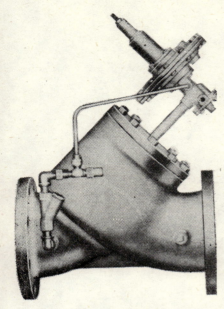

Fig. 7.54. Altitude valve (Rockwell Co.).

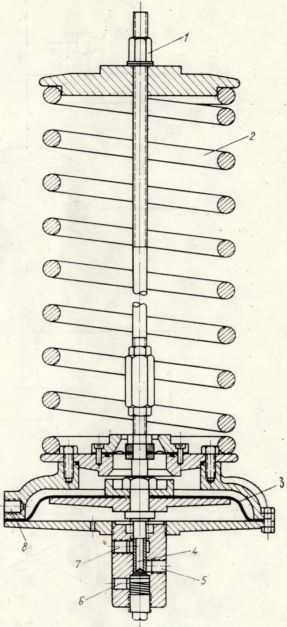

Fig. 7.55. Level sensing control:
1 — adjusting screw; *2* — spring; *3* — diaphragm; *4* — stem; *5* — connection to main valve cover; *6* — upstream supply connection; *7* — exhaust to drain; *8* — sensing connection elevated tank.

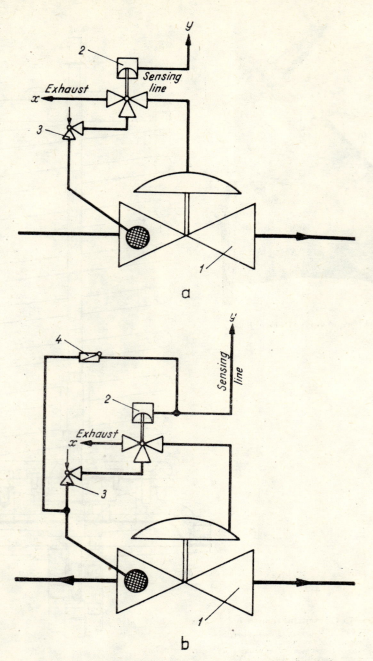

Fig. 7.56. Schematic piping arrangement of altitude valves:
a — one-way flow; b — two-way flow; 1 — main valve; 2 — level sensing control; 3 — opening speed control (needle valve); 4 — auxiliary check valve; x — exhaust to drain; y — sensing connection to water tower.

Control of System Operation

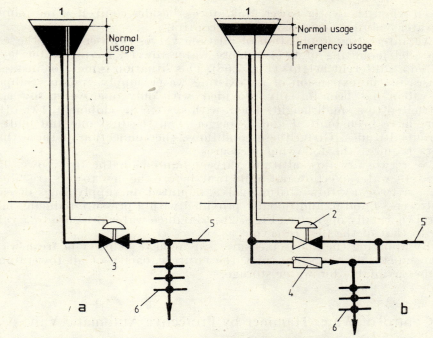

Fig. 7.57. Typical water tower systems with automatic level control:
a — normal plant water storage; *b* — combined normal plant water storage and fire protection regime; *1* — elevated tank; *2* — one-way altitude valve; *3* — two-way altitude valve; *4* — swing check valve; *5* — supply source; *6* — distribution system.

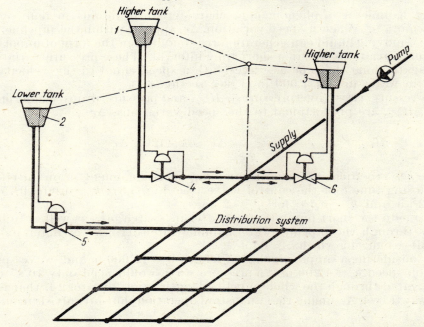

Fig. 7.58. Parallel connection to a system of three water towers with different heights:
1 and *3* — new water towers; *2* — old tower; *5* — two-way altitude valve; *4* and *6* — combined two-way flow control and pressure-reducing valves.

be used when the water tower tank stores, besides compensation volume, the water volume necessary for fire control.

An interesting application of altitude valves is observed in the case of parallel mounting of two or more water towers, having water levels situated at different heights (Fig. 7.58). This situation is met, for instance, in cases of the expansion of an existing water supply system. Suppose the system has been initially provided with only one water tower, *2*, constructed to a small height. Then, with system development, two other towers have been built, *1* and *3*, bigger than the first one and built at different altitudes. Given these conditions, the connection of these three towers requires the following measures:

— a two-way flow altitude valve, mounted on the first tower line to prevent water overflow out of its tank before the new towers are filled;

— a two-way flow altitude valve, mounted on supply lines of each new tower, *4* and *6*, additionally provided with a pressure-reducing pilot control (normally open) (see Fig. 7.26) to reduce water pressure to a value close to that of the first reservoir.

By applying the above measures, we achieve delivery of fresh water to consumers. At the same time, the entire capacity of all tower tanks becomes available for water storage.

7.3 Control of Water Hammer by Protective Automatic Valves

7.3.1 Effect of Valve Operation on Water Hammer

Water hammer is a phenomenon, characteristic of liquids in non-continuous motion. When a speed variation, Δv, appears within the pipeline, the mass of moving liquid can generate forces manifest in the form of important pressure variations within a short time interval. These pressure variations propagate along the pipe as waves with a speed equal to the velocity of ound in water (about 1,000 m/s) (see section 6.1).

Pressure rises (overpressures), $H_{o \cdot p}$, and pressure drops (underpressures), $H_{u \cdot p}$, are proportional to the speed variations, Δv

$$H_{o \cdot p} = H_{u \cdot p} = k \times \frac{a}{g} \times \Delta v \approx k \times 120 \times \Delta v \qquad (7.57)$$

where k is a coefficient depending on pump inertia moment, or on the effective closing time t_e of the control valve (see Fig. 5.3); $k = 1.00 - 0.75$ for $t_e \leqslant 2 \, l/a$ and $k \leqslant 0.75$ for $t_e > 2 \, l/a$.

Due to the short pump rotation after its switch-off, as well as to head losses through the pipeline, water hammer intensity is lowered, but only to a small extent, for which $k \approx 0.75$.

Considering a simple pipeline of length $l = 12,000$ m and wave propagation speed $a = 1,200$ m/s, a pressure wave would need only 10 s to be propagated through the whole pipeline length, and 20 s for 2 l, that is for two-way travel. Assuming that an instant operation shut-off valve is mounted near the pump, and that this valve is suddenly closed within a time interval of 20 s or even less, a flow speed variation $\Delta v = 0.3$ m/s occurs, and consequently, a pressure rise $H_{o \cdot p} = 36$ m, while if $\Delta v = 3$ m/s, $H_{o \cdot p} = 360$ m. The overpressure $H_{o \cdot p}$ is independent of the static head value $(H_{st} + h_{st})$ of the pipe (Fig. 7.59), but the pipeline is subjected to a total pressure head given by

$$H = H_{o \cdot p} + H_{st} + h_{st} \tag{7.58}$$

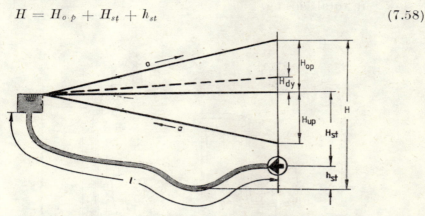

Fig. 7.59. Sketch showing overpressure and underpressure within a pumping station.

If instead of an instant shut-off valve we mount a slow closing device (whose opening and closing controls are individually adjustable and thus reduce the liquid flow speed progressively over a longer time than the reflection time ($t = 2\ l/a$), then pressure variations within the pipeline are much smaller than those taking place in the case of instant stopping of the liquid mass. The longer the closing time, that is the time taken to slow down the liquid flow as compared to the reflection time, the smaller the pressure variations within the pipe. A complete reduction of these variations cannot be obtained. Instead, we could achieve an almost linear slow-down of the flow by adequate adjustment of the closing valve. The linear head loss, produced within the pipeline, plus the local head loss within the valve give a total head loss which, at progressive narrowing of the valve orifice, becomes ever greater, which reduces the liquid speed progressively and linearly down to zero.

In Chapter 6 (see Section 6.1.3), we showed that it is not the total closing time, t_t, of a valve which is important, but the real time, which we have called "effective closing time", and which depends on the used valve type. As (Fig. 6.3) shows, the gate valve is not recommended as a closing device, since it has a very small t_e/t_t ratio ($t_e/t_t = 0.392$). This means that its local head loss rises significantly just before complete closing, very quickly reaching the maximum value. The same is valid for swing check valves, too. Consequently, closing these valves produces a reduction in flow speed only shortly before their complete closing, and with an extraordinary quickness, which is equivalent to an instantaneous cessation of flow, generating water hammers of high intensities.

Although the operating characteristic of a rapid-closing check valve is very disadvantageous, it can be improved by mounting this valve in parallel with another check valve, but of the non-slam type, which has a much smaller diameter than the first one. This valve system achieves successive closing characterized by two times. The first time corresponds to a rapid but partial closing of the cross-section, and the second time to a slow but total closing.

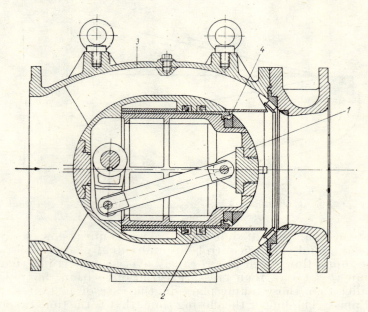

Fig. 7.60. Needle valve as a throttling device (cross-section):
1 — piston; *2* — guidance ball; *3* — casing; *4* — sealing ring.

An absolutely linear closing characteristic cannot be obtained with known valve types. The linear shape is almost obtained with the globe valve with wee ports. It shows an effective closing time of $t_e = 0.65\, t_t$ (see Fig. 6.3).

An excellent throttling device is the needle valve shown in Fig. 7.60. Its closing characteristic is much closer to the linear characteristic, having a closing time of $t_e = 0.72\, t_t$ (Fig. 7.61).

These valves have the role of preventing water hammers generated by normal pump-starts or stops, and thus it is reasonable to consider them efficient only as long as pumps are still running. Generally, needle valve operation is correlated to pump starting and stopping controls. Thus, at first the respective valve is closed, and then the pump is started or stopped (see Section 7.3.4).

Figure 7.62 shows a pumping installation whose pumps are equipped with needle valves correlated to pump cut-in or cut-out controls.

Given an almost linear operation of the needle valve, water hammer intensities are substantially reduced, but show a great dependence on the value of effective closing time. Some usual values are given below [7.5]:

when $t_e = N = 2\,l/a \leqslant 1$, we get instant closing and $H_{o.p} = H_{u.p} = 100$ per cent;

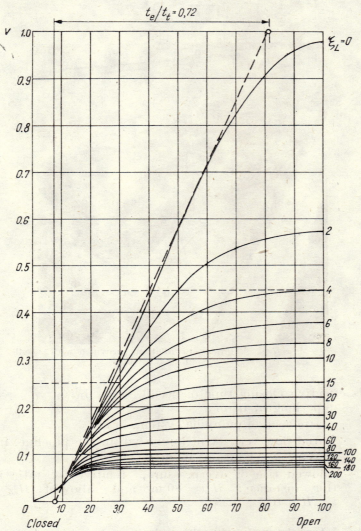

Fig. 7.61. Throttle curves for different pipeline resistances.

when $t_e = 2N$, we get 65 per cent of the value corresponding to instant closing;

when $t_e = 3N$, pressure variations represent only 40 per cent of the value corresponding to instant closing;

when $t_e = 5N$, pressure variations stand for 15 per cent of the maximum value.

The closing time of a needle valve with an approximately linear characteristic can be computed by means of the Voith relation [7.13]

$$t_e = \frac{1}{g} \times \frac{l}{z} \times \frac{v_0}{H_{st}} \sqrt{1+z} \qquad (7.59)$$

Fig. 7.62. Pumping installation, provided with needle closing valves.

where t_e is effective closing time of needle valve, in s;
l — pipe length, in m;
H_{st} — static head of pumping system, in m;
v_0 — initial flow velocity of liquid within the pipeline, in m/s;
g — gravity acceleration, in m/s²;
z — allowed rise of overpressure, relative to static head H_s, (for instance: $H_{st} = 80$ m and allowed $H_{o.p} = 20$ m $z = H_{o.p}/H_{st} = 0.25$).

Example 7.7. **Computation of needle valve closing time.** Consider a discharge pipeline made of steel (Fig. 7.63) having a length of $l = 4{,}310$ m, rated diameter $D_r = 0.90$ m, and wall thickness $\delta = 0.01$ m. In normal operating conditions the pipe has an initial flow rate $Q_0 = 0.75$ cu.m/s, to which an initial flow velocity $v_0 = 1.18$ m/s corresponds. We want to find the total closing time of a needle valve, mounted on the pump discharge line, so that the valve should secure a progressive reduction of flow velocity, in order to keep water hammer within permitted limits.

Control of Water Hammer

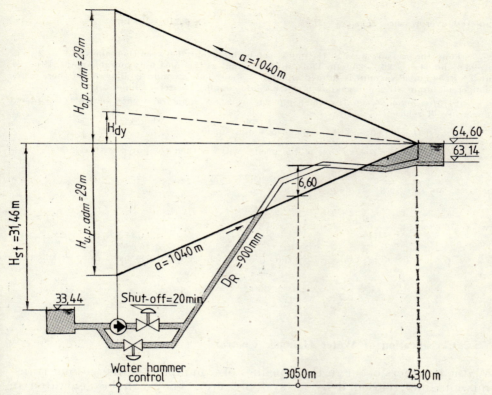

Fig. 7.63. Sketch for constructional and operational characteristics of the pipeline in example 7.7.

Solution

(1) *Wave celerity or propagation speed* is computed by means of equation (6.7)

$$a = \sqrt{\dfrac{9.81}{\dfrac{1}{210{,}000} + \dfrac{1}{22{,}000{,}000} \times \dfrac{0.90}{0.01}}} = 1{,}040 \text{ m/s}$$

(2) *Reflection time* is determined with relation (6.12)

$$t = \dfrac{2 \times 4{,}310}{1{,}040} = 8.3 \text{ s}$$

(3) *Maximum overpressure/underpressure* is computed making use of equation (7.57).

$$H_{o \cdot p} = H_{u \cdot p} = \dfrac{1{,}040}{9.81} \times 1.18 = 125 \text{ m}$$

(4) *Admissible value of overpressure/underpressure* ($H_{o \cdot p \cdot a}/H_{u \cdot p \cdot a}$) within the pipeline is determined. In the case shown in Fig. 7.63 the underpressure is limited at a vacuum equal to -6 m, admitted in the highest point, located at $3 + 200$ km. At that point, the underpressure can drop to $63.14 - 6.00 = 57.14$ m. Underpressure transmission to the pump is made lin early.

383

allowed overpressure $H_{o.p.a} = H_{u.p.a} = \dfrac{64.60 - 57.14}{4,310 - 3,200} \times 4,310 = 29$ m.

Thus, the pressure wave at the pump can oscillate only between the values $64.6 + 29 = 93.6$ and $64.6 - 29 = 35.6$ m. Thus, it follows that at the most unfavourable point, situated at $3 + 0.50$ km a vacuum is formed equal to -6.60. This vacuum is allowed, so there is no danger of liquid stream breakdown which could amplify water hammer.

(5) *Maximum pressure* to which pipeline walls are *subjected is checked*. To this end we make use of relation (7.88)

$$H = 93.60 - 33.14 = 60.46 \approx 6 \text{ daN/sq.cm}$$

(6) *The effective closing time* of the needle valve is then computed with relation (7.59)

$$t_e = \dfrac{1}{9.81} \times \dfrac{4,310}{0.92} \times \dfrac{1.18}{31.46} \times \sqrt{1 + 0.92} = 24.8 \text{ s}$$

where $0.92 = z = \dfrac{29.00}{31.46}$

(7) A *needle valve is chosen with a ratio* $t_e/t_t = 0.72$ (see Fig. 7.61) and a total closing time

$$t_t = \dfrac{t_e}{0.72} = \dfrac{24.8}{0.72} = 34.4 \text{s} \approx 35 \text{ s}.$$

7.3.2 Classification of Water Hammer Controls

Water hammers occur in the pipelines of a pumping system (long transmission lines, closed delivery systems, etc). They can be generated in pumping lines by two causes:

(1) *Normal pump operation with successive starting and stopping*. In this case, preventive measures are necessary, and protection is by means of speed control closing devices, allowing the adjustment of closing and opening speeds within a time interval, longer than the wave reflection time $(2\ l/a)$. These devices can be: *no-slam check valve, electric check valve, electric recirculation valve*, etc. They are mounted on each pump discharge line and are effective as long as the pump is still running.

(2) *Instantaneous and simultaneous stopping of all running pumps due to a general power failure*. In this case, measures of dissipation of water hammers are required, and protection is by means of surge control valves with very fast opening and slower closing times, which allow water discharge within an adjustable time interval, longer than the wave reflection time $(2\ l/a)$. These valves can be: *surge relief valve, surge anticipator (arrestor) valve, surge sensor valve* etc. They are mounted on common pump discharge lines, and discharge is generally into a suction tank.

Consequently, water hammer controls can be classified as follows:

(1) *Controls for surge prevention*:
1.1. Reverse-flow prevention controls
1.2. Controls associated with pump starting and stopping

(2) *Controls for surge dissipation*:
2.1. Controls for discharge at overpressure
2.2. Controls for discharge at underpressure
2.3. Controls for discharge at power failure.

7.3.3 Reverse-Flow Prevention Controls

These controls secure a one-way flow of the liquid through the pump discharge line. Control is by means of a hydraulic check valve (Fig. 7.64). This hydraulic check valve is operated automatically by means of a check pilot, and its closing and opening speeds are regulated by a needle valve. The closing and opening times, as adjusted by the needle valve, are in the range 6 to 60 s. Two connection schemes of their pilots are possible, resulting in the following types of check valves: (*1*) with rapid closing and slow opening (Fig. 7.65, *a*), and (*2*) with slow closing and opening (Fig. 7.65, *b*). The latter scheme shows the pilots connected in anti-parallel.

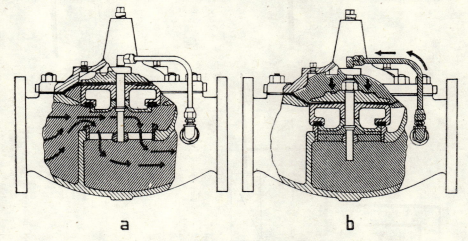

Fig. 7.64. No-slam check valve (sequence of operation):
a — open position; *b* — closed position.

An improved mounting scheme shows the parallel mounting of two check valves (Fig. 7.66). This system secures two alternative closing times of the line section. The first time corresponds to a rapid but partial closing of the reversed flow section, while the second corresponds to a total but slow closing. This case is approached as a combination of reversed flow and surge relief valve. A similar case is the Johnson type of cylindrical valve (Fig. 7.67).

A system of two closing times can be efficient, according to the graph shown in Fig. 6.26, only when $\rho > 0.5$ and $\tau > 1$ (their meaning has been explained in Chapter 6). These conditions are specific to medium head pumping installations whose pumps have medium inertia moments.

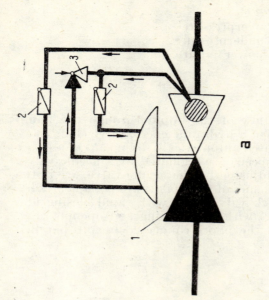
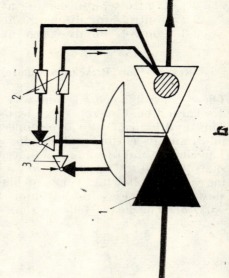

Fig. 7.65. Schematic piping arrangement of non-surge check valve:

a — with rapid closing and slow opening; b — with slow closing and opening; 1 — main valve; 2 — auxiliary check valve; 3 — speed control needle valve.

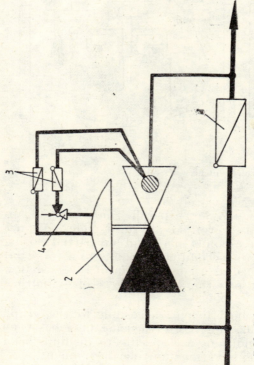

Fig. 7.66. Schematic piping arrangement of check valve with dual closing speed control:

1 — swing check valve; 2 — hydraulic check valve; 3 — auxiliary check valve; 4 — closing speed control needle valve.

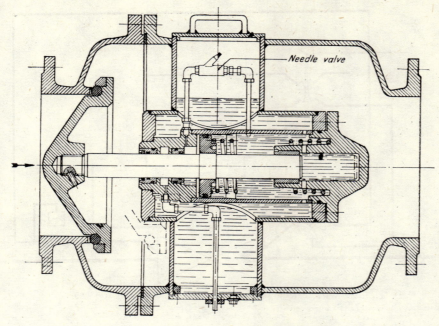

Fig. 7.67. Johnson-type cylindrical valve.

7.3.4 Controls Associated with Pump Starting and Stopping

These controls have the role of preventing water hammer formation due to normal pump starting and stopping.

Two types of control valve are associated with pump starting and stopping: (a) electric check valve; and (b) electric recirculation valve.

(a) *Electric check valve.* This is a hydraulically-operated valve provided with electrical commands correlated with pump starting and stopping (Fig. 7.68). Its controls are arranged so that the pump cannot be started or stopped unless the respective valve is in the *closed* position.

The electric check valve is mounted on the pump discharge line and comprises a main throttling valve and its controlling pilots.

The main throttling valve (Fig. 7.69) is hydraulically operated by a double-acting servomotor, whose operation is based on the differential piston principle.

The main throttling valve is controlled by a four-way solenoid pilot valve (Fig. 7.70). When the pilot solenoid is energized, the pilot commands the main valve opening, and vice-versa. Two needle valves (Fig. 7.71) are mounted into supply lines going from the pilot valve to the servomotor operation chambers. These needle valves adjust the closing and opening speeds of the electric check valve. The electric check valve is also provided with a limit switch assembly (Fig. 7.72) which secures an electric interlock of the pump motor starting switch.

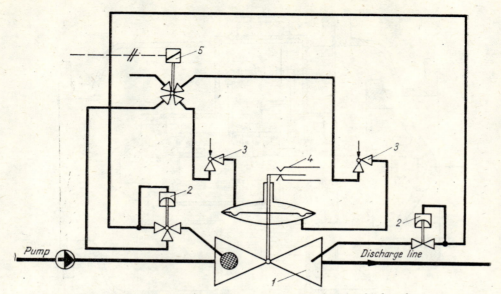

Fig. 7.68. Schematic piping arrangement of electric check valve:
1 — power (main) valve; *2* — auxiliary hydraulic check valve; *3* — speed control needle valve; *4* — limit switch; *5* — solenoid pilot.

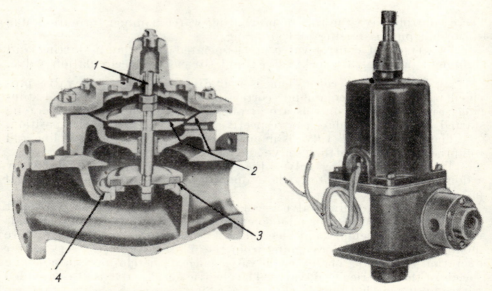

Fig. 7.69. Power (main) valve:
1 — stem; *2* — servomotor; *3* — disk; *4* — seat.

Fig. 7.70. Four-way solenoid pilot valve.

The electric check valve has three operational cycles:

(1) *Normal pump starting cycle* (Fig. 7.73). When the pump starting command is acted, the protection valve is in the closed position, and is

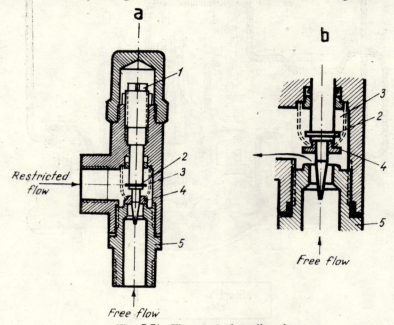

Fig. 7.71. Flow control needle valve:
a — assembly; *b* — layout; *1* — adjusting screw; *2* — snap ring; *3* — spring; *4* — disk; *5* — seat body.

opening slowly only after the pump picks up speed, thus progressively allowing liquid into the discharge line.

(2) *Normal pump stopping cycle* (Fig. 7.74). At the normal pump stopping command, the check valve is slowly closing, thus gradually reducing water velocity and admission into the pipe. Near the closing position, a valve extension is acting like a micro-switch, de-energizing the electric motor and thus stopping the pump. The stopping takes place at the same time as total closing of the check valve, resulting in minimum pressure variations within the pipeline.

(3) *Abnormal pump stopping cycle* (Fig. 7.75). When an accidental power failure takes place, the pump is stopped and the static pressure of the existing water in

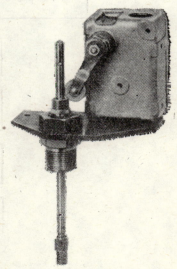

Fig. 7.72. Limit switch assembly.

Automated Control of Pumping Systems

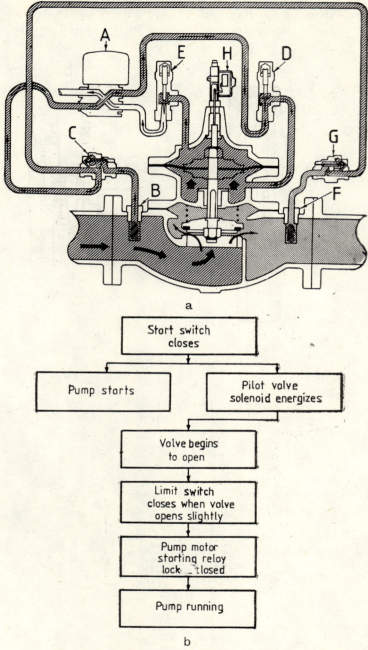

Fig. 7.73. Electric check valve; normal pump starting cycle: a — schematic piping arrangement; b — sequences of operation.

Control of Water Hammer

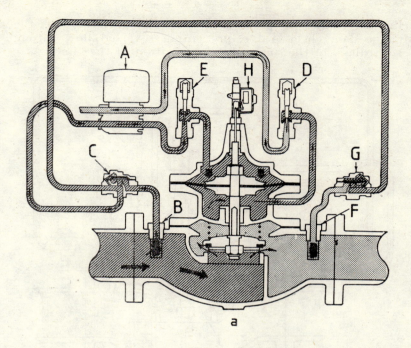
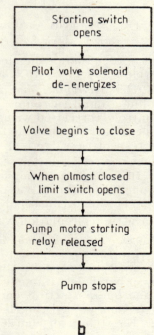

Fig. 7.74. Electric check valve; normal pump stopping cycle:
a — schematic piping arrangement; *b* — sequences of operation.

the pipeline determines the protection valve closing. In this case, the valve acts like a no-slam check valve, preventing reversed flow through the pump, regardless of the solenoid pilot position at the time.

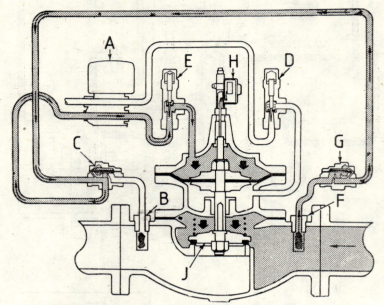

Fig. 7.75. Electric check valve: abnormal pump stopping cycle.

Figure 7.76,*a* shows the electric scheme, correlated with pump operational cycles, and Fig. 7.76,*b*, a view of an electric check valve.

(b) *Electric recirculation valve*. This valve (Fig. 7.77) is mounted on a by-pass line around the pump. It is a hydraulically-operated valve with electric controls, correlated with pump starting and stopping. These electric controls are arranged so that the pump cannot be started or stopped unless the respective valve is in the *open* position.

The electric recirculation valve has a four-way solenoid pilot valve (see Fig. 7.70) whose operation determines the closing or opening of the main valve. The main valve is also provided with two needle valves (see Fig. 7.71) which can separately adjust closing and opening speeds.

The electric recirculation valve has three distinct operation cycles:

(1) *Normal pump starting cycle* (Fig. 7.78). When the pump is stopped, the valve is in the *open* position. After starting the pump, the whole flow is directed backwards to the suction pipe, or out to the atmosphere in the case of boreholes. Immediately afterwards, the flow rate is diminished, since the pump starting energized the solenoid pilot control which in turn has commanded the electric recirculation valve closing. As it is closing, the pump discharge pressure rises until the existing static pressure in the pipeline is exceeded, and the pump check valve starts

Control of Water Hammer

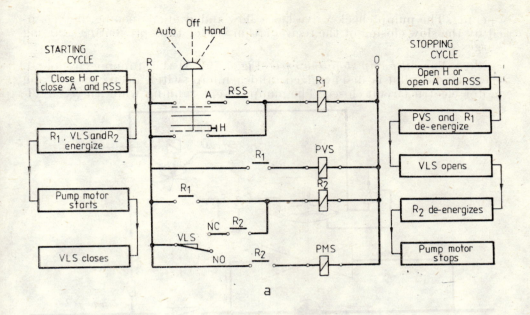

Fig. 7.76, *a*. Wiring diagram for electric check valve, with the pump in the off position and valve closed:

Auto-off-hand-selector switch; R_1, R_2 — double pole, normally-open relay; RSS — automatic switch (pressure switch, etc.); VLS — valve limit switch; PVS — pilot valve solenoid; PMS — pump motor starter.

Fig. 7.76, *b*. Electric check valve (Clayton Co.)

opening. The pump check valve has a slow and gradual opening, determined by the slow closing of the recirculation valve, thus preventing starting surges.

(2) *Normal pump stopping cycle* (Fig. 7.79). At the stopping command, the solenoid pilot is de-energized, and a micro-switch keeps the driving motor electric circuit closed. The pump is kept running while the recircul-

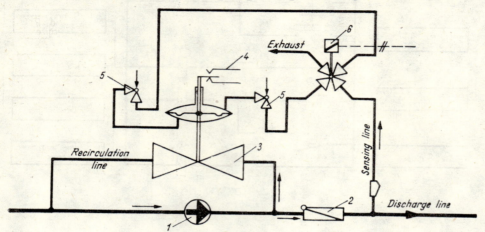

Fig. 7.77. Schematic piping arrangement of electric recirculation valve:
1 — pump; *2* — pump check valve; *3* — main valve; *4* — limit switch; *5* — closing speed needle valve; *6* — solenoid pilot valve.

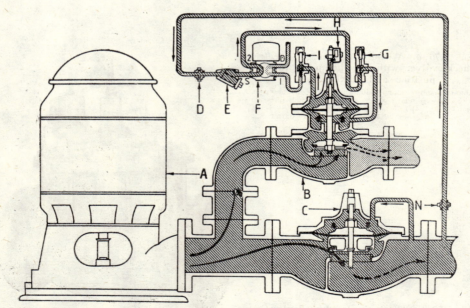

Fig. 7.78. Electric recirculation valve; normal pump starting cycle.

Control of Water Hammer

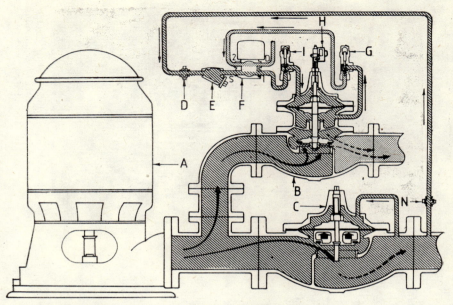

Fig. 7.79. Electric recirculation valve; normal pump stopping cycle.

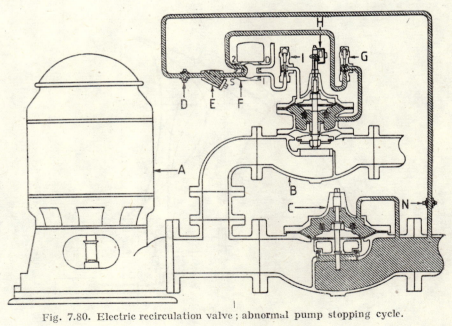

Fig. 7.80. Electric recirculation valve; abnormal pump stopping cycle.

Automated Control of Pumping Systems

Fig. 7.82. Pump, provided with electric recirculation valve, mounted in a pumping station, in Bucharest.

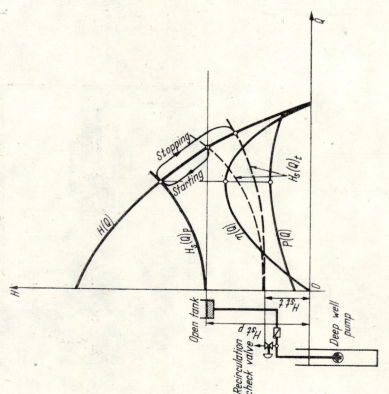

Fig. 7.81. Head-flow-rate curves of a pump, provided with electric recirculation valve.

ation valve is opening, and thus the pump pressure falls gradually as the pump check valve is slowly closing. The pump is stopped only when the recirculation valve is completely open.

(3) *Abnormal pump stopping cycle* (Fig. 7.80). When an accidental power failure takes place, the pump is instantly stopped, and consequently the static pressure of water in the discharge line opens the pump check valve, while the recirculation valve is gradually opening, regardless of the solenoid pilot position at that moment.

Since during this transitory regime the installation duty point is displaced very much to the right of the maximum efficiency point (Fig. 7.81), an electric recirculation valve is recommended mainly on mixed-flow pumps, whose power curve is non-overloading (see Section 2.6.3). On the other hand, there is a danger of overloading the electric motor.

The analyzed regulation method has the advantage of using a much smaller check valve (Fig. 7.82) than in the case of mounting it on the pump discharge line. At the same time, we do not introduce in this way additional head losses during continuous duty. A check valve of smaller diameter is cheaper and, at the same time, it secures the required closing and opening speeds much easier.

7.3.5 Controls for Discharge at Overpressures

The best surge control is prevention, but this is not always possible. For instance, in the case of a sudden power failure of all pumps, the water column in the discharge line is suddenly halted, producing high intensity overpressure waves. In this situation, wave dissipation measures are necessary.

The control for discharge at overpressures is by means of a surge relief valve (Fig. 7.83), connected laterally to the common discharge line of the pumps.

The surge relief valve comprises an angle body main valve and its controlling pilot.

The main valve (Fig. 7.84) is hydraulically actuated by a hydraulic servomotor with differential piston. This piston allows a larger area to be exposed to closing than to opening forces. The main valve function is exclusively determined by the function of its selected pilots.

The surge relief valve operates automatically. Its automatic command is given by an overpressure control pilot (Fig. 7.85), normally closed, that is closed when not under a pressure higher than the set one. The pressure pilot is provided with extra large ports to permit very fast opening of the main valve. The value of pressure, set on this pilot, is higher by 1.10—1.15 than the duty pressure in the main line.

The closing speed control of the surge relief valve is achieved by a needle valve (see Fig. 7.71), which gradually closes the surge relief valve within an adjustable preset time interval (0—30 s).

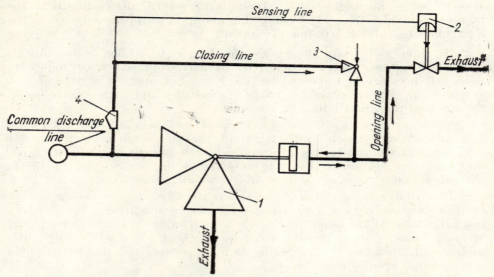

Fig. 7.83. Schematic piping arrangement of surge relief valve:
1 — main valve; *2* — overpressure pilot; *3* — closed speed control needle valve; *4* — strainer.

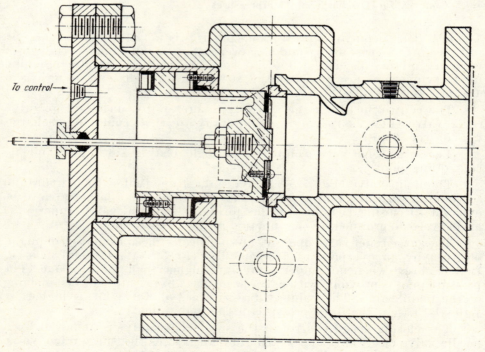

Fig. 7.84. Angle body main valve (cross-section).

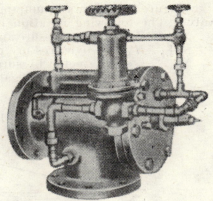

Fig. 7.86. Surge relief valve (Golden-Anderson Co.).

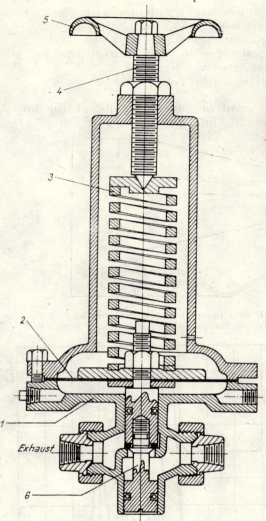

Fig. 7.85. Overpressure control pilot:
1 — body; 2 — membrane; 3 — spring; 4 — adjusting screw; 5 — star; 6 — stem.

Figure 7.86 shows the view of a surge relief valve. This valve is built according to the scheme shown in Fig. 7.83, having a two-phase operation:

— in the first phase when an overpressure wave of value higher than the set one occurs, the surge relief valve is rapidly opened to start the discharge action.

— in the second phase, when the pressure in the system is back below the set value due to water hammer damping, the surge relief valve is gradually closing within an adjustable time interval, thus concluding the discharge action.

Figure 7.87 shows a pumping installation, provided with a surge relief valve. The following notation has been used:

$H_{o.p}$, $H_{u.p}$ is overpressure and underpressure head, respectively;

$H'_{o.p}$, $H'_{u.p}$ — overpressure/underpressure head into a pipeline, protected by surge relief valve;

H_{st}, h_{st} — positive/negative static head.

$H = H_{o.p} + H_{st} + h_{st}$ — total pressure head on pipeline walls at the lowest point of the longitudinal profile.

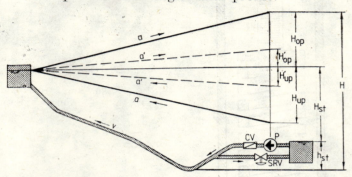

Fig. 7.87. Piezometric profile of a pumping installation, provided with surge relief valve on the pump discharge line.

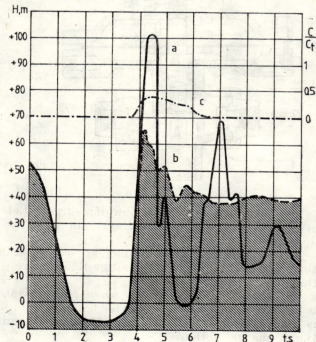

Fig. 7.88. Graph showing pressure variations at a pumping station, provided with surge relief valve:

a — for an unprotected system; b — for a system protected by means of a surge relief valve; c — valve stroke variation during opening.

Figure 7.88 shows pressure variation plots of an experimental pumping station. They include:
— pressure head variation curves near the pump for the case of an unprotected pipeline (curve *a*) and of a pipeline protected by a surge relief valve (curve *b*);
— variation curve of control valve stroke within the time interval necessary for the effective water discharge from the pipeline (curve *c*).

We notice that, when the pipeline is unprotected, the pressure head at the pump reaches a value of 120 m, and when protected, this head does not exceed 66 m. We also observe that the relief valve starts opening at $t = 3.8$ s, corresponding to a pressure head $H = 35$ m, which is smaller by 10.5 m than the pilot preset pressure. It follows that the surge relief valve can only partially anticipate the surge, and opens only 40 per cent (that is, its effective stroke *c* is 0.4 of the total stroke), and then slowly closes. During the time interval necessary for discharge ($t = 6.5 - 3.8 = 2.7$ s), the pressure recovers to its normal value in the pipeline.

The computation for surge relief valve selection is done in two stages. During the first, the discharge flow rate Q_{ds} is determined by means of the graph shown in Fig. 7.89, and during the second the rated diameter D_R of the surge relief valve is computed, making use of the graph shown in Fig. 7.90.

Protection of pipelines against surges by means of surge relief valves is efficient when $\rho < 0.5$ and $\tau < 1$ for the pipelines (the meaning of these coefficients has been explained in Section 6.4.2) and the pipeline length does not exceed 3,500 m. These conditions can be met in the case of short pipelines with medium or high pumping heads, and whose pumps have small inertia moments.

Example 7.8. **Computation for surge relief valve selection.** Consider a discharge line of length $l = 2,000$ m which has to be protected. Assuming as known: initial flow rate $Q_0 = 300$ dm³/s, initial pumping head $H_0 = 80$ m, and initial velocity $v_0 = 1.10$ m/s, we have to select the necessary surge relief valve.

Solution. The computation has two stages:
(1) *Discharge flow rate* Q_{ds} is easily computed, making use of the graph shown in Fig. 7.89. By introducing there the values $H_0 = 80$ m and $v_0 = 1.10$ m, we get a discharge flow rate $Q_d = 1.2\, Q_0$. Thus, $Q_{ds} = 1.2 \times 300 = 360$ l/s.
(2) *The surge relief valve rated* diameter is determined. To this end we use Fig. 7.90. Thus, for $Q_{ds} = 360$ l/s and $H_0 = 80$ m we get $D_R = 200$ mm.

7.3.6 Controls for Discharge at Underpressures

Under certain conditions it becomes necessary to have the surge control valve start to open on a low pressure wave. This wave is created by the pump stopping. With the surge valve opening on the initial low wave, the returning high wave is diverted from the system. In effect the valve has anticipated the returning high wave and is open. Consequently, the high wave never occurs.

A. **Controls opening at negative underpressures.** When surge computations show an underpressure wave going below the pipe centre-line,

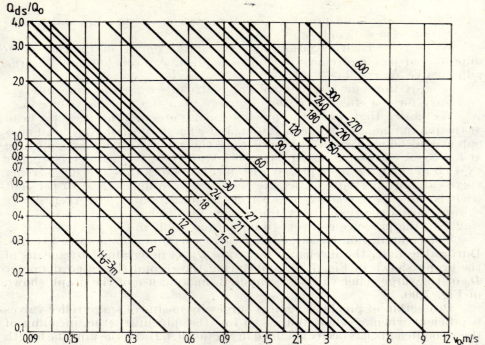

Fig. 7.89. Discharge output Q_{ds}, plotted against initial velocity, v_0, and initial head, H_0.

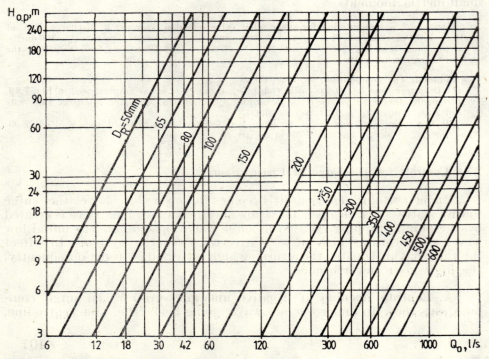

Fig. 7.90. Rated diameter D_R of the surge discharge valve versus initial head, H_0, and initial output, Q_0.

a vacuum may occur during the first stage of the surge. For preventing underpressure waves during the second stage, a protection with controls opening at vacuum (negative underpressure) should be used. This protection is achieved by means of a surge anticipator valve (Fig. 7.91, *a*), connected laterally to the pumps common discharge line (as in Fig. 7.87). This valve is controlled by a swing check pilot valve, opening at the occurrence of vacuum, and in its turn opening the main valve.

The operation of a surge anticipator valve takes place in two stages, as follows :

— during the first stage, vacuum is created in the pipeline by surges, and the check pilot control rapidly opens the main valve for discharge ;

— during the second stage, the vacuum disappears due to damping of the surges and the check pilot control is closed, also slowly closing the main valve within an adjustable time interval, thus cutting off the discharge.

Surge anticipator valves can also have the function of surge relief valves (Fig. 7.91, *b*) when provided additionally with an overpressure pilot control (normally closed) (see Fig. 7.85).

Figure 7.92 shows the effect of pipeline protection by means of a surge anticipator valve [7.12]. Curve *a* is the pump pressure head variation in the case of an unprotected pipeline, and curve *b* for a pipeline protected by a surge anticipator valve. The same diagram includes the curve of variation of surge anticipator valve stroke (curve *c*). The graphs show that in an unprotected pipeline the maximum pressure head rises up to 102 m while in a pipeline, protected by a surge anticipator valve, this value does not exceed 22 m. We also notice that the valve starts opening as soon as the pressure becomes zero, and corresponds to a time interval $t=1.1i$ s. After a time interval of 0.8 s, the valve is completely open, and remains in the position as long as there is vacuum within the pipeline, that is for one second. As the vacuum disappears, that is at $t = 3$ s, the main valve starts closing slowly over 7 s. Finally, the pressure wave is smoothly damped, becoming equal again to the static pressure, without generating any overpressure wave.

The design computation for a surge anticipator valve is done in two stages : during the first, we determine the discharge flow rate, making use of the graph shown in Fig. 7.89, and during the second, the rated diameter of the valve is found, by means of the graph shown in Fig. 7.93.

Protection by a surge anticipator valve, opening at negative underpressure, is efficient when $\rho < 0.5$ and $\tau < 1$ (see Section 6.4.2) and the protected pipeline length exceeds 3,500 m. These conditions are specific to long transmission lines, with high pumping heads and coupled to pumps with small inertia moments.

Example 7.9. **Design computation for a surge anticipator valve.** Consider a pumping line 5,000 m long which has to be protected. Assuming as known : initial flowrate $Q_0=1,171$ dm³/s initial pumping head $H_0 = 60$ m, and initial velocity $v_0 = 1.49$ m/s, we have to select the necessary surge anticipator valve.

Solution

(1) *Discharge flow rate* Q_{ds} is determined by means of the graph shown in Fig. 7.89. For $H_0 = 60$ m and $v_0 = 1.49$ m/s, we get $Q_{ds} = 0.6\ Q_0 = 0.6 \times 1,171 = 700$ l/s.

(2) *Rated diameter of surge anticipator valve* D_R is determined by means of Fig. 7.93. For $Q_{ds} = 700$ l/s and $H_0 = 60$ m we get $D_R = 250$ mm.

Automated Control of Pumping Systems

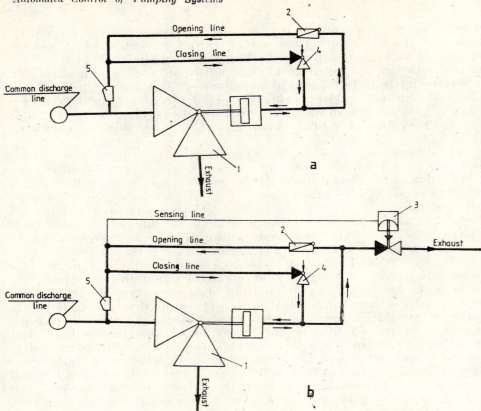

Fig. 7.91. Schematic piping arrangement of surge anticipator valve opening on negative underpressure:

a — simple arrangement; *2* — combined arrangement; *1* — main valve; *2* — auxiliary swing check valve; *3* — overpressure control pilot; *4* — needle valve for closing speed control; *5* — strainer.

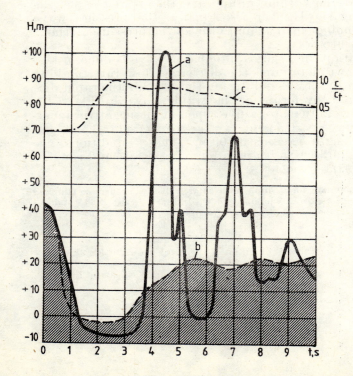

Fig. 7.92. Graph showing pressure variations at a pumping station, provided with a surge anticipator valve:

a — for an unprotected system; *b* — for a system protected by means of a surge anticipator valve; *c* — valve stroke variation during opening.

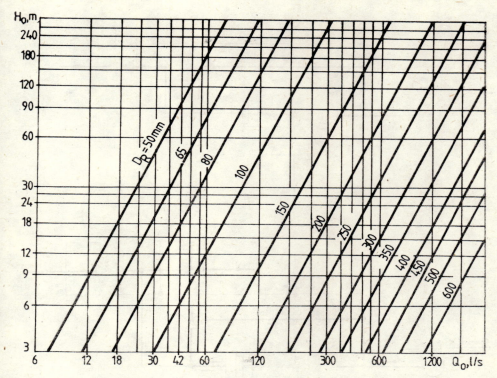

Fig. 7.93. Rated diameter D_R of surge anticipator valve versus initial head H_0 and initial output Q_0.

B. **Controls opening at positive underpressures.** These controls have the role of discharging protected pipelines by means of surge anticipator valves opening when positive underpressure waves, produced during the first stage of surges, are signalled. By positive underpressure wave we mean those waves which do not lower under pipes summit and thus no vacuum can be created inside the respective pipelines.

For protecting these pipelines against surges, we recommend the surge anticipator valve built according to the scheme shown in Fig. 7.94, a. It is provided with an underpressure pilot control (Fig. 7.95), normally open, that is, open when not underpressure, with extra large ports to permit very fast opening of its main valve. The preset value of minimum pressure head for the underpressure pilot is chosen at 60 per cent of the initial pressure value.

The surge anticipator valve can also function as a surge relief valve (Fig. 7.94, b) when provided additionally with an overpressure pilot (normally closed) (see Fig. 7.85).

The selection, mounting, and operation of the surge relief valves, shown in Fig. 7.94 are similar to those shown in Fig. 7.91.

Protection of pipelines by the valves mentioned above is efficient when underpressure waves produced within these pipelines are of small

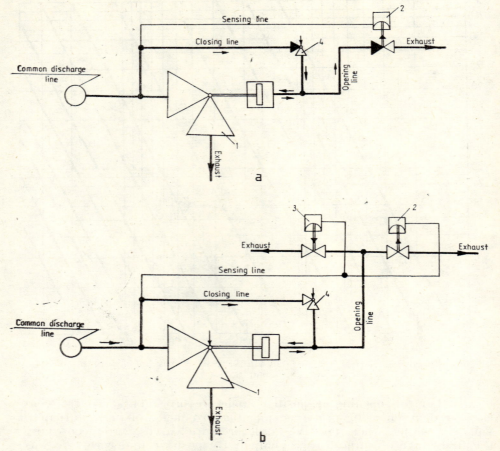

Fig. 7.94. Schematic piping arrangement of surge anticipator valve with opening at positive underpressure:
a — simple arrangement; *b* — combined arrangement; *1* — main valve; *2* — underpressure pilot; *3* — overpressure pilot; *4* — needle valve for closing speed control.

amplitude. These conditions are characteristic of pumping systems, provided with pumps of high inertia moments.

Under certain conditions, the pressure within the pipeline cannot recover to a value equal to, or higher than, the one set. Consequently, the controlling pilot can no longer be energized and the main valve stays open, uselessly discharging liquid from the system. In such conditions, additional measures should be taken for a forced hydraulic closing of the main valve (Fig. 7.96). This means that the surge anticipator valve should be provided with a hydraulic time-delay pilot (Fig. 7.97). This pilot produces the forced hydraulic closing of the main valve depending on time, that is, after a predetermined time interval, and not on pressure, as in the case of the other protection valves previously presented.

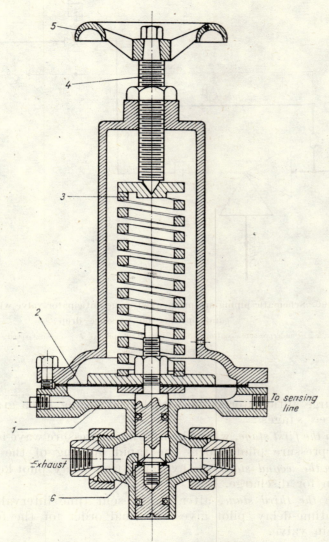

Fig. 7.95. Underpressure control pilot:
1 — body; *2* — membrane; *3* — spring; *4* — adjusting screw; *5* — star; *6* — stem.

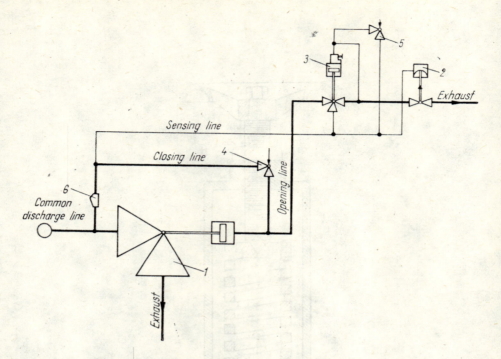

Fig. 7.96. Schematic piping arrangement of surge anticipator valve with forced closing and opening on pressure drop:

1 — main valve; *2* — underpressure control; *3* — time-delay pilot; *4* — closing speed control; *5* — needle valve; *6* — strainer.

The surge anticipator valve with forced closing has a gradual operation in three stages:

(1) *In the first stage*, when the first underpressure wave is intercepted, the underpressure pilot orders the rapid opening of the main valve.

(2) *In the second stage*, the hydraulic time-delay pilot keeps the main valve open for discharge.

(3) *In the third stage*, after the present time interval is over, the hydraulic time-delay pilot gives the final order for the forced closing of the main valve.

The surge anticipator valve can also function as a surge relief valve when the main valve is additionally provided with an overpressure pilot (normally closed) (see Fig. 7.80), thus obtaining a combined valve with opening at underpressure and overpressure (Fig. 7.98).

Figure 7.99 shows such a protection valve, mounted in a pumping station in the city of Bucharest.

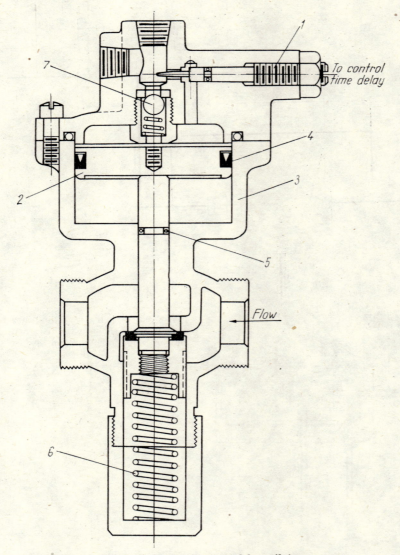

Fig. 7.97. Hydraulic time-delay pilot:
1 — metering stem to control time-delay; *2* — piston; *3* — body; *4* — piston seal; *5* — steam seal; *6* — spring; *7* — ball.

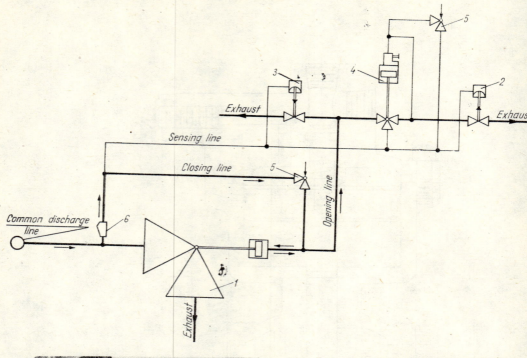

Fig. 7.98. Schematic piping arrangement of surge anticipator valve with forced closing and opening on pressure drop and rise:

1 — main valve; *2* — underpressure pilot; *3* — overpressure pilot; *4* — time delay pilot; *5* — closing speed control; *6* — strainer.

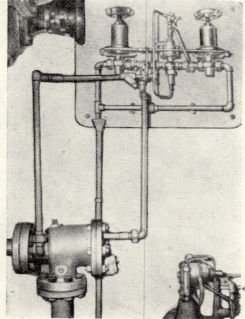

Fig. 7.99. Surge anticipator valve with forced closing and opening on pressure drop and rise (Golden-Anderson Co.).

7.3.7 Controls for Discharge at Power Failure

These controls have the role of protecting a pumping system against surges by means of a protection valve, opening at power failures. Its command (Fig. 7.100) is given by means of two pilots: a solenoid control pilot normally open (that is, open when not energized) (Fig. 7.101), and a time-delay relay, hydraulically actuated (see Fig. 7.97).

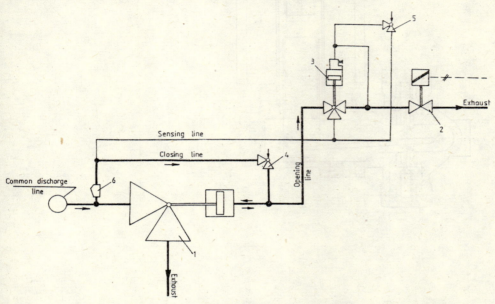

Fig. 7.100. Schematic piping arrangement of surge anticipator valve with forced closing and opening on power failure:
1 — main valve; *2* — solenoid pilot; *3* — time-delay pilot; *4* — closing speed control; *5* — needle valve; *6* — strainer.

The check valve for power failure operates in three stages as follows:

(1) *During the first stage*, at power failure, the solenoid control pilot is de-energized and thus orders fast opening of the main valve.

(2) *During the second stage*, the time-delay relay keeps the main valve open to discharge liquid from the pipeline.

(3) *During the third stage*, after the present time interval is over, the time-delay relay orders slow closing of the main valve.

The control valve shown in Fig. 7.100 can also have other functions, when provided with adequate supplementary pilot controls. Thus it can be fitted with:

— overpressure pilot controls (see Fig. 7.85) for a combined valve opening at power failure and overpressures (Fig. 7.102);

— underpressure pilot controls (see Fig. 7.85) and another overpressure pilot control (see Fig. 7.95), for a combined valve opening at power failure and under/overpressure (Fig. 7.103).

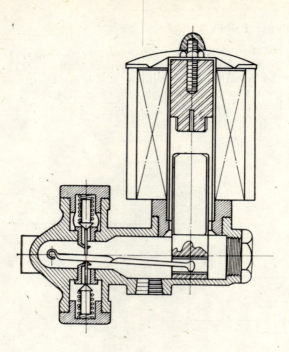

Fig. 7.101. Solenoid control pilot.

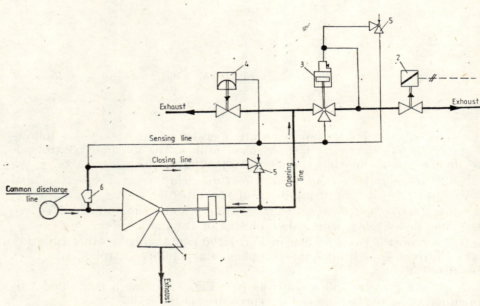

Fig. 7.102. Schematic piping arrangement of surge anticipator valve with forced closing and opening on power failure and pressure rise:

1 — main valve; *2* — solenoid pilot; *3* — time-delay pilot; *4* — overpressure pilot; *5* — closing speed control; *6* — strainer.

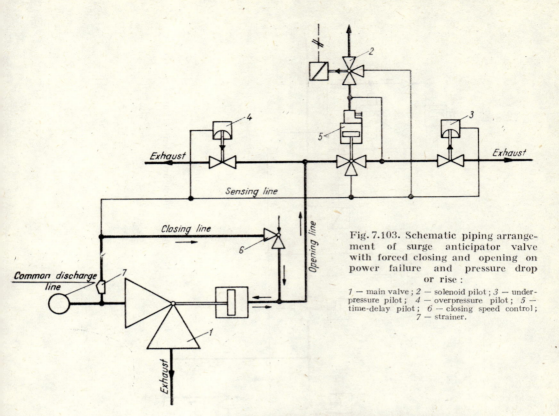

Fig. 7.103. Schematic piping arrangement of surge anticipator valve with forced closing and opening on power failure and pressure drop or rise:

1 — main valve; *2* — solenoid pilot; *3* — underpressure pilot; *4* — overpressure pilot; *5* — time-delay pilot; *6* — closing speed control; *7* — strainer.

Fig. 7.104. Surge anticipator valve according to scheme shown in Fig. 7.103. (Golden-Anderson Co.).

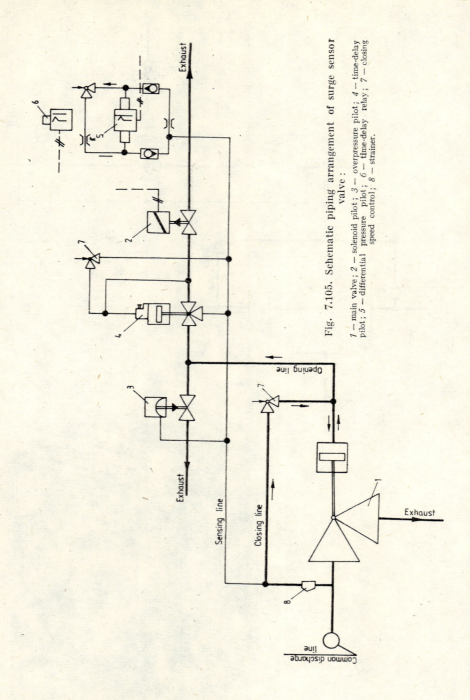

Fig. 7.105. Schematic piping arrangement of surge sensor valve:

1 — main valve; *2* — solenoid pilot; *3* — overpressure pilot; *4* — time-delay pilot; *5* — differential pressure pilot; *6* — time-delay relay; *7* — closing speed control; *8* — strainer.

Figure 7.104 shows such a protection valve, built according to the scheme shown in Fig. 7.103.

The computation for selecting the protection valves shown in Figs. 7.100, 7.102, and 7.103 is identical to that for the selection of the valve shown in Fig. 7.91.

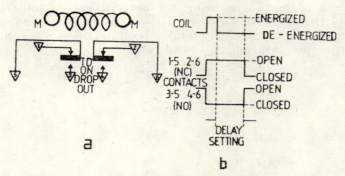

Fig. 7.106. Time-delay relay:
a — wiring diagram; *b* — operation diagram.

In the case of a long discharge line with a very skewed system curve, head losses increase proportionally with the square of the flow rate, and the same applies for the pressure at the pump. Consequently, in that case we should use several fixed-speed pumps connected in series, or connected to a variable-speed pump, fed by a static convertor. Such a pipeline can be protected against surges by a certain control valve, provided with special automatic controls (Fig. 7.105), called a *surge sensor valve* [7.6].

Fig. 7.107. Special pilots of surge sensor valve (Golden-Anderson Co.).

415

The main valve command is so arranged that its pilots open it only at a certain pressure variation within a certain time interval. For example, such a valve will open on a 35 m pressure change in two sec, or a 70 m pressure change in five sec, and so forth.

The pressure and time modes are individually adjustable with the pressure-time adjustments (Fig. 7.106) so that a normal pump start-up or shut-down or normal valve closure will not activate the controls.

Figure 7.107 shows a panel with two special pilots (a differential pressure switch and a time-delay relay), both mounted on a surge sensor valve at a pumping station in Bucharest.

REFERENCES

7.1. BOGER, W. H., *Flow characteristics for control valve installations*. ISA Journal, No. 10 (1966).
7.2. DRISKELL, R. L., *Practical guide to control valve sizing*. Instrumentation Technology, No. 10 (1966).
7.3. GUREVICH, I. P., *Raschiot i constructzia turboprovodnyih armatury*. Mashgyz, Moscow, 1964.
7.4. IONEL, I. I., *Stations de pompage a régulation débitmetrique*. La technique de l'eau et de l'assainissement, No. 1 (1972).
7.5. KERR, L. S., *Effect of valve operation on water hammer*. Jour. AWWA. No. 1 (1960).
7.6. LESCOVICH, E. I., *The control of water hammer by automatic valve*. Jour. AWWA, No. 5 (1966).
7.7. MAMRELLI, S. E., *Pressure zoning and maintenance in the Los Angeles Distribution System*. Jour. AWWA, No. 2 (1961).
7.8. MARINOIU, V., POSCHINA, I., *Robinete de reglare*. Technical Publishing House, Bucharest, 1972.
7.9. ROSENBAUM, A. A., MARYNICK, M. S., *Water hammer control in Los Angeles distribution pumping plants*. Jour. AWWA, No. 6 (1967).
7.10. SCHUDER, B. B., *Control valve characteristics*. Instruments and control systems, No. 3 (1967).
7.11. VOLK, W., *Absperrorgane in rohleitungen*. Springer Verlag, Berlin, 1959.
7.12. WEAVER, L. D., *Surge control*. Jour AWWA, No. 7. (1972).
7.13. WENZEL, H., *Werteilung des wassers*. Verlag technic, Berlin, 1960.

8

Automated Control of Pumps

The fundamental control functions of a pump can be subdivided into two main groups : (1) the interruption and resumption of flow, and (2) the modification of the pump-head capacity-curve. The interruption and resumption of flow can be accomplished by starting and stopping a fixed-speed pump-driver, regulated by means of an on-off control such as a pressure switch, or an on-off control by a flowmeter. The pump-head capacity-curve can be modified by adjusting the speed of a variable-speed pump-driver which is regulated by means of a modulating control such as a constant-pressure control, or a parabolic pressure control.

8.1 Hydraulic Accumulator Theory and Computation Methods

Hydraulic accumulators are indispensable appliances to pump installations, equipped with fixed-speed pumps, regulated by means of a pressure switch on-off control. They limit the maximum frequency of turbopumps, starting automatically (thus having a corrective function), as well as the adverse effects of water hammer in the pumping system (having a protective function). These accumulators are used for starting the installation for compensating the minimum flow rate, and whenever the required consumption is not within the flow range of the turbopumps.

Depending on the way the energy is stored, the hydraulic accumulators are classified in two principal categories : hydromechanical accumulators, and hydropneumatic accumulators.

8.1.1 Hydromechanical Accumulators

Hydromechanical accumulators store the hydraulic energy resulting from the compression of an elastic body (spring or rubber). Considering their construction, these accumulators are of two types : *with piston or with membrane*. They comprise a tank with a mobile inner wall, which is either a piston on which a steel spring is acting, or a cylindrical rubber membrane with thickened wall.

Piston-type accumulator. The piston creates two chambers of variable volume. The force developed by the liquid under pressure (discharged by the pumps) acts upon the lower face of the piston, while the counter-force provided by the metallic spring acts in a reverse sense (Fig. 8.1).

For determination of the size of a piston-type accumulator, the equation of balance of forces acting on the piston (assuming the dynamic parameters of the piston, and hence the damping factor, are neglected) is

$$S(p_a - p_0) = k(x_0 - x) \tag{8.1}$$

where S is piston surface, in cm²;
k — spring elasticity constant, in mm/kg;
$x - x_0$ — spring deformation, in mm;
$p_a - p_0$ — pressure variation, in daN/cm².

Differentiating equation (8.1)

$$S \frac{dp_a}{dt} = -k \frac{dx}{dt} \tag{8.2}$$

The total volume, V_t, of the accumulator depends on the construction characteristics of the piston and spring, and can be determined from

$$V_t = \frac{S}{k} \frac{dp_a}{dt} \tag{8.3}$$

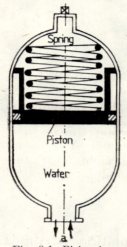

Fig. 8.1. Piston-type accumulator.

Memberane-type accumulator. This hydraulic accumulator is known in technical literature as "Hydrocel" (Fig. 8.2). It consists of an elastic tube, 1, of high quality rubber, with a thickened membrane. This tube is superposed on a metallic pipe, 2, (which has circulation orifices), and is sealed at its both ends by two packing rings, 3. The tube is protected on its exterior by a pipe, 4 (made of aluminium or plastic), closed with covers at both ends, 5. The membrane is thickened so that it should balance the forces corresponding to the working liquid pressure, exerted on the inner surface of the flexible tube.

A membrane-type accumulator operates like an ornamental baloon: it is filled with pressurized liquid by membrane tensioning, and throws the liquid out by membrane contraction. The graph of Fig. 8.3 shows the pressure head curve against volume (curve a) for the case of a Hydrocel membrane-type accumulator of 12 dm³. The same graph shows the characteristic operation curve of a hydropneumatic (conventional) accumulator of 36 dm³ (curve b). Analysis of the two curves reveals that curve a is flatter than curve b. This means that the membrane-type accumulator maintains a constant pressure over a large range of volume variation, that is, during most of the time it is emptying to meet consumption needs without help from pumps. A membrane-type accumulator stores a bigger useful volume, and thus it has an excellent efficiency ($\eta = V_{useful}/V_{total} = 0.8$ to 0.9). However, these accumulators have an extremely small volume of 10 to 12 dm³ and are therefore mounted in batteries of 2 to 30 units (Fig. 8.4). In general, they are fitted at the highest point of the network (under the roof, or on the top floor of a building) since their membranes do not stand excessive pressures (maximum working pressure $\leqslant 5$ daN/cm²). The cut-in and cut-out pressures also have relatively low values. For instance, the Jacuzzi

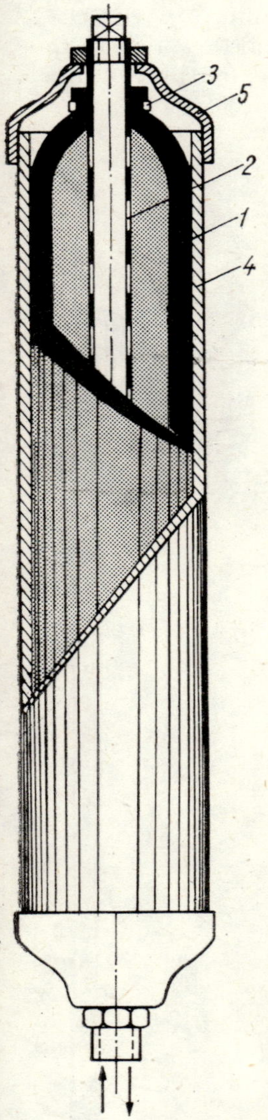

Hydraulic Accumulator Theory

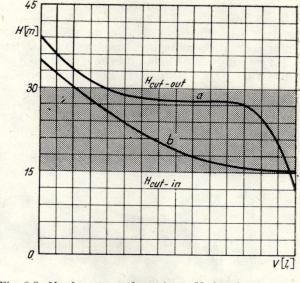

Fig. 8.3. Head versus volume for a Hydrocel accumulator:
a — Hydrocel (12 dm³); b — hydropneumatic tank (36 dm³).

Fig. 8.2. Hydrocel membrane-type accumulator, (Jacuzzi Co.).

Fig. 8.4. Hydrocel accumulators mounted in compact batteries (Jacuzzi Co.).

Company (U.S.A.) produces membrane-type hydraulic accumulators in two variants. The first includes accumulators having 1.5 daN/cm² cut-in pressure and 3.0 daN/cm² cut-out pressure, and the second accumulators of 2.0 daN/cm² cut-in pressure and 4.0 daN/cm² cut-out pressure.

The number of accumulators used depends on the unitary values of pump flow rate and power. For instance, eight accumulators of 10 dm³ each are required for a pump having $Q = 2.5$ dm³/s and $P = 2.2$ kW, and 28 accumulators of the same capacity, for a pump with $Q = 4.5$ dm³/s and $P = 5.5$ kW. In the first case, the total flow rate of the system (maximum per hour) is 7.5 dm³/s and is provided by three pumps, while in the second it is 18 dm³/s and is provided by four pumps.

Figure 8.5 shows a package pumping installation, fitted with a Hydrocel membrane-type accumulator. This applies to the case of

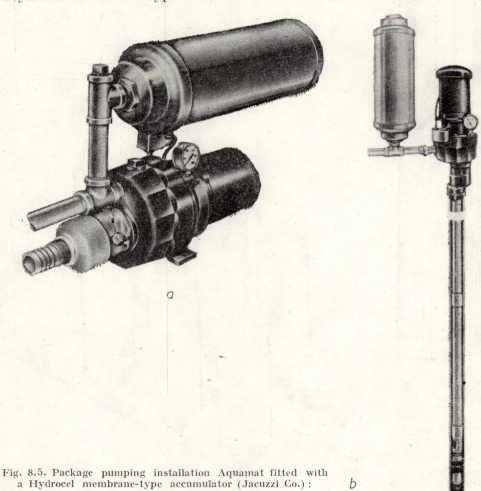

Fig. 8.5. Package pumping installation Aquamat fitted with a Hydrocel membrane-type accumulator (Jacuzzi Co.):
a — horizontal mounting; *b* — vertical mounting.

extremely small pumping installations ($Q \approx 2$ dm³/s and $H \approx 30$ m), used for the water supply plants of rural or urban individual farms (water supply for house, gardening, car washing, etc.).

8.1.2 Hydropneumatic Accumulators

Hydropneumatic accumulators, or hydrophore tanks, store the pressure resulting from gas compression (air or nitrogen). Two construction types are known: with separation bladder, or conventional.

Bladder hydropneumatic accumulator. This type of accumulator (Fig. 8.6) has been patented by the Charlatte Company (France) under the name of Hydrofort [8.4]. It comprises a flexible, alimentary butyl separation bladder (or chamber), *1*, (playing the role of separation chamber between water and air), whose orifice, *2*, is fitted to the discharge pipe of the pump, operating with the accumulator. The bladder is placed into a metallic oval outer cover *3* (the tank) whose ends have access-hole covers, *4*, necessary both for inside control of the tank and for bellows replacement in a case of rupture. The tank is also fitted with a pressure switch, *5*. The bladder is filled with drinking water, with air outside. Thus, the bladder separates the drinking water from a gaseous volume, which brings the following advantages:

— it avoids loss of the air cushion in the event of an undesired cut-in power supply to the pumps;

— it avoids the danger of polluting drinking water with oil vapour (which has escaped from the compressors) or by corrosion of the metallic tank).

As water fills the separation chamber, the gas volume contained in the tank is diminished, and its pressure rises. This pressure variation with volume develops according to the Boyle-Mariotte law: $pV = p_1 V_1 =$ constant, temperature variation being negligible. When the cut-out pressure is reached, the pressure switch stops the pump, operating with the relevant tank. As the water is consumed, the compressed gas volume presses the chamber, which discharges the stored water into the supply system. Thus, the pressure decreases as the chamber is emptied, and when cut-in pressure is reached, the pressure switch re-starts the pump. This cycle is automatically repeated whenever cut-in or cut-out pressures are reached.

Figure 8.7 shows the family of characteristic curves for tank volume variation with pressure. They are real curves, and characterize a Hydrofort-type tank of *1* cu. m [8.4].

Example 8.1. **Computation of useful volume for a Hydrofort tank of 1 cu.m.** Consider a pumping installation receiving drinking water at 1 m pressure head from a municipal system. This station is designed for the supply of a building with a 9 m static head and a 1 m dynamic head (head losses). The water supply to the last floor implies a minimum pressure head of 20 m, and allows a maximum pressure head of 50 m.

Solution

(1) *The cut-in and cut-out head of the pump* are computed:

— cut-in head $H_{cut-in} = 20 + 9 + 1 = 30$ m

— cut-out head $H_{cut-out} = 50 + 9 + 1 = 60$ m

Automated Control of Pumps

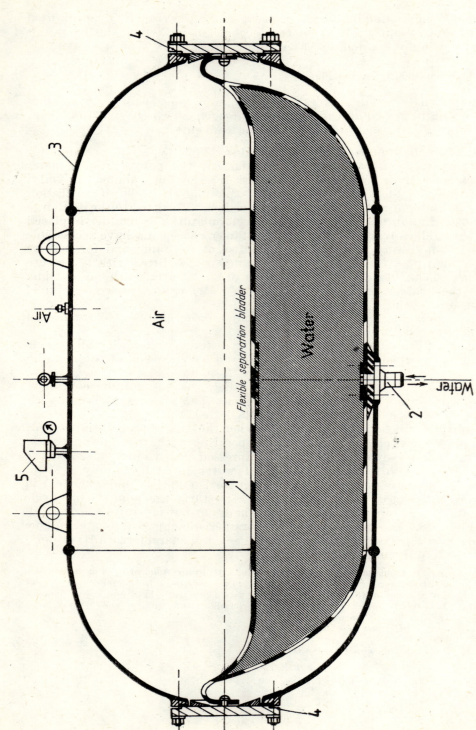

Fig. 8.6. Hydrofort hydropneumatic accumulator with separation bladder (Charlotte Co.):
1 — separation bladder; *2* — outlet pipe; *3* — tank; *4* — manicle; *5* — pressure switch.

Hydraulic Accumulator Theory

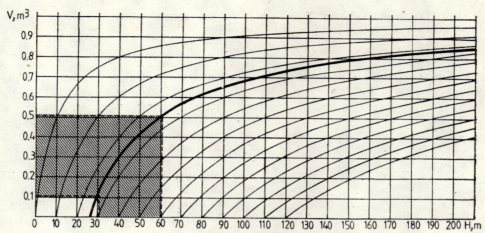

Fig. 8.7. Volume versus head for Hydrofort-type tank of one cu.m.

(2) *The data obtained above* are plotted on the graph shown in Fig. 8.7. and we thus get :
— a water reserve of 0.1 cu.m at starting, for a 30 m pressure head ;
— a water reserve of 0.5 cu.m at stopping, for a 60 m pressure head.
(3) *The useful volume of water in the tank* results from the following difference
$$V_{useful} = V_{cut\text{-}out} - V_{cut\text{-}in} = 0.5 - 0.1 = 0.4 \text{ cu.m}$$

It turns out that a Hydrofort tank with a total volume of one cu.m is able to store, under the given operation conditions, a useful volume of 0.4 cu.m, that is 40 per cent of its nominal capacity.

Figure 8.8 shows a package pressure-boosting plant, produced by the KSB Company (F. R. of Germany). The installation is provided

Fig. 8.8. Hyamat package pressure-boosting installation, in compact form provided with separation bladder and submersible pumps (KSB Co.).

with a Hydrofort tank of 1 m³ and two horizontal pumps. This installation shows the following performances

$$Q = 3{,}5 \text{ dm}^3/\text{s}, \; H_{cut\text{-}in} = 90 \text{ m}, \; H_{cut\text{-}out} = 115 \text{ m}.$$

Conventional hydropneumatic accumulator. This is a water tank under pressure, consisting of a metallic cover, *1*, made of welded steel, fitted at its lower side with a bottom valve with a floating body, *2* (necessary for keeping a minimum water volume and thus, implicitly, for maintaining the air cushion), and with a laterally-located water gauge glass, *3*, and a manometer, *4*, (for monitoring water level and pressure in the tank) as shown in Fig. 8.9.

Operating principle. Consider a hydropneumatic tank, coupled to a constant-speed pump according to the scheme shown in Fig. 8.10. The notations are: $H_{cut\text{-}out}$ — the cut-out head, $H_{cut\text{-}in}$ — the cut-in head, Q — the pumped flow rate, and Q_d — the demand flow rate. The hydropneumatic accumulator operates in accordance with the variation of the demand flow rate Q_d. As Fig. 8.10,*b* shows, the accumulator operates as follows: if $Q_d < Q$, the accumulator is charged with the filling-in flow rate $Q_f = Q - Q_d$, and thus, during this time interval, the pumping head H developed by the turbopump rises until the cut-out head, fixed on the pressure switch, is reached. Then the pump is stopped. If $Q_d > Q$, the accumulator discharges with an emptying flow rate $Q_e = Q_d - Q$, and the pump head falls. This decrease continues until the pumping head H reaches the cut-in head, fixed on the pressure switch, and the pump starts again.

The diagram of Fig. 8.10,*b* shows two operating curves of the turbopump: the $H(Q)$ curve, corresponding to the minimum value

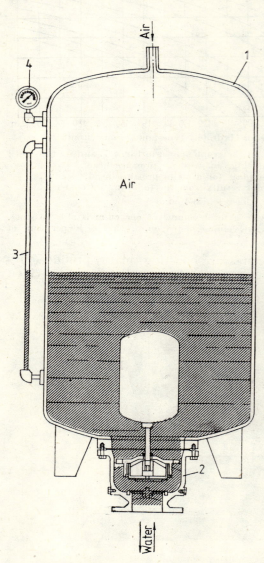

Fig. 8.9. Conventional hydropneumatic tank provided with bottom float-valve (Vögel Co.).

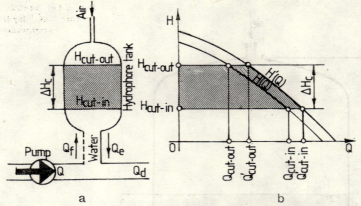

Fig. 8.10. Operating principle of a hydrophore tank:
a — schematic piping arrangement; *b* — working diagram.

of suction head, and determining the cut-out and cut-in flow rates, $Q_{cut-out}$ and Q_{cut-in}, and the $H'(Q)$ curve, corresponding to the maximum suction head value, and determining the $Q'_{cut-out}$ and Q'_{cut-in}. The computation of accumulator volume requires the following flow rates: $Q'_{cut-out}$ and Q'_{cut-in} for the case in which the suction head value is kept constant (suction by means of a free-level buffer tank), and $Q'_{cut-out}$ and Q'_{cut-in} for the case in which the suction head value is variable (direct suction by the mains of a town water supply system).

Volume computation. Two methods may be used for the computation of tank volume: (a) the analytical method for approximate estimations, and (b) the graphical method for exact estimations.

(a) *Analytical method.* This method has been developed for rapid computation of the hydropneumatic tank volume. This is a fast method, leading to approximate results (a maximum error of two per cent relative to the graphical method). It starts from the fact that the total tank volume, V_t, is made of partial volumes of air, V, and water, ΔV, (as shown in Fig. 8.11). The relation between these volume is given by

$$V_t = V_{cut-out} + \Delta V_c + \Delta V_s + \Delta V_{dead} \tag{8.4}$$

or

$$V_t = V_{cut-in} + \Delta V_s + \Delta V_{dead} \tag{8.4, a}$$

Noting the sum, $V_{cut-in} + \Delta V_s = V_{nt}$, we may write

$$V_t = V_{net} + \Delta V_{dead} \tag{8.4, b}$$

where V_t is the total volume;
$V_{cut-out}$ — the cut-out volume;
V_{cut-in} — the cut-in volume;
V_{net} — the net volume;
ΔV_c — the compensation volume, or useful volume;
ΔV_s — the succession volume;
ΔV_{dead} — the dead volume;
all expressed in cu.m.

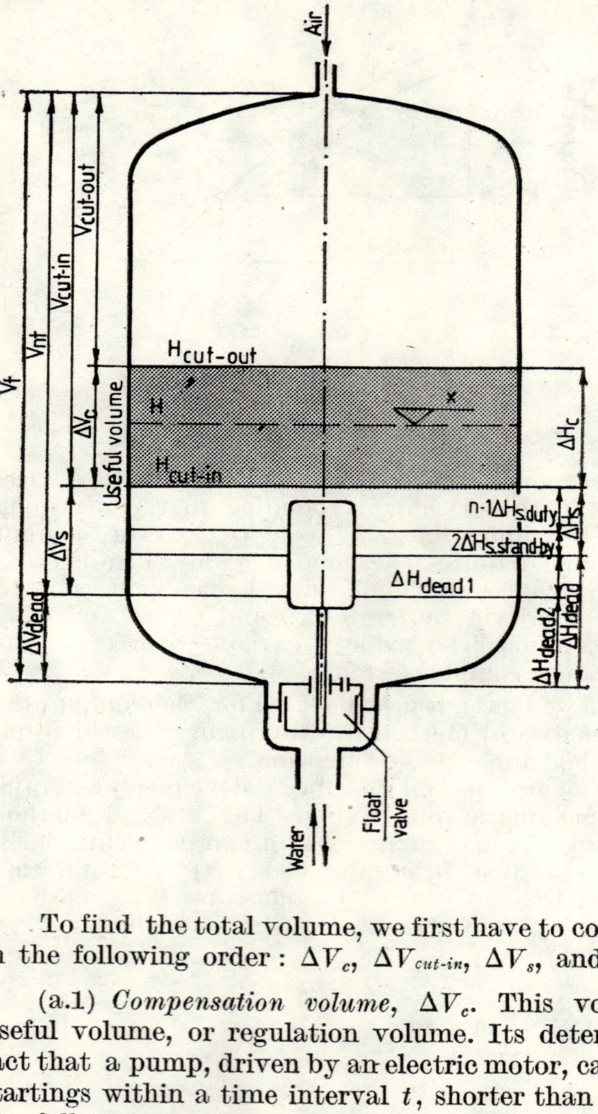

Fig. 8.11. Sketch for deriving volume of hydropneumatic accumulator.

To find the total volume, we first have to compute the partial volumes in the following order: ΔV_c, ΔV_{cut-in}, ΔV_s, and ΔV_{dead}.

(a.1) *Compensation volume*, ΔV_c. This volume is also called the useful volume, or regulation volume. Its determination starts from the fact that a pump, driven by an electric motor, cannot stand two successive startings within a time interval t, shorter than the critical time t_k. Thus, the following condition should be met: $t \geqslant t_k$.

The necessary time between two successive startings, t, is given by

$$t = t_e + t_f \tag{8.5}$$

where t_e is the emptying time corresponding to the pump pause time;
t_f is the filling time, which is the same as the pump operation time.

Assuming that the pumped flow rate, Q, and the required flow rate, Q_d, are invariant during a time interval t, we may write

$$t_e = \frac{\Delta V_c}{Q_d} \tag{8.6}$$

$$t_f = \frac{\Delta V_c}{Q - Q_d} \tag{8.7}$$

where the difference $Q - Q_d$ is the accumulator filling flow rate.

Substituting relations (8.6) and (8.7) into (8.5), we get

$$t = \Delta V_c \left(\frac{1}{Q_d} + \frac{1}{Q - Q_d} \right)$$

and solving, we have

$$t = \frac{\Delta V_c Q}{Q Q_d - Q_d^2}$$

whence it follows that

$$\Delta V_c = \frac{t}{Q}(Q Q_d - Q_d^2) \tag{8.8}$$

Contrary to the previous assumption (that Q and Q_d are invariant), in reality both flow rates are variable: Q_d varies due to consumption modifications, while Q varies due to changes of pumping head H. Thus, equation (8.8) has two variables, Q and Q_d.

Substituting the mean pumped flow rate Q_m, which has a constant value, for the pumped flow rate Q, we get an equation with only one variable, Q_d

$$\Delta V_c = \frac{t}{Q_m}(Q_m Q_d - Q_d^2) \tag{8.9}$$

The volume ΔV_c reaches its maximum value when $d\Delta V_c/dQ_d = 0$. Deriving, we get

$$\frac{d\Delta V_c}{dQ_d} = \frac{t}{Q_m}(Q_m - 2Q_d)$$

Thus, the derivative is zero for $Q_d = Q_m/2$. Introducing this equality equation in (8.8), we get the maximum value of the compensation volume

$$\Delta V_c = \frac{t}{4} Q_m \tag{8.10}$$

or, by replacing time t with the maximum permitted frequency f of pump startings ($t = 1/f$)

$$\Delta V_c = \frac{Q_m}{4f} \tag{8.11}$$

Given the pumped mean flow rate Q_m, as

$$Q_m = \frac{\int_{H_{cut-in}}^{H_{cut-out}} Q \, dH}{H_{cut-out} - H_{cut-in}}$$

and regarding the pump $H(Q)$ characteristic as a quadratic (parabola) whose path crosses the co-ordinate points ($H_{cut-out}$, $Q_{cut-out}$) and (H_{cut-in}, Q_{cut-in}), we get

$$Q_m = \frac{2}{3} \frac{(Q_{cut-in} + Q_{cut-out})^2 - Q_{cut-in} Q_{cut-out}}{Q_{cut-in} + Q_{cut-out}} \tag{8.12}$$

where $H_{cut-out}$, $Q_{cut-out}$ are *cut-out* head, in m, and flow rate, in cu.m;

H_{cut-in}, H_{cut-in} — the cut-in head, in m, and flow rate, in cu.m.

(a.2) *Cut-in volume*, V_{cut-in}. This can be determined starting from the premise that the air volume in the accumulator changes isothermally, and that the pressure head varies between two discrete values, $H_{cut-out}$ and H_{cut-in}. Applying the Boyle-Mariotte law to these conditions, we may write

$$V_{cut-out}(H_{cut-out} + H_b) = V_{cut-in}(H_{cut-in} + H_b) \tag{8.13}$$

and by substituting the difference $V_{cut-in} - \Delta V_c$ for $V_{cut-out}$, we get

$$(H_{cut-out} + H_b)(V_{cut-in} - \Delta V_c) = (H_{cut-in} + H_b)V_{cut-in} \tag{8.14}$$

Working out equation (8.14), we get

$$V_{cut-in} = \Delta V_c \frac{(H_{cut-out} + H_b)}{H_{cut-out} - H_{cut-in}} = \Delta V_c \frac{H_{cut-out} - H_b}{\Delta H_c} \tag{8.15}$$

Combining equation (8.15) with (8.11), we have

$$V_{cut-in} = \frac{Q_m}{4f} \frac{H_{cut-out} + H_b}{\Delta H_c} \tag{8.16}$$

where f is the maximum allowed frequency for the driving motor of the pump, in h^{-1};

ΔH_c — compensation head, or optimum useful head, in m;

H_b — barometric pressure head, in m ($H_b \approx 10$ m).

The value of maximum permitted frequency is established by the constructors of electric motors in accordance with their power. For instance, the Vögel Company (Austria) produces electric motors having the following values for f

for $P \leqslant 2.2$ kW, $f = 15-20$ h^{-1}

for $3 \leqslant P \leqslant 13$ kW, $f = 12-10$ h^{-1}

for $P > 13$ kW, $f = 10-6$ h^{-1}.

The optimum value of the compensation head, ΔH_c, is equivalent to the value of the pressure head set on the pressure switches. It is established in accordance with the number of duty pumps, z, as follows:

for $z = 1$, $\Delta H_c = 14$ m

for $z = 2$, $\Delta H_c = 12$ m

for $z = 3$, $\Delta H_c = 10$ m.

(a.3) *Succession volume*, ΔV_s. This helps the programming of two or more pumps in successive starting. Considering a successive starting programme, we may write the following equation for tank filling and emptying

$$t = \Delta V_c \left[\underbrace{\frac{1}{Q_d - (z-1)Q_m}}_{t_e} + \underbrace{\frac{1}{zQ_m - Q_d}}_{t_f} \right] \qquad (8.17)$$

where $Q_d - (z-1)Q_m$ is the flow rate side, securing the tank emptying after the last pump z has stopped;

$zQ_m - Q_d$ — the flow rate side, securing the accumulator filling after the last pump z has started.

The minimum value of time results for $dt/dQ_d = 0$, occurring for

$$Q_d = \frac{(2z-1)}{2} Q_m \qquad (8.18)$$

Combining equations (8.18) and (8.17), we get

$$\Delta V_c = \frac{t[zQ_m - (z-1)]Q_m}{4} \qquad (8.19)$$

or

$$\Delta V_c = \frac{zQ_m - (z-1)Q_m}{4f} \qquad (8.19\text{ a})$$

Comparing the above equations with (8.10) and (8.11), we notice that *volume ΔV_c is not dependent on the number of pumps. But when the pumps are not identical, the volume ΔV_c is computed for the turbopump with the highest flow rate and the flattest $H(Q)$ characteristic*.

The succession volume ΔV_s is necessary for automated starting of duty and stand-by pumps when they start in descending or ascending pressure cascade (see section 8.2.1). The magnitude of this volume is influenced by the pressure switch allowance, used in the system of pumps for automated control. The determination of compensation volume starts from the same considerations as above [see (a.2)], namely, that the air in the acumulator follows an isothermal law, and that pressure heads vary within two discrete limit values, $H_{cut\text{-}out}$ and $H_{cut\text{-}in}$. Given these conditions, and applying the Boyle-Mariotte law, we get

$$V_{cut\text{-}in}(H_{cut\text{-}in} + H_b) = (V_{cut\text{-}in} + \Delta V_s) \times (H_{cut\text{-}in} + H_b - \Delta H_s)$$

From this relation, we obtain

$$\Delta V_s = V_{cut-iu} \frac{\Delta H_s}{H_{cut-in} + H_b - \Delta H_s} \qquad (8.20)$$

where ΔH_s denotes the necessary pressure head for successive starting of $(z-1)$ duty pumps and of the stand-by pump.

Combining equation (8.20) with equations (8.15) and (8.19 a), we get

$$V_{net} = \frac{zQ_m - (z-1)Q_m}{4f} \times \frac{H_{cut-out} - H_b}{\Delta H_c} \times \frac{H_{cut-in}}{H_{cut-in} + H_b - \Delta H_s} \qquad (8.21)$$

Equation (8.21) is the general form of the fundamental equation for the computation of net hydropneumatic tank volume. This relation can be found in various particular forms, depending on the characteristics of automated control system used. The term ΔH_s should be replaced by a suitable one, dependent on the control system characteristics (see Section 8.2.1).

(a.4) *Dead volume*, ΔV_{dead}. This volume depends on the construction characteristics of the tank and can be computed by means of the equation:

$$V_{dead} = m_{(w+a)} \omega \left(\Delta H_{dead_1} + \frac{2}{3} \Delta H_{dead_2} \right) \qquad (8.22)$$

where $m_{(w+a)}$ is the number of tanks, containing both air and water, since most tanks are filled only with air;
    ω — tank cross-section, in sq.m.;
    $\Delta H_{dead\,1}$ — distance between water level which closes the bottom tap and lower welding of the tank, in m;
    $\Delta H_{dead\,2}$ — distance between lower welding of the tank and its bottom, in m.

Out of the total number m_t of tanks, only some contain both air and water, that is, those ones included with $m_{(w+a)}$.

The distribution is:

for $m_t = 2$, $m_{(w+a)} = 1$

for $m_t = 3-4$, $m_{(w+a)} = 2$.

The other tanks contain only air.

The distance $\Delta H_{dead\,1}$ depends on the rated diameter D_R of the tap, mounted under the tank bottom (Fig. 8.12).

Thus

for $D_R = 80$ mm, $\Delta H_{dead\,1} = 0.050$ m

for $D_R = 100$ mm, $\Delta H_{dead\,1} = 0.075$ m

for $D_R = 125$ mm, $\Delta H_{dead\,1} = 0.100$ m.

Fig. 8.12. Bottom float valve for flow breaker at power failure (Vögel Co.).

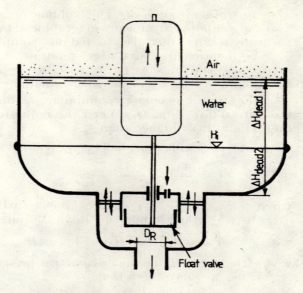

Example 8.2. **Computation of a hydropneumatic tank volume by the analytical method.** Consider a turbopump operating between the points

$$A(H_{cut-out} = 60 \text{ m}, \quad Q_{cut-out} = 18 \text{ cm/h})$$

$$B(H_{cut-in} = 46 \text{ m}, \quad Q_{cut-in} = 36 \text{ cm/h})$$

Given the maximum permitted frequency for the starting of its driving motor, $f = 6 \text{ h}^{-1}$, we want to compute the total volume of the water supply tank for the turbopump automated control.

Solution.

(1) *The mean flow rate of the turbopump* is computed from equation (8.12)

$$Q_m = \frac{2}{3} \frac{(36 + 18)^2 - 36 \times 18}{(36 + 18)} = 27.90 \text{ cu.m/h}$$

(2) *The cut-in volume* (V_{cut-in}) is computed by means of equation (8.16)

$$V_{cut-in} = \frac{27.90}{4 \times 6} \times \frac{60 + 10}{14} = 50.80 \text{ cu.m}$$

Two tanks of three cu.m. each are selected, one to be filled with air and water, and the other, with air only.

(3) *The dead volume* (V_{dead}) is computed with equation (8.22) for a three cu.m. tank, coupled to a bottom tap of $D_R = 125$ mm nominal diameter

$$\Delta V_{dead} = 1 \times 1.125 \left(0.075 + \frac{2}{3} 0.15 \right) = 0.136 \text{ cu.m}$$

The total tank volume is computed by relation (8.4, a)

$$V_t = 5.80 + 0.00 + 0.196 = 5.996 \text{ cu.m} < 6.00 \text{ cu.m}$$

(b) *Graphical method.* The volume of a hydropneumatic tank may also be computed by means of graphical methods, leading to extremely accurate results, but also to tedious computations.

As mentioned above [see relation (8.5)], the time between two successive startings is $t = t_e + t_f$, where $t_e = f(Q_a)$ is a descending function, and $t_f = f(Q)$ is a rising function.

The time t reaches a minimum t_k for a certain flow rate called the critical flow rate, Q_k. When the requested flow rate differs from Q_k, then t is always higher than t_k.

(b.1) *Emptying time*, t_e. This is given by the equation

$$t_e = \frac{V_{cut\text{-}in} - V_{cut\text{-}out}}{Q_a}$$

where $(V_{cut\text{-}in} - V_{cut\text{-}out})$ denotes the air volume used for regulation.

On the other hand, the Boyle-Mariotte law gives

$$V_{cut\text{-}in}(H_{cut\text{-}in} + H_b) = V_{cut\text{-}out}(H_{cut\text{-}out} + H_b)$$

where the term H_b is the barometric pressure head.

Thus, we may infer that

$$\frac{t_e}{V_{cut\text{-}in}} = \frac{1}{Q_a}\left(\frac{H_{cut\text{-}out} - H_{cut\text{-}in}}{H_{cut\text{-}out} - H_b}\right) \tag{8.23}$$

(b.2) *Filling time*, t_f. Taking Q and H as the pumping flow rate and head, respectively, x as the water level in the tank, corresponding to pumping head H and to moment t, and ω as the tank cross-section, we may write

$$\omega dx = (Q - Q_a)dt_f$$

$$dt_f = \frac{\omega}{Q - Q_a}dx$$

By applying the Boyle-Mariotte law, we obtain

$$\omega t(H + H_b) = \text{constant}$$

Thus, we find that

$$\frac{dt_f}{V_{cut\text{-}in}} = -(H_{cut\text{-}in} + H_b)\frac{1}{(H + H_b)^2(Q - Q_a)}dH \tag{8.24}$$

and finally

$$\frac{t_f}{V_{cut\text{-}in}} \int_{H_{cut\text{-}in}}^{H_{cut\text{-}out}} \frac{H_{cut\text{-}in} + H_b}{(H + H_b)^2} \times \frac{dH}{Q - Q_a} \tag{8.24, a}$$

Since the function $Q = f(H)$ is given by the pump characteristic curve, the graphical integration gives various values of the $Q_a/Q_{cut\text{-}out}$ ratio.

(b.3) *Flow rate* Q_k *and volume* $V_{cut\text{-}in}$. The two curves are drawn

$$\frac{t_e}{V_{cut\text{-}in}}(Q_a) \quad \text{and} \quad \frac{t_f}{V_{cut\text{-}in}}(Q_a)$$

and the resulting one (representing the summation of the two above mentioned)

$$\frac{t}{V_{cut-in}}(Q_a).$$

The peak (minimum) of this latter curve stands for the critical value of the flow rate Q_k, and the value of the $k_m = t/V_{cut-in}$ ratio, for $Q_a = Q_k$.

The minimum time between two successive startings being t_k, it follows that the volume V_{cut-in} of the tank is given by

$$V_{cut-in} = \frac{t_k}{k_m} \qquad (8.25)$$

Example 8.3. **Computation of hydropneumatic tank volume by the graphical method.** Consider a pump, coupled to a hydrophore tank whose operation parameters are adjusted for the following values: $Q_{cut-in} = 42 \text{ dm}^3/\text{s}$, $H_{cut-in} = 51 \text{ m}$, $Q_{cut-out} = 81 \text{ dm}^3/\text{s}$, and $H_{cut-out} = 66 \text{ m}$ (Fig. 8.13). We have to determine the tank useful volume.

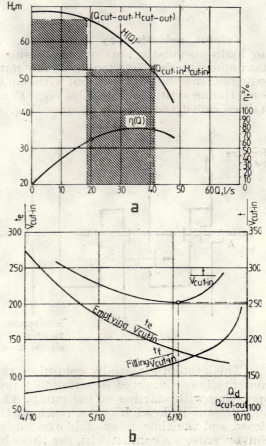

Fig. 8.13. Head (*a*) and volume (*b*) versus flow rate, for a hydropneumatic tank.

Solution. The minimum value of the t/V_{cut-in} ratio occurs for $Q_d = 0.80\ Q_{cut-out}$, when

$$k_m = \frac{t}{V_{cut-in}} = 252 \times 10^{-4}$$

Taking the minimum time between two successive startings as ten minutes, we have

$$V_{cut-in} = \frac{t_k}{k_m} = 23.8 \text{ cu.m}$$

On the other hand, we have

$$V_{cut-out} = V_{cut-in} \frac{(H_{cut-in} + H_b)}{(H_{cut-out} + H_b)} = 19.1 \text{ cu.m}$$

Thus, the water volume securing the control is

$$V_u = V_{cut-in} - V_{cut-out} = 4.7 \text{ cu.m}$$

or, with respect to the minimum water volume in the tank, V_{cut-in} is

$$\frac{V_u}{V_{cut-in}} \times 100 = \frac{4.7}{23.7} \times 100 = 20\%$$

Volume reduction by artifices. As we have demonstrated above, the hydropneumatic tank volume results from the condition of limited maximum starting frequency of the pumping aggregate. Tank volume reduction may be obtained, not withstanding this condition, by means of some timing artifices. Thus, the Vögel Company (Austria) applied a timing artifice called the Evomat system (Fig. 8.14).

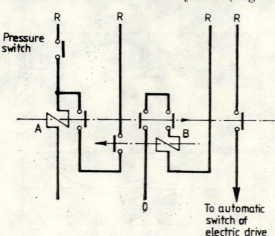

Fig. 8.14. Simplified wiring diagram of an Evomat control system (Vögel Co.).

This system essentially comprises two time-delay relays of a special construction, connected in series to a pressure switch. The system starts operating with starting of the pump. Thus, when the tank pressure falls and the cut-in head H_{cut-in} is reached, the pressure switch contacts are closed and the time relay A is excited, since, at opening, its self-holding contact crosses a contact with the timer belonging to time relay B.

Figure 8.15 shows the operating diagram of the Evomat system. Analysis of this diagram reveals the following:

— when the time between two successive openings of the pressure switch is shorter than the adjusted time t^* (noted on the time relay B), then the pump goes on operating until the end of the adjusted time, irrespective of the command, given by the pressure switch;

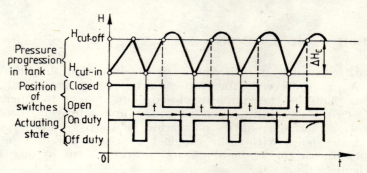

Fig. 8.15. Working diagram of the Evomat system.

— when the time between two successive closings of the pressure switch is longer than the adjusted time t^*, then the time relay B is de-energized and thus the pump is normally stopped by the opening of pressure switch contacts.

Consequently, the time interval between two successive stops will always be longer than, or equal to, the selected adjusted time, and thus the maximum starting frequency of the pumping aggregate is no longer determined by the increase of hydropneumatic tank volume but by the orders given by the Evomat system.

The Evomat system has the following advantages and disadvantages:

Advantages:

— it reduces by 25—33 per cent the hydropneumatic tank volume without altering its operation characteristics:

— it limits the maximum starting frequency of the pumping aggregate, irrespective of the amount of air introduced into the tank.

Disadvantages:

— it makes the pumping head exceed the cut-out head $H_{cut-out}$ adjusted by means of the pressure switch;

— it reduces the global efficiency of the pumping installation when the tank volume is excessively reduced;

— it eliminates the possibility of using pumps with an unstable $H(Q)$ characteristic (with $n_s < 80$).

Marking of computed volumes on the tank. The variation ranges of computed volumes are noted, for easier exploitation, on the tank exterior wall. Thus, the pressure heads ΔH_c and ΔH_s, corresponding to the computed volumes ΔV_c and ΔV_s, are drawn with different colours.

Automated Control of Pumps

The values of these pressure heads are determined by the following relations

$$\Delta H_c = \frac{\Delta V_c}{m_{(\omega+a)}\omega} \quad ; \quad \Delta H_s = \frac{\Delta V_s}{m_{(\omega+a)}\omega} \qquad (8.26)$$

The other notations are mentioned above.

8.2 On-off Pump Controls

The on-off control system is based on the interruption and resumption of flow. A pump driver is therefore running or not, or a valve is open or closed. This is the simplest closed-loop system to operate on-off between two fixed limits such as water level, pressure or flow. The on-off action is at extremities of a wide- or narrow-band type that can be set at any point within the range. For instance, a tank level control may work in an on-off band of 0.25 cm or 25 cm at any level in a 1.5 m. deep tank.

8.2.1 Pressure Switch On-off Controls

Depending on technological parameters, we may distinguish the following types of regulation: (a) simple pressure switch control, based on turbopump pressure measurements, and (b) compensated pressure switch control, based on the simultaneous measurement of turbopump pressure and flow rate.

(a) *Simple pressure switch control system*. The main control element in this method is the conventional pressure switch (Fig. 8.16).

The operating principle of the pressure switch lies in the application of a force F, exerted by the pump pressure head H, on a metallic spring *1*. The force applied on the pressure switch spring is expressed by

$$F = aH. \qquad (8.27)$$

The values of the cut-out pressure head, $H_{cut\text{-}out}$, and the cut-in pressure head, $H_{cut\text{-}in}$, are specified by two adjustable screws, *2*. These values are related to the change of tension F in the pressure switch spring

$$F = \frac{1}{k}(x_0 - x) \qquad (8.28)$$

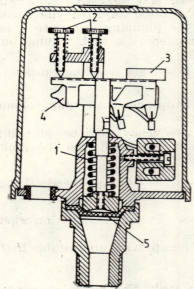

Fig. 8.16. Pressure switch (Vögel Co.):
1 — spring; *2* — adjusting screw; *3* — counterweight; *4* — glass ampoule with mercury; *5* — rubber diaphragm.

Combining equations (8.27) and (8.28) gives the characteristic of the simple (conventional) pressure switch

$$H = b(x_0 - x) \qquad (8.29)$$

where $(x_0 - x)$ is the spring movement, in mm;
k — spring constant, in mm/kg;
a, b — constants.

The graph of this equation is a horizontal straight line, parallel to the flow-rate axis of the $H-Q$ co-ordinate system.

The block diagram of Fig. 8.17 shows the components of a simple pressure switch system.

The diagram of Fig. 8.18 shows the operation of a pump controlled by means of a simple pressure switch regulation system.

The pump operation may take place in either of two ways, depending on the requested flow rate Q_d.

Intermittently: when $0 < Q_d < Q_{cut\text{-}out}$. In this case the pump is repeatedly started and stopped, and the hydraulic accumulator is successively filled and emptied.

Continuously: when $Q_{cut\text{-}out} < Q_d < Q_{cut\text{-}in}$. In this case the pump is operating uninterruptedly, but the operation point is to be found on that section of the $H(Q)$ characteristic curve which determines the flow-rate range ΔQ.

When the requested system flow rate is large, it should be divided between two or more pumps, to reduce the volume of the hydraulic accumulator. Figure 8.19 shows the operation diagram of two pumps also controlled by means of a simple pressure switch regulation system. This time the system includes two pressure switches which fix the individual head ranges, ΔH, for each pump separately, as well as the overall head range ΔH_i of the pumping installation.

Operation of the pumps depends on the flow rate requested and develops in four distinct stages:

(1) When $0 < Q_d < Q_{cut\text{-}out\,1}$, the first pump operates intermittently, and the hydraulic accumulator is successively filled and emptied.

(2) When $Q_{cut\text{-}out\,1} < Q_d < Q_{cut\text{-}in\,1}$, the first pump operates continuously, the duty point being on the $H(Q)$ curve of this pump.

(3) When $Q_{cut\text{-}in\,1} < Q_d < Q_{cut\text{-}out\,2}$, the first pump remains in continuous operation, while the second is intermittently operating together with the hydraulic accumulator.

(4) When $Q_{cut\text{-}out\,2} < Q_d < Q_{cut\text{-}in\,2}$, both pumps operate continuously. The duty point of the installation is on the characteristic curve of the coupled pumps denoted by $2H(Q.)$

Consequently, according to Fig. 8.18, when two or more pumps are controlled by a simple pressure switch regulation system we get *starting in a descending pressure cascade*. In this type of starting, calculation of the net volume, V_{nt} (see equation 8.21), of the hydropneumatic tank is done as follows.

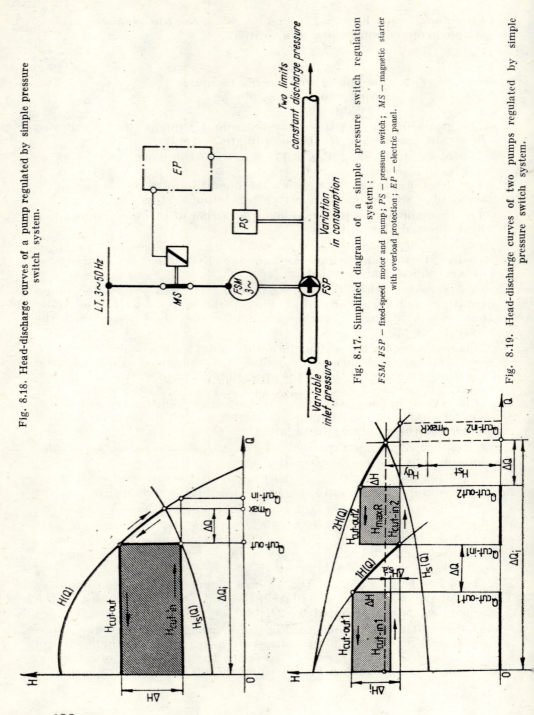

Fig. 8.18. Head-discharge curves of a pump regulated by simple pressure switch system.

Fig. 8.17. Simplified diagram of a simple pressure switch regulation system:

FSM, FSP — fixed-speed motor and pump; PS — pressure switch; MS — magnetic starter with overload protection; EP — electric panel.

Fig. 8.19. Head-discharge curves of two pumps regulated by simple pressure switch system.

The term ΔH_s is determined from equation (8.21) by the relation

$$\Delta H_s = (z-1)H_{sdp} - H_{ssp} \tag{8.30}$$

and then, combining this relation with equation (8.21), we get

$$V_{net} = \frac{zQ_m - (z-1)Q_m}{4f} \times \frac{H_{cut\text{-}out} - H_b}{\Delta H_c} \times$$

$$\times \frac{H_{cut\text{-}in} + H_b}{H_{cut\text{-}in} + H_b - [(z-1)\Delta H_{sdp} + 2H_{ssp}]} \tag{8.31}$$

where $H_{cut\text{-}out}$ is the cut-out head, in m;
        $H_{cut\text{-}in}$ — the cut-in head, in m;
        H_b — the barometric pressure head, in m, ($H_b \approx 10$ m);
        H_c — the compensation pressure head, in m;
        H_{sdp} — the succession head of the duty pump, in m, ($H_{sdp} = 3$ m);
        H_{ssp} — the succession head of the stand-by pump in m, ($H_{ssp} = 2$ m);
        Q_m — the mean pumped flow rate, in cu·m/h;
        f — maximum admissible frequency of startings, in h^{-1};
        z — number of pumps.

System review. Simple pressure switch regulation offers some advantages and disadvantages:

Advantages:
— simplicity of construction;
— cheap apparatus and relays;
— reduced capital cost.

Disadvantages:
— an extremely high global head range ΔH_i;
— reduced overall efficiency η_i (40 per cent $< \eta_i <$ 60 per cent), although the pumps, taken separately, have high proper efficiencies;
— high specific energy consumption, e_s, mainly in the region between the minimum and medium flow rate.

Consequently, the system analyzed above is recommended only for the control of pumps with a sloped $H(Q)$ characteristic curve in a pumping installation with a small flow rate ($Q_d < 100$ cu·m/h), with the number of pumps $z \leqslant 2$.

Example 8.4. **Computation of hydropneumatic tank volume for simple pressure switch regulation.** Consider two identical pumps ($z = 2$), automatically controlled by means of a simple pressure switch regulator. Assuming the computation selected pump has been set to operate with the following parameters

$H_{cut\text{-}out} = 60$ m; $Q_{cut\text{-}out} = 18$ cu.m/h; $H_{cut\text{-}in} = 48$ m.

$Q_{cut\text{-}in} = 36$ cu.m/h; $H_{sdp} = 3$ m; $H_{ssp} = 2$ m, and $f = 6h^{-1}$,

we want to compute the total volume, V_t, of the hydropneumatic tank of the pumping installation.

Automated Control of Pumps

Solution.

(1) *The pump mean flow rate* is obtained from equation (8.12)

$$Q_m = \frac{2}{3} \times \frac{(36+18)^2 - 36 \times 18}{(36+18)} = 27.30 \text{ cu.m/h.}$$

(2) *The net volume of the hydropneumatic tank* is computed by means of equation (8.21)

$$V_{net} = \frac{2 \times 72.90 - (2-1)27.90}{4 \times 6} \times \frac{60+10}{12} \times \frac{48+10}{48+18-(2-1)3+2 \times 2} = 8 \text{ cu.m.}$$

This volume is distributed equally between two tanks, each of 5 cu·m·capacity, of which one is filled only with air.

(3) *The dead volume,* V_{dead} is obtained from equation (8.22)

$$V_{dead} = 1 \times 1.54 \left(0.075 + \frac{2}{3} + 0.20\right) = 0.315 \text{ cu.m.}$$

(4) *The total volume,* V_t, *of the tank* is obtained by means of equation (8.4, b)

$$V_t = 8.00 + 0.315 = 8.315 \text{ cu.m} < 10 \text{ cu.m tank capacity.}$$

Compensated pressure switch regulation system. The main element of this system is the compensated pressure switch (Fig. 8.20).

Two forces act on the pressure switch spring, *1*: a force determined by the pressure head H in the discharge pipe, and an opposite force ΔH

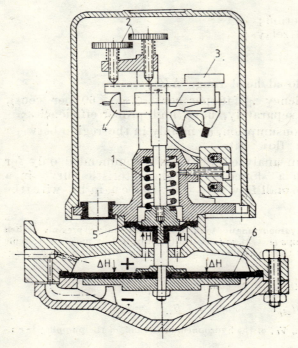

Fig. 8.20. Compensated pressure switch (Vögel Co.):
1 — spring; *2* — adjusting screw; *3* — counterweight; *4* — glass ampoule with mercury; *5* — intermediary diaphragm; *6* — main diaphragm.

governed by the differential pressure head (depression), produced in a throttling device (diaphragm or Venturi tube), mounted in the common discharge line of the pumps.

Thus, the resulting force, exerted on the pressure switch, can be expressed by

$$F = aH - b\,\Delta H \tag{8.32}$$

Since the differential pressure head ΔH is proportional to the square of the flow rate Q, equation (8.32) becomes

$$F = aH - cQ^2 \tag{8.33}$$

The pressure switch has two adjustment screws by means of which two values of the pressure head, $H_{cut\text{-}out}$ and $H_{cut\text{-}in}$, are specified by adjusting the tension F of the pressure switch spring. This tension is given by

$$F = \frac{1}{k}(x_0 - x) \tag{8.34}$$

Combining equations (8.33) and (8.34), we get the characteristic of the compensated pressure switch

$$H = A(x_0 - x) \times BQ^2 \tag{8.35}$$

where $(x_0 - x)$ is the spring distortion, in mm;
k — spring constant, in mm/kg;
a, b, c, A, B — constants.

The graph of this characteristic is a parabola whose path is approximately parallel to that of the pumping system.

Figure 8.21 shows a mounting scheme for compensated pressure switches. In this scheme the pressure switches *1* are connected to the Venturi tube *2* by means of two impulse pipes *3*, of which the one marked with the plus sign feeds both the lower chamber of the small diaphragm and the higher chamber of the large diaphragm; and the one marked with the minus sign feeds only the lower chamber of the large diaphragm.

The scheme of Fig. 8.22 shows the component elements of a compensated pressure switch regulation system, and the operation diagram of Fig. 8.23 shows the operation of a pump, controlled by means of this system.

Unlike the simple pressure switch regulation system, the compensated pressure switch regulation can automatically change the limits of its cut-in and cut-out heads in accordance with flow rate (demand) variations in the pumping system. Because of this peculiarity, the cut-in heads of all pumps can be regulated in the immediate vicinity of the path, travelled by the $H(Q)$ system-head curve of the pumping system (Fig. 8.24). Consequently, we may use pumps with a flat $H(Q)$ characteristic curve, starting in an *ascending pressure cascade*. This leads to a rise of global efficiency in the pumping installation and, consequently, a reduction of electric power consumption in the small and medium range of flow rates

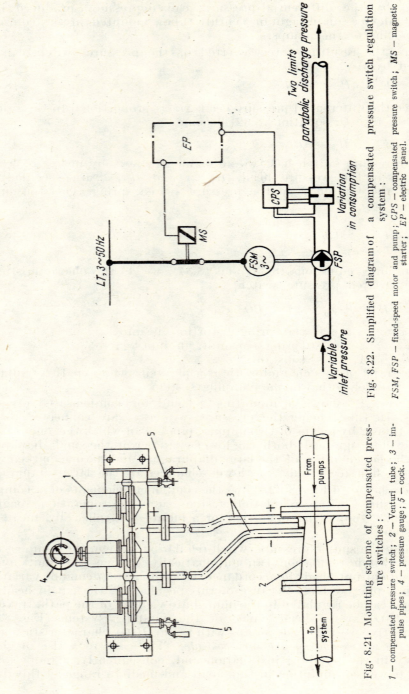

Fig. 8.22. Simplified diagram of a compensated pressure switch regulation system:

FSM, FSP — fixed-speed motor and pump; *CPS* — compensated pressure switch; *MS* — magnetic starter; *EP* — electric panel.

Fig. 8.21. Mounting scheme of compensated pressure switches:

1 — compensated pressure switch; 2 — Venturi tube; 3 — impulse pipes; 4 — pressure gauge; 5 — cock.

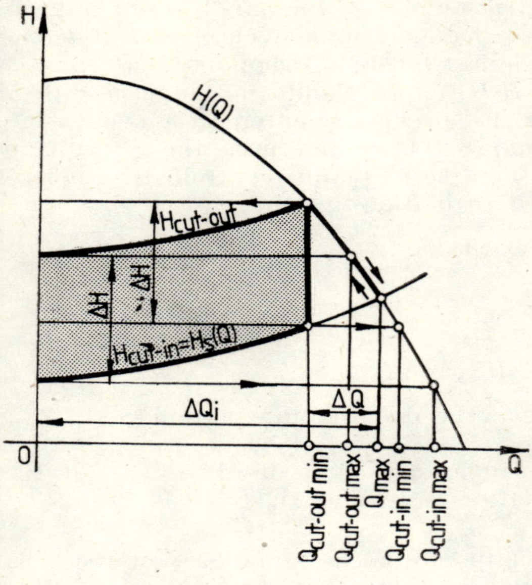

Fig. 8.23. Head-discharge curves of a pump with regulation by compensated pressure switch system.

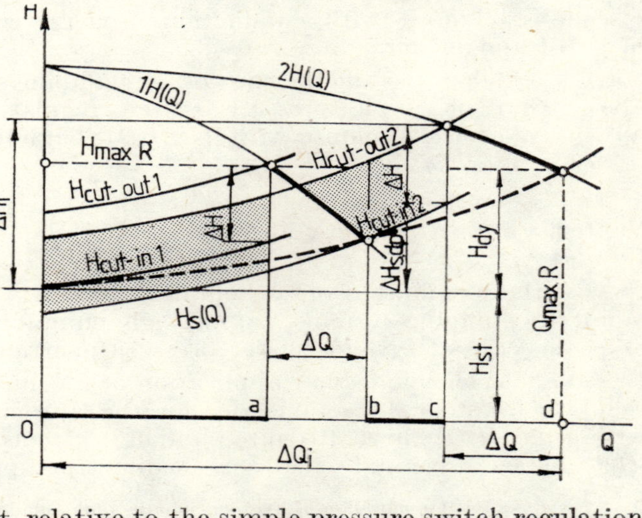

Fig. 8.24. Head-discharge curves of two pumps regulated by compensated pressure switch system.

(by some 25—30 per cent, relative to the simple pressure switch regulation system).

In this case, the volume of the hydropneumatic tanks is computed for the pumps starting in ascending cascade.

The term ΔH_s from equation (8.21) has the following components

$$\Delta H_s = \Delta H'_{sdp} + \Delta H_{ssp} \qquad (8.36)$$

of which the first component is determined graphically by extracting it from the operation diagram. It represents the difference between the cut-in

443

head of the last turbopump (selected for *computation*) and the cut-in head of the first turbopump that entered the operational cycle. By *computation turbopump* we mean the last, or the last-but-one turbopump that entered the operational cycle, but whose $H(Q)$ curve should not superpose over the $H(Q)$ curve of the turbopump that previously entered the cycle.

Combining equations (8.36) and (8.21), we determine the net volume of the hydropneumatic tanks in the case of pumps controlled by means of a compensated pressure switch regulation system

$$V_{net} = \frac{zQ_m - (z-1)Q_m}{4f} \times \frac{H_{cut-out} + H_b}{\Delta H_c} \times$$

$$\times \frac{H_{cut-in} + H_b}{H_{cut-in} + H_b - (\Delta H'_{sdp} + 2H_{ssp})} \tag{8.37}$$

The notation here is the same as the one used in equation (8.31).

Review of the system. The advantages and disadvantages of the compensated pressure switch regulation system are the following:

Advantages:

— the starting points of pumps with the same characteristic as that of the pumping system become closer, so that starting in ascending cascade is achieved with a reduction of energy consumption in the range of small and medium flow rates;

— a higher overall efficiency of the pumping system (some 10 per cent above that of simple pressure switch regulation) is achieved, since advantageous use of pumps with flat $H(Q)$ characteristic curves is possible.

Disadvantages:

— it implies additional use of a throttling device (diaphragm or Venturi tube), which may also be used for the measurement of pumped flow rate;

— the membranes of consumption pressure switches are difficult to adapt to pumping systems, having high pumping heads ($H_s \approx 80-100$, as is the case of systems under pressure in irrigation networks).

We may conclude that application of a compensated pressure switch regulation system is recommended for the automated control of pumps with a flat $H(Q)$ curve, fitted to pumping systems with medium flow rate ($Q \leq 200$ cu.m/h) and small or medium pumping head ($H_s \leq 65$ m).

Example 8.5. **Computation of hydropneumatic tank volume for the case of a compensated pressure switch regulation system.** Consider a pumping installation, equipped with three equal pumps ($z = 3$). Given the operating parameters of the pumps as:

$H_{cut-out} = 58$ m, $Q_{cut-out} = 22.5$ cu.m/h

$H_{cut-in} = 48$ m, $Q_{cut-in} = 45$ cu.m/h

$H'_{sdp} = 6$ m

(from the operation diagram), and $f = 8$ h^{-1}, we want to compute the volume of the hydropneumatic tanks.

Solution

(1) *The mean pumped flow rate* is found by means of equation (8.12)

$$Q_m = \frac{2}{3} \times \frac{(45 + 22.5)^2 - 45 \times 22.5}{(45 + 22.5)} = 35 \text{ cu.m/h}$$

(2) *The tank net volume*, V_{net}, is given by equation (8.21)

$$V_{net} = \frac{3 \times 35 - (3-1)35}{4 \times 8} \times \frac{58 + 10}{10} \times \frac{48 + 10}{48 + 10(6 + 2 \times 2)} = 9 \text{ cu.m}$$

(3) *The dead volume*, V_{dead}, is computed by means of equation (8.22), taking the rated diameter of the bottom as $D_R = 125$ mm

$$V_{dead} = 1 \times 1.54 \left(0.1 + \frac{2}{3} \times 0.20 \right) = 0.354 \text{ cu.m}$$

(4) *The total tank volume* is determined with equation (8.4, b)

$$V_t = 9 + 0.354 = 9.354 \text{ cm} < 10 \text{ cu.m}$$

8.2.2 On-off Regulation by Flowmeters

Depending on the measured parameter, we may distinguish (a) simple on-off regulation by flowmeters, in which the pumps are used in accordance with the flow rate, and (b) regulation by associated pressure switch and flowmeter, in which the pumps receive control signals in accordance with pressure, while their number is established with respect to flow rate.

(a) *Simple on-off regulation by flowmeter.* This regulation type has been the subject of a detailed paper, published by the author in "La technique de l'eau et de l'assainisement" (Paris, [8.9]).

Introductory concepts. On-off regulation by flowmeter requires at least two pumps in the pumping installation. Depending on their entrance sequence into the operation cycle, one pump is called *the lead* while the other is *the lag* pump (or pumps). The lag pumps are identical, but the lead pump may have a flow rate, equal to or half that of the others. When the lead pump has a smaller flow rate, it is also called a *jockey* pump, while the others are called *main* pumps.

Basic principle. The on-off regulation by flowmeters is based on meeting all consumption needs by utilising only the pump operation ranges. Consequently, for each requested flow rate in the system, we have a corresponding stable characteristic point on the $H(Q)$ curves of one or more pumps, operating in parallel.

The individual pump operation ranges are settled by preliminary adjustment of flowmeter contacts. These contacts control the pumps, that is, their successive cut-in and cut-out, when the requested flow rate implies a passage from one operation range to another.

The displacement of the duty point, on the $H(Q)$ characteristic of one or more pumps operating in parallel, is achieved by the throttling effect produced by closing or opening the progressive down-stop valves in the distribution network, in response to variation of consumption. This displacement may be pursued by means of Fig. 8.25, where the curve OG represents the natural characteristic curve of the pumping system (corresponding to a completely open position of all down-stop valves), while the curves OA and OF are the artificial system-head curves (corresponding to various positions of progressive and successive closing of down-stop valves).

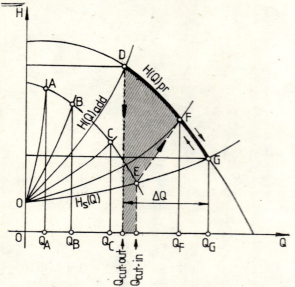

Fig. 8.25. Diagram explaining self-stabilization effect.

Figure 8.25 also shows the $H(Q)$ curves of two unequal pumps, the first of which is a jockey pump, having the role of lead pump (with $Q = Q_a/2$), and the other is the main pump having the role of leg pump (with $Q = Q_a$). The working conditions (duty points) of the first pump are determined by the intersection of $H(Q)_{Add}$ curve with intermediary curves OA, OB, OC, and OE. This pump could meet consumption needs up to the operating point E, where $H_s = H$, where the flowmeter gives the starting order to the main pump (and, at the same time, the stopping order to the jockey pump). Thus another throttling curve OF will be established, which leads to the displacement of the duty point from E to F, and thus to a rapid rise of pressure up to the value corresponding to the new flow rate requested in the system. The operation point displacement along the path EF, inclined to the pressure head axis H, is explained by the self-stabilizing effect [1]. This effect consists of simultaneous rising of flow rate through the valves ($Q_F > Q_E$) and of pressure head ($H_F > H_E$).

---

[1] This effect has been observed by the author at pumps, starting within a pumping installation without hydropneumatic accumulator, and has been theoretically explained in [8.9].

Number of pumps. Regulation by flowmeter is only possible in a pumping system, showing a minimum water consumption Q_{min} of at least 15—25 per cent Q_{max}. This condition is imposed by the leading flowmeter, which makes an inaccurate measurement, and implicitly gives a wrong signal in the range of small flow rates ($Q < 15-25$ per cent). Therefore, for a better overall efficiency of the pumping installation, its total flow should be distributed between several pumps. The bigger the number of pumps, the higher the global efficiency, and the better adapted is the installation to consumption variations. However, equipment costs also rise. Consequently, an optimum number of pumps has to be established. In this sense, we suggest the following three combinations:

(1) *Two equal main pumps*, each with $Q = 55$ per cent Q_{max}, and a jockey pump having $Q = 25$ per cent Q_{max}, securing a variation range of pumping flow rate of $Q = 12.5-110\% \; Q_{max}$ (Fig. 8.26).

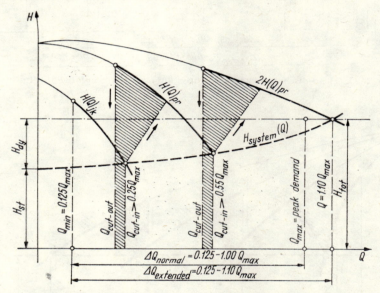

Fig. 8.26. Head-discharge curves of two duty pumps regulated by flowmeter.

(2) *Three equal main pumps*, each having $Q = 34$ per cent Q_{max} and, in certain cases, another jockey pump with $Q = 18$ per cent Q_{max}, securing a variation range of pumping installation flow rate of $Q = 9-102$ per cent Q_{max}. When started again (at the end of a cycle), the auxiliary pump could also provide $Q = 9-120$ per cent Q_{max}, but only in the case of an extremely flat $H(Q)$ curve for the pumping system (the case of land sprinkling irrigation systems), (Fig. 8.27).

(3) *Four equal pumps*, each having $Q = 26$ per cent Q_{max}, securing together a variation range of pumping installation flow rate of $Q = 13-104$ per cent Q_{max} (Fig. 8.28).

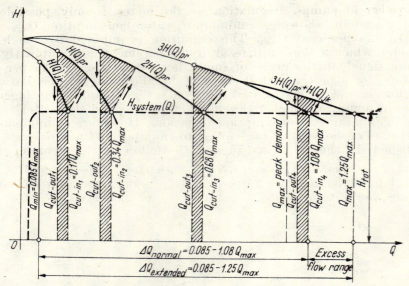

Fig. 8.27. Head-discharge curves of three duty pumps regulated by flowmeter.

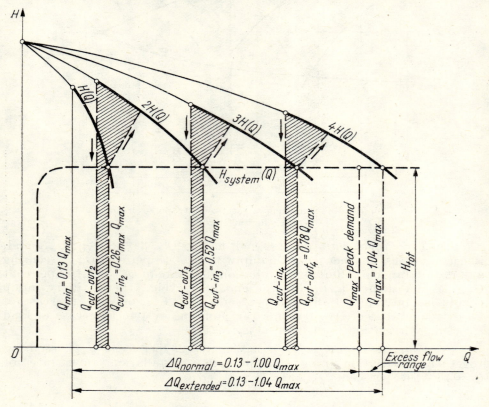

Fig. 8.28. Head-discharge curves of four equal pumps.

Control elements. The following sensing elements, for pump control in accordance with flow rate, are widely used in practice:

(1) *Flow rate switch* (Fig. 8.29). The basic principle of this element is the application of force F, produced by the differential pressure head ΔH, by means of a throttling device (diaphragm or Venturi tube) on a metallic spring *1*.

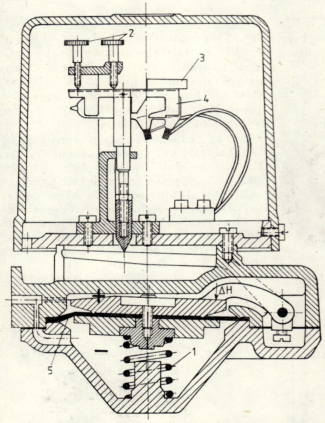

Fig. 8.29. Flow-rate switch (Vögel Co.):
1 — spring; *2* — adjusting screw; *3* — counterweight; *4* — glass ampoule with mercury; *5* — diaphragm.

The force exerted on the spring is expressed by
$$F = a\,\Delta H \tag{8.38}$$
but, since the differential pressure head ΔH is proportional to the square of the flow rate, equation (8.38) becomes
$$F = bQ^2 \tag{8.39}$$

The flow-rate switch specifies two pressure values, corresponding to cut-out and cut-in flow rates, respectively, controlling spring tension by two adjustment screws *2*. Spring tension F is

$$F = \frac{1}{k}(x_0 - x) \tag{8.40}$$

Combining equations (8.39), and (8.40), it gives the flow-rate switch characteristic

$$Q = \sqrt{c(x_0 - x)} \qquad (8.41)$$

where $(x_0 - x)$ is spring distortion, in mm;
k — spring constant;
a, b, c — constants.

(2) *Alarm volumeter.* (Fig. 8.30). This control element, also known as the Pacomonitor flow control, is made of hydraulic and electrical parts.

Fig. 8.30. Alarm volumeter or Pacomonitor flow control (Pacific Inc.).

The hydraulic part, *1*, consists of an orifice and a propeller. The liquid flowing through the orifice makes the propeller rotate proportionally to the volume of liquid passed.

The propeller rotation is transmitted directly, by a special extension, to the electric part outside. The electric part, *2*, comprises an axle with cams and several microswitches. The cams may be staggered, for successive starting of pumps for various pre-determined flow rates. The alarm volumeter is also fitted with a graded dial, *3*, specifying instantaneous flow rate.

The alarm volumeter is mounted horizontally on the common pump discharge line (Fig. 8.31).

 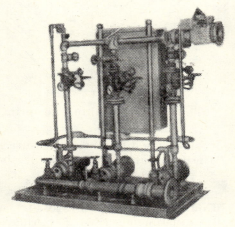

Fig. 8.31. Package constant pressure system, provided with fixed-speed pumps and Pacomonitor flow control (Pacific Inc.):
a — front view; *b* — back view.

The Pacific Pumping Inc. (U.S.A.) produces Pacomonitor volumeter apparatus for the following rated diameters and flow rates:
$D_R = 100$ mm, $Q = 100$ m³/h;
$D_R = 150$ mm, $Q = 270$ m³/h;
$D_R = 200$ mm, $Q = 540$ m³/h;
$D_R = 300$ mm, $Q = 950$ m³/h.

(3) *Alarm rotameter* (Fig. 8.32). This is a flow-rate switch with constant pressure difference, employing a float with linear motion. It is of simple and cheap construction, with a small pressure loss. The alarm rotameter comprises a calibrated meter tube *1*, within which a cylindrical float *2* moves vertically, connected by a rod *3*, to a main magnet *4*, sliding into an extension *5*, which contains the control switches *6*. Each switch (Fig. 8.33) is made of two blades *1*, and an auxiliary magnet *2*, all of them sealed into a plastic cage *3*, placed in the field of main magnet *4*.

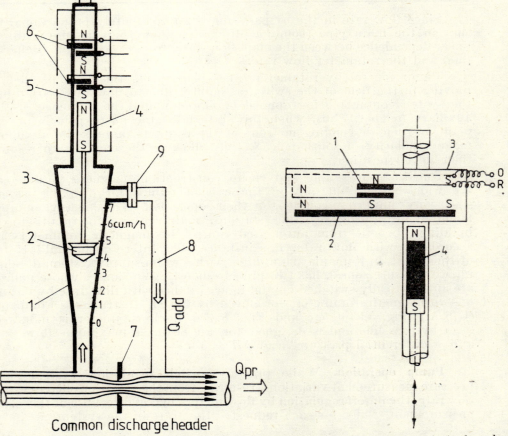

Fig. 8.32. Schematic piping arrangement of an alarm rotameter, mounted in by-pass.

Fig. 8.33. The control contact of an alarm rotameter.

Measurement of medium flow rates ($Q > 18$ m³/h) is achieved with the alarm rotameter, mounted on a by-pass pipe 8, avoiding the diaphragm 7, placed on the common discharge line of the pumps (see Fig. 8.32). In this case, in the region where the diaphragm is to be mounted, the diameter of the common discharge line depends on the maximum flow rate exhausted. This diameter is chosen in accordance with Table 8.1.

TABLE 8.1 **Sizing the common discharge line in zone of orifice plate**

Plant flow rate in dm³/s	5.55	8.33	15.27	23.61	31.11	52.77	94.44	144.44	211.11	291.66	375.00	583.00
Line diameter in mm	50	65	80	100	125	150	200	250	300	400	450	500

Since flow rate in the by-pass line is approximately proportional to that in the main pipe (common discharge line), there is practically a linear dependence between the main flow rate (from the common discharge line) and the rotameter flow rate.

Pump control by rotameter takes place when the main magnet is moving in the field of the switch assembly (due to flow rate variation), the switch contacts being opened or closed under the influence of the auxiliary magnet. Thus, when flow rate falls, the north pole N of the main magnet, descending in front of the contacts, annuls the auxiliary magnet polarity of south pole S, thus allowing the contacts to resume their initial position.

Due to their simple and cheap construction, magnetic rotameters are widely used, but only in the case of pumps for drinking water, since any impurity could stop their operation. Thus, many pumps, mounted in pressure-boosting stations, mainly those for multi-storey buildings, benefit from automated control by magnetic rotameters in accordance with flow rate. For instance, the "Inter-Continental" Hotel of Bucharest, in Romania (90 m high, with a water supply system, divided in two pressure zones), has two pressure-boosting stations [1] whose pumps are automatically switched on and in accordance with flow rate by means of two magnetic rotameters, produced by the SK Instruments Company (U.S.A.) (Fig. 8.34 a, b, c and Fig. 8.35). Each rotameter is mounted on a by-pass line and is designed for an 80 m³/h maximum flow rate and a differential pressure head $\Delta H = 5$ m.

Pump operation. While pressure switch regulation systems use pressure measurement variations, resulting from the variations of requested flow rate, the on-off regulation by flow-meters makes direct use of flow rate measurement. Thus, for each requested flow rate in the system, we have

---

[1] The two installations have been designed by the author as head of project, at "Proiect București" Designing Institute, Romania.

On-off Pump Controls

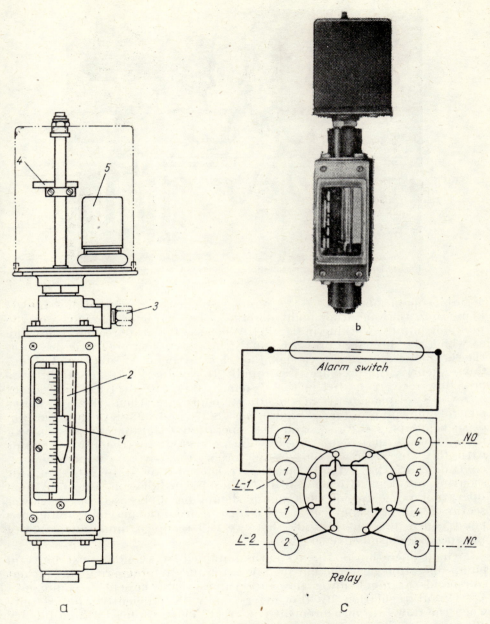

Fig. 8.34, a. Alarm rotameter:
1 — float assembly; 2 — meter tube; 3 — adaptor and range orifice; 4 — switch assembly; 5 — relay.
Fig. 8.34, b. View of an alarm rotameter (SKI Co.).
Fig. 8.34, c. Relay and alarm switch diagram of alarm rotameter.

Fig. 8.35. By-pass alarm rotameters, mounted in a pressure-boosting installation with fixed-speed pumps.

a stable operation point at the pumps. When the requested flow-rate is increasing, the operation point describes the characteristic curve of the pump or pumps, working in parallel. When the flow rate reaches the value at which the discharge pressure equals the system characteristic pressure, the flowmeter always gives a signal for the starting of an additional pump. Consequently, the pressure rises rapidly to a value, corresponding to the new flow rate, requested in the system. A flow rate decrease determines a reverse displacement of the duty point, and then the flowmeter controls the pump stopping. Figures 8.26, 8.27, and 8.28 show the operation of such pumping stations, and the operation diagram analysis, shown there, reveals that regulation by flowmeter requires the use of pumps with flat characteristic curves in order to cover, completely and continuously, the overall variation range of flow rate ΔQ_i. Yet, for securing optimum efficiency of the pumping installation, the individual operation range ΔQ of each pump should be superposed on the lower section of the efficiency curve $\eta(Q)$, starting from the maximum efficiency point, since it is on this side that we can secure a minimum value of the specific consumed energy.

Good operation of the system is obtained by avoiding instability of pump operation near the cut-in flow rate, and by staggering the cut-out commands at much lower flow-rate values than the cut-in commands.

On-off regulation by flowmeters may be efficiently applied only when the flow-rate measurement is correct. Thus, we need a high fidelity measurement apparatus (over the whole measurement range). The great majority of flowmeters used are differential pressure switches, coupled to throttling devices (diaphragms or Venturi tubes) and mounted on the common pump discharge line. It is known that in the zones of minimum flow rate for flowmeters, the diameter measurement is very inaccurate, unless we choose a throttling device with a much smaller orifice and thus a higher head loss at maximum flow rate. This orifice narrowing is a disadvantage (determined by energy loss in the throttling device) in the selection of an adequate differential flowmeter. The most common flowmeters show good measurement only for flow rates of $Q > 15$ to 20 per cent Q_{max}. Thus for automatic control in the region of minimum flow rate, one or two jockey pumps with small flow rates should be used, in conjunction with a hydrophore tank, and controlled by a pressure switch rather than a flowmeter. These pumps may be stopped by the main pumps coming into operation and started again beyond the operation range of the main pumps for meeting the requirements of exceptional consumption points.

A characteristic feature of regulation by flowmeters is the necessity of having a stand-by pump that can instantaneously take the place of any duty pump in trouble. In the case of pressure switch regulation systems, when a pump fails, the pressures fall and the next pump can immediately take its place. In case of on-off regulation by flowmeters, the failure of one pump leads to the progressive breaking of the flowmeter chain, and consequently, to a decreased flow rate, down to the complete stopping of the installation. Thus a stand-by pump, which should automatically start working, is essential. This may be achieved by means of an adequate electric relay, or by a stand-by pump control by means of a pressure switch, whose cut-in pressure should be fixed a little below the pressure described by the system characteristic curve. This latter case implies the use of a pressure switch with a characteristic curve similar to that of the pumping system. Thus, for systems having a horizontal characteristic curve (sprinkling irrigation systems), the simple pressure switch is quite adequate, while the compensated pressure switch is fitted for those having a parabolic characteristic curve (drinking water supply systems).

System review. On-off regulation by flowmeters has advantages and disadvantages.

Advantages:

— the great advantage of this system is the continuous operation of the pumps at all flow rates, higher than the cut-out flow rate of the first pump, which reduces equipment fatigue and extends the life span of machines (pumps and electric motors);

— the system achieves a high efficiency of the pumping installation due to the use of pumps with a flat $H(Q)$ characteristic, which considerably reduces the pressure range ΔH.

Disadvantages:

— the system implies the use of flowmeters whose throttling device leads to high head losses;

— it requires the use of a stand-by pump with automated control by pressure switch, which should take the place of any duty pump in trouble and thus avoid the breaking of the flowmeter chain.

Consequently, the system analyzed is suitable for automated control of pumps with a flat $H(Q)$ characteristic curve, mounted on medium flow rate pumping installations ($Q < 720$ cu.m/h).

On-off regulation by associated pressure switch and flowmeter. This system, like pressure switch on-off regulation, gives an intermittent operation of pumps, the consumption range being covered by pumps with hydropneumatic tanks. These tanks are brought into use when the requested flow-rate is beyond the individual operation ranges of the pumps with which they work. The pressure head in the tanks varies between two limits, H_{cut-in} and $H_{cut-out}$, which are rather close to each other.

This system combines pressure switch T_p, with a flow transducer T_q (Fig. 8.36). The pressure switch determines the two limits and gives the cut-in or cut-out signals accordingly.

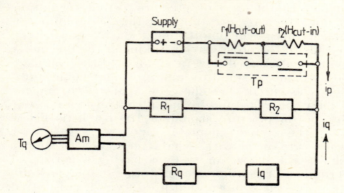

Fig. 8.36. Simplified wiring diagram of regulation by associated pressure switch and flowmeter:

T_p — pressure switch (pressure transducer); T_q — flow transducer; A_m — amplifier; R_1, R_2 — relays; R_q — flow recorder; I_q — integrator.

On the other hand, the flow transducer analyzes the consumption variation in the system and establishes the necessary number of pumps for meeting this consumption. In other words, we may say that the flow transducer is preparing the cut-in and cut-out commands for a pump, which are then confirmed by the pressure switch.

Actually, a command signal is achieved by summing two currents:

— a current i_p, from the pressure switch T_p, ranging between two discrete values i_{p_1} and i_{p_2}, with the pressure switch within a circuit having two resistances, r_1 and r_2, mounted in series or not, allowing sudden variation of the current i_p;

— a current i_q from flow transducer T_q, varying linearly with flow rate, derived via amplifier A_m.

The function $(i_p + i_q)$ is represented by two straight lines: the first one $(i_{p_1} + i_q)$, corresponding to the cut-in pressure (H_{cut-in}), and the second $(i_{p_2} + i_q)$, corresponding to the cut-out pressure ($H_{cut-out}$) (Fig. 8.37).

The sum of the two currents $(i_p + i_q)$ passes through the circuits of relays R_1 and R_2, associated with the pumps P_1 and P_2. The working ranges of the relays are to be found between the two straight lines $(i_{p_1} + i_q)$ and $(i_{p_2} + i_q)$.

The system is characterized by the fact that pumps start without pressure cascade, since cut-in pressures of all pumps are located on a horizontal line, parallel to the flow-rate axis (see Fig. 8.37). In the particular case of sprinkling irrigation systems, the cut-in pressure is equal to the characteristic pressure of the pumping system (see Fig. 8.37).

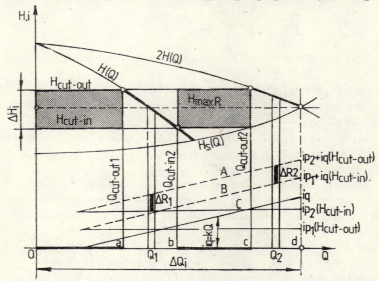

Fig. 8.37. Head-discharge curves of two pumps, regulated by associated pressure switch and flowmeter.

This feature should be taken into consideration when establishing the necessary relations for computation of hydropneumatic tanks, present in the associated pressure switch and flowmeter regulation system. Since cut-in pressure is the same for all pumps, we remove the succession head and, hence, the $H_s = 0$ term too. In this case, the relation (8.21) becomes

$$V_{net} = \frac{zQ_m - (z-1)Q_m}{4f} \times \frac{H_{cut-out} - H_b}{\Delta H_c} \times \frac{H_{cut-in}}{H_{cut-in} - H_b} \quad (8.42)$$

where the symbols have the same meaning as those used in equation (8.31).

Example. 8.6 **Computation of hydropneumatic tank volume for on-off regulation by associated pressure switch and flowmeter.** Consider a pressure-boosting station, provided with four pumps and fixed for

$H_{cut-out} = 90$ m; $H_{cut-in} = 75$ m; $Q_{cut-out} = 135$ cu.m/h;

$Q_{cut-in} = 270$ cu.m/h, and $f = 6$ h^{-1}.

We want to determine the volume of hydropneumatic tanks.

Solution

(1) *Pump mean flow rate* is computed with equation (8.12)

$$Q_m = \frac{2}{3} \times \frac{(270 + 135)^2 - 270 \times 135}{(270 + 135)} = 200 \text{ cu.m/hour}$$

(2) *The net tank volume* is obtained by means of equation (8.21)

$$V_{net} = \frac{4 \times 209 - (4-1)209}{4 \times 6} \times \frac{90 + 10}{16} \cdot \frac{75}{75 + 10} = 48.30 \text{ cu.m.}$$

This volume is divided between two tanks of 25 cu.m each.

(3) *The dead volume*, V_{dead}, is obtained by means of equation (8.22)

$$V_{dead} = 1 \times 5.95 \left(0.10 + \frac{2}{3} 0.3\right) = 1.70 \text{ cu.m.}$$

(4) *The total volume of tanks* is computed with equation (8.4, b)
$V_t = 48.30 + 1.70 = 50$ cu.m

System review. On-off regulation by associated pressure switch and flowmeter has the following advantages and disadvantages:

Advantages:

— it is very simply constructed and has a simplified relay;

— the installation pressure head error, ΔH_i, is reduced, which leads to a high overall efficiency, η_i;

— the flowmeter used must not necessarily be of high accuracy, since we are interested only in correct estimates of flow-rate range ΔQ;

— it reduces the total volume of the hydropneumatic tank by an amount equal to the succession volume, ΔV_s.

Disadvantages:

In certain cases, namely when number of pumps z is too large ($z > 5$), the system implies additional time relays, whose role is to complete pressure switch action.

Thus, we may conclude that the analyzed system is all the more adequate as it has a flatter characteristic curve. Therefore, it was developed considerably for the case of automated control of pumping installations for putting under pressure spraying irrigation systems, whose $H_s(Q)$ characteristic is a horizontal line, parallel to the flow-rate axis. Consequently, the cut-in pressure of all pumps is settled equal to the system-head curve.

On-off regulation by associated pressure switch and flowmeter has been applied to a pumping installation for raising the pressure in the irrigation system of Băneasa-Giurgiu experimental station (Romania). In that case, the installation was provided with five pumps, each having $Q = 540$ cu.m/h, $H = 74$ m, and $P = 143$ kW. Regulation by associated pressure switch and flowmeter includes, as main automation elements, a differential flowmeter with $Q = 2700$ cu.m/h and $H = 3$, and a pressure switch with metallic contacts, having $H = 20$ m. and rated pressure of 16 daN/cm².

8.2.3 On-off Regulation by Levelmeters

Chronometric system by levels. Unlike pressure switch regulation systems, where each pump has its pressure switch, the chronometric systems make use of a single pressure indicator. This indication is actually obtained by measuring the levels, that is the minimum and maximum levels, of the hydropneumatic tank. At the minimum level, the pumps receive a start command, while at the maximum level, they receive a stop command. The start command entails a programme of successive startings in a pre-determined order and at regular time intervals. The commands are transmitted by time relays with the same timing (for instance, all of them are timed at seven seconds). The stop command generates a programme of pumps successively stopping, in reverse order to their starting.

For reducing the hydropneumatic tank volume, we should consider two quantities: the mean flow rate, Q_m, and the minimum accepted time between two startings, t. We may influence Q_m by making use of two pump types: jockey pumps and main pumps. The jockey pumps have half of the flow rate of main pumps. Thus, jockey pumps come into operation in the region with intermittent operation, and the main pumps, only in the stable operation region.

The quantity t may be influenced by the permutation of jockey pumps start commands when they have to cover flow rates in the intermittent region.

Figure 8.38 shows the operating diagram of a pumping installation, provided with two main pumps of flow rate Q, and three jockey pumps of flow rate $Q/2$. Denoting the jockey as A, B and C pumps and the main ones as P and Q, their programme of entering operation will be

$$O/A/A + B/P/P + A/P + A + B/P + Q/P + Q + A/P +$$
$$+ Q + A + B/P + Q + A + B + C$$

Besides the normal programme of entering operation, the chronological system has also an additional preparation programme. This has the role of ensuring accurate commands and of keeping the pumps within the imposed operation range. To this end, the levels should be determined very accurately and they should correspond to the demanded pressures.

The preparation programme is achieved by making use of the following equipment:

— a by-pass valve, for isolating the installation from the system;
— an injection valve, for introducing compressed air into the tanks;
— a zero flow rate starting device, independent of normal control.

This equipment comes into operation in the following order: when the pumping installation is again put into operation, either voluntarily, or after a current failure, the by-pass valve is closed for isolating the installation from the system. A jockey pump raises the water level in the tank, and when the maximum level is reached, it stops. Then, the injection valve is opened and at the same time, the compressor starts, keeping on working until the pressure in the tanks equals the pressure corresponding to the maximum level. As soon as the preparation programme is over,

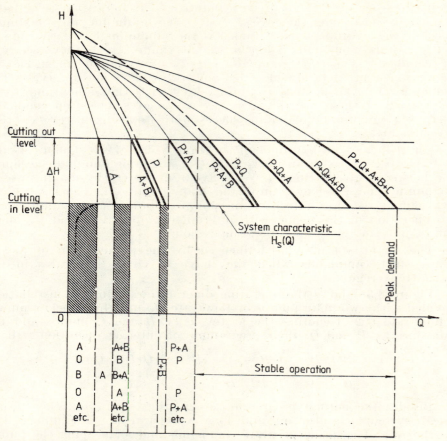

Fig. 8.38. Head-discharge curves of five pumps, regulated by chronometric system.

the by-pass valve is progressively opened and the normal control programme starts.

System review. Chronometric on-off regulation has a series of advantages and disadvantages:

Advantages:

— the global head range of the installation, ΔH_i, is very much reduced without influencing normal operation of the pumps. Thus, pump pressures hardly exceed the system characteristic pressure and, thus, the installation has a high global efficiency.

Disadvantages:

— the automation is rather complex, requiring an extremely accurate record of levels and the use of many relays which increase risks of failure of the installation.

8.3 Modulating Pump Controls

The modulating control system is based on modification of the pump head curve by means of speed variation. It adjusts the speed of pump driver to current needs of the pumping system. This is the best type of closed-loop control.

The commands given by these controls are continuous, since their action has a character, leading to continuous changes in the operation of some machines (for instance, motor speed variation) which continuously change the installation parameters (flow rate, pressure, etc.).

In modulating regulation systems, the final control element is the pump with variable speed. These systems are used for automatically creating a certain dependence law between pump pressure and flow rate so that their $H(Q)$ pump-head curves should be ideally placed on the path followed by the $H_s(Q)$ system-head curve.

Before analyzing modulating regulation systems it is worth noting the evolution of the two characteristic curve paths, $H(Q)$ and $H_s(Q)$, respectively. The pumping system-head characteristic curve, $H_s(Q)$, is represented by a parabola with its concavity upwards (in the case of water supply systems), or by a horizontal line, parallel to the flow-rate axis (in the case of land irrigation systems). Unlike these curves, the natural characteristic curve of a turbopump with constant speed, $H(Q)$, is represented by a parabola with downward concavity, that is opposite to that of the system. This is a great disadvantage in the case of pump flow rate regulation from the nominal point up to the region of minimum flow rates, since in this region the turbopump pressure far exceeds the necessary pressure demanded by the system-head characteristic curve. This leads to useless hydraulic energy losses, and thus, to a rather low global efficiency of pumping installation. For preventing this shortcoming, one should use variable-speed pumps.

Changing the speed of these pumps, we get a family of pump-head artificial curves, $H(Q)$. The hull of these curves could be (depending on imposed speed law) either a parabola with upward concavity, or a horizontal line, parallel to flow rate axis, which should superpose exactly over the system-head characteristic curve. Thus, ideally speaking, one may obtain a result in which pressure, developed by turbopump, is equal to system characteristic pressure, and the global efficiency of pumping installation is excellent.

After these considerations, we may pass to the analysis of modulating regulation systems as applied to turbopump controls.

8.3.1. Pressure Modulating Regulations

These regulations have the role of creating a dependence law between turbopumps pressure and flow rate when the demand in the system is changed. Essentially, two types of regulations are used:
    (a) *constant pressure regulations*, and
    (b) *parabolic pressure regulations*.

(a) *Constant pressure regulation*. This regulation measures a single hydraulic parameter: the outlet pressure of pumping station. The role of this regulation is to keep constant the pump outlet pressure, regardless of variations in consumption or/and inlet pressure.

System arrangement. Figure 8.39 shows the schematic diagram of a constant pressure system, highlighting its components. This system uses

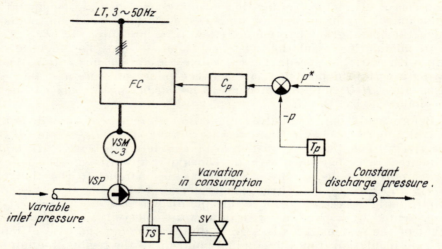

Fig. 8.39. Schematic diagram of constant pressure regulation with variable-speed pump:
VSM, VSP — variable-speed motor and pump; T_p — pressure transducer; C_p — pressure controller; *TS* — temperature sensor; *SV* — solenoid valve; *FC* — frequency converter.

the continuous pressure transducer, T_p, as an element of transmitting information which converts the hydraulic pressure signal into an electronic one, p, and, at the same time, informs presure controller, C_p, on the real value of pressure. Pressure controller compares the real value, p, with the setting value, p^*, and gives the order to actuation element, *FC*, (frequency converter). This order is transmitted to the *VSM* motor as a speed variation, so that the *VSP* pump should adequately change its flow rate, thus keeping constant discharge pressure.

Pumps operation. Let us consider two cases separately: a) with only one variable-speed pump, and b) with two pumps, of which the first with variable speed and the second with fixed speed.

In the first case, the pump operation depends on the disturbances occurring in discharge pressure, and takes place as follows: when pressure falls under the preset value, p^*, due to demand increase in pumping system, the controller makes the driving motor rise its speed, and the pump, its flow rate, so that discharge pressure could be kept constant. Conversely, when discharge pressure rises, tending to exceed the preset value, p^*, due to demand falling down, the controller orders a speed lowering, and implicitly, a lowering of pumped flow rate.

A few component elements from the diagram of a constant pressure regulation are shown in Figs. 8.40 and 8.42. Figure 8.41 shows the operation characteristics of a continuous pressure regulator.

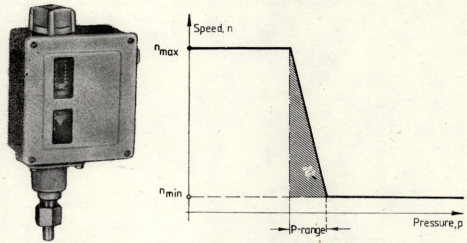

Fig. 8.40. Continuous pressure regulator (Danfoss Co.)

Fig. 8.41. Operation characteristic of a continuous pressure regulator.

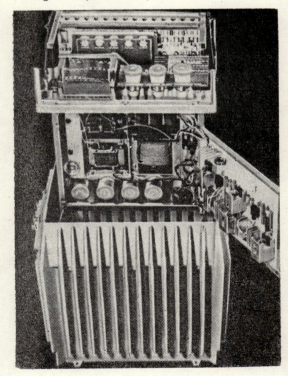

Fig. 8.42. Frequency converter (Danfoss Co.).

Figure 8.43 shows the operation diagram of a pump with constant pressure regulation. The artificial characteristic curve $H(Q)_{VSP}$, obtained by regulation, is represented by the AB segment and has the form of a straight line with constant ordinate $H_{ds} = H_{max} = $ constant (parallel to flow rate axis). The speed variation range, Δn, is narrow, and the control range of flow rate, ΔQ_i, is rather limited. The limitation of ΔQ_i range is

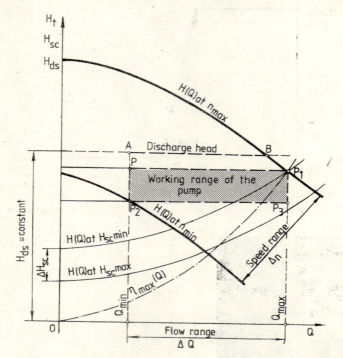

Fig. 8.43. Head-discharge curves of a variable-speed pump, regulated by constant pressure system.

determined by the fact that in minimum flow rate region the overall efficiency of the installation is rather low, since in that region the pump-head artificial curve, $H(Q)_{VSP}$, is moving away both from the system-head curve, $H_s(Q)$, and from that of pump maximum efficiency, $\eta_{max}(Q)$.

For widening the control range of installation flow rate, ΔQ_i, we should distribute the peak demand of pumping system to two or more pumps, connected in parallel, of which one with variable speed. In this case, the resulting artificial characteristic of parallel clutch is obtained by the continuous speed modification of variable speed pump, VSP, and by starting and stopping the pump or pumps with fixed speed, FSP.

Figure 8.44 shows the operation diagram of two pumps, connected in parallel ($VSP + FSP$) and regulated by constant pressure system. The artificial characteristic obtained (the AB segment) is also represented by a horizontal line, parallel to flow rate axis. These pumps operation depend also on the disturbances, occurred in discharge pressure as a result of demand modification, and takes place as follows: the first pump with variable speed, *VSP*, covers by itself the left part of demand variation range, ΔQ_i, keeping constant its pressure by changing pumped flow rate through speed variation. When the demand exceeds the pumped flow rate, secured by this pump at its maximum speed, then the second pump with fixed speed, *FSP*, is started by means of a maximum voltage relay, connected to the electric circuit of a voltage tahogenerator. When this latter pump starts running, the *VSP* one is immediately and adequately

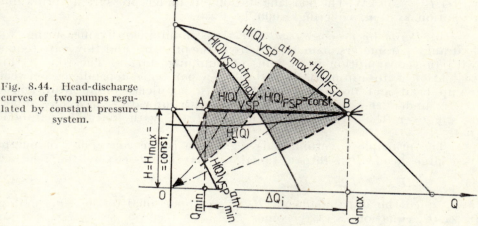

Fig. 8.44. Head-discharge curves of two pumps regulated by constant pressure system.

lowering its speed, meeting the new demand. For covering the right side of the ΔQ_i range, the *FSP* pump settles a certain duty point (for instance, corresponding to maximum efficiency), while the *VSP* pump duty point is displaced (for coping up with consumption variations) by the modification of pumped flow rate and speed, but keeps its pressure constant. When the demand falls down under a smaller value than that of the flow rate, pumped by the two pumps, connected in parallel clutch (the *VSP* pump, working with minimum speed) then, according to speed, the *FSP* pump is stopped by means of a minimum voltage relay, the *VSP* pump rising immediately its speed for meeting the new demand.

A review of the system. The presented regulation has various advantages and disadvantages:

Advantages:

— it keeps a constant outlet pressure, regardless of variation in consumption or/and inlet pressure.

Disadvantages:

— for a small number of pumps, it covers a reduced flow rate range, since the characteristic resultant due to pump regulation has a path (in region of minimum demand) far away from that of maximum efficiency characteristic curve belonging to variable-speed pump.

Consequently, the constant pressure regulation is recommended for the automatization of pumps operating on horizontal system-head curve, $H_s(Q)$ (the case of water supply systems for high-rise buildings and sprinkling irrigation systems).

A constant pressure regulation system has been used, for instance, to control four pumps, working on the horizontal system-head curve (one of which with variable speed) in a pressure boosting station in Lagruère sprinkling irrigation system (France).

Each pump had flow rate $Q = 750$ cu.m/h, head $H = 90$ m, and power $P = 250$ kW. The boosting station put under pressure a sprinkiling irrigation system, covering 1,000 hectares.

(b) *Parabolic pressure regulation*. It is simultaneously measuring two hydraulic parameters, namely, the discharge pressure and flow rate (equal with the consumption in system) of pumping station. The role of this regulation is to automatically establish a parabolic dependence between pump head and flow rate, to obtain an artificial pump-head curve, $H(Q)$, under the form of a parabola with upward concavity, superposing over the system-head curve, $H_s(Q)$, with natural parabolic paths.

The principle governing this regulation is the conversion of pumps discharge head, H_{ds}, into a signal as that expressed by the relation.

$$H_{ds} = H_{ds.st} + H_{ds.dy} \tag{8.43}$$

and, since the dynamic head, $H_{ds.dy}$, is proportional to the square flow rate, Q^2, equation (8.43) becomes

$$H_{ds} = H_{ds.st} + kQ^2 \tag{8.44}$$

Thus, pump discharge head can ideally follow a parabolic dependence law, identical to that followed by system-head curve.

Figure 8.45 shows the simplified diagram of a parabolic pressure system. This system combines a pressure transducer, T_p, with a flow-rate transducer T_q, both connected to an adder instrument. The hydraulic signals of pressure and flow rate are thus converted into a resulting electronic signal of compensated pressure, by flow rate, p_c.

Thus, the transducers inform the controller, C_p, on compensated pressure real value. This time, the controller compares real value, p_c, with preset value, p^*, and gives the adequate order to the actuation element FC (that is to frequency converter). This command is transmitted to the VSM motor as a speed variation, so that the VSP pump should achieve a reversed parabolic dependence between pressure and flow rate, similar to that of the pumping system.

The artificial characteristic curve of a variable-speed pump, obtained by parabolic pressure regulation, is graphically represented in the diagram shown in Fig. 8.46 by the AB curve segment. This segment is part of the path of a parabola with upward concavity, superposing itself over the system-head characteristic curve, $H_s(Q)$. Thus, the pressure developed by the pump is equal to system characteristic pressure throughout the variation range of pumped flow rate, ΔQ_i. An exact superposition could

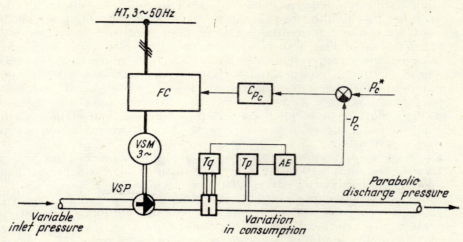

Fig. 8.45. Simplified diagram of a parabolic pressure system :

VSM, VSP — variable-speed motor and pump; T_p, T_q — pressure and flow transducers; AE — added element; C_{pc} — compensated pressure controller; FC — frequency converter.

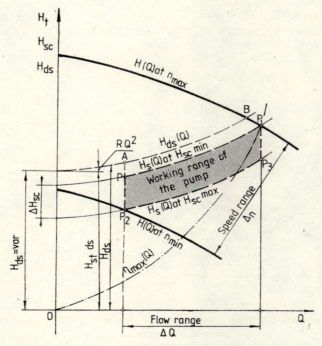

Fig. 8.46. Head-discharge curves of a variable-speed pump, regulated by a parabolic pressure system.

be obtained by an adequate choice of throttling device (diaphragm or Venturi tube) to which the flow rate transducer, T_q, is connected (obviously when this transducer is a differential manometer).

The analysis of Fig. 8.46 also shows that flow rate variation range, ΔQ_i, is larger when the system-head curve, $H_s(Q)$, is more inclined, since pump-head artificial characteristic, $H(Q)_{VSP}$, obtained by regulation (confounded with the first one), has a path much closer to that of maximum efficiency curve $\eta_{max}(Q)$. Thus, we could say that the parabolic pressure regulation is most recommended for pumps working on parabolic system-head characteristic curve, since it secures the highest overall efficiency to pumping installation.

The number of pumps depends on system static head. Thus, when, in extremis, the pumping system is lacking static head, then the artificial pump-head characteristic curve, $H(Q)_{VSP}$, is totally superposed over maximum efficiency curve, $\eta_{max}(Q)$, and therefore, one single pump is enough for securing an excellent overall efficiency to pumping installation. When we have static head, and the system-head curve starts flattening, the peak demand should be secured by two or more pumps, connected in parallel, one of which, or mostly two, should have variable speed. An optimization estimation should be done for the determination of both total number of pumps, and of those with variable speed.

The resulting pump-head characteristic of the two pumps has been drawn in Fig. 8.47 and is represented by the AB curve segment. We may notice that the resulting characteristic curve $H(Q)_{VSP} + H(Q)_{FSP}$ also has a parabolic shape, superposing exactly over the system-head curve $H_s(Q)$.

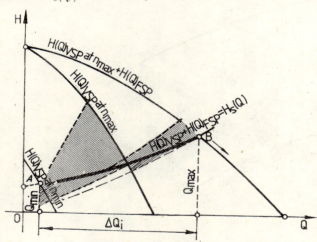

Fig. 8.47. Head-discharge curves of two variable-speed pumps, regulated by parabolic pressure system.

The connected pumps operation is similar to that described in case of constant (horizontal) pressure regulation.

A review of the system. The parabolic pressure regulation has a seris of advantages and disadvantages:

Advantages:
— the artificial pump-head characterstic curve is exactly superposed over the system-head curve, thus securing an excellent efficiency of pumping installation;
— the range of covered flow rate is very large, since the artificial pump-head characteristic has a common or close path to maximum efficiency characteristic, $\eta_{max}(Q)$.

Disadvantages:
— it makes use of a rather complex regulation system, since it requires two transducers and an adder instrument.

Consequently, the parabolic pressure regulation is recommended for the pump operation on inclined system-head characteristics. These are the cases of distribution pumps, connected to long main lines with high dynamic head (from water supply systems for localities, and water recirculation systems). For instance, a parabolic pressure regulation system has been used for the automatization of pumps from distribution pumping station of Toulouse town (France). In that case, distribution station has been provided with four pumps, two of which with variabe speed, and the other two with fixed speed. Each pump had the following parameters: flow rate, $Q = 1,300$ cu.m/h, head $H = 75$ m, and power $P = 430$ kW.

REFERENCES

8.1. ADAM, H., BENGHOUZI, L., *Commande automatique des stations de pompage pour réseau de plaines*, La Houille blanche, No. 5.(1966).
8.2. BRAXTON, S. J., *Design of water systems for high-rise buildings*. Journal American Water Works Association, No. 7.1966.
8.3. DANFOSS, CO., *Regulation of pressure-increasing plant*. The Danfoss Journal No. 4. (1974).
8.4. GOGUELL, L. R., *Nouveaux réservoires sous pression pour stations de pompage*. Travaux communaux, No. 6/7(1968).
8.5. HELLER, P., *Installations de la reprise pour alimentation directe d'un réseau*. Revue technique Sulzer, No.2. 1965.
8.6. HUMMEL, G. H., *Zur Bemessung von drunkvindkessel für automatische Betriebe von Pumpenwerken der Wasserversorgung*. Gas Wasser Wärme, No. 10 (1964).
8.7. IONEL, I. I., *Instalații de pompare reglabile*. Technical Publishing House, Bucharest, 1976.
8.8. IONEL, I. I., *Acționarea electrică a turbomașinilor*. Technical Publishing House, Bucharest, 1980.
8.9. IONEL, I. I., *Stations de pompages à regulation debitmetrique*. La technique de l'eau et de l'assainisement, No. 1.(1972).
8.10. IONEL, I. I., *Stații de pompare cu reglare automată în regim discontinuu*, Hidrotehnica, No. 3. (1969).
8.11. JOST, J., *Alimentation en eau potable d'un réseau de distribution par pompage à la demande, sans réservoirs en charge sur réseau*. Travaux, No. 12, 1969.
8.12. MELVILL, W. A., *Description of continuous pumping*. Journal American Water Warks Association, No. 5. (1965).
8.13. MÎNDRU, S., IONEL, I. I., *Unele aspecte privind proiectarea instalațiilor sanitare și utilajelor de bucătărie și spălătorie la hotelul ,,Inter-Continental București"*, Constructions, No. 2. (1973).
8.14. VALIBOUSE, B., *Les réservoirs d'eau sous pression et leur utilisation pour irrigation*. L'eau, No. 2 (1961).
8.15. VERDIER, J., BAGNÈRES, J., *Différents types des stations de pompage automatique*. La Houille Blanche, No. 5.(1966).

9

Speed Torque Characteristic Curves of Turbopumps

One of the main characteristics of turbopumps, regarded as working machines or as loading devices for electric motors, is the *speed-torque characteristic* $T = f(n)$. It shows the variation of load torque T, developed at the shaft as a function of rotation speed n. The graphical shape of this characteristic is highly influenced by the characteristic of the pumping system, $H_s = f(Q)$.

9.1 System Characteristic Effect on Turbopump Characteristics

In the transitory starting period, turbopump acceleration by means of an electric motor from zero speed to the full rate is possible only if the driving torque T_m, developed by the motor, is always above the load torque T, developed by the turbopump at its shaft

$$T_m > T \tag{9.1}$$

As the load torque is a function of speed, we may write

$$T = f(n) \tag{9.2}$$

Each working machine has its own speed-torque characteristic. However, the $T = f(n)$ characteristics of the majority of working machines, met in practice, may be expressed, in general, by the relation

$$T = T_f + pn^q \tag{9.3}$$

where T_f is the load torque opposed at starting for the $n = 0$ speed, and p and q are parameters depending on the machine type.

When we know the rated value of power P_R, consumed at the rated speed n_R, the rated load torque is

$$T_R = 955 \frac{P_R}{n_R} \tag{9.4}$$

By combining the relations (9.3) and (9.4), we get

$$T = T_f + (T_R + T_f)\left(\frac{n}{n_R}\right)^q \tag{9.5}$$

Denoting the relative values of the load torque and speed by

$$T^* = \frac{T}{T_R} \; ; \quad n^* = \frac{n}{n_R} \; ; \quad T_f^* = \frac{T_f}{T_R} \tag{9.6}$$

the relation (9.5) becomes

$$T^* = T_f^* + (1 - T_f^*) n^{*q} \tag{9.7}$$

where the values of parameter q depend on the machine type (pump, blower, compressor, etc.).

The turbopump, as a working machine, has a load torque, varying with the square of the rotation speed, so that $q = 2$, and equation (9.7) becomes

$$T^* = T_f^* + (1 - T_f^*) n^{*2} \quad (9.8)$$

By giving values between zero and unity to T_f^*, we get the function $T^* = f(n^*)$ shown in Fig. 9.1.

Since the mechanical power, absorbed at the shafts of turbopumps, varies with the third power of speed

$$\frac{P_2}{P_1} = \frac{\eta_1}{\eta_2} \left(\frac{n_2}{n_1}\right)^3 \quad (9.9)$$

and the efficiency is relatively independent of speed variations.

$$\eta_1 \approx \eta_2 \quad (9.10)$$

it follows that

$$\frac{P_1}{P_2} = \left(\frac{n_2}{n_1}\right)^3 \tag{9.11}$$

and

$$\frac{T_2}{T_1} = \frac{P_2}{P_1} \times \frac{n_1}{n_2} = \left(\frac{n_2}{n_1}\right)^2 \tag{9.12}$$

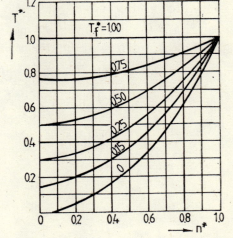

Fig. 9.1. Load torque variation T^* with speed n^* for different values of braking torque T_b^* (case $q = 2$).

whereby we see that $q = 2$. These equations are valid only for particular cases in which turbopumps do not have to overcome a static head. Such cases are met, for instance, with turbopumps mounted in recirculation installations, whose systems have only dynamic heads.

For the calculation of speed-torque characteristics in the general case, where the pumping system has both dynamic and static heads, we should consider the characteristic curves of turbopumps $H(Q)$, $P(Q)$ and $\eta(Q)$. When turbopumps operate at constant speed, we can get economic operation only near the point where $\eta \approx \eta_{max}$, and when they operate at variable speed, economic operation occurs around the bisector of the family of efficiency curves shown in the topogram of the respective turbopump.

Let us denote by

$$H^* = \frac{H}{H_R} \; ; \quad Q^* = \frac{Q}{Q_R} \; ; \quad n^* = \frac{n}{n_R} \tag{9.13}$$

the values of pumping head, flow rate, and speed, relative to their rated values. We also write the equation for relative values of the system pumping head

$$H^* = H_{st}^* + H_{dy}^* \tag{9.14}$$

where H_{st}^* and H_{dy}^* stand for the relative values of static and dynamic head, respectively $\left(H_{st}^* = \dfrac{H_{st}}{H_R} \text{ and } H_{dy}^* = \dfrac{H_{dy}}{H_R} \right)$.

At variable speed, the dynamic head is proportional to the flow rate, so it follows that

$$\frac{H_{dy2}}{H_{dy1}} = \left(\frac{Q_2}{Q_1} \right)^2 \tag{9.15}$$

Also, when we take

$$H_{dy1} = H_{dyR} \text{ and } Q_1 = Q_R \tag{9.16}$$

then

$$H_{dy} = H_{dyR} \left(\frac{Q}{Q_R} \right)^2 \tag{9.17}$$

and thus equation (9.14) becomes

$$H^* = H_{st}^* + H_{dyR}^* Q^{*2} \tag{9.18}$$

When the pumping system does not have a static head, then $H_{st}^* = 0$ and $H_{dy}^* = 1$ and hence

$$H^* = Q^{*2} \tag{9.19}$$

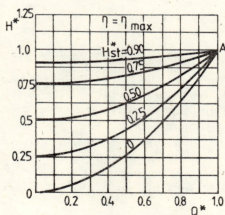

Fig. 9.2. Pumping head H^* variation with flow rate Q^* for different values of the system static head H_{st}^*.

which is a parabola, passing through the points zero and A of the graph in Fig. 9.2.

When the pumping system has also a static component, then $H_{st}^* \neq 0$ and we get a family of parabolae, starting from values of H^* other than zero, but all passing through the point A, having the co-ordinates $H^* = 1$ and $Q^* = 1$, since for $Q = Q_R$, $H = H_R$. Thus

$$H^* = 1 = H_{st}^* + H_{dyR}^*$$

and consequently, equation (9.14) becomes

$$H^* = H_{st}^* \left[1 + \left(\frac{1}{H_{st}^*} - 1 \right) Q^{*2} \right] \tag{9.20}$$

By giving H^* in turn the values 0.00, 0.25, 0.50, 0.75, and 0.90, we get the other four parabolae shown in Fig. 9.2.

With the known pumping head H^* (according to equation 9.20), we can easily find the turbopump power using the equation

$$P = \frac{QH}{\eta} \qquad (9.21)$$

or, in relative units

$$P^* = \frac{Q^* H^*}{\eta^*} \qquad (9.22)$$

where $\eta^* = \eta/\eta_{max}$; the values η^* are taken from the topogram of the turbopump.

For different values of H_{st}, we get the curves $P^* = f(Q^*)$ shown in Fig. 9.3.

Knowing that

$$T^* = 955 \frac{P^*}{n^*} \qquad (9.23)$$

it follows that we can calculate the turbopump load torque for various values of H_{st}^*. By giving H_{st}^* in turn the values 0.00, 0.25, 0.50, 0.75, and 0.90, we get the $T^* = f(n^*)$ curves shown in Fig. 9.4.

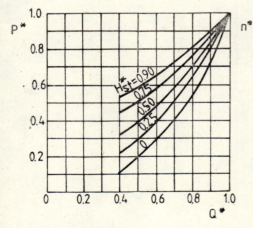 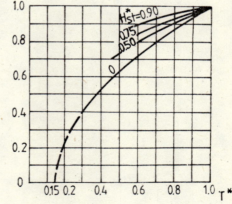

Fig. 9.3. Pump power variation P^* with flow rate Q^* for different values of the system static head H_{st}^*.

Fig. 9.4. Pump load torque variation T^* with speed n^* for different values of system static head H_{st}^*.

In this way, we may easily determine the variation of load torque with speed for all machines. The analysis of Fig. 9.4 shows that the load torque variation with speed does not follow a rigorously parabolic law for all machines, but only for those with $H_{st}^* = 0$. In other cases, for instance for $H_{st}^* = 0.90$, the parabola is of about sixth order.

TABLE 9.1 Pump performance parameters Q^*, H^*, η^*, P^*, n^*, M^* versus static head H^*_{st}

Parameter	$H^*_{st} = 0.25$							
Q^*	0.40	0.50	0.60	0.70	0.80	0.90	1.00	1.10
H^*	0.29	0.46	0.54	0.65	0.76	0.86	1.00	1.10
η^*	0.64	0.75	0.86	0.93	0.97	0.98	1.00	0.98
P^*	0.21	0.30	0.39	0.50	0.63	0.79	1.00	1.30
n^*	0.53	0.65	0.72	0.78	0.86	0.92	1.00	1.07
M^*	0.40	0.46	0.56	0.64	0.73	0.86	1.00	1.21

Parameter	$H^*_{st} = 0.50$							
Q^*	0.40	0.50	0.60	0.70	0.80	0.90	1.00	1.10
H^*	0.58	0.62	0.68	0.75	0.82	0.91	1.00	1.10
η^*	0.72	0.82	0.90	0.95	0.97	0.98	1.00	0.98
P^*	0.32	0.37	0.45	0.55	0.68	0.84	1.00	1.24
n^*	0.72	0.75	0.78	0.83	0.88	0.95	1.00	1.07
M^*	0.43	0.50	0.58	0.67	0.77	0.88	1.00	1.16

Parameter	$H^*_{st} = 0.75$							
Q^*	0.40	0.50	0.60	0.70	0.80	0.90	1.00	1.10
H^*	0.77	0.82	0.84	0.86	0.90	0.94	1.00	1.05
η^*	0.70	0.80	0.89	0.95	0.96	0.99	1.00	0.98
P^*	0.44	0.51	0.58	0.67	0.77	0.87	1.00	1.15
n^*	0.82	0.85	0.88	0.89	0.92	0.98	1.00	1.03
M^*	0.54	0.60	0.66	0.75	0.84	0.89	1.00	1.11

Parameter	$H^*_{st} = 0.90$							
Q^*	0.40	0.50	0.60	0.70	0.80	0.90	1.00	1.10
H^*	0.91	0.93	0.93	0.94	0.95	0.98	1.00	1.03
η^*	0.68	0.79	0.88	0.93	0.96	0.98	1.00	0.98
P^*	0.52	0.59	0.66	0.74	0.82	0.90	1.00	1.08
n^*	0.89	0.92	0.92	0.93	0.94	0.98	1.00	1.04
M^*	0.58	0.64	0.72	0.80	0.87	0.92	1.00	1.04

By extrapolating the load torque characteristic in accordance with the speed for $H_{st}^* = 0$, we get from Fig. 9.4, for $n = 0$, $T_f = 0.15$. In this case, the characteristic established by equation (9.8) becomes

$$T^* = 0.15 + 0.85\, n^{*2} \tag{9.24}$$

This equation helps us find the turbopump characteristics for cases in which they operate in pumping systems without static head.

Table 9.1 has been drawn up for rapid computations, since it gives the turbopump load torque in accordance with its main parameters (Q^*, H^*, P^*, T^* η^* and n) for the following values of pumping system static head:

$$H_{st}^* = 0.25,\ 0.50,\ 0.75,\ \text{and } 0.90.$$

9.2 Application of Similarity Equations to Turbopumps

By varying the rotation speed of a turbopump, we modify its characteristic $H(Q)$ and its flow rate. The operating points $A \ldots A_x$ are found to be at the intersection of the family of characteristics $H(Q)$ with the system characteristic $H_s(Q)$, as shown in Fig. 9.5.

The hydraulic coordinates Q_A and H_A of these points can be found by solving the system of equations

$$\begin{cases} H = A_2 n^2 + B_2 nQ + C_2 Q^2 \\ H = RQ^2 \end{cases}$$

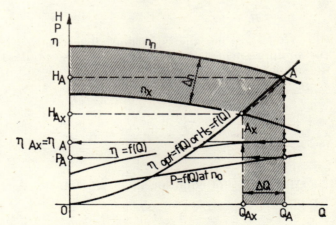

Fig. 9.5. Diagram for deriving the similarity equations.

By solving this system for the flow rate Q, we get

$$(R - C_2)Q^2 - B_2 nQ - A_2 n^2 = 0$$

which gives

$$Q = \left[\frac{\pm B_2 + \sqrt{B_2^2 + 4A_2(R - C_2)}}{2(R - C_2)} \right] n. \tag{9.25}$$

Thus, we notice that *the flow rate of a turbopump is proportional to its rotation speed*. Consequently, we may write

$$\frac{Q}{Q_x} = \frac{n}{n_x} \tag{9.96}$$

or

$$Q = k_Q n \tag{9.27}$$

By solving the equation (9.25) in accordance with the pumping head H, we get

$$\left(1 - \frac{C_2}{R}\right) H - \frac{B_2}{\sqrt{R}} n \sqrt{H} - A_2 n^2 = 0$$

which gives

$$H = \left[\frac{+\frac{A_2'}{\sqrt{R}} + \frac{B_2^2}{R} + 4A_2\left(1 - \frac{C_2}{R}\right)}{2\left(1 - \frac{C_2}{R}\right)}\right]^2 n^2$$

and which leads to the conclusion that the *turbopump head is proportional to the square of its rotation speed*. Hence we may write

$$\frac{H}{H_x} = \left(\frac{n}{n_x}\right)^2 \tag{9.28}$$

or

$$H = k_H n^2 \tag{9.29}$$

After deriving the equations (9.26) and (9.28), and finding that the turbopump effective power is given by the product of flow rate and pumping head, it follows that *the effective power is proportional to the cube of speed*, so that we may write

$$\frac{P_u}{P_{ux}} = \left(\frac{n}{n_x}\right)^3 \tag{9.30}$$

or

$$P_u = k_P n^3 \tag{9.31}$$

Making use of equation (2.90, a) and considering also equation (9.27), we get

$$\eta = \frac{\rho}{102} \cdot \frac{A_2 n^2 k_Q n + B_2 n k_Q^2 n^2 + C_2 k_Q^3 n^3}{A_3 n^2 k_Q n + B_3 n k_Q^2 n^2 + D_3^2 n^3}$$

whence

$$\eta = \frac{\rho}{102} \cdot \frac{(A_2 k_Q + B_2 k_Q^2 + C_2 k_Q^3)}{(A_3 k_Q - B_3 k_Q^2 + D_3)} \tag{9.32}$$

showing that *turbopump efficiency is independent of speed*.

Thus
$$\eta = \text{constant} \tag{9.33}$$

or, according to Fig. 9.5, we may write

$$\eta_{Ax} = \eta_A \tag{9.34}$$

Since the turbopump power output is proportional to the cube of speed, and its efficiency does not depend on speed, the power absorbed by the turbopump is proportional to the cube of speed

$$\frac{P}{P_x} = \left(\frac{n}{n_x}\right)^3 \tag{9.35}$$

or

$$P = k_P n^3 \tag{9.36}$$

Since $T = k_T \dfrac{P}{\eta}$, it follows that the hydraulic *load torque*, T, developed by the turbopump, *is proportional to the square of speed*.

Thus, we may write

$$\frac{T_h}{T_{h_x}} = \left(\frac{n}{n_x}\right)^2 \tag{9.37}$$

or

$$T_h = k_T n^2 \tag{9.38}$$

The above equations (9.26), (9.28), (9.30), (9.35), and (9.37) are called *similarity equations* and are adequate for cases in which the turbopump is operating in a pumping system without static head. In such cases the static torque of the turbopump, referred to the driving motor shaft, consists of the hydraulic load torque T_h, given by equation (9.38), and the friction torque T_f, produced by the mechanical resistances (friction in bearings, seals, reducer, etc.).

Thus,

$$T = T_f + k_T n^2 \tag{9.39}$$

Friction occurrence can be explained by the fact that, before starting, the turbopump is at rest and the oil film, present between the shaft and the bearings, being compressed, allows for closer contact of the two metals. Thus, a relative sticking of the turbopump shaft is produced, resulting in the occurrence of friction torque T_f. Its magnitude depends on the construction characteristics of the bearings and seals, as well as on the duration of the turbopump rest state. Thus, turbopumps fitted with rolling bearings have $T_f = 0.10 \, T_R$, and those with sliding bearings have $T_f = 0.15 \, T_R$. These values apply to a zero speed ($n = 0$). As soon as the speed rises ($n = 0.1 \ldots 0.2 \, n_R$), the friction torque decreases to zero.

In view of these observations, the turbopump friction torque can be expressed in terms of the speed as follows:

— for turbopumps with rolling bearings

$$T = 0.10\ T_R + 0.90\ T_R \left(\frac{n}{n_R}\right)^2 \tag{9.40}$$

— for turbopumps with sliding bearings

$$T_r = 0.15\ T_R + 0.85\ T_R \left(\frac{n}{n_R}\right)^2 \tag{9.41}$$

9.3 Turbopumps Load Torque Computation Methods

As shown in Section 9.2, when turbopump speed varies within a certain range Δn, its flow rate is changed in the domain Q and its efficiency remains constant within the domain ΔQ (Fig. 9.6).

Due to speed variation the turbopump characteristic $H(Q)$ is also changed, resulting in the family of characteristic curves $H(Q_1) - H(Q_3)$. The operation points $B_1 - B_3$ are found at the intersection of these curves with the characteristic curve of the pumping system. The efficiencies

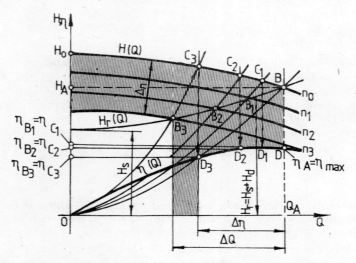

Fig. 9.6. Graphical computation of variable-speed pump efficiency.

corresponding to these points may be found by drawing the parabolae $OB_1C_1 - OB_3C_3$, which cross the characteristic $H_1(Q)$, corresponding to nominal speed n_R, and give the points $C_1 - C_3$. The perpendiculars drawn from these points on the flow-rate axis OQ intersect the efficiency curve $\eta(Q)$ in the points $D_1 - D_3$. Drawing parallels through these points to the flow-rate axis, we finally get the values of the efficiencies corresponding to the operation points $B_1 - B_3$.

Hence, it follows that

$$\eta_{B_1} = \eta_{C_1}; \quad \eta_{B_2} = \eta_{C_2}; \quad \eta_{B_3} = \eta_{C_3}. \tag{9.42}$$

The flow rate $Q_B = f(n)$ and the pumping head $H_B = f(n)$ are found by solving the system of equations

$$\begin{cases} H = A_2 n^2 + B_2 n Q + C_2 Q^2 \\ H = H_{st} + RQ^2 \end{cases} \tag{9.43}$$

By solving the above system for the flow rate Q, we get

$$Q = \frac{B_2 n - \sqrt{B_2^2 n^2 + 4(R - C_2)(A_2 n^2 - H_{st})}}{2(R - C_2)} \tag{9.44}$$

where, if $H_{st} = H_0$, the flow rate $Q = 0$. When solving this system for the pumping head, we get

$$H = H_{st} + \frac{R}{2(R - C_2)^2}[B_2^2 n^2 + 2(R - C_2)(A_2 n^2 - H_{st}) -$$

$$- B_2 \sqrt{B_2^2 n^2 + 4(R - C_2)(A_2 n^2 - H_{st})}] \tag{9.45}$$

Thus, we notice that equations (9.44) and (9.45) have an extremely complex form. Moreover, when we consider the variation of turbopump efficiency with flow rate, the situation is all the more intricate. For these reasons, we do not develop these equations here, but make use of more simple equations. Thus, by comparing the turbopump characteristic $H(Q)$ with a parabola, we can express it analytically by means of an equation [similar to equation (2.85)] having the form

$$H = H_0\left(\frac{n}{n_R}\right) - C_4 Q^2 \tag{9.46}$$

where H_0 is the shut-off head, in m;

$C_4 = \dfrac{H_0 - H_R}{Q_R^2}$ —a coefficient, whose value derives from the nominal values of the turbopump operation parameters.

The work condition (that is the duty point) of a turbopump is found at the intersection of the turbopump characteristic curve $H(Q)$, defined by equation (9.46), with the curve of the pumping system, defined by equation

$$H = H_{st} + H_{dy} = H_{st} + RQ^2 \tag{9.46, a}$$

By solving the system formed by these two equations for the flow rate Q, we get

$$Q = \frac{\sqrt{H_0\left(\dfrac{n}{n_R}\right)^2 - H_{st}}}{R + C_4} \tag{9.47}$$

Speed Torque Characteristic Curves

For the great majority of turbopumps, the power characteristic curve $P = f(Q)$ can be approximated by a straight line, defined by the equation

$$P = P_0 \left(\frac{n}{n_R}\right)^3 + A_4 Q \left(\frac{n}{n_R}\right)^2 \qquad (9.48)$$

where P_0 is the shut-off power, in kW;

$A_4 = \dfrac{P_R - P_0}{Q_R}$ — a coefficient whose value derives from the nominal values of the turbopump parameters.

By substituting equation (9.47) in (9.48), we get the expression for power as a function of speed for the case in which the turbopump is coupled to a pumping system with invariable operation parameters

$$P = P_0 \left(\frac{n}{n_R}\right)^3 + A_4 \left(\frac{n}{n_R}\right) \sqrt{\frac{H_0 \left(\frac{n}{n_R}\right)^2 - H_{st}}{R + C_4}} \qquad (9.49)$$

From equation (9.49), we get the load torque equation

$$T = \frac{955\, P_0}{n_R} \left(\frac{n}{n_R}\right)^2 + \frac{955\, A_4}{n_R} \frac{n}{n_R} \sqrt{\frac{H_0 \left(\frac{n}{n_R}\right)^2 - H_{st}}{R + C_4}} \qquad (9.50)$$

or

$$T = T_0 \left(\frac{n}{n_R}\right)^2 + (T_R - T_0) \frac{n}{n_R} \sqrt{\frac{H_0 \left(\frac{n}{n_R}\right)^2 - H_{st}}{R Q_R^2 + H_0 - H_R}} \qquad (9.51)$$

where

$$T_0 = 955 \frac{P_0}{n_R} \qquad (9.52)$$

is the shut-off torque, at nominal speed, and

$$T_R = 955 \frac{P_R}{n_R} \qquad (9.53)$$

is the nominal load torque, corresponding to nominal speed.

By substituting $H_{st} = 0$ in equation (9.51), we find a similar equation to (9.38).

By further expressing the static head H_{st} as a function of the minimum speed by means of equation $n_{min} = n_R \sqrt{\dfrac{H_{st}}{H_0}}$ and then using equation (9.51), we get

$$T = T_0 \left(\frac{n}{n_R}\right)^2 + (T_R + T_0) \frac{n}{n_R} \sqrt{\frac{\left(\frac{n}{n_R}\right)^2 - \left(\frac{n_{min}}{n_R}\right)^2}{1 - \frac{H_R}{H_0} + \frac{R Q^2}{H_0}}} \qquad (9.54)$$

9.4 Transitory Regimes, Occurring at Turbopump Start

During the starting of a turbopump, from the moment when its flow rate is zero until it reaches the rated flow rate, starting transitory regimes are manifest. They may show much higher pumping heads and load torque than their nominal values. This requires that, in certain conditions, the pump and its driving motor be selected in accordance rather with the starting conditions than with those corresponding to the rated flow rate. In this sense, radial-flow and mixed-flow pumps with small and medium specific speed ($n_s \leqslant 320$) show advantageous starting characteristic curves. The pumping head at shut-off is not much higher than that corresponding to the rated flow rate, and the shut-off torque is even small, relative to that for nominal flow rate conditions ($T_{shut-off} = 0.33$ to $0.80\ T_R$). On the other hand, high speed pumps ($n_s > 325$), such as the axial-flow type and some of the mixed-flow pumps, generate higher shut-off pumping heads and torques, relative to those for the rated flow rate ($T_{shut-off} = 1.10$ to 2.10). This requires that special attention be given to the time interval necessary for starting.

We give below several graphs, representing the function $T = f(n)$ for various starting conditions specific to turbopumps (starting with closed or open discharge valve, starting against a check valve, and starting in the reverse direction) and with a view to the shape of pumping system characteristic curves [9.1], [9.2], [9.3].

9.4.1 Turbopumps Starting Against a Closed Discharge Valve

A turbopump starting against a closed discharge valve represents a reduced load on its electric driving motor. This mode leads to starting powers and torques approximately 50% below their full rate. When turbopumps start against an open discharge valve, they show higher shut-off head values than the full-rated ones. These values vary with the turbopump specific speed, in that the higher the specific speed, the higher the values of shut-off head (see Fig. 2.19, a). Turbopump power and shut-off torque also increase with the specific speed of the pump (Table 9.2).

We see from Table 9.2 that low and medium-specific speed pumps of the radial and mixed-flow types ($n_s < 325$) have favourable characteristics. The power and shut-off torque are less than at normal flow. High-specific speed pumps of mixed and axial-flow types ($n_s > 325$, develop relatively high power and shut-off torque. These characteristics of high-specific speed pumps require special attention during the starting period.

For radial-flow pumps starting against a closed discharge valve, the pumping head and torque, in general, follow the curves $H = f(Q)$ and $T = f(n)$ shown in Fig. 9.7.
We see that during the pump starting phase, the pumping head varies proportionally with the square of rotation speed (curve OA in Fig. 9.7, a). The load torque also shows a parabolic variation (section OA of curve OB

in Fig. 9.7, b). On the AB section, the load torque variation depends on the shape of the characteristic curve of turbopump power.

As seen in Fig. 9.7, b, the turbopump torque is not zero at zero speed since the influence of friction torque is manifest in this region (due to

TABLE 9.2 **Comparison of power** (P/P_R) **and torque** (T/T_R) **versus flow rate** (Q/Q_R) **for different specific-speed** $(n_s\text{-}n_q)$ [9.3]

n_s	Q/Q_R					n_q
	0	0.25	0.50	0.75	1.10	
	P/P_R or T/T_R					
40	0.33	0.51	0.70	0.81	1.05	10.0
50	0.35	0.53	0.70	0.81	1.05	12.5
60	0.37	0.54	0.70	0.81	1.05	15.0
70	0.42	0.56	0.71	0.81	1.05	17.5
100	0.50	0.62	0.73	0.82	1.05	25.0
150	0.60	0.67	0.76	0.83	1.05	37.0
200	0.73	0.84	0.82	0.86	1.00	50.0
300	0.80	0.87	0.94	0.94	1.00	75.0
325	1.00	1.00	1.00	1.00	1.00	81.2
350	1.10	1.05	1.05	1.05	0.96	87.5
400	1.30	1.20	1.10	1.05	0.94	100.0
500	1.50	1.30	1.18	1.10	0.90	125.0
600	1.70	1.45	1.25	1.15	0.81	150.0
700	2.10	1.70	1.40	1.21	0.79	175.0

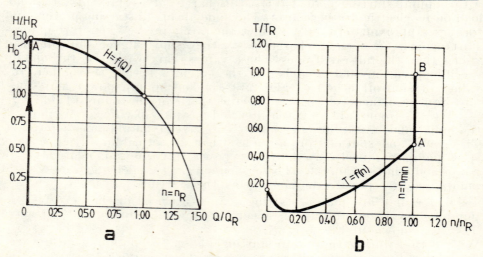

Fig. 9.7. Turbopump starting against a closed discharge valve:
a — head-discharge characteristic; b — torque-speed characteristic.

the static friction in bearings and seals). This is the explanation for curve $T = f(n)$ within the speed range $n = 0$ and $n = 0.10\ n_R$. We can explain the speed-torque characteristic of a turbopump, $T = f(n)$, by analogy with equation (9.39) as

$$T = T_f + T_0 \left(\frac{n}{n_R}\right)^2 \tag{9.55}$$

The graphs of Fig. 9.8 show variation of load torque with speed, n, both for pumps with low specific rotation speed (Fig. 9.8, a), and for those with high specific rotation speed (Fig. 9.8, b) [9.3]. The same figures

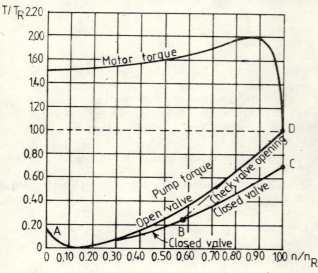

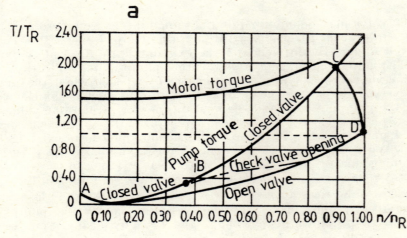

Fig. 9.8. Torque variation during the starting of a pump with closed valve, with open valve, and with check valve: a — a low-specific-speed pump; b — a high-specific-speed pump.

include the driving torque curve of a squirrel-cage induction motor. Analysis of Fig. 9.8, *a* shows there is a substantial difference between the motor and pump torques, representing the necessary excess torque for pump acceleration from rest to the full-rated speed. Consequently, when establishing the turbopump shaft size, we should consider not only the turbopump load torque (curve *ABC*), but also the excess torque at the motor shaft.

Pumps with high specific speed, mainly the axial-flow type, require a higher shut-off torque than the full-rated one. They are therefore never started against a closed discharge valve, since to do so would require a driving motor of higher power which would be more expensive. In Fig. 9.8, *b*, the curve *ABC* represents the variation of load torque for a pump of high specific speed, started against a closed discharge valve. Superposed is the driving torque curve of a squirrel-cage induction motor, able to overcome a normal load torque (its full-rated value). We see that the motor has an insufficient torque for accelerating the pump from rest to its full-rated speed. This motor remains overloaded until the discharge valve opens. Thus, we begin opening the valve before starting the electric motor, but not so early that it might produce a reverse flow, leading to the motor starting in reverse flow conditions (see Section 9.4.4).

When a synchronous motor is used for starting the pump then, beside the driving torque necessary for overcoming the existing pumping head and for accelerating the pump and motor masses, we should also consider the additional driving torque necessary for achieving synchronous rotation of the motor. We should also bear in mind the low value of asynchronous driving torque when the synchronous motor is provided with a starting squirrel-cage.

9.4.2 Turbopump Starting Against a Check Valve

A check valve may be mounted on individual discharge pipes of the pump in order to prevent reverse flow through the pump. Reverse flow could be produced under the influence of the existing static head of the system, or of the pumping head, generated by other pumps, coupled to the system. In principle, a check valve is automatically opened when the pumping head exceeds the system head. When a pump is started against a check valve, the pumping head and the load torque have the same values as in in the case of a closed discharge valve until the full-rated speed is reached, and the shut-off head exceeds the pumping head existing in the system. When the check valve is opened, the pump head keeps growing, overcoming (for any flow rate) the static head existing in the system and the head, generated by other pumps, operating simultaneously, as well as the hydraulic resistances of the pipes and fittings, and the inertia of the pumped liquid mass. Torque variation curves for pumps with check valves are shown in Figs. 9.8, *a* and 9.8, *b* by means of curves *ABC*. Both curves represent the case in which the pumping system has both static and dynamic head.

9.4.3 Turbopump Starting Against an Open Discharge Valve

The load torque, generated at turbopumps starting against open discharge valves, is considerably influenced by the shape which the static or dynamic head represents of the total pumping head of the system. We give in the following some detailed analyses.

Pumping system without static head. In the case of a system without static head, the pumping head and the load torque of a turbopump couple to it follows the $H = f(Q)$ and $T = f(n)$ curves shown in Fig. 9.9.

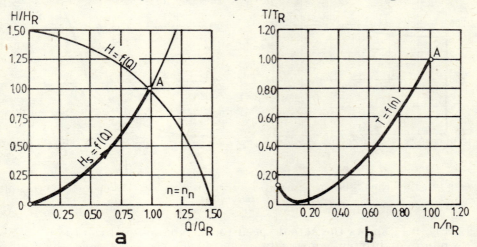

Fig. 9.9. Turbopump starting against an open discharge valve. Pumping system without static head:
a — head-discharge characteristic; b — torque-speed characteristic.

Analysis of Fig. 9.9, b shows that the load torque (curve OA) is a parabola, in accordance with equation (9.39). A curve of this shape is obtained only when the discharge line is not too long. Otherwise, the curve has a similar shape to that shown in Fig. 9.7, since the acceleration time of the mass of liquid in the discharge line is much longer than the turbopump starting time. This effect is explained by the fact that the large mass of liquid at rest, all along the pipe, acts like a discharge valve in the closed position.

Pumping system with static head. When the total pumping head of the system includes a static component, the curves of the pumping head $H = f(Q)$ and of load torque $T = f(n)$ follow the graphs of Fig. 9.10.
Analysis of these graphs leads to the following conclusions:
— for $0 < n < n_{min}$, the turbopump pumping head is smaller than that existing in the system and, thus, the load torque curve (section OA) has the same shape as that in Fig. 9.7, b and the load torque values could be determined by means of equation (9.46);

— for $n_{min} < n < n_R$, the turbopump head is higher that the static head of the system. In this case, the turbopump operates at a point whose co-ordinates of flow rate and pumping head are given by equations (9.44) and (9.45) respectively, and the resulting load torque characteristic (section AB) has the form of a very inclined curve with downward concavity (see Fig. 9.10, b) and torque values, determined by equation (9.51).

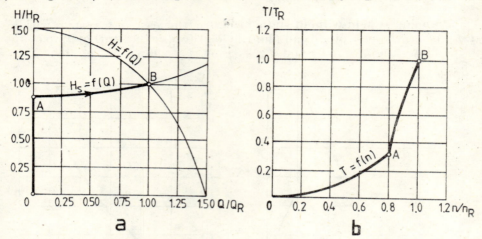

Fig. 9.10. Turbopump starting against an open discharge valve. Pumping system with static head:
a — head-discharge characteristic; b — torque-speed characteristic.

Fig. 9.11 shows the actual speed-torque characteristics of two radial-flow pumps of different construction, one of which is single stage (Fig. 9.11, a) and the other multistage (Fig. 9.11, b) design.

Also tested were two submerged pumps, of radial-flow and mixed-flow types, started in different conditions, with the discharge line either empty or filled with water. The results are shown in Fig. 9.12.

9.4.4 Turbopump Starting in Reverse Running

Suppose a turbopump discharges into a system with static head, or together with other turbopumps into a common discharge header, and then is stopped. Unless the discharge valve is closed, or there is a check valve in the system, or a broken siphon in a siphon system, the flow would be reversed, turning the turbopump in a reverse direction and thus making it operate like a no-load running turbine. The turbopump operates in a turbine condition in the following situations:

— when it is coupled to a system with a static head. In this case, the turbine head applied to the turbopump will equal the difference between the system static head and the dynamic head corresponding to the reversed flow (the head losses due to friction);

— when it discharges, together with other turbopumps, into a common discharge header, even if the pumping system does not have a static

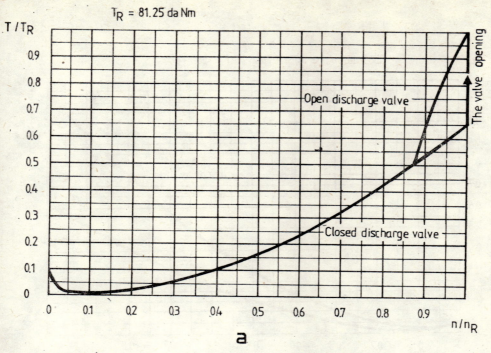

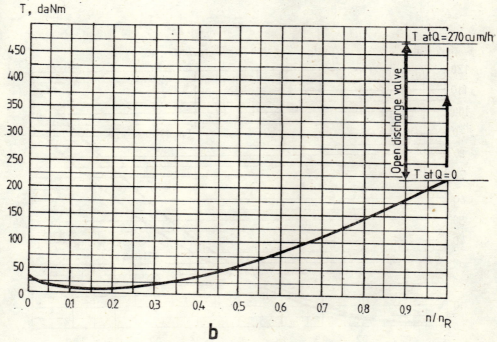

Fig. 9.11. Speed-torque characteristics of two radial-flow pumps:
a — single stage pump; b — multistage pump.

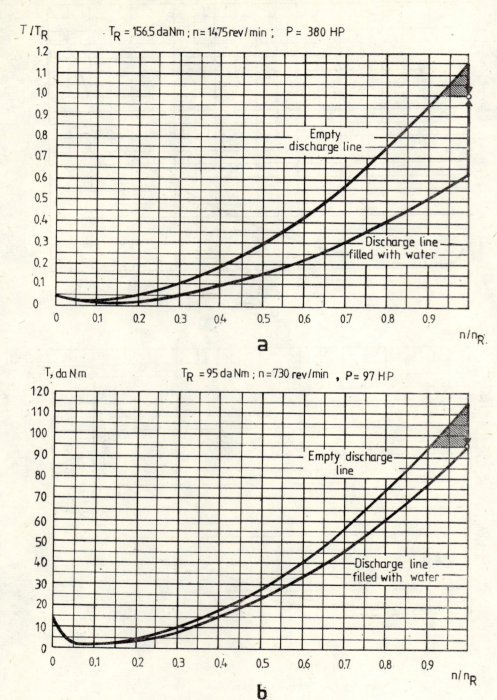

Fig. 9.12. Starting of submerged pumps with empty and water-filled discharge pipelines.

head. In this case, the turbine head (the net load), which rotates the turbopump in the reverse direction, is equal to the pumping head corresponding to the pumping regime, or to the pumping head developed by the other turbopumps, minus the head losses corresponding to the reversed flow rate.

Figure 9.13 shows the speed-torque characteristics (reverse speed-torque) of both low and high specific speed turbopumps. Analysis of

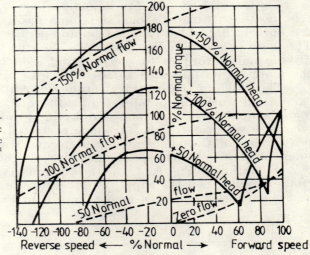

Fig. 9.13. Typical reverse speed-torque characteristics of pumps:
a — a radial-flow pump with low specific speed; b — an axial-flow pump with high specific speed.

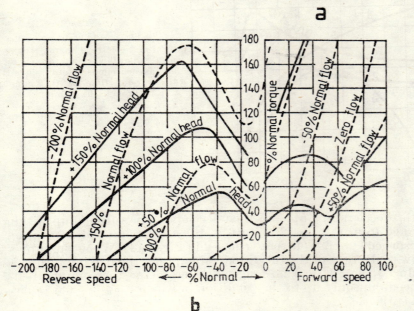

these graphs shows that when the flow rate is reversed through the pump (its motor also shows a load torque, but of extremely small value), this reaches a higher speed in the reversed direction than in the normal one. This runaway speed is directly proportional to the pump specific speed and to the system pumping head. At the same time, we should notice that the parameter values, indicated on these graphs (speed, torque, flow rate, and pumping head), are expressed as percentages of their full-rated values when the pump is rotating normally.

The attempt to start a turbopump which is already in reverse rotation requires that its motor should develop a positive torque, equal to the initial one, during which time the pump operates in the negative direction. This may be seen in Fig. 9.13. These graphs also show the necessary driving motor torques for decelerating, stopping, and then accelerating the pump to its full-rated speed.

The value of the contrary torque, generated by a reverse flow, may exceed the turbopump full-rated load torque. Consequently, the electric motor should develop an increased driving torque for overcoming the augmented load torque of the turbopump started from reverse running. The peak value of the augmented load torque corresponds to the stopping moment, that is, to zero speed (the curve shown in Fig. 9.14).

According to Fig. 9.14, the turbopump may be started from reverse running only when the supply voltage to the electric motor is constant and equal to its full-rated level (curve C in Fig. 9.14). Only in this case

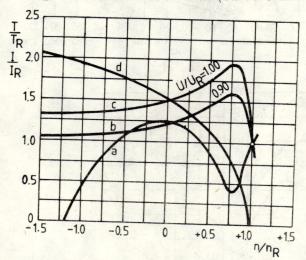

Fig. 9.14. The hydraulic, mechanical and electrical characteristics of a radial-flow pump, driven by an asynchronous motor during the reverse flow of the liquid through the pump:

a — load torque curve; b — motor torque curve for $U/U_R = 0.9$; c — ditto for $U/U_R = 1.0$; d — current curve.

can the motor decelerate, stop, and then accelerate the turbopump within the speed range from $n = -1.2\, n_R$ to $n = 1.0\, n_R$, but in conditions of an increased absorbed starting current (curve d in Fig. 9.14). When the supply voltage is low (only 10% of the full-rated value), then the driving torque is so much reduced that it can no longer accelerate the turbopump to its full-rated speed, but only up to the condition in which $n/n_R = -0.5$ (curve b in Fig. 9.14) [9.5].

9.4.5 Turbopump Starting, Provided with a Broken Siphon

Axial-flow pumps are usually provided with individual discharge pipes. Each of these pipes is fitted with a rising-broken siphon (Fig. 9.15) whose role is to prevent the flow reversing after the pump is stopped. To this end, the siphon is fitted with an automatic vacuum breaker, located at its summit. Siphons are often used only for the elimination of any shut-off or

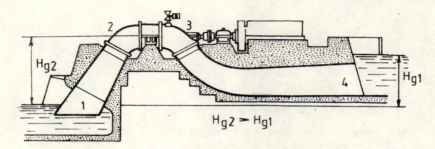

Fig. 9.15. A pumping installation with broken siphon.

check valves, etc., from the discharge circuit, and hence of local hydraulic resistances, which leads to significant savings of electrical energy.

Since the power characteristic curve $P = f(Q)$ of axial-flow pumps shows a maximum P_0 for $Q = 0$, this leads to a ratio $P_0/P_R \geqslant 2$ and hence, to an extremely unfavourable speed-torque characteristic $T = f(n)$. Consequently, when the pump is coupled to a discharge pipe with rising-broken siphon, it generates several problems at the starting of its driving motor. These problems arise because the siphon summits much above the normal discharge level, which obliges the pump to overcome (until the siphon is filled and primed) a much higher transitory static head H_{g2} than the normal one H_{g1} resulting after the siphon priming (Fig. 9.16).

The graphs shown in Fig. 9.17 represent the pumping head curve $H = f(Q)$ and the load torque curve $T = f(n)$ for the case of an axial-flow pump coupled to a discharge pipe with a broken siphon. Analysis of these graphs reveals that the load torque value (curve OA in Fig. 9.17, b) exceeds the nominal one by a value, equal to the segment AB of the same graph. The size of the segment AB depends on the shape of the power characteristic curve, $P = f(Q)$, and on the turbopump specific speed. Thus, the more inclined the power curve and the higher the specific speed, the greater the size of this segment, and vice-versa.

When a pump and driver are to be selected to prime a siphon system, it is necessary to estimate the pumping head and the power, required to produce the minimum flow, needed to start the siphon in the conditions of a transitory static head, H_{g2}. The minimum required flow increases with the length and diameter of the down-leg pipe and decreases with its slope.

Speed Torque Characteristic Curves

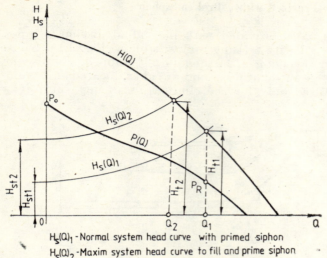

Fig. 9.16. Transient system of total head priming a siphon.

$H_s(Q)_1$ — Normal system head curve with primed siphon
$H_s(Q)_2$ — Maxim system head curve to fill and prime siphon

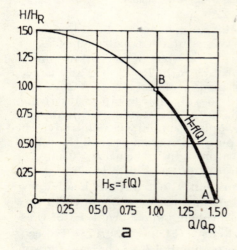
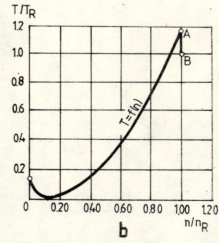

Fig. 9.17. Starting of a turbopump, provided with broken siphon:
a — head-discharge characteristic; b — torque-speed characteristic.

9.5 Conclusions on Turbopump Behaviour Features

Assembling the features of turbopump behaviour and their requirements with respect to driving motors, the conclusions below may be drawn:

(a) *The turbopump flow rate* is proportional to the speed and pumping head, the load torque to the square of the speed, and the absorbed power to the cube of the speed. The lines joining the points of constant efficiency represent a family of symmetrical parabolae with respect to the pumping head axis, whose peaks are to be found at the origin of co-ordinate axes $H-Q$ (Fig. 9.18).

(b) *The shut-off torque* T_0, corresponding to the flow rate $Q = 0$ and the full-rated speed n_R, obtained with the discharge valve closed, varies with

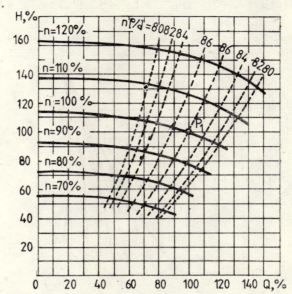

Fig. 9.18. The complete (universal) characteristic of a turbopump.

the pump construction and specific speed. The higher the specific speed, the higher the shut-off torque. On average, the shut-off torque is about 0.55 as to the full-rated torque (with full load) (Fig. 9.19).

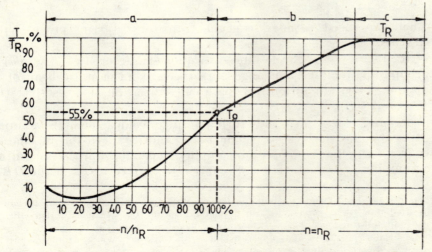

Fig. 9.19. Turbopump starting torque in various conditions:
a — with closed check valve; b — with open check valve; c — with a check valve with progressive opening.

(c) *Turbopumps show a marked interdependence* between the load torque and the speed, since they generally operate in systems having also static heads. This interdependence is influenced by the form of the pumping system characteristic curve $H_s = f(Q)$, that is, by the ratio between the static head H_{st} and the total head H_t (Fig. 9.20).

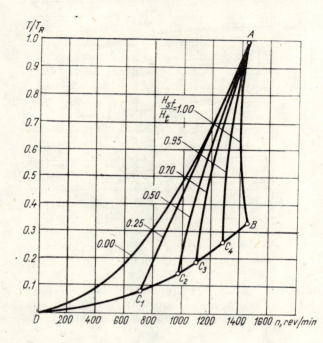

Fig. 9.20. Modification of torque-speed characteristic of a turbopump with the static system head variation.

When the speed varies during the transitory starting process, it is noticed that:

— if $0 < n < n_{min}$, the speed-torque characteristic $T = f(n)$ of the turbopump remains unchanged, irrespective of the static head of the system and shows an absolutely parabolic variation (curves OC_1, OC_2, OC_3, OC_4 and OB);

— if $n_{min} < n < n_R$, the turbopump develops enough pumping head to overcome the system static head, and its speed-torque characteristic $T = f(n)$ changes, becoming highly dependent on the magnitude of the pumping system static head (curves OC_1A, OC_2A, OC_3A and OBA).

Example 9.1. **Load torque computation for a turbopump.** Let us consider a turbopump with full-rated flow $Q_R = 0.70$ cu.m/s, rated pumping head $H_R = 42.50$ m, rated efficiency $\eta_R = 0.89$ and full-rated speed $n_R = 730$ rev/min. This turbopump is coupled to a system with static head $H_{st} = 25$ m and pipe resistance $R = 36$ s²/m⁵. We want to compute the turbopump load torque value for an intermediate speed of $n = 585$ rev/min, corresponding to the operation point of the pumping installation.

Solution

(1) *We first draw the turbopump characteristic* $H = f(Q)$, corresponding to the given speed. To this end we use equation (9.38). Then we draw the system characteristic $H_s = f(Q)$ by means of equation (9.46, a).

(2) *The two operation points, A and B* (Fig. 9.21), are found at the intersection of the turbopump characteristic $H = f(Q)$ (corresponding to the two given speeds, $n = 730$ rev/min, and $n = 585$ rev/min) with the system characteristic $H_s = f(Q)$.

(3) *We then determine the turbopump efficiency for point B* (see Fig. 9.21, a). A quadratic parabola is drawn through point B, thus obtaining point C at its intersection with curve $H = f(Q)$, corresponding to the speed $n = 730$ rev/min. Then the flow rate Q_C is determined by means of equation (9.28)

$$Q_C = Q_B \frac{n}{n_B} = 0.38 \frac{730}{585} =$$

$$= 0.475 \text{ cu. m/s}$$

From point C, we then draw a vertical line crossing the curve $\eta = f(Q)$ and thus get point D. From point D, we draw a horizontal line crossing the efficiency axis to the efficiency value corresponding to the intersection point B. Thus

$$\eta_B = \eta_C = 0.77$$

(4) *We then calculate the turbopump absorbed power corresponding to point B*

$$P_B = \frac{\rho Q_B H_B}{102 \, \eta} =$$

$$= \frac{1000 \times 0.38 \times 30}{102 \times 0.77} = 145 \text{ kW}$$

(5) *Finally, we compute the turbopump load torque corresponding to point B* by means of equation (9.28)

$$T = 955 \frac{P_B}{n_B} = 955 \frac{145}{585} =$$

$$= 237 \text{ daNm.}$$

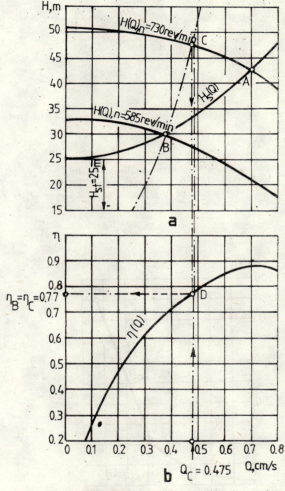

Fig. 9.21. Graphical computation of efficiency for a variable-speed turbopump (for $H_{st} = 25$ m).

Example 9.2. **Drawing a turbopump speed-torque characteristic.** Let us consider a radial flow pump having full-rated flow $Q_R = 0.70$ cu.m/s, nominal pumping head $H_R = 42.50$ m, shut-off head $H_0 = 51$ m, rated power $P_R = 350$ kW, shut-off power $P_0 = 150$ kW, friction torque (produced in turbopump bearings) $T_f = 9$ daNm, and full-rated speed $n_R = 730$ rev/min. This pump is coupled to a pumping system with static head $H_{st} = 25$ m and pipeline resistance $R = 36$ s²/m⁵. We want to draw the speed torque characteristic of the turbopump for the case of its starting with an open discharge valve.

Solution

(1) *We first determine the minimum speed* (by which we understand here that speed at which the turbopump develops a pumping head, higher than the static head of the system, and at which the check valve opens) by means of equation

$$n_{min} = n_R \sqrt{\frac{H_{st}}{H_0}} = 730 \sqrt{\frac{25}{51}} = 510 \text{ rev/min}.$$

(2) *We compute the turbopump shut-off torque* by means of equation (9.52):

$$T_0 = \frac{955 \times 150}{730} = 200 \text{ daNm}$$

(3) *We construct the speed-torque characteristic* for the interval of turbopump acceleration within the speed range $0 < n < 510$ rev/min (OA curve) by equation (9.55),

$$T = 9 + 200 \frac{n^2}{730^2} = 9 + \frac{n^2}{2660} \text{ daNm}$$

We then take several values for the speed n, within the range from zero to 510 rev/min, and thus we get the necessary values for drawing the curve OA in Fig. 9.22.

(4) *We compute the full-rated load torque of the turbopump* by means of equation (9.53)

$$T_R = 955 \frac{350}{730} = 466 \text{ daNm}$$

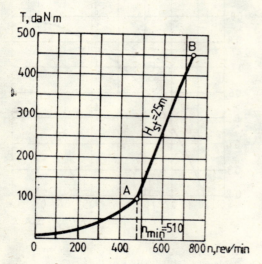

Fig. 9.22. Speed-torque characteristic of a radial-flow pump, derived for the case of its coupling to a system with $H_{st} = 25$ m.

(5) *We build the speed-torque characteristic* corresponding to turbopump acceleration within the speed range 510—730 rev/min (curve AB). To this end, we make use of the equation

$$T = 200 \left(\frac{n}{n_R}\right)^2 + 266 \frac{n}{n_R} \sqrt{1.95 \left(\frac{n}{n_R}\right)^2} = 0.95 \text{ daNm}$$

By assigning various values to n within the range $n = 510$—730 rev/min, we get a series of values for T by means of which we may draw the AB section of the curve, shown in Fig. 9.21.

REFERENCES

9.1. IONEL, I. I., *Acționarea electrică a turbomașinilor*. Technical Publishing House, Bucharest, 1980.
9.2. IONEL, I. I., *Instalații de pompare reglabile*. Technical Publishing House, Bucharest, 1976.
9.3. KARASSIK, I. J., KRUTZSCH, W. C., FRASER, W. H., *Pump Handbook*. McGraw-Hill Book Co., New York, 1976.
9.4. ONISCHENKO, G. B., IUNKOV, M. G., *Electroprivod turbomehanizmov*. Energya, Moscow, 1972.
9.5. STEPANOFF, A. I., *Centrifugal and Axial Pumps*. McGraw-Hill Book Co., New York 1948.

10

Speed Control of Turbopumps, Run by Means of Variable Voltage D.C. Motors

The finest speed control is obtained by means of d.c. motors, fed from variable voltage supplies. These supplies can be either external (natural) commutation converters (semiconductor controlled rectifiers), generally used for the supply of classic commutator d.c. motors, or load (motor.) controlled converters, particularly designed for the supply of modern d.c motors, without a commutator, also known under the designation of *static converter synchronous motors*. Voltage controlled electric drives confer an economic speed control over a wide enough speed range ($\Delta n = 1 : 200$) and yet they are used with turbopump drives only in special cases, because they are expensive and, furthermore, are difficult to operate (the last drawback refers to d.c. commutator motors).

The high cost of these driving machines is due both to the d.c. motor which is more expensive than the a.c. one, and to the static converter. Indeed, these static converters, through which the motors are supplied with a variable voltage, have to be designed for the whole power of the driving motor, whereas with static converter asynchronous cascades, the converter is designed only for the slip power, which represents some 15% of the driving motor rated power.

Since turbopumps are generally characterized as working machines, requiring comparatively only narrow ranges of speed control, the use of d.c. motors that offer wide speed control ranges as drives for such machines is hardly a wise proposition. However, there are certain very special cases when the use of such drives is justified, for instance with recycling pumps in nuclear stations, or with feed pumps of steam generators. As may be inferred, such cases imply either adverse environmental conditions, or wide speed control ranges as a consequence of the fact that turbopumps are operating in pumping systems without a static head (i.e. with no back pressure). Another special case worth mentioning is that of turbopumps of high and very high power (over 8,000 kW) and generally of a high speed too (3,000 rpm), when d.c. motors without a commutator (static converter synchronous motors) represent the sole possible solution to the problem, because in such conditions only synchronous motors can be used.

For turbopump drives, the voltage sources mentioned above can serve only to control the armature of d.c. motors, the control action being achieved only by changing the voltage applied across the armature while keeping the excitation current at a constant value. This control procedure is preferred with turbopumps, because these are machines with a parabolic mechanical characteristic (i.e. their resistive torque varies in direct proportion to the square of the speed) and consequently they require the highest torque at the highest speed.

The converters, used to supply a variable voltage to d.c. motors, are converters with external (natural) commutation. Their basic function is to convert the electric energy, of a variety of kinds and parameters, by means of natural commutation. This external commutation can be achieved either from the main supply or from the load (the motor), and hence converters are classified in two groups, either as phase-controlled converters (in rectifier or inverter duty) or as load-controlled converters (inverter duty). The adjustable drives of d.c. motors were developed along these general lines.

10.1 Adjustable Electric Drive Systems, Using Commutator D.C. Motors, Supplied by Phase-Controlled Converters

10.1.1 Operation Principles of Phase-Controlled Converters (Controlled Rectifiers)

Single-phase full-wave rectifier (with centre-tap). One of the most representative systems is the two-pulse (full wave) centre-tap rectifier (Fig. 10.1), which is generally used with low powers and voltages. The rectifier arrangement includes two thyristors in the secondary of a centre-tapped single-phase transformer, while the load is connected between the central tap and the thyristor cathodes.

We shall now explain the operation of this arrangement by asuming that the load circuit has a strongly inductive character ($L = \infty$), thus permitting a continuously rectified current. The waveforms of currents and voltages for various control angles are shown in Fig. 10.2. As can be seen in Fig. 10.2, *a*, each thyristor is triggerred on when the crossing voltage becomes positive in sign; in such a case, the controlled thyristor operates as a non-controlled one, made up of diodes. The thyristors are triggered off when the voltage passes through zero. Each thyristor conducts

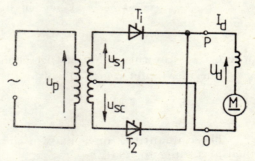

Fig. 10.1. Diagrammatic representation of the controlled rectifier with centre tap.

during one half-oscillation of the secondary alternating voltage, so that the rectified voltage has two pulses during each period. The rectified voltage mean value (U_d) is maximum. The current in the transformer primary has a square shape, while its fundamental harmonic I_{1p} is in phase with the primary voltage, U_p. The wave forms for a control angle from 0° to 90°

are shown in Fig. 10.2, *b*. The ignition pulse of each thyristor is phase-shifted (delayed) by an angle α with respect to the point where the secondary voltage passes through zero in the positive direction. As can be seen, the mean value of the rectified voltage decreases with increase in the control angle, while the phase shift between the fundamental harmonic I_{1p} of the primary current and of the voltage becomes larger. When the control angle reaches 90°, the rectified voltage mean value, U, becomes nil. When the control angle exceeds 90° (Fig. 10.2, *c*), the rectified voltage mean value becomes negative. In such a case, the rectifier can operate as an inverter, provided the load circuit includes a d.c. motor, operating as a generator with an electromotive voltage, higher than the mean value of the rectified voltage.

We shall now estimate a few basic parameters, as a tentative design analysis of the rectifier components [10.5]. To this end, let us consider a transformer ratio of unity (hence $U_p = U_s$).

The rms value of the secondary current is

$$I_s = \sqrt{\frac{1}{T} \int_0^{T/2} I_d^2 \, dt} = \frac{I_d}{\sqrt{2}} \qquad (10.1)$$

where I_d is the value of the rectified d.c. current.

The rms value of the primary current is

$$I_p = I_d \qquad (10.2)$$

The mean value of the rectified voltage is

$$U_d = \frac{1}{\pi} \int_{-\frac{\pi}{2}+\alpha}^{\frac{\pi}{2}+\alpha} U_{s\,max} \cos \omega t \, d\omega t = \frac{2\sqrt{2}}{\pi} U_s \cos \alpha \qquad (10.3)$$

where U_s is the rms value of the secondary voltage.

The maximum rectified voltage is

$$U_{do} = \frac{2\sqrt{2}}{\pi} U_s \qquad (10.4)$$

These quantities once known, the transformer output can be determined according to relation

$$P_{TR} = \frac{1}{2}(I_p U_p + 2 I_s U_s) = 1.34 \, U_{do} I_{dR} = 1.34 \, P_{do} \qquad (10.5)$$

where P_{do} is the actual transformer output. It will be seen that the transformer output should be 34% higher, this reduction in output being due to the fact that, at any given moment in time, only one half of the transformer winding is operative.

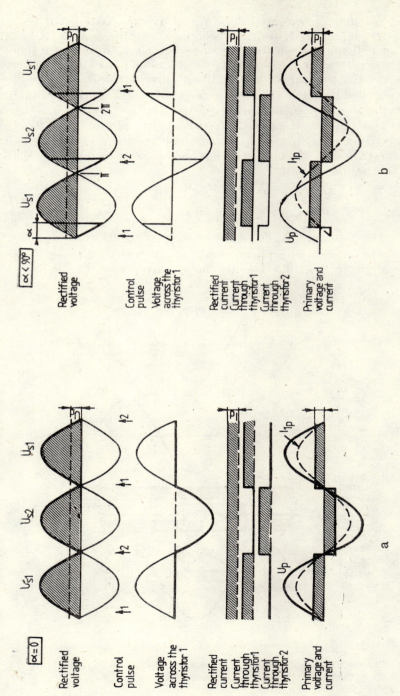

Fig. 10.2, *a* and *b*. Ideal waveforms for the currents and voltage of the centre top rectifier at various control angles: *a* — $\alpha = 0$; *b* — $\alpha < 90°$.

Actual rectifier operation is determined by the existence of the leakage reactance as referred to the secondary winding ($X_s + X'_p = X_k$), this reactance being responsible for the fact that the current changeover from one thyristor to another cannot take place instantaneously. The angle, needed to transfer the rectified current from one thyristor to the other, is called *the angle of anodic overlapping* (γ). Fig. 10.2, *c* illustrates the commutation phenomenon, that is, the change over of the current from I_d to zero, in the triggerred-off thyristor, and from zero to I_d, in the triggerred-on thyristor. Thus, a common circuit is established, built up out of the two arms of the rectifier. These current changes bring about a drop in the inductive voltage

$$u_k = L_k \frac{di_k}{dt} = X_k \frac{di_k}{d\omega t} = u_{sc} \frac{U_s}{I_{dR}} \frac{di_k}{d\omega t} \tag{10.6}$$

where u_{sc} is the short-circuit voltage of the transformer.

The anodic overlapping phenomenon is illustrated, for four values of the control angle, in Fig. 10.3. During the overlap period, the value of the

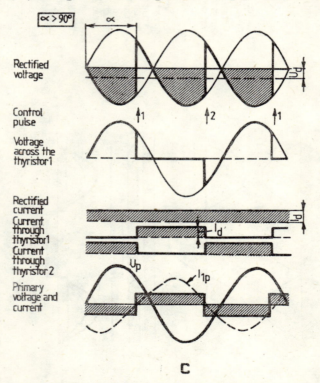

Fig. 10.2, *c*. Ideal waveforms for the currents and voltages of the centre tap rectifier at various control angles:
$\alpha > 90°$ (*c*).

rectified voltage is $u_d = \frac{1}{2}(u_{s1} + u_{s2})$ and for this reason the rectified mean voltage is lower. The anodic overlapping surface ψ_k is found with relation (10.6), thus

$$\psi_k = \int_\alpha^{\alpha+\gamma} u_k \, d\omega t = \sqrt{2U_s} \int_\alpha^{\alpha+\gamma} \sin \omega t \, d\omega t = u_{sc} \frac{I_d}{I_{dR}} U_s \qquad (10.7)$$

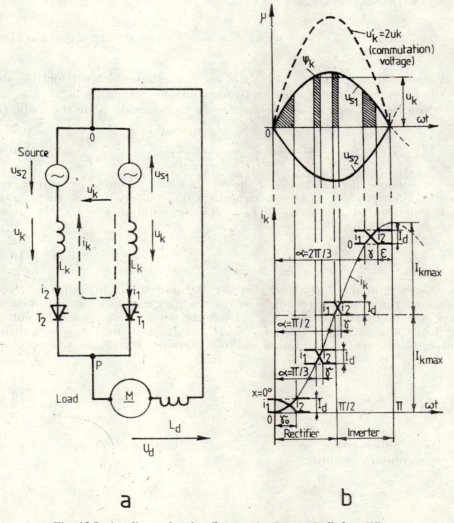

Fig. 10.3. Anodic overlapping for a centre tap controlled rectifier:
a — commutation wiring; b — commutation voltage and current for four values of the control angle.

Integrating gives

$$\cos \alpha - \cos(\alpha + \gamma) = K_x u_{sc} \frac{I_d}{I_{dR}} \qquad (10.8)$$

where $K_x = \dfrac{1}{\sqrt{2}}$ is a coefficient which depends on the rectifier, while the remainder of the expression is generally common to all typical rectifiers.

For the real rectifier, the mean value of the rectified voltage is estimated with relation

$$U_d = \frac{\sqrt{2}\,U_s}{\pi} \int_{\alpha+\gamma}^{\alpha+180°} \sin \omega t \, d\omega t = \frac{U_{do}}{2}[\cos(\alpha+\gamma)+\cos\alpha] \qquad (10.9)$$

and the resulting value is always lower than that obtained for the ideal rectifier, as determined with relation (10.3).

Now let us define a fractional ratio for the drop in the inductive voltage, thus

$$d_x = \frac{U_{do} - U_d}{U_{do}} = \frac{\psi_k}{\pi U_{do}} = \frac{u_{sc} I_d}{2\sqrt{2}\, I_{dR}} = u_{sc}\frac{K_x}{2}\frac{I_d}{I_{dR}} \qquad (10.10)$$

We choose, for the design of the rectifier components, the least favourable moment, which is the starting of a d.c. motor, where the starting current is limited at a value of $I_{d\,max}$. Then we define a new ratio for the voltage drop across load resistor (R_d)

$$d_{RR} = \frac{R_d I_d}{U_{dR}} \qquad (10.11)$$

Bearing in mind that on starting the motor $E = 0$ and $I_d = I_{d\,max}$, it follows that the control angle of the rectifier cannot be zero but should have a minimum value of α_{min}. Assuming that on starting, the voltage drop in the a.c. network is $\Delta U_p = \Delta U_s$ and using the ratios established above, the following maximum value U_{do} results for the rectified voltage

$$U_{do} = \frac{U_{dR}\left(1 + d_{RR}\dfrac{I_{dmax} - I_{dR}}{I_{dR}}\right)}{\dfrac{U_{sR} - sU_s}{U_{sn}}\cos\alpha_{min} - d_{xR}\dfrac{I_{d\,max}}{I_{dR}}} \qquad (10.12)$$

With this relation, we can now find the voltage developped in the rectifier secondary winding

$$U_s = \frac{\pi}{2\sqrt{2}} U_{do} \qquad (10.13)$$

The above relations hold true for the design analysis of other typical rectifiers as well.

An important feature of the converter is *the control characteristic*, that is, the change in the rectified mean voltage as a function of the thyristor ignition control angle (Fig. 10.4). Using relation (10.3) as well as the considerations in Fig. 10.4, we conclude that for $0 < \alpha < 90°$ $U_d/U_{do} > 0$, the converter operates as a *rectifier*, while for $90° < \alpha < 180°$, $U_d/U_{do} < 0$,

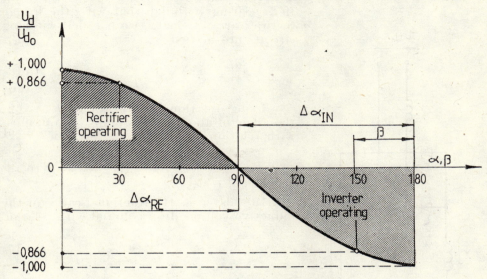

Fig. 10.4. Control characteristic of the phase-controlled converter.

the converter can operate as an *inverter*. Bearing in mind the existence of the anodic overlapping phenomenon, the rectifier control range is reduced to $180° - \beta$, where β is a spare angle needed to switch the current between the thyristors; the value of this angle is controlled by the value of the transformer leakage reactance. For all practical purposes, one takes $\beta \approx 20-30°$.

The single phase full-wave bridge rectifier. The drawback, represented by the centre tap in the transformer secondary winding of the rectifier circuit shown in Fig. 10.3, can be eliminated by using a bridge arrangement, as shown in Fig. 10.5. In such a case, the transformer is better used, because its secondary winding is fully effective (under load) during both half-cycles of the secondary voltage. The transformer rated power is found in a way similar to that used for the centre tap rectifier, i.e.

$$P_{TR} = \frac{\pi}{2\sqrt{2}} U_0 I_{dR} = 1.11 \ U_{do} I_{dR} = K_{TR} U_{dc} I_{dR} \qquad (10.14)$$

Here it can be seen that for the same useful power $U_{do}I_{dR}$, the transformer used with the bridge rectifier is only 11% bigger whereas for the centre tap rectifier, the transformer is 34% bigger (see relation 10.5).

According to the single-phase full-wave bridge arrangement, two opposite thyristors are conducting during each half-period. Current changeover from one group of thyristors to the other (T_2, $T_3 \rightarrow T_1$, T_4) is accompanied by the anodic overlapping phenomenon. The voltage area corresponding to anodic overlapping ψ_k is established as for the centre tap rectifier, bearing in mind that with the bridge rectifier arrangement, two parallel switching circuits are needed

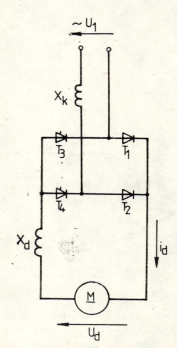

Fig. 10.5. Diagrammatic representation of the single-phase bridge rectifier.

$$\psi_k = 2\, u_{sc} U_s \frac{I_d}{I_{dR}} \qquad (10.15)$$

Using relation (10.8), where $K_x = \sqrt{2}$, the angle of anodic overlapping is obtained with relation

$$1 - \cos\gamma = \sqrt{2}\, u_{sc} \frac{I_d}{I_{dR}} \qquad (10.16)$$

In this case, the fractional ratio for the inductive voltage drop is found with relation

$$d_x = \frac{U_{d0} - U_d}{U_{d0}} = \frac{\psi_k}{\pi U_{d0}} = \frac{1}{\sqrt{2}} u_{ss} \frac{I_d}{I_{dR}} \qquad (10.17)$$

It should be mentioned that for the single-phase full-wave bridge arrangement, relations (10.12) and (10.13) are still valid.

Looking at the voltage and current wave forms, as established for the single-phase full-wave bridge rectifier arrangement, one can see that they are most distorted for a control angle of $\alpha = 90°$.

Now, if the armature of the d.c. motor is connected in the external circuit of the rectifier, the only electrically useful quantities will be the rectified mean voltage U_d and the rectified mean current I_d, which are responsible for the voltage balance and for the electromagnetic torque in the armature circuit. Because the rectified voltage does not have a continuous waveform, the rectified current has a pulsating character, that is, in addition to the continuous component I_d, it has an alternating component $i_{d\sim}$ (Fig. 10.6). This component, which brings no contribution to the torque development, must be damped out, and hence a smoothing coil is introduced in the armature circuit. Such a situation is illustrated in Fig. 10.6 for a control angle $\alpha = 30°$. Figure 10.6, c represents the waveform of the alternating component, $i_{d\sim}$, while Figure 10.6, d, shows the full rectified current.

Two solutions can be applied for the design of the smoothing coil:

(1) *Restrict the ripples of the rectified current to a given value* $I_{d\sim} = K_{1m} I_{dR}$, where $I_{d\sim}$ is the rms value of the alternating component. This

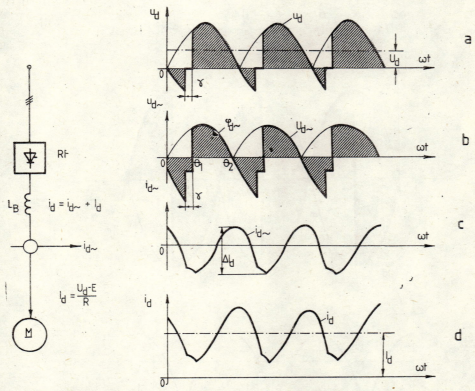

Fig. 10.6. Symbolic representation and the d.c. and a.c. components of the rectified voltage and current.

restriction is imposed to avoid thermally loading the motor and hence impairing the commutation phenomenon.

(2) *Restrict the mean value of the rectified current to a minimum value*, equal to the no-load(idle) current of the d.c. motor. This is suggested because on no-load running, I_d being of a low value, the rectifier can be operated in interrupted conductive conditions (Fig. 10.6, d).

The limiting case of rectifier operation, according to the two solutions above, is illustrated in Fig. 10.7. For $\alpha = 90°$, the voltage surface ψ_d is found with relations

$$\psi_d = \int_{\theta_1}^{\theta_2} U_{d\sim} \, d\omega t \qquad (10.18)$$

where $U_{d\sim}$ is the alternating (ripple) component of the rectified voltage, equal to the actual rectified voltage at $\alpha = 90°$, because $U_d = 0$. This component is given by relation

$$U_{d\sim} = L_{B_1} \frac{di_{d\sim}}{dt} = X_{b1} \frac{di_{d\sim}}{d\omega t} \tag{10.19}$$

where L_{B_1} is the inductance of the smoothing coil.

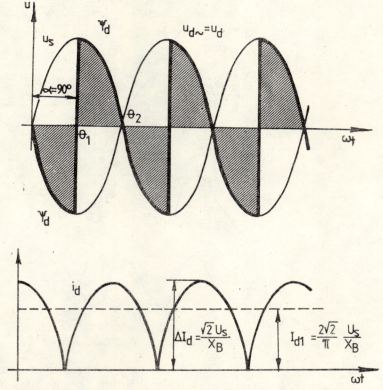

Fig. 10.7. Design analysis of the smoothing coil for the limit case of rectifier operation ($\alpha = 90°$).

Applying this relation to the alternating (ripple) component of the rectified current results in

$$i_{d\sim} = \frac{1}{X_{B_1}} \int U_{d\sim} \, d\omega t \tag{10.20}$$

from which the maximum value of the rectified current can be derived by means of Fig. 10.7 and relation (10.18)

$$\Delta I_d = \frac{1}{X_{B_1}} \int_{\theta_1}^{\theta_2} U_{d\sim} \, d\omega t = \frac{\psi_d}{X_B} \tag{10.21}$$

According to *the first solution* applied in the smoothing coil design analysis, the e.m.f. value of the alternating component of the rectified current is computed for the limit situation shown in Fig. 10.7

$$I_d = \Delta I_d \sqrt{\frac{1}{\pi}\int_0^\pi \left(\sin \omega t - \frac{2}{\pi}\right)^2 d\omega t} = 0.32\, \Delta I_d = 0.32\, \frac{\psi_d}{X_B} \tag{10.22}$$

from which the value of the smoothing coil inductance follows

$$L_{B_1} = 1.6\, \frac{U_{do}}{I_{d\sim}} = \frac{1.6}{K_{1m}}\, \frac{U_{do}}{I_{dR}} = \frac{K_{B1} U_{do}}{K_{1m} I_{dR}} \tag{10.23}$$

where

$$K_{1m} = \frac{I_{d\sim}}{I_{dR}}$$

and the quantities are expressed thus

$$L_{B_1}\, [\text{mH}];\ U_{do}\, [\text{V}];\ I_{dn}\, [\text{A}]$$

According to *the second solution* of the smoothing coil design analysis, the mean value of the alternating component of the rectified current is estimated, knowing that for $\alpha = 90°$, $\psi_d = \psi_{d\,max} = 0.5\, \pi U_{do}$ (relation 10.18)

$$I_{d1} = \frac{2}{\pi}\, \frac{\psi_{d\,max}}{X_B} \int_0^{\pi/2} \cos \omega t \cdot d\omega t = \frac{U_{dR}}{X_1} \tag{10.24}$$

from which the smoothing coil inductance results

$$L_{B_2} = 3.18\, \frac{U_{do}}{I_{d1}} = \frac{3.18\, U_{do}}{K_{2m} I_{dR}} = \frac{K_{B_2} U_{do}}{K_{2m} I_{dR}} \tag{10.25}$$

where

$$K_{2m} = \frac{I_{d1}}{I_{dR}}$$

and the quantities are expressed thus

$$L_{B_2}\, [\text{mH}];\ U_{do}\, [\text{V}];\ I_{dn}\, [\text{A}]$$

Design analysis computations for the smoothing coil according to the two solutions described above generally lead to $L_{B1} < L_{B2}$; in practice, however, both values are calculated and the higher is chosen.

The relations established above for the smoothing coil apply also for other typical rectifiers, with change of coefficients K_{B1} or K_{B2}, as the case may be, whose values are tabulated in Table 10.1. The same table gives the characteristic constants for other typical rectifiers as well.

TABLE 10.1 Characteristic coefficients K_{B_1} and K_{B_2} for the typical rectifiers

Characteristic coefficients	Single-phase rectifier with nill	Single-phase bridge rectifier	Single-phase half-wave bridge	Three-phase bridge with nill	Three-phase bridge rectifier	Three-phase half-wave bridge
K_{B_1}	1.60	1.60	1.05	0.58	0.13	0.32
K_{B_2}	3.18	3.18	1.79	1.25	0.30	0.76

10.1.2 Behaviour of D.C. Motors Supplied by Phase-Controlled Converters

D.C. motors, used as drives in semiconductor rectifying arrangements, have some operational peculiarities that are reflected in their design, owing to the ripple component of the voltage and current. These peculiarities will now be reviewed.

The voltage u_A, applied across the rotor circuit as taken from the output of a semiconductor rectifier, is not constant in time, even in stationary conditions, but pulsates around the mean value U_A. The rotor winding has a certain resistance R_A and a self inductance L_{AA} (which actually includes that of the auxiliary polar windings and the magnetic reaction compensating windings, in series with the rotor winding). It was found that the natural time constant of the rotor circuit L_{AA}/R_A is comparatively very low, since current i_a shows comparatively high pulses around the mean value I_A. Moreover, with a low I_A, the phenomenon of an interrupted current is apparent at low speeds, which leads to even higher pulses around the mean value. To limit these pulses and restrict as much as possible the area of the interrupted current in the mechanical characteristic of the motor, induction or smoothing coils are introduced in the rotor circuit to provide additional inductance L and thus to increase artificially the rotor time constant $\tau_A = (L_{AA} + L)/R_A$. It was shown that, if the value of the total inductance or the rotor circuit is practically infinite, current i_A is practically free of pulses and hence it is constant in time. Of course, a coil with a very high inductance L is not only bulky, but also very expensive. For the sake of economy, a certain oscillation of i_A around the mean value is therefore allowed in practice. The instantaneous current will be

$$i_A(t) = I_A + \sum_{\nu=1}^{\infty} I_{Am}^{(\nu)} \sin(\nu m \omega t + \varphi_\nu)$$

where ν represents the harmonic order; m — the number of pulses of rectified voltage u_A per cycle of the a.c. network;

ω — the oscillation frequency of this network; $I_{Am}^{(\nu)}$ — harmonic ν amplitude;

φ_ν — original phase of harmonic ν. Current i_A is represented in Fig. 10.8.

The modulation coefficient of current i_a is the quantity

$$w = \frac{\sqrt{\frac{1}{2} \sum_{\nu=1}^{\infty} I_{Am}^{(\nu)2}}}{I_A}$$

that is, the ratio of the rms value of the a.c. current superimposed on the d.c. component I_A, to the latter component. This modulation coef-

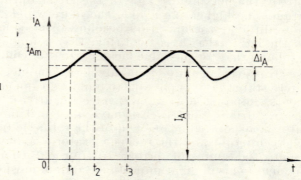

Fig. 10.8. Pulses of the rectified current.

ficient **w** is a measure of the i_A oscillation. Unfortunately, even with the same rectifier, this coefficient varies as a function of the lag in the thyristor ignition time and of the change in the additional inductance L. Of course, with an increased number of pulses m in the a.c. network that supplies the rectifier, yet keeping the same I_A, the current modulation coefficient **w** will be lower. This is why, particularly where higher powers are expected, the three-phase bridges ($m = 6$) are preferred to the single-phase bridge ($m = 2$). The explanation is simple enough, for with higher m, the frequency oscillation of the first harmonic of current i_0 is greater, and so is the coil reactance $m\omega L$; hence amplitude $I_{Am}^{(1)}$ will be lower for the same amplitude of the applied voltage $U_A^{(1)} m$ (and furthermore, with higher m, $U_{Am}^{(1)}$ drops too). As a general rule, the modulation coefficient w, allowed at the rated load (U_{An}, I_{An}), falls between 0.10 and 0.20.

A first result of oscillatory current i_A is to increase the Joule losses in the rotor loop

$$P_j = R_A I_{Aef}^2 = R_A I_A^2 (1 + w^2)$$

while the additional losses are proportional to w^2. Actually, the losses by Joule effect are increased also by the eddy currents, generated in the leads of the notches by i_a current harmonics.

Another consequence of oscillatory currrent i_A is the increase in the commutation current. At the commutator end, the commutation frequency is much higher than the current oscillation frequency. Therefore, during commutation, in any section of the rotor winding, the current varies within limits that differ according to the moment when commutation takes place. Referring to Fig. 10.8 for instance, in the commutation section, the current around time t_1 will change from $+I_A$ to $-I_A$. For the section where commutation takes places around time t_2, the current will change from $+I_{Am}$ to $-I_{Am}$, the latter being the maximum value of current i_A. And since $I_{Am} > I_A$, in the section where commutation takes place around time t_2, this operation will be more difficult and the self induction opposing electromotive voltage will have a higher mean value, as will the opposing electromotive voltage induced by the reaction field. Commutation is therefore affected not by the mean value I_A of current i_A, but by the maximum value of the latter during one period of its oscillation. The higher the ripple, the more will I_{Am} exceed I_A, and hence the more difficult commutation becomes. And, since the ripple depends both on the lag in the thyristor firing, that is angle α, and on the number m, it follows that for the same mean current I_A commutation will vary with α.

However, in spite of all adverse phenomena decribed above, one can safely state that it is possible to build a d.c. machine to meet all the requirements imposed by modern electric drives. Thus, for automated d.c. electric drives, constructors have designed motors with the tachogenerator incorporated at one end of the rotor shaft. In this way, the coupling between motor and the tachogenerator is avoided, such a coupling being a potential source of errors. In transient conditions, the motor inertial moment (J) is of great importance and, in most cases, represents the major inertial component of a drive system. To reduce this moment, the constructors have changed the classic geometry of electric machines in favour of machines of a smaller diameter but greater length (the moment of inertia J is a function of the fourth power of the diameter and only of the first power of length). Owing to these operational peculiarities, nowadays d.c. motors are specially built to meet the demands of the modern automatically controlled electric drives.

10.1.3 Speed Control of Commutator D.C. Motors by Phase-Controlled Converters

Figure 10.9 shows the principles of a d.c. motor drive with the commutator fed from a thyristor controlled rectifier, while in Fig. 10.10 the most frequently used arrangement for high-power adjustable drives of turbopumps employing a rectifier is shown, i.e. the three-phase bridge. This arrangement includes six thyristors that are fired by pulses recurring at $\pi/3$ radian phase intervals in the following sequence: a_1, c_2, a_2, c_1, b_2, b_1. One thyristor in the left-hand group and one in the right-hand group are always conducting at the same time (of course, from different phases). In Fig. 10.10, thyristors a_1 and b_2 are simultaneously conducting. Voltage u_A across the motor terminals has the same value as the secondary voltage across phases $u_{ab} = u_a - u_b$, until the next thyristor is fired, that is c_2, provided the current has no interruptions. At this moment, voltage $u_{ac} = u_a - u_c$ being increased, the current changes direction and flows through thyristors a_1 and c_2; voltage u_A becomes equal to u_{ac}. Then thyristor b_1 is fired and takes over the current from thyristor a_1 of the same

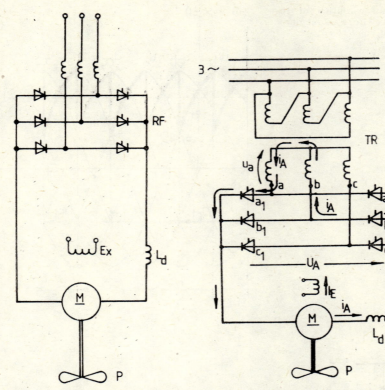

Fig. 10.9. Electrical diagram of a d.c. motor with the commutator fed from a thyristor controlled rectifier.

Fig. 10.10. Electrical diagram of the three-phase converter in a bridge arrangement.

group; voltage u_A will be equal to u_{bc}, while the current is closed through thyristor and so on.

In Fig. 10.11 it is assumed that the ignition lag angle α is 45°. Thyristor a_1 is shown at time $\alpha = 0$, which is coincident with the moment when voltage u_{ab} is higher than other voltages between phases. This is taken as the reference time for the ignition lag of thyristor a_1.

The automatic speed control arrangement, using a d.c. motor with a commutator fed through a three-phase bridge controlled rectifier, is shown in Fig. 10.12. As can be seen, this arrangement uses two cascade controllers, R_n for speed, and R_i for current, as well as a d.c. motor with a classic commutator with separate excitation. The motor drives turbopump P, speed n is measured by means of the speed transducer T_n (speed-voltage generator), while current i_a in the rotor circuit is measured by a shunt resistor f.

The voltage u_{in}, corresponding to the rotational speed required, as obtained across the potentiometer P_{ot} output, is compared with voltage u_{rn}

513

Speed Control by D. C. Motors

Fig. 10.11. Time variation of the rectified voltage in a three-phase bridge converter.

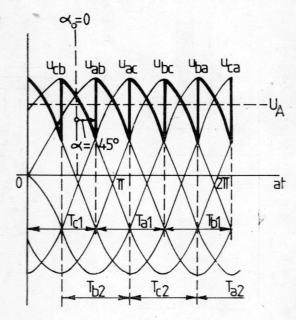

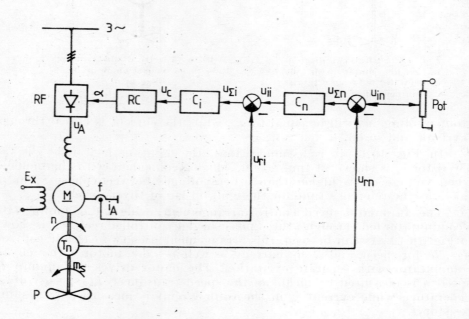

Fig. 10.12. Automatic speed control using a d.c. motor with a commutator fed through a three-phase bridge controlled rectifier.

across the speed-indicating generator, whereas at the input of current controller R_i, the output voltage u_{ii} of the speed controller is compared with the feedback voltage u_{ri}.

The arrangement described above is called a multiple-loop control system or cascade arrangement, and uses one controller for each variable. The main variable n is controlled through an external loop, and the secondary variable i_A through an internal loop (see Fig. 10.12). The speed controller output represents the reference quantity for the current controller. The internal loop is chosen to be at least twice faster than the external one, so that a dynamical separation of the two loops is achieved, which particularly facilitates the controller design analysis and optimization, and hence the improvement in the performance of the arrangement as a whole. This dynamical separation permits the secondary quantity to be limited during the transient conditions by using the "saturation" of the external loop controller.

Let us consider in more detail this "saturation" phenomenon by choosing as transient conditions the starting in no-load conditions of the motor, which subsequently reaches a given speed. We further assume that a comparatively high level is imposed on u_{in} at the speed controller input, when the motor is idle. Since $n = 0$, the feedback voltage $u_{rn} = 0$ and hence $u_{\Sigma n} = u_{in} - u_{rn}$ is comparatively high, which saturates controller R_n and generates at its output a signal u_{ii}, whose value is prescribed at the design stage. This value is kept constant even during starting, until the speed controller is no longer saturated and starts feeding back, when its output is not constant any more. As long as voltage u_{ii} across the current controller input is kept constant at the saturation level, the operation of the internal current loop is completely independent. Like any other automated control system, the internal loop will feed back in an attempt to bring the controlled variable (current i_A) to, and keep it at, the level imposed by the saturated voltage u_{ii}, so designed as to correspond to the limit current allowed in the rotor circuit. Therefore, starting takes place with a practically limited current i_A as shown in Fig. 10.13, a, when the limit current is $3\,I_{An}$. For comparison purposes, on the same figure one can see current i_A in the case of a natural starting, with $u_A = U_{An}$, with no current limitation. It will be seen that with natural starting, the current shock reaches the approximate value of $9\,I_{An}$, which can be very dangerous for the machine, particularly because of the commutation phenomenon.

The change in the rotational speed of the automated system on no-load starting is shown in Fig. 10.13, b. As long as current i_A is practically constant during the transient conditions prevailing during the start-up period, speed n will increase linearly with time, according to the motion equation

$$M_{EA} I_A i_A = J \frac{dn}{dt} = \text{constant}$$

(the load torque being assumed as nil $m_s = 0$).

Now, when speed n reaches a certain level during the run-up period, the feedback voltage u_{rn} will be high enough and close to u_{in}, so that controller R_n no longer saturates, its output voltage will drop and eventually will reach zero when $u_{in} = u_{rn}$, and the current i_A will tend towards zero (Fig. 10.13, b) and the speed towards the set value n_i. Also on Fig. 10.13, b) one can see, for comparison purposes, how the speed varies during natural starting, with $u_A = U_{An} =$ constant and $m_s = 0$. The areas

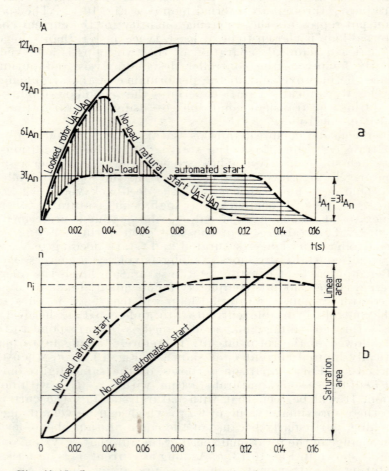

Fig. 10.13. Current and rotational speed (a and b) change during starting, with and without intervention of the automated system.

shaded (with vertical and horizontal lines) should be equal, since from the motion equation it follows that

$$M_{EA} I_E \int_0^{t_p} i_A \, dt = j \int_0^{t_p} dn = j n_i$$

that is, the area limited by curve $i_A(t)$ and the time axis is the same, whether the automated or natural arrangement is used for the no-load starting. Hence the important conclusion that the start-up time (the time response for any disturbance in the system) will be the lower, the higher the allowed current limit and, of course, the lower the inertial moment J.

10.2 Adjustable Drives with D.C.Motors, without a Commutator Supplied by Load-Controlled Converters

Operation principle. On such drives, the mechanical rotary commutator of the d.c. machine is replaced by an electronic arragement, of which the main component is the static converter. This machine-converter assembly is called in the literature a *static converter synchronous motor*, or a *d.c. motor with electronic commutator*. Being a synchronous motor, fed through a static converter, the machine takes on the operating properties of a d.c. machine since the commutation energy is supplied to the converter by the motor itself. At the same time, the synchronous machine also supplies the converter with current pulses. The trigger-pulses are generated by a position generator, coupled to the machine rotor. This generator is a probe (FET or Hall) responsive to the magnetic field generated by permanent magnets, or may feature an optoelectronic positional transducer. The maximum width allowed for the trigger-pulses from the generator is limited by the inverter duty threshold (which is, as for the phase-controlled converters, about 150° phase angle).

From an operational aspect, the unit built up from the armature of the synchronous machine and the converter, controlled as a function of the machine shaft position, corresponds to the unit built up with the d.c. machine armature and the commutator. Indeed, they show the same dynamical behaviour, and hence the risks of the machine going into oscillation (which is characteristic of the synchronous machine) or going outside the rated operational conditions in case of overloads, are eliminated. From a structural point of view, the d.c. motor without commutator and fed by a load-controlled converter is, in fact, a synchronous machine, whereas from an operational point of view it should be considered as a d.c. machine with an electronic commutator.

The static converter. As a general rule with external (natural) commutation converters, the reactive power needed for commutation is supplied from the mains supply, and for this reason such converters are also called "phase-controlled converters". However, the reactive power of commutation can be supplied from the motor as well, provided certain conditions are fulfilled, so we can speak of motor-controlled converters or load-controlled converters. The condition for this kind of commutation, i.e. from the motor (load), is that the load current must have a capacitive component. This condition can be met by an over-excited synchronous machine.

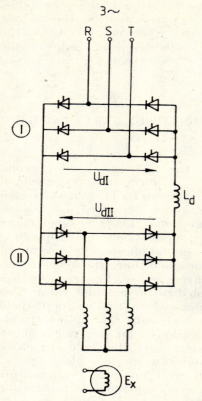

Fig. 10.14. Diagram of the load-controlled converter with an over-excited synchronous motor.

Let us now explain the structure of a load-controlled converter (or inverter). With these converters, the commutation power is obtained from an adequately excited synchronous machine. Such arrangements also allow reverse energy transfer. Combining a phase-controlled rectifier with a load-controlled inverter results in a motor-driven converter (Fig. 10.14). As a general rule, the d.c. circuit includes an energy storing component (coil) to separate the phase-controlled rectifier from the load (motor)-controlled inverter. Converter I, when actuating the synchronous machine, operates as a phase-controlled rectifier. By changin its control angle, it generates the adjustable voltage U_{dI}. The current rectified by the intermediate circuit I_d is smoothed by coil L_d. Converter II works as a motor-controlled inverter and generates the d.c. voltage U_{dII}. In inverter duty, the mean value of this voltage is of negative sign. In stationary conditions, $U_{dII} = -U_{dI}$. When reversing the sign of energy transfer, converter II should operate as a rectifier, and converter I as an inverter. D.C. voltages U_{dII} and U_{dI} change their signs while the sign of current I_d stays unchanged, and the synchronous machine works like a generator. The synchronous machine can lead converter II only if its current has a capacitive component.

The phasor-diagram of the synchronous machine is shown in Fig. 10.15; in a generator mode, U_s is the stator voltage and I_s is the curent through the stator, leading U_s by an angle φ_s.

Speed control. The block diagram for the speed control of a d.c. motor without commutator, fed by a load-controlled converter, is shown in Fig. 10.16. Automatic speed control is achieved through converter I (rectifier) which is controlled from the mains (phase-control). The control diagram includes an electronic control arrangement, similar to that of the rectifier normally used to supply the classic commutator d.c. motor, and designed according to similar criteria (see Section 10.1.3). More often than not, the controlled variables are first the speed, then the current, the latter being also limited. Bearing in mind that with turbopumps the load torque varies parabolically (i.e. the load torque varies in direct propor-

Adjustable Drives with D.C. Motors

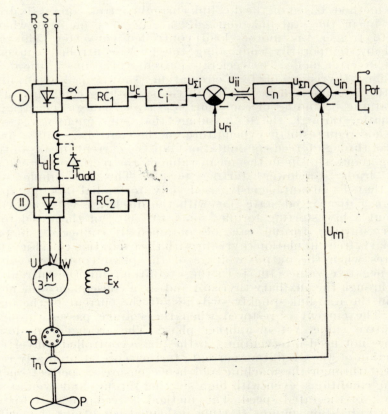

Fig. 10.15. Phasor diagram of the synchronous machine.

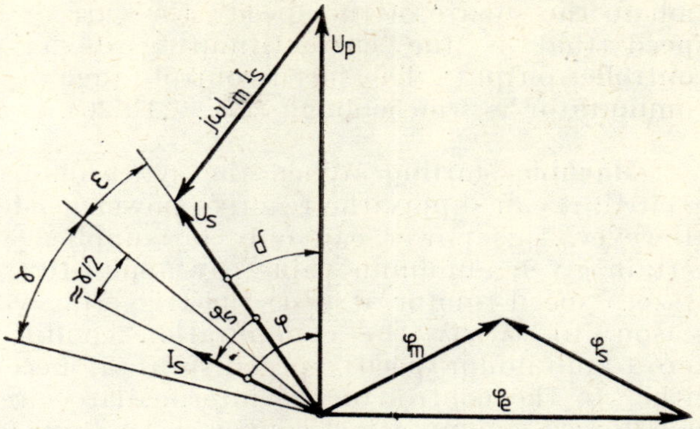

Fig. 10.16. Block-diagram for the speed control of a d.c. motor without commutator, fed from a load-controlled converter:

RC_1, RC_2 — ring counter; C_i — current controller; C_n — speed controller; P_{ot} — potentiometer; T_θ — angular transducer; T_n — speed-voltage generator; I — controller rectifier; II — inverter; M — over-excited synchronous motor; P — pump; E_x — excitation coil; T_{add} — stand-by free-flow thyristor.

tion to the square of the speed), the current is limited as a function of speed (that is the current-limiting device is placed at the speed controller output). The speed control range of a d.c. motor drive without commutator is wide enough ($\Delta n = 1 : 20$).

Machine starting. When the synchronous machine is slightly overexcited, it can supply the reactive power needed to control the converter. However, this power can only be supplied when the speed exceeds a certain given minimum value (by about 10% of the rated speed), since at zero speed (motor at rest) the triggering voltage is zero too. For this reason, to satisfy the commutation conditions during starting (from zero to minimum speed), special steps are necessary. These steps can refer either to the control of the intermediate circuit current, or to the use of a forced-commutation converter (except for the starting through a motor or transformer).

The method, based on the d.c. link current control, consists in dropping the current in the said link markedly at the very moment when it is desired to trigger the motor (load)-controlled converter. This can be achieved by temporarily controlling the phase-controlled converter in a perfect inverter mode. To accelerate current variations in the d.c. link, an auxiliary thyristor, of free circulation (shown in broken lines on Fig. 10.16), is connected in parallel to the smoothing coil; this thyristor is fired at the beginning of each current break (pause). The triggerring is automatic, through the d.c. voltage that subsequently rises again. This method requires little expenditure for the control circuits (as against the control through forced commutation) but the current breaks in the d.c. link bring about a drop in the mean value of the rectified current which, in turn, leads to a lower starting torque. This is a simple starting method that is advantageously used only for radial-flow pumps that start from a discharged state (i.e. with the discharge valve closed) and require but a low starting torque (some 40—50% of the rated torque). A better starting torque can be obtained by connecting in parallel the auxiliary, free-circulation thyristor with the excitation coil (Fig. 10.17,a) which fires when the output voltage of the phase-controlled converter shows a negative value [10.5]. In the excitation coil, the current keeps flowing through the auxiliary thyristor and as a consequence the negative voltage in the d.c. link quickly switches off the current in the machine winding. The current is restored when the voltage passes through the next positive value, yet in another phase. The arrangement described above does not need interventions in the phase-controlled converter, nor the reduction of the excitation current; furthermore, it promptly restores the current through the machine and hence ensures a perfect operation in starting conditions, even with high starting torques and even at speeds up to 10% of the rated speed. This method is recommended for drives used with radial-flow pumps starting in loaded conditions (i.e. with the discharge valve open), in which case a high starting torque is needed (some 90—110% of the rated torque).

Another method used to obtain an independent starting of the static converter motor is the temporary operation of the supply converter as an internal (forced) commutation converter, until the minimum required speed is reached. This can be achieved by using the forced-commutation converter, as shown in Fig. 10.17, b. Apart from the main thyristors $T_{11}-T_{16}$ and the attached extinction circuits built up with capacitors

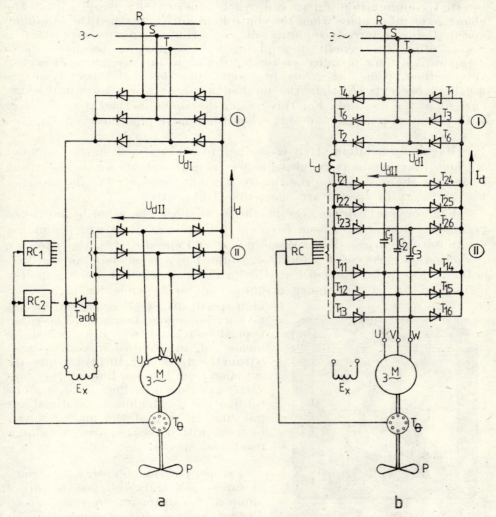

Fig. 10.17. Starting of the synchronous motor with a static converter:
a — arrangement including an auxiliary thyristor connected in parallel with the excitation winding; b — arrangement including extinction circuits built up with auxiliary capacitors and thyristors.

C_1, C_2, C_3, this converter also includes the blow-out auxiliary thyristors $T_{21}-T_{26}$, which all require expansion and, implicitly, more complexity of the control circuit. The starting trigger-frequency is also taken from a position generator or from an external generator, and rises from zero until the minimum speed is reached. When the motor reaches the speed at which the inverter current can be commutated from the machine voltage, the blow-out thyristors are locked off by the control pulses. During starting, the time duration of pulses sent by the position generator, in internal commutation duty, can reach almost 180° phase angle. The changeover takes place when the control circuit has reached the minimum speed. Using a forced commutation converter as an auxiliary starting means offers the benefit of applying a high starting torque with a comparatively low installed power for the blow-out thyristors. However, this method, while being advantageous from the load aspect, leads to much higher expense than the method using as a control variable the current of the d.c. link. For this reason the above method is rarely used with turbopump drives and only when these have high inertial moments.

Machine excitation. If it is desired to build synchronous machines without rotary contacts, requiring no servicing attendance for the excitation system in the case of the d.c. motor without commutator, the designer has to resort to special arrangements [10.1], [10.2].

For low power drives, the machines used are energized by means of permanent magnets, fed from frequency static converters with simplified commutation devices. For average power and speed drives, the machine is excited by means of an alternator, coupled to a semiconductor rectifier, both devices being fastened to the machine shaft. For average-power high-speed drives, reluctance machines are used, while for high-power high-speed drives, the present trend favours homopolar machines. These two typical machines are d.c.-excited by means of an excitation coil, placed within the stator (Fig. 10.18) and directly supplied from the d.c. link of the converter (as shown in Fig. 10.17, a). The rotor of such machines is solid and has no winding, so that the machines can be used with drives requiring very high rotational speeds.

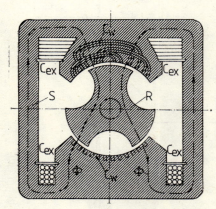

Fig. 10.18. Variable reluctance electric motor, d.c. — excited by means of a stator excitation winding:
S —stator; R —rotor; C_w — working winding; C_{ex} — excitation winding.

Scope. D.C. machines without commutator are mainly designed for turbopump drives operating in special conditions, and are recommended in all cases where the operating conditions, speed and power parameters restrict the use of classic commutator d.c. machines.

According to their use, namely according to the turbopump power, d.c. machines without a commutator can be classified into three characteristic groups.

522

(a) *For low power turbopump drives* (from 20 to 200 kW), operating in heavy environmental conditions (turbopumps used in chemical industry), one can use reluctance machines or those excited by permanent magnets, supplied from separate converters and fitted with simple commutation devices. A magneto-electric machine is simple enough; it has a stator winding, similar to that of a three-phase motor, while its rotor encloses only permanent magnets, where no losses are produced. Siemens, for instance, a well known constructor in F.R. of Germany, uses 12-pole permanent magnets for their motors with rated parameters speed — 3,000 rpm; powers — 7.5, 11, 15, 22 and 30 kW [10.2].

(b) *For medium power and speed turbopump drives* (from 500 to 1,500 kW), operating in special environments (e.g. cycling pumps in nuclear stations), one can use machines excited by an alternator and a rectifier, mounted on the rotor shaft. For such applications, Siemens has built two typical d.c. motors without a commutator (static converter motors), one for 660 kW and the other for 1,500 kW, both being used in experimental Na-cooling loops. Both designs proved to meet the requirements perfectly [10.1, 10.2].

(c) *For high and very high power turbopump drives* (from 2,000 to 20,000 kW), operating as a rule at very high speeds (in feed and extraction pumps, used with steam generators), the most suitable are homopolar motors with a rotor that has neither steel-sheets nor windings and permits very high speeds to be reached. In this field we can cite only a few research works aimed at two well-defined goals: (1) the development of multi-phase arrangements, whose number of phases is ever increasing and allows a considerable reduction in the torque ripple (a drawback that affects the static converter motor); (2) the development of high-voltage static converters that permit the use of high-voltage motors, as such motors are advantageous in view of their electromechanical structure.

REFERENCES

10.1. HABÖCK, A., KÖLLENSPERGER, D., *Développement du moteur à convertisseur statique*. Revue Siemens, No. 6, (1971).
10.2. HABÖCK, A., KÖLLENSPERGER, D., *Application et perfectionnement du moteur à convertisseur statique*, Revue Siemens, No. 6, 1971.
10.3. IONEL, I. I., *Acționarea electrică a turbomașinilor*. Technical Publishing House, Bucharest, 1980.
10.4. IONEL, I. I., *Instalații de pompare reglabile*. Technical Publishing House, Bucharest, 1980.
10.5. KELEMEN, A., *Acționări electrice*. Didactic and Pedagogic Publishing House, Bucharest, 1976.

11

Speed Control of Turbopumps, Run by Means of Stator-controlled Asynchronous Motors

The asynchronous motor has a hard (flat) speed-torque characteristic, like the d.c. shunt motor. In many applications, electric motors used as turbopump drives are required to change their mechanical characteristics within quite wide limits, and the control means need not be very accurate and should not be too expensive.

With the d.c. motor, change in excitation current represents an economic solution that can achieve a fine control action, within wide enough limits, provided the motor is adequately designed and built. Another efficient solution for the speed control problem is the more or less recently developed technique of power electronics, which offers comparatively easy solutions to the continuous change, within broad limits, of the supply voltage. Such means are not available with the asynchronous motor; indeed, with such motors it may be that the synchronism speed (that is, a direct function of the mains frequency) imposes an upper limit that can be exceeded only by resorting to special facilities, which are, of course, very expensive. This is why, even though the asynchronous motor is sturdy, cheap and easy to operate, the d.c. motor is supreme when the specifications impose a fine adjustment within a wide range of voltages, with no excessive sophistication nor too high costs.

With stator-controlled asynchronous motors, the speed control action is achieved by varying the motor supply voltage or by varying frequency of the supply source.

11.1 Adjustable Electric Drives by Varying the Supply Voltage

11.1.1. Speed Control through Variation of Stator Voltage

The phase rms voltage U_1 in the stator can be changed by using an adjustable three-phase autotransformer (Fig. 11.1, a). According to relation

$$T_{max} = \frac{3U_1^2}{2c\Omega_1[R_1 + \sqrt{R_1^2 + (X\sigma_{12} + cX'\sigma_{21})^2}]}$$

the maximum electromagnetic torque, T_{max}, depends on the squared value of the applied voltage, while the critical slip s_{max} is independent of U_1. Hence, if a squirrel-cage asynchronous motor is used, the family of mechanical characteristics shown in Fig. 11.1, b will be obtained, the parameter being U_1. Now, assuming that the working machine is a turbopump with a parabolic torque, it follows straightforwardly that speed can be varied within very narrow limits; furthermore, the flatness

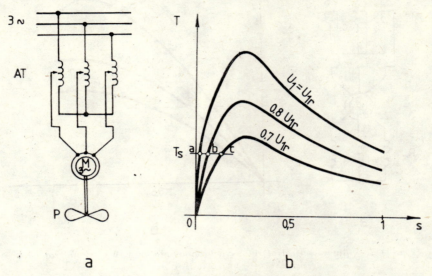

Fig. 11.1. Voltage control of the asynchronous motor with a short-circuited rotor:
a — electric wiring; b — speed-torque characteristics for various voltages.

of the speed-torque characteristic drops with decreasing U_1 and, more importantly, the rotor losses increase while the loading capacity falls quickly (owing to the reduction in the maximum torque T_{max}).

The speed control range can be broadened only on asynchronous motors with a phase-wound rotor, by introducing additional resistances in the rotor circuit (Fig. 11.2, a). In such a case, however, the critical slip s_{max} is considerably increased (Fig. 11.2, b) and the overall efficiency of the arrangement is even lower, but on the other hand the most important losses take place in the external rotor resistances.

The sign of the control action can only be negative, since voltage U_1 can only be decreased with respect to the rated voltage U_{1n}. The precision of the control action is higher when an autotransformer with a continuous voltage control action is used. Of course, this is much more expensive, and raises operational problems owing to the sliding contacts.

The power P_1 taken from the mains by the asynchronous motor, approximately equal in value to the electromagnetic power P (if one neglects the iron and Joule losses in the stator), is spent as mechanical power $(1-s)P$ and as slip power $P_s = sP$, that is lost in the rotor through the Joule effect. The amount of slip power depends on speed and reaches the maximum value of $P_{s\,max} = 0.15\,P_R$ for a speed $n = 0.66 n_0$ i.e. for a slip $s = 0.33\,s_R$; at $n = 0.66\,n_0$, the slip power drops (Fig. 11.3).

525

Speed Control of Stator Asynchron Motors

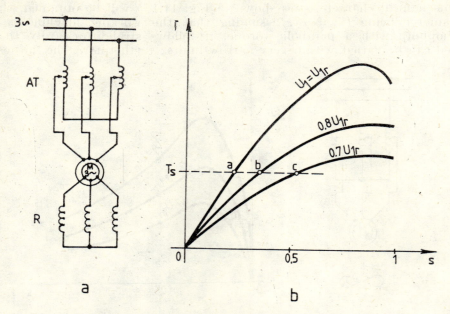

Fig. 11.2. Voltage control of the asynchronous motor with a wound rotor:
a — electrical diagram; *b* — speed-torque characteristics at various voltages.

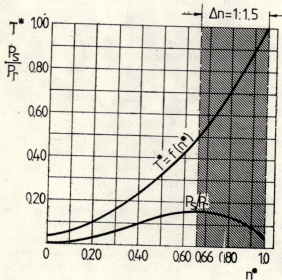

Fig. 11.3. Dependence of the motor torque $T^*(T/T_R)$ and of the P_s/P_R ratio, on the rotational speed $n^*(n/n_R)$.

While the slip power varies as shown above, the motor torque varies parabolically as a function of the supply voltage (see Fig. 11.3). Owing to this fact, below 66% of the synchronism speed, the torque developed by the synchronous motor is substantially reduced and, as a consequence, the speed control range is limited at a value $\Delta n \approx 1:1.5$ (that is the interval from 66% to 100% of the synchronism speed).

Since the voltage controlled speed adjustment is always a cause of losses, it follows that adjustable drives based on voltage control are economic only for low power turbopumps. At present, for such drives thyristor variators (i.e. triacs) are used. The power of a synchronous motor supplied through a thyristor variator is chosen with due consideration for the particular operating conditions in view (the parabolic variation of the motor torque as a function of the supply voltage, and the speed control that operates over a range $\Delta n \approx 1:1.5$, where the slip reaches a maximum value). It is known that in ordinary operating conditions the rated value of slip power $P_{sR} \approx (0.04-0.05)\, P_R$. If the speed is controlled over the range $\Delta n \geqslant 1:1.5$, the slip power increases and reaches the maximum value of $P_{s\,max} = 0.15\, P_R$, and so $P_{s\,max}/P_{sR} = 3$ to 4. Therefore, when the speed is controlled by means of a thyristor variator, one can use a mass-produced asynchronous motor, provided however that its power is 3 to 4 times higher than in the standard case (when the motor is supplied at the rated voltage). Considering also that motor cooling is impaired as the fan speed drops, and that non-sinusoidal

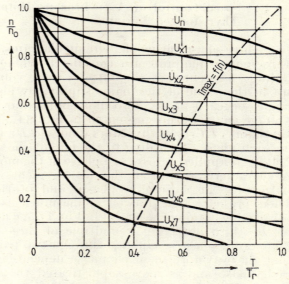

Fig. 11.4. Speed-torque characteristics of a disk-rotor asynchronous motor with an axial gap, as obtained by changing the feed voltage.

voltages and currents are likely to develop (owing to the motor being supplied through a thyristor variator), it clearly follows that with asynchronous machines an additional power capability should be provided, right from the design stage of the machine. Thus, owing to drive oversizing and to the losses inherent in the control feature, speed control based on change in the supply voltage is economical only when applied to low power turbopump electric drives ($P \approx 5$ kW).

Another, more economical solution, that does not call for an oversize motor, consists in employing a disk-rotor (axial-gap) asynchronous motor of a special design. The resistance of the disk-rotor varies in inverse proportion to the speed. This rotor tolerates overheating while offering at the same time good cooling conditions which permit the heat arising from the conversion of the motor slip power to be readily released. Another benefit of this typical motor is that, having a higher power factor, the stator current requirements are lower for the same speed and torque performance, which permits reduction of the limit current flowing through the thyristors of the voltage variator that supplies the motor.

When the supply voltage of the disk-rotor (axial-gap) motor is changed, the family of speed-torque characteristics shown in Fig. 11.4 results. The shape of these characteristics also show the square law that relates the motor torque to the supply voltage.

The disk-rotor asynchronous motor, supplied through a thyristorized variator, represents an economic solution for the adjustable drives of low power turbopumps ($P \leq 10$ kW), operating in adverse environmental conditions (such as chemical and petrochemical plants).

11.1.2. Speed Control of Asynchronous Motors through Thyristorized Variators

Three-phase principle circuit diagrams. In what follows, only those three-phase diagrams are considered whereby a symmetrical system of voltages is secured at the output; these voltages can be varied from zero up to the value of the mains supply, by changing the thyristor ignition angles on each phase. Two more important thyristor circuits designed for voltage control are represented in Fig. 11.5. Each includes three pairs of thyristors (TT) or of thyristors-diodes (TD), in an antiparallel arrangement on each phase. According to the circuit shown in Fig. 11.5, b, the supply voltage is applied across each phase of the motor via one pair of thyristors. A characteristic trait of this arrangement is the presence of third-order harmonics in the motor phases and of those of a multiple of three. Their presence is a drawback, inasmuch as it impairs the efficiency of the driving action. The circuit in Fig. 11.5, a does neither include harmonics of the third order nor a multiple of three. According to this diagram, each motor phase is supplied through two pairs of thyristors, which means that the waveform of each phase depends on the thyristor condition in each phase and is more sophisticated than the arrangement shown in Fig. 11.5, b. We shall therefore limit our attention to the operation of the diagram shown in Fig. 11.5, a.

The drive employing a thyristorized voltage variator with a disk-rotor (axial-gap) asynchronous motor is diagrammatically represented in Fig. 11.6, where the stator circuit of the motor includes three pairs of thyristors in an antiparallel arrangement. Voltage, and hence motor speed-control is achieved by changing the thyristor angle of ignition over a broad range, from 0 to 150°. The ring counter (RC) sends control pulses phase-shifted by an angle α with respect to the beginning of the positive half-periods of the anodic voltages of each thyristor.

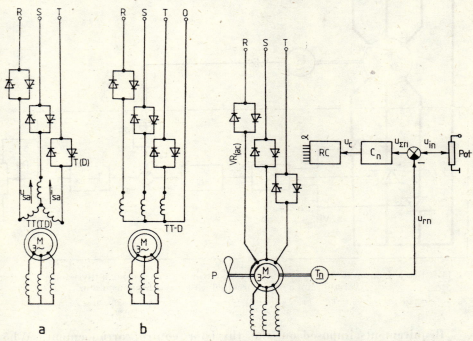

Fig. 11.5. Thyristorized three-phase arrangements for the feed voltage control.

Fig. 11.6. Block diagram for the automatic speed control, using a thyristorized voltage variator and a disk-rotor, axial-gap motor:

$VR(ac)$ — a.c. voltage regulator; RC — ring counter; C_n — speed regulator; Pot — potentiometer; Tn — speed-voltage generator; M — disk-rotor, axial-gap motor; P — pump.

An improved circuit, specially designed by Siemens (F.R. of Germany) for adjustable drives of low power turbopumps by means of squirrel-cage asynchronous motors, is shown in Fig. 11.7 [11.1]. According to this diagram, the voltage variator is built with triacs, placed in the star-connections of the asynchronous motor. Such an arrangement considerably reduces the load imposed on semiconductor devices as a result of overvoltages, developed during starting. Triacs are controlled by means of a ring counter RC, driven by a speed controller Rn, whose control parameter are the reference value n, set by means of potentiometer Pot, and the real value — n, as given by the speed-voltage generator Tn. The speed controller (used with this arrangement) is designed as a continuously variable governor with a non-linear characteristic, as a result of the square law that relates the stator voltage to the motor torque.

Speed Control of Stator Asynchron Motors

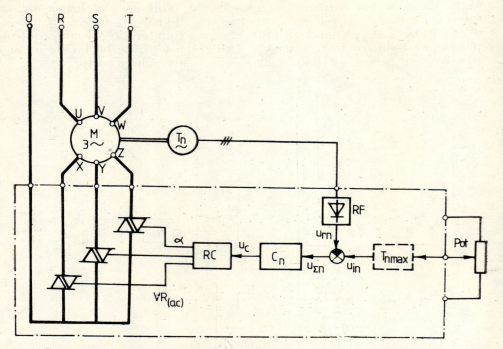

Fig. 11.7. Block diagram for the automated speed control of a drive, using a thyristorized voltage variator and a short-circuited rotor asynchronous motor :

$VR(ac)$ — voltage regulator (a.c) ; RC — ring counter ; C_n — speed controller ; Tn — speed-voltage generator ; M — asynchronous motor with a short-circuited rotor ; RF — rectifier.

Requirements imposed on the thyristor control arrangement. When the thyristors are in the fully open position, the voltages applied across the motor phases are sinusoidal in shape and can be determined with the following relations

$$\left. \begin{array}{l} U_R = U_m \sin \omega t \\[4pt] U_S = U_m \sin \left(\omega t - \dfrac{2\pi}{3} \right) \\[4pt] U_T = U_m \sin \left(\omega t - \dfrac{4\pi}{3} \right) \end{array} \right\} \qquad (11.1)$$

For this reason, the phase current is sinusoidal in shape too and phase-shifted with respect to the voltage by an angle $\varphi = \arctan(I\omega/R)$.

The current through the thyristor is triggerred (commutated) at the moment $\omega t = \varphi$. The ignition is delayed by an angle φ, determined by the inductive character of the load, which imposes special conditions on the thyristor control arrangement. If the control pulse to be applied on the thyristor control electrode is lagging by an angle α with respect to the beginning of the positive half-period of the mains voltage, then for $\alpha < \varphi$ the pulse front should be broader than $\varphi - \alpha$ so that when $\omega t = \varphi$, a positive ignition voltage is applied across the control electrode. And, since φ can be varied within wide limits, the width of control pulses at the beginning of the change in the control angle α should not be less than 96° phase angle.

With this arrangement, connection to the mains is only possible when at least two thyristors are simultaneously ignited (fired), but this condition can be only met if, over the range where angle α is varied, the width of control pulses is not smaller than 60° phase angle.

The voltage of phase R for a variety of ignition and active-inductive load angles, i.e., α and φ respectively, is shown in Fig. 11.8; the conduction angle resulting is λ. As can be seen, the motor is supplied with sine-wave portions that are functions of several parameters. The current is not sinusoidal in shape, but is determined by the impedance which, in turn, is a function of the rotational speed.

With inductive loads (see Fig. 11.8), owing to the self-induction e.m.f, the conduction angle λ of thyristors increases by δ as compared with the active load. For control angles α smaller than a limit value α_{lim} (to be determined below), the phase-voltage curves will show both three-phase and two-phase working portions (see Fig. 11.8, a—d).

For $\alpha > \alpha_{lim}$, the three-phase working portions disappear and each thyristor is ignited twice during one half-period.

The shape of the curve shown in Fig. 11.8 is fully determined by the control angle α and by angle δ, which depends on the phase-load angle φ. The thyristor angle of conduction λ is related to α and δ through the following expression

$$\lambda = \pi - \alpha + \delta$$

The shape of the load current pulse is found by solving the differential equation written for the active-inductive circuit for each of the five voltage portions (see Fig. 11.8, a). Taking the time origin as the beginning of each portion of the voltage curve, the current equation can be written, in relative units:

For the first portion

$$\left.\begin{array}{l} \alpha < \omega t < \dfrac{\pi}{3} + \delta \\[6pt] i_1 = -\sin(\alpha - \varphi)\, e^{-\dfrac{\tau}{\operatorname{tg}\varphi}} + \sin(\tau + \alpha - \varphi) \\[6pt] 0 \leqslant \tau \leqslant \dfrac{\pi}{3} + \delta - \alpha \end{array}\right\} \quad (11.2)$$

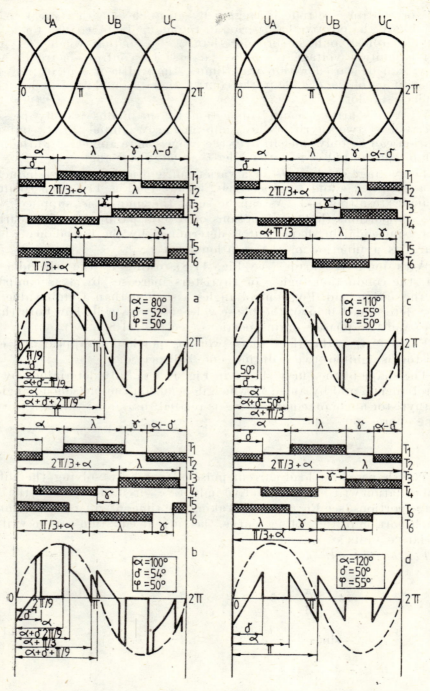

Fig. 11.8. Voltage change in one stator phase for various loads and ignition angles of the thyristors.

For the second portion

$$\frac{\pi}{3} + \delta \leqslant \omega t \leqslant \frac{\pi}{3} + \alpha$$

$$i_2 = \left[-\frac{\sqrt{3}}{2} \cos(\delta - \varphi) + i_{2int} \right] e^{-\frac{\tau}{\operatorname{tg}\varphi}} + \frac{\sqrt{3}}{2}(\tau + \delta - \varphi) \quad \quad (11.3)$$

$$0 \leqslant \tau \leqslant \alpha - \delta.$$

The initial value of the current in the second portion, i_{2int}, should equal the end value of the current in the first portion, i_{1end} thus

$$i_{2\,int} = i_{1\,end} = i_1\left(\frac{\pi}{3} + \delta - \alpha\right)$$

For the third portion

$$\frac{\pi}{3} + \alpha \leqslant \omega t \leqslant \frac{2\pi}{3} + \delta$$

$$i_3 = \left[-\sin\left(\frac{\pi}{3} + \alpha - \theta\right) + i_{3\,int} \right] e^{-\frac{\tau}{\operatorname{tg}\varphi}} + \sin\left(\tau + \frac{\pi}{3} + \alpha - \varphi\right)$$

$$0 \leqslant \tau \leqslant \frac{\pi}{3} + \delta - \alpha \quad \quad (11.4)$$

$$i_{3\,int} = i_{2\,end} = i_2(\alpha - \delta) \quad \quad (11.5)$$

For the fourth portion

$$\frac{2\pi}{3} + \delta \leqslant \omega t \leqslant \frac{2\pi}{3} + \alpha$$

$$i_4 = \left[-\frac{\sqrt{3}}{2} \cos(\delta - \varphi) + i_{4\,int} \right] e^{-\frac{\tau}{\operatorname{tg}\varphi}} + \frac{\sqrt{3}}{2} \cos(\tau + \delta - \varphi) \quad (11.6)$$

$$0 \leqslant \tau < \alpha - \delta$$

$$i_{4int} = i_{3\,end} = i_3\left(\frac{\pi}{3} + \delta - \alpha\right) \quad \quad (11.7)$$

For the fifth portion

$$\frac{2\pi}{3} + \alpha \leqslant \omega t \leqslant \pi + \delta$$

$$i_5 = \left[-\sin\left(\frac{2\pi}{3} + \alpha - \varphi\right) + i_{5\,int} \right] e^{-\frac{\tau}{\operatorname{tg}\varphi}} + \sin\left(\tau + \frac{2\pi}{3} + \alpha - \varphi\right)$$

$$0 \leqslant \tau \leqslant \frac{\pi}{3} + \delta - \alpha \quad \quad (11.8)$$

$$i_{5\,int} = i_{4\,end} = i_4(\alpha - \delta) \quad \quad (11.9)$$

in these equations, as the basis of the relative units, the value I_m of the current is taken

$$I_m = \frac{U_m}{Z} = \frac{U_m}{\sqrt{R^2 + (\omega L)^2}}$$

The end value of the current in the fifth portion should be equal to zero

$$\left[-\sin\left(\frac{2\pi}{3} + \alpha - \varphi\right) + i_{5\,int}\right] e^{\frac{-\frac{\pi}{3} + \delta - \alpha}{\mathrm{tg}\,\varphi}} - \sin(\delta - \varphi) = 0$$

Introducing the initial value of the current in the above expression and making a few trigonometric changes, the equation relating α, φ, and δ results thus

$$k(\varphi)\sin(\alpha - \varphi) e^{\frac{-\alpha - \delta}{\mathrm{tg}\,\varphi}} - \sin(\delta - \varphi) = 0 \tag{11.10}$$

where

$$k(\varphi) = \frac{\frac{1}{2} e^{-\frac{\pi}{3\mathrm{tg}\,\varphi}} - \frac{1}{2} e^{-\frac{2\pi}{3\mathrm{tg}\,\varphi}} - e^{-\frac{\pi}{\mathrm{tg}\,\varphi}}}{\frac{1}{2} e^{-\frac{\pi}{3\mathrm{tg}\,\varphi}} - \frac{1}{2} e^{-\frac{2\pi}{3\mathrm{tg}\,\varphi}} + 1}$$

The dependence of δ on α for various values of φ is obtained by graphically and analytically solving equation (11.10) by means of a computer; this relationship is shown in Fig. 11.9.

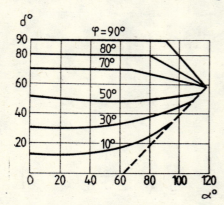

Fig. 11.9. Change in the values of angle δ as a function of the ignition angle (with the load out-of-phase angle as a parameter).

Estimation of speed-torque characteristics. We shall now describe a method for estimating the torque of the asynchronous motor as a function of slip (s) and of angle (α), representing the delay in thyristor ignition: this method is based on space phasors [11.4].

First, the stator current is determined in the form of space phasors, then the instantaneous torque and its average value are estimated, the latter being the analytical expression of the speed-torque characteristic.

Analytical expression of the stator current. The equivalent diagram of the asynchronous motor using space phasors can be represented, in a simplified version, as in Fig. 11.10, where

R_s, R_r' — are the stator and rotor resistances respectively, as referred to the stator;

s — the slip;

$L' = L_s + \dfrac{L_m L_r}{L_m + L_r}$ — the transient inductance;

u, i — space phasors associated with the voltage and current respectively, and expressed in a co-ordinate system fixed in the space

For the equivalent diagram given, the following equation can be written, which holds true only during the conduction period of the thyristors

$$\underline{u} = R_s \underline{i} + L' \frac{d\underline{i}}{dt} + \frac{R_s'}{s} \underline{i} \quad (11.11)$$

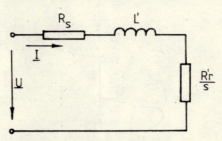

Fig. 11.10. Equivalent dagiram of the asynchronous motor using space phasors.

Denoting by $X' = \omega_1 L' = 2\pi f_1 L'$ the transient reactance and by $\tau = \omega_1 t$ the new variable, results in

$$\underline{u} = \left(R_s + \frac{R_r'}{s}\right) \underline{i} + X' \frac{d\underline{i}}{dt} \quad (11.12)$$

In a co-ordinate system that rotates at a speed ω_1, $u = \underline{U} e^{j\tau}$, where $\underline{U}$ is the vector, as expressed in the new co-ordinate system. By introducing the critical slip $s_k = \dfrac{R_r'}{X'}$, equation (11.12) becomes

$$\frac{d\underline{i}}{dt} + \left(\frac{R_s}{X'} + \frac{s_k}{s}\right) \underline{i} = \frac{\underline{U}}{X'} e^{j\tau} \quad (11.13)$$

The transient solution of equation (11.13) is found by integrating the homogeneous differential equation, thus:

$$\underline{i} = \underline{A} e^{-\left(\frac{s_k}{s} - \frac{R_s}{X'}\right)\tau} \quad (11.14)$$

The particular, stationary solution is of the form

$$\underline{i}_0 = \underline{B} e^{j\tau} \quad (11.15)$$

where constant $\underline{B}$ is found by replacing solution (11.13) in the homogeneous equation. It follows

$$\underline{i}_0 = \underline{B} e^{j\tau} = \frac{\underline{U}}{X'} \frac{1}{\dfrac{s_k}{s} + \dfrac{R_s}{X'} + j} e^{j\tau} \quad (11.16)$$

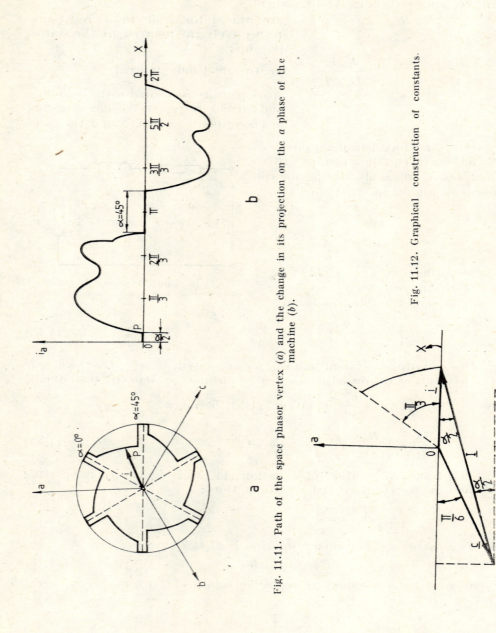

Fig. 11.11. Path of the space phasor vertex (*a*) and the change in its projection on the *a* phase of the machine (*b*).

Fig. 11.12. Graphical construction of constants.

Thus, the general solution of equation (11.13) is

$$\underline{i}_0 = \underline{A}\, e^{-\left(\frac{s_k}{s} + \frac{R_1}{X'}\right)\tau} + \frac{U}{X'} \frac{1}{\frac{s_k}{s} + \frac{R_s}{X'} + j} e^{j\tau} \qquad (11.17)$$

As can be seen, this expression is of the form $\underline{i} = \underline{C} + \underline{I} e^{j\tau}$ where

$$\underline{C} = A\, e^{-\left(\frac{s_k}{s} + \frac{R_1}{X'}\right)\tau} \qquad (11.18)$$

$$\underline{I} = \frac{U}{X'} \frac{1}{\frac{s_k}{s} + \frac{R_s}{X'} + j} \qquad (11.19)$$

If follows therefore that constant $\underline{A}$ will necessarily depend on angle α.

If $\underline{C} = 0$, then the spatial phasor $\underline{i}$ would describe a circle of a radius I which does not depend on angle α (the lag in the thyristor ignition). As known, according to the theory of space phasors, current waveforms on the three stator phases represent perfect sine waves. The existence of term $\underline{C}$ is responsible for the exponential shift of the centre described by vector $\underline{i}$, so that the path of vertex $\underline{i}$ will depend on angle α. Hence, it follows that constant $\underline{A}$ will necessarily depend on angle α. The path of space phasor $\underline{i}$ vertex, as well as its projection on axis (a) of the machine, i.e. the instantaneous current i_a in phase a, are presented in Fig. 11.11, a—b.

Constant $\underline{A}$ is geometrically determined (Fig. 11.12).

When $\tau = \frac{\alpha}{2}$, the thyristor on phase a is still in the off state, which means that $i_a = 0$, and therefore vector $\underline{i}$ is perpendicular to axis (a). At this moment, $C \neq 0$ for $\alpha \neq 0$. During one period (2π rad), the conduction time of a thyristor is $\pi/3$ for each conduction direction. For symmetry reasons, it follows that vector C is lying along the bisector of angle $\pi/3$. Thus, the value

$$\underline{C} = -2\underline{I} \sin \frac{\alpha}{2} e^{j\frac{\pi}{6}} \qquad (11.20)$$

is found, which is further used to find the value of constant $\underline{A}$ by putting in relation (11.18), $x = \frac{\alpha}{2}$

$$\underline{A} = -2\underline{I} \sin \frac{\alpha}{2} e^{\left(\frac{s_k}{s} + \frac{R_s}{X'}\right)\frac{\alpha}{2}} \cdot e^{j\frac{\pi}{6}} \qquad (11.21)$$

This gives the analytical expression for current i

$$i = \frac{U}{X'} \frac{1}{\frac{s_k}{s} + \frac{R_s}{X'} + j} e^{j\tau} - \frac{2U}{X'} \frac{1}{\frac{s_k}{s} + \frac{R_s}{X'} + j} \times$$

$$\times e^{\left(\frac{s_k}{s} + \frac{R_s}{X'}\right)\left(\frac{\alpha}{2} - \tau\right)} \cdot \sin\frac{\alpha}{2} \cdot e^{j\frac{\pi}{6}} \quad (11.22)$$

The expression for the instantaneous torque. The instantaneous torque can be determined by means of the known relation [11.4]

$$t = \frac{3}{2} p \frac{L_m^2 \omega_0 s R_r'}{R_r'^2 + s^2 X_r'^2} (i_d^2 + i_q^2) \quad (11.23)$$

where p is the number of pole pairs;
    R_r', X_r' — the rotor resistance and reactance, respectively, as referred to the stator;
    i_d, i_q — the real and imaginary components, respectively, of vector i.

Axes α and q of the complex plane are chosen so that

$$\underline{U} = U_d + jU_q = U_d (U_q = 0)$$

Space phasor i can be also written in the form

$$\underline{i} = \frac{U_d}{X'} - \frac{\frac{s_k}{s} + \frac{R_s}{X'} - j}{\left(\frac{s_k}{s} + \frac{R_s}{X'}\right)^2 + 1} \left[\cos\tau + j\sin\tau - \right.$$

$$\left. - 2\sin\frac{\alpha}{2} e^{\left(\frac{s_k}{s} + \frac{R_s}{X'}\right)\left(\frac{\alpha}{2} - \tau\right)} \times \left(\frac{\sqrt{3}}{2} + j\frac{1}{2}\right) \right] \quad (11.24)$$

the real (i_d) and imaginary (i_q) components of which are squared, then introduced in expression (11.23). The expression for the instantaneous torque becomes thus

$$t = \frac{3}{2} p \frac{L_m^2 \omega_0 s R_r'}{R_r'^2 + s X_r'^2} \cdot \frac{U_d^2}{X'^{1/2}} \frac{1}{\left(\frac{s_k}{s} + \frac{R_s}{X'}\right)^2 + 1} \times$$

$$\times \left[1 + 4\sin^2\frac{\alpha}{2} \cdot e^{2\left(\frac{s_k}{s} + \frac{R_s}{X'}\right)\left(\frac{\alpha}{2} - \tau\right)} - 4\sin\frac{\alpha}{2} \sin\left(\tau - \frac{\pi}{3}\right) \times \right.$$

$$\left. \times e^{\left(\frac{s_k}{s} + \frac{R_s}{X'}\right)\left(\frac{\alpha}{2} - \tau\right)} \right] \quad (11.25)$$

The expression for the average torque. The average torque is the mean value of function $t(\tau)$ for one period, thus

$$T = \frac{1}{2\pi}\int_0^{2\pi} t(\tau)\,d\tau \tag{11.26}$$

One can write as well

$$\frac{s_k}{s} + \frac{R_s}{X'} = \frac{s_k}{s}\left(1 + s\frac{R_s}{R'_r}\right) \cong \frac{s_k}{s}(1+s)$$

and eventually

$$T = \frac{3}{2}\,p\,\frac{L_m^2\,\omega_0\,s\,R'_r}{R'^2_r + s X'^2_r}\cdot\frac{U_d^2}{X'^2}\,\frac{1}{\dfrac{s_k^2}{s^2}(1+s)^2 + 1}\times$$

$$\times\left\{1 + \frac{1}{\pi}\sin^2\frac{\alpha}{2}\,\frac{1}{s_k(1+s)}\,e^{\frac{2s_k}{s}(1+s)\frac{\alpha}{2}}\left[1 - e^{-\frac{2s_k}{s}(1+s)2\pi}\right] - \right.$$

$$\left. - \frac{1}{\pi}\sin\frac{\alpha}{2}\,\frac{\sqrt{3\,s}\cdot s_k(1+s) + s^2}{s_k^2(1+s)^2 + s^2}\,e^{\frac{s_k}{s}(1+s)\frac{\alpha}{2}}\times\right.$$

$$\left.\times\left[1 - e^{-\frac{s_k}{s}(1+s)2\pi}\right]\right\}. \tag{11.27}$$

As can be seen, the torque depends on angle α. For $\alpha = 0$, the well known expression for torque should result, as derived through classical methods; that is, the normal speed-torque characteristic of the asynchronous motor

$$T_0 = \frac{2T_k}{\dfrac{s}{s_k} + \dfrac{s_k}{s}} \tag{11.28}$$

where

$$T_k = \frac{3p\,U_{10}^2}{2\,\omega_0\,C_r}\cdot\frac{1}{C_r\,X_{s\sigma} + X'_{r\sigma}};\quad U_{10} = \frac{U_d}{\sqrt{2}}$$

$$C_r = 1 + \frac{X'_{r\sigma}}{X_m} = \frac{X'_r}{X_m} \tag{11.29}$$

$$X_{s\sigma} = \omega_0 L_{s\sigma};\quad X'_{r\sigma} = \omega_0 L_{r\sigma};\quad X_m = \omega_0 L_m;\quad X'_r = X'_{r\sigma} + X_m.$$

To demonstrate relation (11.28), we take $\alpha = 0$ in expression (11.27) and

$$T_{\alpha=0} = \frac{3}{2}\,p\,\frac{L_m^2\,\omega_0\,s\,R'_r}{R'^2_r + s^2 X'^2_r}\cdot\frac{U_d^2}{X'^2}\,\frac{1}{\left(\dfrac{s_k}{s} + \dfrac{R_s}{X'}\right)^2 + 1}$$

Neglecting R_s and $R_r'^2$ results in

$$T_{\alpha=0} = \frac{3}{2} p \frac{L_m^2}{X_r'^2} \cdot \frac{\omega_0^2 \cdot U_{10}^2}{X'} \cdot \frac{R_{rs}'}{X'} \frac{1}{s_k^2 + s^2}$$

but

$$\frac{\omega_0^2 L_m^2}{X_r'^2} = \frac{X_m^2}{X_r'^2} = \frac{1}{C_r^2}; \quad \frac{R_r'}{X'} = s_k$$

$$X' = X_{s\sigma} + \frac{X_m X_{r\sigma}'}{X_m + X_{r\sigma}'} = \frac{1}{Cr}(X_{r\sigma}' + Cr\, X_{s\sigma})$$

and therefore it follows

$$T_{\alpha=0} = \frac{3}{2} p \frac{U_{10}^2}{\omega_0 C_r} \frac{1}{X_{r\sigma} + C_r X_{s\sigma}} \cdot \frac{2}{\frac{s}{s_k} + \frac{s_k}{s}} =$$

$$= \frac{2 T_k}{\frac{s}{s_k} + \frac{s_k}{s}} = T_0$$

In conclusion, it can be shown that for any angle α, the expression or the torque will be

$$T(s, \alpha) = T_0(s) \cdot K(s, \alpha) \tag{11.30}$$

where $T_0(s)$ represents the natural mechanical characteristic of the asynchronous motor, while

$$K(s, \alpha) = 1 + \frac{1}{\pi} \sin^2 \frac{\alpha}{2} \frac{1}{2s_k(1+s)} e^{\frac{s_k}{s}(1+s)\alpha} \left[1 - e^{-4\pi \frac{s_k}{s}(1+s)} \right] -$$

$$- \frac{1}{\pi} \sin \frac{\alpha}{2} \frac{\sqrt{3s}\, s_k (1+s) + s^2}{s_k^2 (1+s)^2 + s^2} e^{\frac{s_k}{s}(1+s)\frac{\alpha}{2}} \times$$

$$\times \left[1 - e^{-2\pi \frac{s_k}{s}(1+s)} \right] \tag{11.31}$$

represents a correction factor that depends on angle α.

The mechanical characteristics, estimated for a motor having power 10 kW and speed $n = 1540$ rpm, are shown in Fig. 11.13.

The electromagnetic torque, estimated as the average value of the instantaneous torque, also incorporates the effects of the parasite torques of a higher order, as determined by the voltage and current harmonics. The use of the space phasors permits a simpler determination of the torque of asynchronous motors supplied by thyristors.

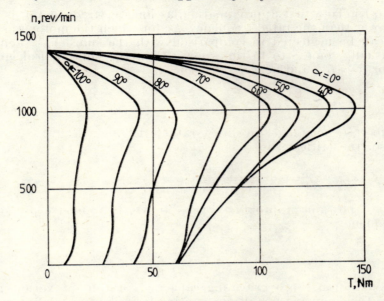

Fig. 11.13. Speed-torque characteristics of a 10 kW motor for various ignition angles.

11.2 Adjustable Electric Drives, Based on Varying Frequecy of the Supply Source

11.2.1 Frequency Control of Asynchronous Motors

Using the standard notation, the angular velocity of the asynchronous motor is

$$\Omega_2 = \Omega_1 (1 - s) = \frac{\omega_1}{p} (1 - s) = \frac{2\pi f_1}{p} (1 - s) \qquad (11.32)$$

From this relation, it readily follows that one possible way of speed control is by changing the frequency f_1 of the voltage, applied across the stator windings. This method implies the use of a frequency converter, since the industrial mains are fed at a constant frequency, and therefore requires significant investment costs, for a converter is generally more expensive than is the asynchronous motor whose speed is to be adjustable.

Let us consider in more detail the problems of operation of the asynchronous motor at a variable frequency. According to expression

$$T_{max} = \frac{3U_1^2}{2c\Omega_1 \left[R_1 + \sqrt{R_1^2 + (X_{\sigma 12} + cX'_{\sigma 21})^2} \right]}$$

the maximum electromagnetic torque depends on frequency f_1 (both the angular frequency $\Omega_1 = \dfrac{\omega_1}{p}$ and the dispersion reactances are proportional to f_1). To a first approximation, as long as frequency f_1 is not too much lower than the rated frequency f_{1R} for which the motor was designed, R_1 can be neglected in comparison with the sum $X_{\sigma 12} + cX'_{\sigma 21}$ (R_1 being usually some 8 to 15 times lower than the sum of leakage reactances for $f_1 = f_{1R}$) and hence

$$T_{max} \sim \frac{U_1^2}{f_1^2} \tag{11.33}$$

the simbol $\sim$, meaning "in direct ratio with".

Also, the critical slip

$$s_{max} = \frac{cR'_2}{\sqrt{R_1^2 + (X_{\sigma 12} + cX'_{\sigma 21})^2}}$$

which, to the same approximation as above, is in inverse proportion to the frequency

$$s_{max} \sim \frac{1}{f_1} \tag{11.34}$$

To maintain the overloading capacity constant at various resistive torques, the following conditions can be derived from relations above

$$\frac{T_{rb}}{T_{ra}} = \frac{T_{max\,b}}{T_{max\,a}} = \left(\frac{U_{1b}}{f_{1b}}\right)^2 \cdot \left(\frac{f_{1a}}{U_{1a}}\right)^2 \tag{11.35}$$

from which one can find the variation of the supply voltage with the change in the speed by modifying the supply frequency with due consideration given to the resistive torque

$$\frac{U_{1b}}{U_{1a}} = \frac{f_{1a}}{f_{1b}} = \sqrt{\frac{T_{rb}}{T_{ra}}} \tag{11.36}$$

Knowing that the resistive torque with turbopumps is a parabolic function of speed, the relationship between the motor torque, i.e. the power, or the supply voltage, on the one hand, and the frequency, on the other, can be found as follows

$$\frac{T_a}{T_b} = \left(\frac{f_{1b}}{f_{1a}}\right)^2 ; \quad \frac{P_b}{P_a} = \left(\frac{f_{1b}}{f_{1a}}\right)^3 ; \quad \frac{U_{1b}}{U_{1a}} = \left(\frac{f_{1b}}{f_{1a}}\right)^2 \tag{11.37}$$

If frequency f_1 is variable then, with voltages whose rms value is constant, the amplitude $\Phi_{\mu m}$ of the resulting flux will be variable too. Indeed, according to relation

$$U_1 \approx E_{\mu 1} = \frac{2\pi}{\sqrt{2}} f_1 w_1 kw_1 \Phi_{\mu\,max}$$

one can write

$$U_1 \sim f_1 \Phi_{\mu max}$$

With a descrease of f_1, for $U_1 = $ constant, a higher flux $\Phi_{\mu max}$ results, which brings about the saturation of the magnetic core (which is easily saturated even at the rated frequency), as well as an important increase in the magnetizing current I_μ. To avoid any change in the saturation state of the magnetic circuit when $f_1 < f_{1R}$, the condition

$$\frac{U_1}{f_1} = \text{constant} \qquad (11.38)$$

should be strictly obeyed. Returning now to relation (11.33) and bearing in mind the above condition, the conclusion is reached that the maximum torque is invariable for all practical purposes. Taking also into consideration relation (11.34), one can see that the critical slip increases with $f_1 < f_{1R}$, hence the speed-torque characteristics are less hard. Now, bearing in mind that at lower frequencies R_1 cannot be neglected any more in comparison with the sum of the leakage reactances, we find the family of speed torque characteristics $\Omega_2 = f(T)$ by using equation (11.38) (see Fig. 11.14). It will be seen that for frequencies much lower than f_{1R}, the maximum torque drops considerably and the flatness of the speed-torque characteristic visibly drops too, which restricts the speed control range. Of course, for $f_1 > f_{1R}$, condition (11.38) is not justified any more and the condition $U_1 = $ constant is maintained.

To broaden the control range without reducing the loading capacity of the machine at low and very low frequencies (sometimes values as low

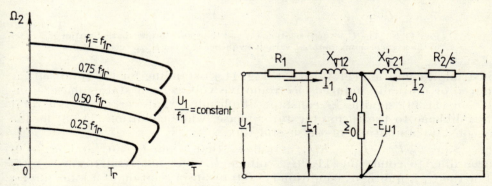

Fig. 11.14. Speed-torque characteristics for different frequencies with $U_1/f_1 = $ constant.

Fig. 11.15. Equivalent circuit, used to bring out total e.m.f., E_1 of the stator.

as 1 or 2 Hz are reached for machines with $f_{1R} = 50$ Hz), the procedure described below is applied, which introduces a correction to condition (11.38). This procedure will be better understood if we refer to the equivalent diagram, shown in Fig. 11.15. As can be seen, should this diagram fail to include R_1, both the maximum torque T_{max} and the maximum flux would be kept strictly constant with the varying f_1. To reach this performance, and bearing in mind that R_1 cannot be ignored, the equivalent diagram itself suggests the condition

$$\frac{E_1}{f_1} = \text{constant} \tag{11.39}$$

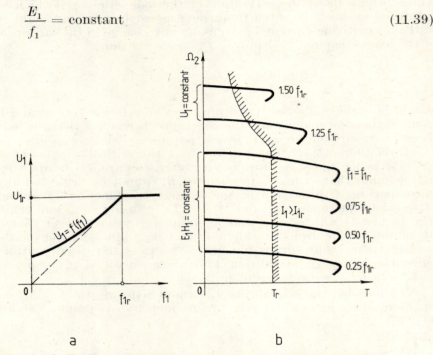

Fig. 11.16. The $U_1 = f(f_1)$ relation (a) and the speed-torque characteristics of the asynchronous motor at various frequencies, when E_1/f_1=constant (b).

instead of condition (11.38), E_1 in Fig. 11.15 standing for the total self-induced or mutually induced electromotive voltage in a stator phase. Condition (11.39) signifies a constant amplitude, whatever f_1, of the total flux linkage to the turns of a stator phase, while condition (11.38) means a constant amplitude of the useful flux.

Since $-E_1 = U_1 - R_1 I_1$, it follows that when $f_1 \to 0$ [and $E_1 \to 0$, according to condition (11.39)], voltage U_1 does not tend towards zero. For the U_1/f_1 quotient, condition (11.39) means a relation of a kind similar to that represented in Fig. 11.16, a, the representative curve departing from a straight line in the vicinity of the origin. If U_1/f_1 obeys this condition, then the speed-torque characteristics will show as in Fig. 11.16, b, having the same maximum and practically parallel torques over the stable operation interval. For $f_1 > f_{1R}$, the rated voltage is not to be exceeded, and hence the speed-torque characteristics cannot have a constant maximum torque; in fact, the torque drops in inverse proportion to the squared frequency. Besides, at $f_1 > f_{1R}$, the speed is higher than the ratted level and the dominating condition is to avoid exceeding the machine rated power: therefore, the load through the resistive torque of the machine should in any case be reduced. As a consequence, curves shown in Fig. 11.16, b are to be used as follows: for $f_1 < f_{1R}$, the machine operates with a maximum resistive torque, equal to the rated one; for $f_1 > f_{1R}$, the machine operates at a maximum active power, equal to the rated one. The operational limit is shown by the broken line in Fig. 11.16, b.

To sum up, speed control through changing the frequency f_1 has the following benefits: a broad control range, (sometimes from 1 to 100 Hz), a fine control action in both directions (below and above the rated speed); and no energy losses (except for the converter); on the other hand, it has the drawback that investment costs substantially exceed the cost of the asynchronous motor.

Both conditions (11.38) and, particularly, (11.39) lead to complications for the motor supply (converter supply). As will be shown in the following paragraph, modern power electronics practice has developed perfectly satisfactory static converters. When the cost of these converters becomes acceptable, speed control through power converters will have widespread application, since the speed-torque characteristics, achieved by means of this control method, are similar to those achieved by the separately excited d.c. motor with an adjustable rotor voltage. Thanks to the additional benefits, the asynchronous motor is more sturdy, has a lower inertial moment, a lower overall size, and does not require special maintenance, for it has no sliding contacts (in the case of the squirrel-cage asynchronous motor).

11.2.2 Speed Control by Varying Frequency with Static Converters

The phase-controlled converter (cycloconverter). The cycloconverter is an electrical device that directly converts a.c. energy of given parameters into another a.c. energy of other parameters. Its operation is based on the same principle as that of the *four-quadrant converters* (the double converter), used to supply the d.c. motors, applied in reversible electric drives, except that in this case the armature of the d.c. motor is replaced by one phase of the induction motor. Being a double converter, the four-quadrant converter is conductive in both directions of the load circuit. The phases of the ignition pulses are controlled as a function of the a.c. supply voltage, and the result is a rectified average voltage that can be continuously controlled in both directions, giving a possibility of obtaining an a.c. voltage across the load terminals. When the control action covers the commutating frequency of the two converters as well, and the frequency of the voltage across the load terminals can be varied, the converter is called a *cycloconverter*. In other words, the cycloconverter can supply the windings of induction motors with an a.c. voltage whose amplitude and frequency are adjustable. The amplitude of the output voltage is adjusted by changing the converter control voltage U_c, while its frequency is adjusted by acting on the commutation frequency between the two converters. As can be easily seen, if at the input of the double converter an a.c. reference voltage (U_c is applied sinusoidal, trapezoidal etc.), of a given frequency and amplitude, then across the load terminals an a.c. voltage of the same frequency but of an amplitude multiplied by the converter transfer factor will be obtained. The advantage offered by the double converter, of operation in four quadrants, confers to the cycloconverter the property to supply loads with any power factor, while the energy can flow in any direction through the cycloconverter. The output voltage has inevitably a certain content of harmonics, apart from the fundamental harmonic needed. The distortions in the output voltage are explained by the very mechanics of the cycloconverter operation, the output voltage being built up out of portions of the supply voltage of the a.c. mains. By using filters where needed, one can get a good quality of the waveform for the output voltage. Distortions in the output voltage increase with the increased output/input frequency ratio. Conversely, the distorsions drop with the increased number of converter pulses. The need for a natural commutation of the current from one wave of the supply voltage to another causes the maximum frequency of the output voltage to be lower than the frequency of the supply voltage, whatever the number of converter pulses. This is one of the fundamental limits of the cycloconverter.

The output voltage of a phase-controlled cycloconverter is single-phased. To obtain at the output a three-phase voltage (of a low frequency), three reversible cycloconverters are connected in a three-phase bridge arrangement, one converter on each phase (Fig. 11.17); the converters are controlled by pulses, phase-shifted by 120° from one phase to another. For a better understanding of the cycloconverter operating principle, we refer to Fig. 11.18 which is a simplified circuit of it. Each converter consists of an a.c. voltage supply (equal to the fundamental voltage component desired) and a diode that controls the direction of the current through the converter. The harmonic components of the voltage are neglected. Between the two converters, there is no current flow. The voltages of the two generators have the same amplitude, frequency, and phase, while the voltage across the output terminals of the cycloconverter is equal to the voltage of each cycloconverter. The impedance of the a.c. voltage generator is zero (as measured across the output terminals) and therefore the current can flow in any direction, at any moment; thus the converter can work at any power factor of the load.

Because in each converter the current flows in one direction only, each converter should build up one half of the load current output cycle, whatever the phase of the load current as referred to the voltage. One can therefore safely state that during one half-period of the load current cycle, each converter can generate voltages of both polarities. Hence, in two quadrants, each converter can operate both as a rectifier and as an inverter. This condition is illustrated in Fig. 11.19 for several phase-

Adjustable Electric Drives by Frequency

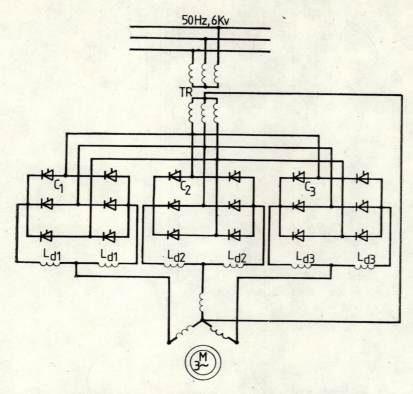

Fig. 11.17. Principle diagram of an adjustable electric drive, using an asynchronous motor with a short-circuited rotor and fed through a cycloconverter:
M — motor; TR — transformer; $C_1 \ldots C_3$ — reversible frequency converters; $L_{d_1} \ldots L_{d_3}$ — filter coils.

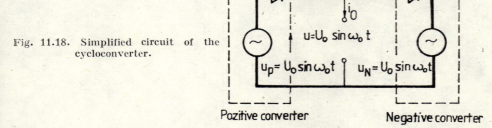

Fig. 11.18. Simplified circuit of the cycloconverter.

shifts of the load current with respect to the voltage. In Fig. 11.19, *a*, the load phase-shift angle is $\varphi_0 = 0$. Each converter is conducting as long as it operates as a rectifier, and is locked off when it operates as an inverter. In Fig. 11.19, *b*, the phase-shift is $\varphi_0 = 60°$, inductive. During a 120° period of each half-cycle, the associated converter works as a rectifier, and over the last 60°, as an inverter, feeding the power back to the input. In Fig. 11.19, *c*, the load phase-shift angle is $\varphi_0 = 60°$, capacitive.

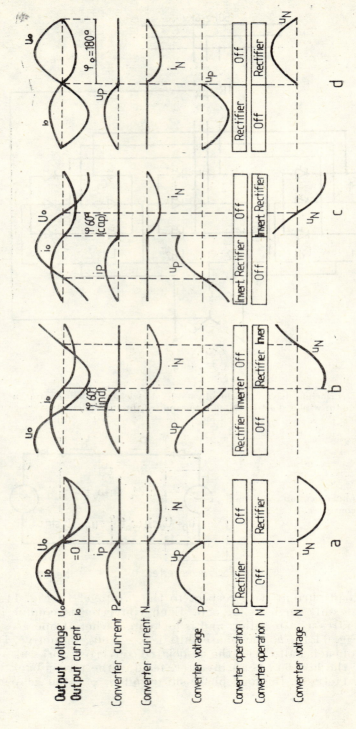

Fig. 11.19. Voltage and current wave forms for:
a) — $\varphi = 0°$; *b*) — $\varphi = 60°$; *c*) — $\varphi = 60°$.

During the first 60° of the half-cycle, the associated converter works as an inverter, and over the other 120° as a rectifier. In Fig. 11.19, d, the phase-shift angle is $\varphi_0 = 180°$, and in this case the converters are working only as inverters. For the cycloconverter, the no-load (idle) condition can be considered that situation when the ignition angles of the two converters are at 90°. This happens when no control signal exists at the input of the circuits that control the synchronizing of the ignition pulses.

A positive mean voltage is obtained by shifting forwards the ignition angles of the positive converter and by delaying the ignition angles of the negative converter, as referred to the static point of 90°C. To obtain a negative mean voltage, the ignition angles of the positive converters should lag behind, while those of the negative converter should lead the static point of operation, i.e. 90°. Therefore, to obtain at the output an a.c. voltage, the ignition angles of each converter should vary continuously around the static point of oscillation. The oscillations of the ignition angles of the two converters should be phase-opposed and their sum should always be 180°. If the ignition pulses are simultaneously applied to both converters and if the sum of ignition angles is 180°, then the voltage components, while being equal, will nevertheless show some instantaneous disparities between them. These voltage differentials will cause a flowing current whose value is theoretically infinite and must be limited by means of the current limiting reactance coil.

Owing to a spectrum rich in harmonics, both for the output voltage and for the load current, and paying due consideration to the control dynamics, a direct converter with three-phase bridges can yield a maximum output frequency of $f_2 = (0-0.3) f_1$. Higher output frequencies can also be obtained, but then the number of phases on each bridge converter should be increased, say from 3 to 6, or else one should increase the converter supply frequency. Thus, a maximum output frequency of $f_2 = (0-0.7)f_1$ can be obtained.

As for the installed power P_i of the components included in the diagram shown in Fig. 11.17, the approximate value is 5.80 P_R for the thyristors and 1.75 P_R for the transformer, P_R being the rated power of the asynchronous motor fed through the cycloconverter.

The cycloconverter diagram represents an economic solution for the supplies of high-power asynchronous motors (from 400 up to 10,000 kW).

As yet, the speed control of asynchronous motors through a cycloconverter has not been applied in the case of turbopumps, for the driving equipment is rather complex and the control action restricts the frequencies to the lower side of the range; and a great majority of turbopumps are characterized as fast working machines. However, we know the performance of a direct frequency converter can be valuable when studying the adjustable drives through a double-fed asynchronous motor, as described in Section 12.2.3.

The frequency converter (d.c. link converter). This converter is characterized by a double conversion of the electric energy: (1) conversion of the mains a.c. voltage of a constant frequency into a d.c. voltage by means of a rectifier; (2) conversion of the d.c. link voltage into a single or a three-phase voltage of an adjustable frequency, by means of a static inverter.

D.C. link converters fall into two groups, the classification criterion being the component whereby the frequency and the output voltage are adjusted.

(a) *The voltage and frequency are adjusted by the same component*, that is, by the inverter, which in this case is fed at a constant d.c. voltage; thus, the converter includes a non-controlled rectifier and an inverter, fed with a constant voltage (Fig. 11.20, *a*).

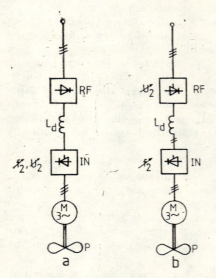

Fig. 11.20. Link-converters:
a — with a constant voltage;
b — with a variable voltage.

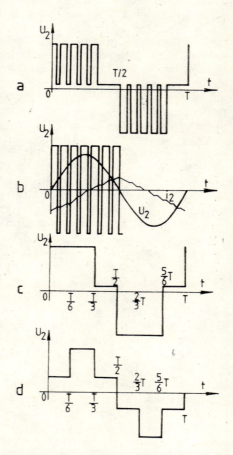

Fig. 11.21. Specific forms of the output voltage of various typical converters.

(b) *The output voltage and frequency are controlled through two different components*, namely, the output voltage in the rectifying circuit and the frequency in the inverter. In this case, the frequency converter features a controlled rectifier and an inverter fed with a variable voltage (Fig. 11.20, *b*).

The characteristics of the output voltages follow the classification above (Fig. 11.21). On constant voltage d.c. link converters, the conduction is permanently interrupted to change the output voltage mean value as a function of the input voltage of the converter. The output voltage mean value is changed by using static regulators of the chopper type, i.e. the a.c. voltage is pulse-modulated at the output by a constant d.c. of voltage. When the relative on-time interval of the conduction period of a phase is constant, the output voltage is built up in equivalent rectangular voltage steps (Fig. 11.21, *a*). When the relative on-time interval is modulated with respect to a sine-wave reference quantity (variable), then the output voltage will have an equivalent sine-wave shape owing to the sine variation of the average value (Fig. 11.21, *b*). On variable voltage d.c. link converters, supplied from a rectifier controlled through an a.c. voltage of a rectangular waveform, characterized by continuous conductive time intervals (Fig. 11.21, *c — d*), the inverter complexity grows with the number of steps of the output voltage, which is higher. The amplitude of the output voltage is equal to the converter d.c. link voltage. For three-phase outputs, inverters can be designed for conductive time-intervals of 120° or 180°.

As a general rule, the complexity of converter arrangements is determined by the inverter. In the case of constant voltage d.c. link converters, the rectifying function is very simple, but the inverter that controls the output frequency and voltage is more complex. In the case of variable voltage d.c. link converters, the inverter is simpler because it is associated only with the frequency control, whereas the rectifying circuit is more complex owing to the voltage control function. On the whole, converters with a d.c. link, while being complex arrangements, still offer the advantage that output quantities U_2, f_2 are independent of U_1, f_1, over their whole range.

Several versions of arrangements using d.c. link frequency converters have been developed already as economic solutions for the adjustable drives of low and high powered turbopumps. According to the inverters used, these can be classified as inverters with autonomous extinction, characterized by the fact that the extinction of a major thyristor is due to the ignition of the following major thyristor, or inverters with independent extinction, characterized by the fact that the extinction of a major thyristor is done by separate circuits, by means of auxiliary triggering-off thyristors.

a. *Constant voltage d.c. link converters.* These are economic solutions for adjustable drives of low powered turbopumps (1 to 40 kW). Two such arrangements are shown in Fig. 11.22, *a—b*, which are designed for supplies of low-power three-phase asynchronous motors. The operating principle of such a converter consists in the alternate commutation, at a high frequency, of each motor phase between the positive and the negative pole of a d.c. source. This action is achieved by a forced triggering of the thyristors of an inverter, wired in a three-phase bridge arrangement and supplied with a constant d.c. voltage. The shape of the output current is a function of the commutation times of each phase between the positive and the negative poles.

Forced commutation converters do not have available a reactive (rms) voltage or power, external to the converter, to switch the load current from one thyristor over to the other, so that this triggering action can be achieved only in a forced way. Forced commutation presupposes the existence of an extinction circuit C_{ex}, which reduces the current through the thyristor to zero, while supplying the voltage needed for the commutation phenomenon. With such an arrangement, the extinction voltage is taken from an extinction capacitor that feeds the main thyristors of the inverter at the appropriate time, through an auxiliary thyristor T_{add}. As a result of this action, the inverter is switched off from the d.c. link by an induction coil, and thus the d.c. side is locked off from the effects of the extinction process. In addition to the extinction circuit, the converter also includes a static control device that keeps the motor flux $\Phi = \dfrac{U_1}{f_1}$ constant, whatever the frequency changes. In order to do this, the inverter input voltage should be continuously matched to the frequency. Thus a d.c. chopper (d.c. voltage-control converter, or VCC) is introduced between the non-controlled rectifier and the inverter. This d.c. chopper is a static device that converts the voltage applied at the input into rectangular voltage pulses at the output; in principle, the d.c. chopper is a circuit breaker (interruptor) whose on-off action can be time-controlled. The average voltage of the output is proportional to the relative on-time duration in the conductive interval, and can be modified from zero to the value of the supply voltage. The d.c. chopper is controlled by a main frequency, equal to that of the output, and by a much higher auxiliary frequency so as to change the relative on-time duration and hence the output voltage. In other words, the output voltage is built up by alternating impulse sequences of a variable or constant width, the average value of which determines the amplitude of the output voltage (see Fig. 11.21, $a-b$). The control signal for the width modulation of voltage pulses is subordinated to the useful output frequency control signal. The d.c. chopper is combined with a current limiter that restricts the maximum working current at the converter output, and thus protects the motor and the converter against overcharges and short-circuits. As a general rule, for low powers (from 4 to 40 kW), the d.c. chopper is built up with thyristors as shown in Fig. 11.22, a, and it needs additional extinction features (including, apart from the main thyristor T, a leakage diode D and an auxiliary extinction thyristor T_{add}) because it does not have available reactive power, external to the converter, to achieve commutation. For very low powers in particular (from 1 to 2 kW), the d.c. chopper can be built with power transistors, as shown in Fig. 11.22, b, or with new semiconductor devices such as the gate-extinction thyristor, which permits the power side of the arrangement to be considerably simplified.

The speed control of an asynchronous motor, supplied by a constant voltage d.c. link converter, is achieved through an electronic control

Adjustable Electric Drives by Frequency

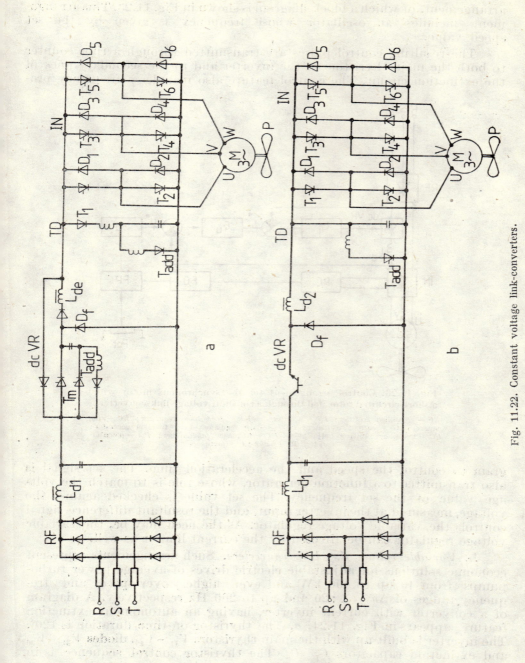

Fig. 11.22. Constant voltage link-converters.

553

arrangement, of which a block diagram is shown in Fig. 11.23. This arrangement includes an oscillator whose frequency is given by the set speed value.

The resulting control pulses are transmitted through a ring counter to both the main thyristors of the inverter and to the secondary ones of the extinction circuit. The control feature also includes a frequency pro-

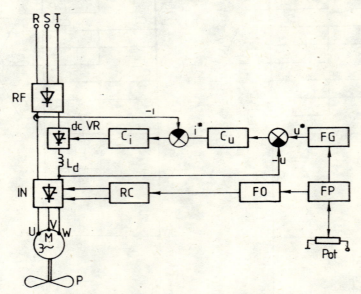

Fig. 11.23. Control arrangement for the synchronous motor with a short-circuited rotor, fed through a constant voltage link-converter:

RF — controlled rectifier; *IN* — inverter; *dcVR* — *dc* voltage regulator; *RC* — ring counter; C_i — current regulator; C_u — voltage regulator; *FG* — function generator; *FO* — frequency oscillator (adjustable); *FP* — frequency programmer; *Pot* — potentiometer; *M* — motor; *P* — pump.

gram to control the speed and the acceleration time. The set signal is also transmitted to a function generator, whose role is to match the voltage value of the set frequency. The set value is checked against the voltage, measured at the inverter input, and the resultant difference signal controls the variable voltage regulator. As the need may be, the variable voltage regulator can be affected by the current limiter as well.

b. *Variable voltage d.c. link converters.* Such arrangements represent economic solutions for adjustable electric drives of average power turbo-pumps (from 40 up to 400 kW and even higher) over speed and frequency ranges of $\Delta n = 1 : 20$ and up to 200 Hz respectively. A diagram of a converter with current inverter, having an autonomous extinction feature, appears in Fig. 11.24, *a*. The thyristor on-time duration is 120°. The inverter is built up with the main thyristors $V_{11} - V_{16}$, diodes $V_{21} - V_{26}$ and extinction capacitors $C_1 - C_6$. The thyristor control sequence is in the order of their numbering, which means that at a certain moment

Adjustable Electric Drives by Frequency

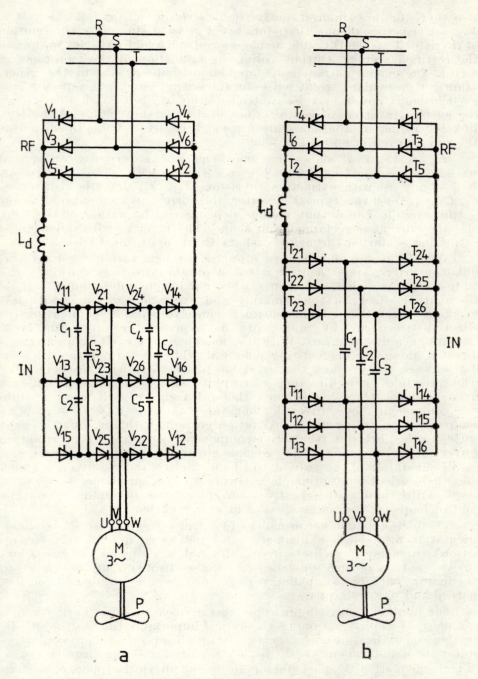

Fig. 11.24. Link-converter arrangements with a variable voltage.

two thyristors are simultaneously conductive on 60°, i.e.: V_{11}, $V_{12}-V_{12}$, V_{13}... Extinction of one thyristor, say V_{11}, is produced by the ignition of thyristor V_{13}. The D.C. link includes a coil and a high capacity to smooth the rectified voltage. Output voltage is a function of the motor power factor. The output current is rectangular in shape. Owing to the generation of rectangular current pulses for the output circuit, the speed of the asynchronous motor may be controlled step by step. With this circuit, the current inverter capacitor is not loaded by the voltage of the D.C. link but rather by the load current, which confers reliable operation on the circuit even in no-load running.

Figure 11.24, b shows the converter with a current inverter that has independent extinction (in the case shown, with phase extinction); it is built up with extinction thyristors $T_{21}-T_{26}$ and the capacitors C_1, C_2, C_3. With this typical inverter, the current is commutated in the output circuit. The output voltage depends on the nature of the load, and the current is rectangular in shape. The inverter with independent extinction is started through auxiliary thyristors to load the capacitors.

With asynchronous motors supplied through variable voltage d.c. link converters, speed control is based on the "field-orientation" principle [11.4]. As with d.c. machines, where the torque is determined without the mutual effects of the armature and excitation currents, with a.c. machines a rotational field control is conceivable, thus eliminating the effects between current components that generate the torque, and those that produce machine flux. With d.c. machines, the excitation current is directed along the magnetizing pole axis while the armature current is led between the brushes, that is at 90° phase angle with respect to the pole direction. These directions of currents and of the armature flux are fixed in space. On a.c. machines, the stator current will break up into two components, one directed along the rotational flux (the reactive component) according to the excitation current, with the other at right angles to this direction (the active component) according to the armature current of d.c. machines. The principle governing current resolution, that is, the separation of the two control loops (active and reactive), is called the "field-orientation principle". Owing to this principle, the control loops of flux and of the reactive current can be superimposed on the control loops of the motor speed and active current.

The high dynamic performance of control systems based on the field-orientation principle is explained by the fact that actually the control action is carried out by a d.c. current, although it is applied to a.c. machines. An analysis of such sophisticated speed control systems, as applied to a.c. motors with high dynamic performance, requires an adequate mathematical and physical apparatus.

The method, consisting in applying the space phasors (Park vectors) in a matrix form has not only a theoretical importance (due to the simplification and unification of the analysis) but it affords rather a physical interpretation of phenomena, both of the stabilized and the transient ones. In fact, the method of space phasors forms the physical support of modern control systems and offers a wide variety of solutions.

Adjustable Electric Drives by Frequency

The idea of applying the field-orientation principle to three-phase machines is also justified by the fact that in such a machine, supplied by a symmetrical system of three-phase voltages, it does not only rotate the space phasor of the rotational field at a synchronous speed, as determined by the supply frequency, but also the space phasors of currents fed into the stator windings and voltages applied across the motor terminals (Fig. 11.25). All these space phasors, having the

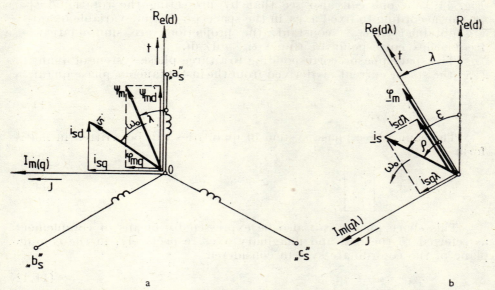

Fig. 11.25. Method of coordinate axes selection in case of "field orientation principle":
a — components of space phasors in fixed axes are sinusoidal; b — components of space phasors in the axes of "field oriented" system are constant quantities.

same angular velocity with reference to a co-ordinate system determined by the position of the rotational field, will take a comparatively stationary position; therefore, they cannot be considered as d.c. quantities although all phase quantities are periodic and even of a sinewave form. Thus the control action is carried out by means of d.c. quantities that are obtained by reference to a system of rotational axes whose real axis is directed along the space phasors of the gap rotational flux. These d.c. quantities are not directly measurable; they rather have to be derived from the real (natural) parameters of the motor, that is, from those corresponding to the periodical variation. In order to separate the flux and the speed control loops, the phasor of the stator current, which is stationary with respect to the space phasor of the rotational flux, will split along the two directions determined by the latter, as shown in Fig. 11.25,b. The three stator phases of the three-phase motor are denoted by a_s, b_s and c_s, respectively. The reference system (d, q) has the real axis aligned to phase a_s of the stator. The position of the rotational flux, i.e. of space phasor ψ_m, is given by angle λ; this angle also determines the position of the coordinate system (dλ, qλ), to which also the space phasor of the stator current i_s will be referred. The position of the phasor is determined by angle ε. Since the axis system (dλ, qλ) has the same angular velocity as the current space phasor, that is $d\lambda/dt = d\varepsilon/dt = \omega_0$, by solving the latter we shall get the quantities $i_{sd x}$ and $i_{sd x}$ which are constant in time. As $i_{sd\lambda}$ is directed according to the magnetizing flux, it will represent the reactive portion, while $i_{sq\lambda}$, being normal to this direction, will represent the active portion of the stator current. From Fig. 11.25, a one can also see that by projecting the rotational space phasor according to fixed axes in the space (d, q), time variable quantities are obtained (if ω_0 = constant, the projections are sinusoidal) whose frequencies correspond to ripple $\omega_0 = d\lambda/dt$.

The space phasor, corresponding to a three-phase system of quantities (e.g. the stator current), is derived from the instantaneous phase quantities

$$\underline{i}_s = \frac{2}{3}(i_{sa} + \underline{a}i_{sb} + \underline{a}^2 i_{sc}) \tag{11.40}$$

The same three-phase system of quantities can be written in matrix form

$$[i_s] = \begin{bmatrix} i_{sa} \\ i_{sb} \\ i_{sc} \end{bmatrix} \tag{11.41}$$

This space phasor can also be expressed by means of complements as referred to the real and imaginary axes respectively, in the complex plane of the co-ordinate system considered

$$\underline{i}_s = i_{sd} + j i_{sq} \tag{11.42}$$

that is, in matrix form

$$[i_s]_\perp = \begin{bmatrix} i_{sd} \\ i_{sq} \end{bmatrix} \tag{11.43}$$

Since the phase quantities that assist the determination of the space phasor in relation (11.40) are expressed in a system of axes fixed in the space, the components of this space phasor will be referred to the same system (d, q).

To change over to the system of rotational axes, the rotational operator of an angle λ should be applied; the matrix form of this operator is $[D(\lambda)]$, the elements being directly derived from Fig. 11.25, thus

$$[i_s]_{\perp\lambda} = \begin{bmatrix} i_{sd\lambda} \\ i_{sq\lambda} \end{bmatrix} = [D(+\lambda)][i_s]_\perp = \begin{bmatrix} +\cos\lambda & +\sin\lambda \\ -\sin\lambda & +\cos\lambda \end{bmatrix} \begin{bmatrix} i_{sd} \\ i_{sq} \end{bmatrix} \tag{11.44}$$

The change in the co-ordinate axes is carried out by means of axis conversion elements AC shown in Fig. 11.26, a, the operations being carried out as in relation (11.44) by applying matrix $[D(\lambda)]$; for the reverse conversion, matrix $[D(-\lambda)]$ is applied, that is, the reverse rotational operator, i.e. the reverse of the matrix defined above. The determination of angle λ, needed for the co-ordinate axis transformation, imposes the knowledge of the rotational field in the gap, that is, of the

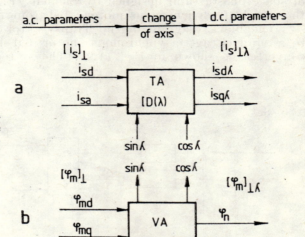

Fig. 11.26. Element of the transformation of axes coordinate system (*a*) and vector analyzer (*b*).

corresponding space phasor. By measuring or by computing the two components of this phasor in the system of fixed axes in the space (see Fig. 11.25, *a*), its modulus and its angle can be found, thus

$$\psi_m = \sqrt{\psi_{mq}^2 + \psi_{md}^2}, \quad \cos \lambda = \frac{\psi_{md}}{\psi_m}, \quad \sin \lambda = \frac{\psi_m}{\psi_{mq}} \tag{11.45}$$

The components of the space phasor of the rotational flux in the complex plane $(d\lambda, q\lambda)$, "field orientated" as in Fig. 11.25, *b*, are

$$\psi_{md\lambda} = \psi_m, \quad \psi_{mq\lambda} = 0 \tag{11.46}$$

or, in a matrix form

$$[\psi_m]_{\perp \lambda} = \begin{bmatrix} \psi_{md\lambda} \\ 0 \end{bmatrix} = \begin{bmatrix} \psi_m \\ 0 \end{bmatrix} \tag{11.47}$$

The element that takes care of the operations described by relations (11.45) is called the vector analyser and is denoted by *VA* in Fig. 11.25,*b*; it solves a particular case of the axis transformation, namely, the case when the direction of the space phasor is coincident with the direction of one of the axes.

Owing to the remarkable advances made in the field of integrated circuits, the method of space phasors (Park vectors) has emerged from the stage of purely theoretical speculation to become the physical support of many modern control systems. It is no coincidence that well-known firms have invested great material and brain efforts for bringing this theory to practical purposes, and so to work out sophisticated speed control systems that could be applied to a.c. motors. In what follows, we shall describe the "Transvecktor control" system, based on the theory of space phasors and designed by Siemens (F.R. of Germany) to control the speed of asynchronous motors, fed from static frequency converters.

The block diagram of the Transvecktor control system, as applied to a squirrel-cage asynchronous motor fed from a current frequency converter, is shown in Fig. 11.27. The system is based on the "field-

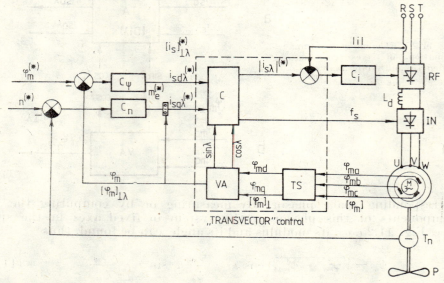

Fig. 11.27. Achievement of "field orientation" by direct measurement of rotational flux.

orientation" principle and directly measures the rotational flux in the gap by means of Hall probes.

For the control circuit in the closed loop, transducers are needed for speed, flux, and current (when the rectifier is supplied from the mains). In order to achieve a control action based on the field-orientation principle, the flux quantities should be converted into the components of the corresponding space phasors, which is done by the SC element with the help of matrix $[A]$. In this case, matrix $[A]$ elements can be derived from relation (11.48) as established by [11.4]

$$[A] = \frac{2}{3} \begin{bmatrix} 1 & -\frac{1}{2} & -\frac{1}{2} \\ 0 & \frac{\sqrt{3}}{2} & -\frac{\sqrt{3}}{2} \\ \frac{1}{2} & \frac{1}{2} & \frac{1}{2} \end{bmatrix} \qquad (11.48)$$

One thus finds the function of element SC for the conversion of three-phase quantities into two-phase quantities

$$[\psi_m]_\perp = \begin{bmatrix} \psi_{md} \\ \psi_{mq} \end{bmatrix} = \frac{2}{3} \begin{bmatrix} +1 & -\frac{1}{2} & -\frac{1}{2} \\ 0 & +\frac{\sqrt{3}}{2} & -\frac{\sqrt{3}}{2} \end{bmatrix} \begin{bmatrix} \psi_{ma} \\ \psi_{mb} \\ \psi_{m\rho} \end{bmatrix} = [A][\psi_m] \tag{11.49}$$

The magnetizing flux is found by applying the principle of field orientation and by separating the two control loops (the active and the reactive), that is, of the speed and of the flux, respectively. To this end, one has to establish position λ of the space phasor within the gap, as well as the module ψ_m, which is done by means of the above described VA vector, an analyser whose function is given by relations (11.45).

Quantities marked by an asterisk (*) are control variables which are d.c. quantities $[i_s]^*_{\perp\lambda}$, as expressed in the system of co-ordinate axes, determined by the position of the rotational field, i.e. by angle λ. As previously stated, $i^{(*)}_{sd\lambda}$ is the reactive component of the current and can be found from the imposed flux (that is, from the working point on the motor magnetizing characteristic), while $i^{(*)}_{sq\lambda}$ is the active component of the current and can be determined from the imposed speed (that is, it corresponds to the mechanical working point of the motor and depends on the load at the shaft of the latter).

As can be readily seen, the control loop of the magnetizing flux ψ_m is superimposed on that of the active current $i_{sq\lambda}$ of the stator. Connection of the corresponding loops is done by means of flux and speed governors, i.e. $R\psi$ and Rn respectively, designed according to the magnetizing and to the speed-torque characteristic of the motor, respectively.

To control the mains rectifier, one should know the amplitude of the stator current, while to control the inverter, the frequency of the current sent by it to the motor has to be known. These two quantities are determined by the working point, established by speed $n^{(*)}$ and by the imposed flux $\psi^{(*)}$, respectively. They are estimated by means of computer C shown in the figure and supplying functions similar to those supplied by a vector analyser. The phase-diagram for the determination of the control quantities described above is shown in Fig. 11.25, b. The amplitude of the stator current i_s can be estimated from the active and reactive components of the current, as established at the output of the corresponding governors

$$|i_{s\lambda}| = i_{s\lambda} = \sqrt{i^2_{sd\lambda} + i^2_{sq\lambda}} \tag{11.50}$$

The position of this current with respect to the magnetizing flux is determined by angle ρ thus

$$\cos \rho = \frac{i_{sd\lambda}}{i_{s\lambda}}, \quad \sin \rho = \frac{i_{sq\lambda}}{i_{s\lambda}} \tag{11.51}$$

The frequency supplied to the stator is determined from the derivative of angle ε, representing the position of the stator current as referred to the fixed real axis

$$f_s = \frac{1}{2\pi} \frac{d\varepsilon}{dt} \qquad (11.52)$$

As can be seen in Fig. 11.25, *b*, $\varepsilon = \lambda + \rho$, hence angle ε can be determined by relations

$$\left.\begin{array}{l}\cos \varepsilon = \sin \lambda \cos \rho - \cos \lambda \sin \rho \\ \sin \varepsilon = \sin \lambda \cos \rho + \cos \lambda \sin \rho\end{array}\right\} \qquad (11.53)$$

The relations (11.50)—(11.53) are solved by a computer program. The stator current thus determined is compared with the amplitude of the fundamental of the current taken from the mains, as in the case of the motor, and so another current governor R_i is required.

The control system described above is sophisticated enough, and furthermore it uses a score of analogue operational components whose accuracy is limited. A promising solution could be the use of microprocessors to convert the analogue control system into a digital one, of much higher accuracy.

REFERENCES

11.1. ETTNER, N., *Simotras Antriebe für Pumpen und Lufter Kleiner Leistung*. Siemens-Zeitschrift, No. 4 (1971).
11.2. FRANSUA, A., MĂGUREANU, R., *Electrical Machines and Drive Systems*. Technical Press, Oxford and Technical Publishing House, Bucharest, 1981.
11.3. IONEL, I. I., *Acționarea electrică a turbomașinilor*. Technical Publishing House, Bucharest, 1980.
11.4. KELEMEN, A., *Acționări electrice*. Didactic and Pedagogic Publishing House, Bucharest, 1976.

12

Speed Control of Turbo-pumps Employing Rotor-Circuit Controlled Asynchronous Motors

The speed and power factor of asynchronous slip-ring motors can be controlled by coupling the rotor circuit, via the slip-rings, to an external variable e.m.f. source. This arrangement, known as an *asynchronous cascade*, allows the slip power of the drive to be recovered and thereby ensures a high efficiency. The slip power sP (s = slip, P = electromagnetic power) can be recovered either electrically, by feeding it back to the mains or to another recovery circuit (an arrangement known as *electrical cascade* or *Scherbius cascade*), or mechanically, by converting it into mechanical power that is fed to the main drive shaft (an arrangement known as *electromechanical cascade* or *Krämer cascade*). Medium and high power turbopumps ($P \approx 400-25{,}000$ kW), that require a comparatively narrow range of speed adjustment ($\Delta n \leqslant 1:2$), can be economically controlled in this way, and are commonly driven by such systems.

The variable required in this form of control can be provided by a frequency converter, consisting either of auxiliary rotating machines (i.e. a machine cascade) or of static devices (rectifiers and inverters — i.e. converters), or of a combination of static and rotating elements. The frequency converter must be capable of working at the maximum voltage likely to be developed in the rotor circuit of the main drive within the required range of speed adjustment, but its power rating need only match the maximum slip power and not the full output power of the main drive. The power rating of the converter is therefore proportional to the required range of speed control; as a general rule, the maximum slip $s_{max} = 0.2 - 0.5$. A cascade connection close to $s = 1$ is not economic, that the power of the auxiliary rotating machines or static devices is close to that of the main drive. Thus start-up of the main motor (until the required range of speed adjustment of the asynchronous cascade is reached) is done by means of an external resistor (whose Ohmic value can be adjusted to change the motor speed-torque characteristics) connected in the rotor circuit, and only when the maximum slip value s_{max} is reached (that is, the minimum speed required, n_{min}) will the main drive be cascade-connected.

For medium and high powered drives of turbopumps, speed control by series-connecting a variable resistance in the rotor circuit cannot be contemplated, because in such cases the power losses are exceedingly high. The major part of slip power, lost through Joule effects, can be recovered, provided the speed control is applied by insertion of a variable e.m.f. in the rotor circuit of the asynchronous motor.

12.1 Operation Principle of Asynchronous Cascades

12.1.1 Speed Control by Insertion of an Additional e.m.f. in the Rotor Circuit

Let us consider a double supply asynchronous machine (i.e. one fed both through the stator and the rotor circuits, see Fig. 12.1).

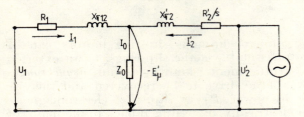

Fig. 12.1. Equivalent circuit when an e.m.f. is inserted in the rotor circuit of an asynchronous motor.

In such a case, the machine will operate as a motor with a constant load (resistive) torque equal to, say, the rated torque. The equations describing such a mode of operation, taken from the theory of the three-phase synchronous motor in stationary conditions will be

$$\underline{U}_1 = R_1 \underline{I}_1 + jX_{\sigma 12}\underline{I}_1 - E_{\mu 1}$$
$$\underline{U}'_2 = \frac{R'_2}{s} \underline{I}_2 + jX_{\sigma 12}\underline{I}_2 - sE_{\mu 1} \tag{12.1}$$

where all rotor quantities are referred to the stator, and hence

$$U'_2 = \frac{U_2}{s} \cdot \frac{w_1 k_{w1}}{w_2 k_{w2}}$$

$$E'_{\mu 2} = \frac{E_{\mu 2}}{s} \cdot \frac{w_1 k_{w1}}{w_2 k_{w2}} = E_{\mu 1}$$

where U_2 is the real voltage of the rotor-circuit supply; U'_2 — the rotor voltage referred to the number of turns and to the stator frequency of the asynchronous motor; $E_{\mu 1}$, $E_{\mu 2}$ — the resulting (useful) stator and rotor e.m.fs. respectively; $E'_{\mu 2}$ — the rms values of the resulting (useful) e.m.f. as referred to the stator (this reference is made assuming that $E'_{\mu 2} = E'_{\mu 1}$); w_1, w_2 — number of the series connected turns on the stator and rotor winding respectively; k_{w1}, k_{w2} — the corresponding winding coefficients; s — slip of the asynchronous motor.

Assuming that the small voltage drops in the stator winding are negligible, we can write

$$E_{\mu 1} \approx -U_1 = \text{constant}$$

where U_1 is the voltage as measured across the terminals.

On the other hand, from the equation of constant torque it follows that

$$T = \frac{3E_{\mu2}I_2(\underline{E}_{\mu2}, \underline{I}'_2)}{\Omega_1 - \Omega_2} = \frac{3sE_{\mu1}I'_2 \cos(\underline{E}_{\mu1}, E\underline{I}'_2)}{\Omega_1 - \Omega_2}$$

$$= \frac{3E_{\mu1}I'_2\cos(\underline{E}_{\mu1}, \underline{I}'_2)}{\Omega_1} = \text{constant}$$

that is, considering that $\Omega_1 = $ constant and $E'_{\mu1} \approx $ constant

$$I'_2 \cos(\underline{E}_{\mu1}, \underline{I}'_2) \approx \text{constant}$$

where Ω_1 is the angular speed of the rotational field and Ω_2 the motor angular speed.

Assuming further that the external voltage U_2 applied to the motor is initially in phase with the e.m.f. $\underline{E}_{\mu1}$, then

$$\cos(\underline{E}_{\mu1}, \underline{I}_2) = \frac{R'_2}{\sqrt{R'^2_2 + s^2 X'^2_{\sigma21}}} \approx 1$$

because generally the slip is low and $sX_{\sigma21} < R'_2$. As a consequence, relation (12.1) can be written

$$R'_2 \underline{I}'_2 = s\underline{U}'_2 + s\underline{E}_{\mu1} = \underline{U}_2 \frac{w_1 k_{w1}}{w_2 k_{w2}} +$$
$$+ s\underline{E}_{\mu1} \approx \text{constant} \qquad (12.2)$$

from which it may be concluded that, if U_2 is varied but kept either in phase or in opposition with $\underline{E}_{\mu1}$, a change in the slip can result for a constant torque.

Figure 12.2, *a* represents the torque of the ordinary asynchronous motor when $s = s_R$ and $U_2 = 0$, and Fig. 12.2, *b* represents the phasor diagram of rotor quantities when $U_2 > 0$, in phase with $\underline{E}_{\mu1}$, with the result that $s < s_R$. If

$$U_2 \frac{w_1 k_{w1}}{w_2 k_{w2}} = \underline{E}_{\mu1}$$

then $s = 0$, with the result that the motor will operate synchronously. In such a case the frequency of the rotor currents is nil, which means that U_2 is a d.c. voltage. If, however, U_2 is in opposition to $\underline{E}_{\mu1}$, then $s > s_R$ (Fig. 12.2, *c*).

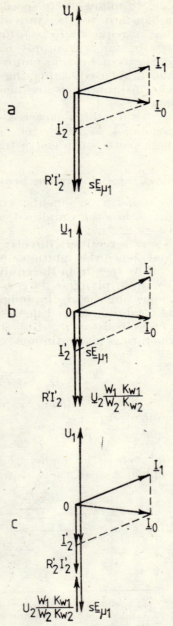

Fig. 12.2. Phasor diagram when an e.m.f. is inserted in the rotor circuit of an asynchronous motor:
a — $U_2 = 0$; *b* — U_2 in phase with I'_2; *c* — U_2 in opposition to the I'_2.

It follows that speed can be controlled in both directions, even for a constant torque, provided that U_2 applied across the rotor winding has a frequency $f_2 = s\, f_1$ for any slip s of the machine. Now the frequency of an a.c. supply is either constant or adjustable (independently of the asynchronous motor control). Therefore a special frequency converter must be used, supplying U_2 of the desired amplitude and yet having the instantaneous frequency of the rotor currents of the asynchronous motor.

Many high powered drives, employing this method, used rotating machines as converters in the past, but with the development of the high-power static converters, rotating machines have been replaced.

12.1.2 Special Actuating States of the Asynchronous Cascade-Connected Motor

Cascade arrangements, with d.c. link frequency converters for the main drive, use the simplified equivalent circuit based on Thévénin's principle (Fig. 12.3).

The rectifiers, directly connected across the slip-rings, are not controlled and hence the ignition angle $\alpha = 0$; therefore, U_2 and I_2 are in phase.

As seen from the equivalent circuit, the effective voltage is obtained by multiplying U_0 by s. Bearing in mind Thévénin's principle and neglecting the overlapping effect and the harmonics, we infer that a voltage sU_0 can be induced in the rotor circuit to produce a current intensity I_2 across resistor R_2. But the value of U_2 is determined by the rectifier operating in continuous conducting conditions, and therefore

$$I_2 = \frac{sU_0 - U_2}{R_2 - jsX} \tag{12.3}$$

For low current values, assuming that $I_2 = 0$, the slip will be a no-load slip, i.e.

$$s_0 U_0 = U_2 \tag{12.4}$$

and hence the higher the no-load slip will be, the higher the voltage applied across the slip-rings. If, for instance, U_2 is constant and the motor speed drops with respect to the no-load speed, then sU_0 will be higher than $s_0 U_0 = U_2$ and, according to relation (12.3), high current intensities (that is, high torques) will be developed. In fact, such developments occur even for small speed or voltage changes, because the impedance $R_2 - jsX$ is very low; we may therefore conclude that in this case the motor will have a hard speed-torque characteristic.

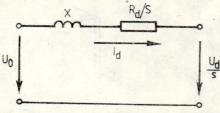

Fig. 12.3. Simplified equivalent circuit of the cascade-connected asynchronous motor.

Cascade-connected asynchronous motors operate in special conditions. Considering the main drive, the most difficult problem is the connection of the rectifier across the slip-rings, knowing that the former generates harmonics both on the d.c. and on the a.c. side. The problem is made even more complex by the overlapping phenomenon. Indeed, in all standard connections where the rectifiers are branched to the mains transformers, the leak reactance is 3—5%; but in our case here, with the motor leak (short-circuit) reactance being 20%, a very strong overlapping effect takes place. Moreover, since overlapping plays an important part (particularly with low slips), the rectifier supply voltage and frequency vary.

Commutation calculations are very arduous because leak reactances are very high, the slip frequency varies over a wide range, and various operating conditions of the rectifying diodes must be considered. The simplest case is obtained by considering the resistances to be negligible and the d.c. current perfectly smooth, which is valid for slips over 30%. In such conditions, the current intensity in the main drive rotor circuit will be independent of slip and will be controlled by the load alone.

Figure 12.4 shows the three-phase equivalent circuit based on Fig. 12.3: resistances are ignored, and the current is assumed to be

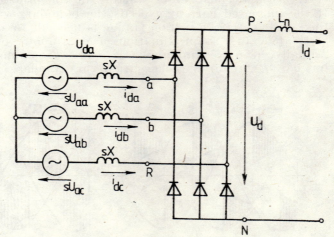

Fig. 12.4. Three-phase equivalent circuit, according to Fig. 12.3.

perfectly smooth, which implies that the smoothing choke will take a very high value.

For a better understanding of this phenomenon, let us assume that, instead of diodes, the rectifier is built with thyristors (*SCRs*). With three-phase bridge controlled rectifiers, the current will be

$$I_d = I_{2sc} \frac{3}{2} [\cos \alpha - \cos(\alpha + \gamma)] \qquad (12.5)$$

and the voltage

$$U_d = U_{s\,max} \frac{\cos \alpha + \cos(\alpha + \gamma)}{2} \qquad (12.6)$$

The maximum value of the rotor short-circuit current will be

$$I_{2\,sc} = \frac{\sqrt{2}\, U_0}{X} \qquad (12.7)$$

while the maximum voltage will be

$$U_{s\,max} = \frac{3}{\pi} \sqrt{3}\, \sqrt{2}\, s \cdot U_0 = s U_{d0} \qquad (12.8)$$

which will depend upon slip.

Actually, three-phase rectifier bridges use only diodes.

With uncontrolled three-phase bridge rectifiers, *two operating ranges* can be identified. *The first* is characterized by $\alpha = 0$ and $\gamma \leqslant 60°$ (Fig. 12.5, *a*), and *the second* by $\alpha \geqslant 30°$ and $\gamma = 60°$ (Fig. 12.5, *b*).

In the *first range*, only angle γ varies with the load; the higher the load current, the larger the commutation area, i.e. the overlapping angle γ needed. Over the so-called first operating range, commutation in one arm of the bridge ends before starting in the other arm (see Fig. 12.5, *a*). If the increased current exceeds the boundary value of the first range, then an even larger voltage range is needed for commutation purposes and, at a first approximation, a continuous overlapping will result. Actually, if for example phases *a* and *b* commutate onto the positive bus-bar (*P*), then on the negative bar (*N*) phase *a* cannot take over the

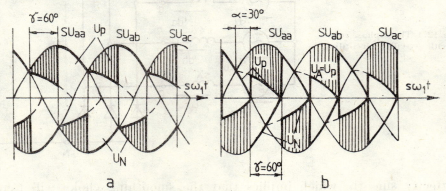

Fig. 12.5. The rectified voltage for the two working ranges of a rectifier in a non-controlled three-phase bridge arrangement:

a — in the first range, for $\gamma = 60°$ and $\alpha = 0$; *b* — in the second range, for $\gamma = 60°$ and $\alpha = 30°$.

conduction from phase *c* while the overlapping on the positive bus-bar is not finished; until this occurs, say at point *A*, the potential will be determined by that of bus-bar *P*. This potential is more positive than that of point *C* since commutation between *c* and *a* on the negative bus-bar can only start when point *A* potential is determined again by phase *a*. In such cases, commutation in one arm of the bridge can only start when it has finished in the other arm (Fig. 12.5, *b*). Over this so-called *second range*, the overlapping angle $\gamma = 60°$ stays constant, while the ignition angle $\alpha = 0$, automatically grows from 0 to 30°. Rectifiers, uncontrolled over the second operating range, perform like those controlled over the first range.

Any estimation of the main drive torque should start with the electromagnetic torque; assuming the losses are negligible, the torque will be the same as that produced by the power on the d.c. side

$$T = \frac{P}{\omega_1} = \frac{P_s}{s\omega_1} = \frac{U_d I_d}{s\omega_1} \tag{12.9}$$

Substituting the values for U_d and I_d in equations (12.5) and (12.6) gives

$$T = T_{k0} \frac{3}{\pi} [\cos^2\alpha - \cos^2(\alpha + \gamma)] \tag{12.10}$$

where

$$T_{ko} = \frac{3\sqrt{2}}{4} \cdot \frac{U_0 I_{2sc}}{\omega_1} = \frac{3}{\omega_1} U_0^2 \frac{1}{X}$$

is the ideal breakdown (pull-out) torque.

After the necessary transformations, equation (12.10) becomes

$$T = T_{ko} \frac{3}{2\pi} [\cos 2\alpha - \cos 2(\alpha + \gamma)] \tag{12.11}$$

which can be also written as

$$T = T_{ko} \frac{3}{\pi} \sin(2\alpha + \gamma) \sin \gamma \tag{12.12}$$

But the speed-torque characteristics $T(s)$ can only be determined if function $s = f(\alpha, \gamma)$ is also known, which is done by starting from the assumption that the relation

$$U_d = s U_{dmo} \frac{\cos\alpha + \cos(\alpha + \gamma)}{2} = s_0 U_{dmR} \tag{12.13}$$

always holds true, provided the resistances are neglected and provided also that the d.c. current is considered to be smooth.

From the above relations, it follows that

$$\frac{s_0}{s} = \frac{\cos\alpha + \cos(\alpha + \gamma)}{2} \tag{12.14}$$

Over the first operating range of the rectifier, $\alpha = 0$ while $0 \leq \gamma \leq 60°$; then, based on equations (12.11) and (12.14), we can write

$$T = T_{ko} \frac{3}{2\pi} (1 - \cos 2\gamma)$$

and
$$\frac{s_0}{s} = \frac{1 + \cos \gamma}{2}$$

After transformation, we obtain

$$\frac{T}{T_{ko}} = \frac{3}{\pi} 4 \frac{s_0}{s}\left(1 - \frac{s_0}{s}\right) \tag{12.15}$$

At the end of the range, for $\gamma = 60°$, $\alpha = 0$, $s_0/s = 3/4$ and $T/T_{ko} = 9/4\pi$.

In the second operating range of the uncontrolled rectifier, i.e. when $0 \leqslant \alpha \leqslant 30°$ and $\gamma = 60°$, the control angle will be forced at high torques. Over this second operating range $\alpha \geqslant 30°$ and $\gamma = 60°$ and, on the basis of relations (12.12) an (12.14) we can write

$$T = T_{ko} \frac{3\sqrt{3}}{2\pi} \sin(2\alpha + 60°) \tag{12.16}$$

and

$$\frac{s_0}{s} = \frac{\sqrt{3}}{2} \cos(\alpha + 30°) \tag{12.17}$$

which, after transformations, lead to

$$\frac{T}{T_{ko}} = \frac{3}{\pi} 2 \frac{s_0}{s} \sqrt{1 - \frac{4}{3}\left(\frac{s_0}{s}\right)^2} \tag{12.18}$$

From relation (12.16) it follows that the maximum torque is obtained for $\alpha = 15°$, i.e.

$$T_{max} = T_{ko} \frac{3\sqrt{3}}{2\pi} = 0.827 \, T_{ko}$$

while the slip from equation (12.15) becomes

$$s_k = \frac{2\sqrt{2}}{3} s_0 = 1.633 s_0$$

Equation (12.17) holds true for $4/3 \, s_0 \leqslant s \leqslant 4/\sqrt{3} \, s_0$ and also at the end of the second operating range, when $\gamma = 60°$, $\alpha = 30°$ and the torque $T/T_{ko} = 0.716$. As can be seen above, the breakdown (pull-out) torque drops by 17% while the speed-torque characteristic is the softer, the higher the U_2 applied.

The natural characteristic of the slip-ring asynchronous motor, with two cascade-actuating characteristics, is illustrated in Fig. 12.6. To illustrate the speed-torque characteristics in the cascade state graphically, all that is needed is to estimate the points of only one characteristic, since from these points one can find the corresponding characteristic $T(s)$

for any s_0. Assuming, for illustrating purposes, that the speed-torque characteristic corresponding to s_{01} is known, then for the same torques the following relation will be valid

$$\frac{s_{01}}{s} = \frac{s_{02}}{s}$$

The above ratios are similar to those written for the changes in the resistances of the rotor circuit, except that in the ratios above, the d.c.

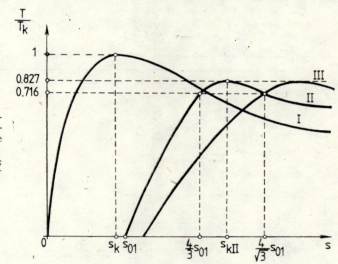

Fig. 12.6. Speed-torque characteristics of the asynchronous motor in cascade arrangement:
curve I — natural characteristic;
curves II and III — cascade-actuating characteristics.

voltage s_0 is considered instead of resistances. The reduced breakdown torque can be explained by the fact that, owing to overlapping, the current of the fundamental harmonic lags behind the voltage. Over the second operating range, the maximal torque drops even lower since the current-to-voltage lag grows with the ignition (control) angle.

Taking account of the overlapping phenomenon, it can be said that the diode rectifier behaves, as regards the motor, as if resistors and inductances were connected across the slip-rings.

12.2. Adjustable-Speed Drives by Asynchronous Cascade Connection

12.2.1. The Asynchronous Electrical Cascade (Scherbius Cascade)

Whenever a variable speed is specified for turbopumps whose speed-torque characteristic is a parabola (that is, their resistive or load torque varies proportionally to the square of the speed), the highest torque is requested for the highest speed. In such a case, a rational solution is to apply an arrangement whereby the electric drive is controlled at a constant torque; this is the so-called *static Scherbius cascade*, also known in the literature as the *sub-synchronous static converter (inverter) cascade*, or the *asynchronous cascade with static elements*. Indeed, this arrangement has, to date, the widest application in adjustable-speed electric drives for turbopumps [12.7], [12.8].

Diagram of the principle. The Scherbius cascade, using a semiconductor rectifier and an inverter in the rotor circuit, is illustrated in Fig. 12.7, where an asynchronous motor and its starting rheostat can be seen. Slip power, sP, is first rectified by rectifier RF (featuring a three-phase

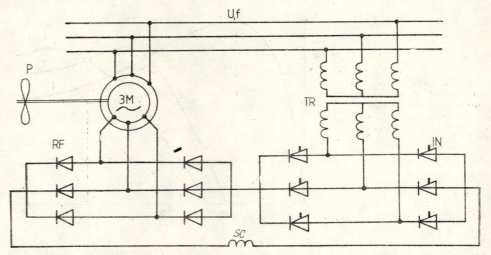

Fig. 12.7. Electric wiring of the static converter electric cascade (Scherbius):
P — pump; M — motor; RF — rectifier; IN — inverter; TR — transformer; SC — smoothing coil.

bridge with semi-conducting diodes), then converted again into an a.c. voltage of a frequency f_1 equal to that of the mains, through a thyristor inverter IN. Transformer TR serves to match the rotor voltage of the asynchronous motor to that of the mains.

Control characteristic of the converter. This is an important factor, representing the change in the rectified mean voltage U_d as a function of control angle α, whereby the thyristor ignition is controlled (Fig. 12.8).

For a three-phase bridge, the mean value of the rectified voltage is given by

$$U_d = \frac{3\sqrt{2}}{\pi} U_s \cos\alpha \qquad (12.19)$$

and its maximum value by

$$U_{do} = \frac{3\sqrt{2}}{\pi} U_s \qquad (12.20)$$

where U_s is the rms value of the secondary voltage. Using relation (12.19) and bearing in mind the considerations made for Fig. 12.8, we conclude that for $0 < \alpha < 90°$, $U_d/U_{do} > 0$ and the converter can operate as a

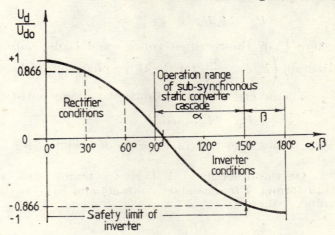

Fig. 12.8. Control characteristic of a converter in an integrally controlled three-phase bridge arangement

rectifier, while for $90° < \alpha < 180°$, $U_d/U_{do} > 0$, hence the converter can operate as an inverter. With due consideration of the anodic overlapping phenomenon, the control range of the rectifier is set by $180° - \beta r$ where β is a safety angle needed to switch the current over between the thyristors whose value is determined by the transformer leak reactance. For all practical purposes

$$\beta \approx 20° - 30°$$

General actuating equations. From the principle diagram of the rotor circuit, shown in Fig. 12.9, the variation of no-load slip (resistive torque nil) of the motor can be inferred. In the above diagram I_2, the d.c. current between the rectifier and the inverter, and the current taken by the primary winding of transformer TR (whose secondary is connected to the mains), are all nil.

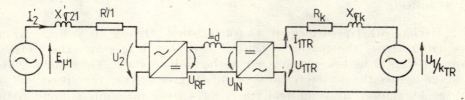

Fig. 12.9. Equivalent circuit of the asynchronous electric cascade (Scherbius-cascade) with a static converter.

According to the same diagram, for the no-load conditions we can write

$$E\mu_1 = U_2' = U_2 \frac{1}{s_0} \cdot \frac{w_1 k_{w1}}{w_2 k_{w2}} \qquad (12.21)$$

where the symbols have the same meanings as those in Section 12.1.1. For the input and output voltages of the rectifier RF, the following relation can be written

$$U_2 = k_u U_d \qquad (12.22)$$

where U_d is the rectified voltage and k_u depends on the rectifier wiring diagram $\left(k_u = \dfrac{\pi}{3\sqrt{6}} \text{ for the three-phase bridge}\right)$.

The inverter input and output voltages satisfy the relation

$$k_u U_1 = U_{1TR} |\cos\alpha|$$

with values over 90° (inverter mode), while $\cos\alpha$ has a negative relative value.

On the other hand, if U_1 is the mains phase voltage (and hence the transformer TR secondary voltage) and U_{1TR} is the transformer primary voltage, then

$$k_t U_{1TR} = U_1 \qquad (12.23)$$

where k_t is the conversion ratio of the matching (recovery) transformer.

However, since $U_d = U_1$ (Fig. 12.9), it follows from equations (12.21) that

$$s_0 = \frac{U_2}{E} \cdot \frac{w_1 k_{w1}}{w_2 k_{w2}} = \frac{U_1}{k_t E \mu_1} \cdot \frac{w_1 k_{w1}}{w_2 k_{w2}} |\cos\alpha|$$

The no-load slip will therefore be proportional to $|\cos\alpha|$. For $\alpha = 90°$, $s_0 = 0$ (i.e. the natural characteristic), assuming that the transformer and the intermediate smoothing choke sc, connected between the rectifier and the inverter, do not induce additional impedances which could be reflected in the rotor circuit of the motor. As a rule, the maximum permitted slip is 0.5, to avoid oversizing of the rectifier, inverter and recovery transformer otherwise needed when too high power values are likely to be met in practice.

Mechanical (speed-torque) characteristics. Owing to voltage drops occurring across the motor in the on-load actuating state, the rectified voltage U_d will be lower than that in the no-load state. Thus, for a given voltage across the inverter, the slip will be greater, i.e. the speed will be lower. Therefore, in an on-load state, for a constant control angle α, the voltage drop across the motor will cause its speed to change relative to that of a no-load state. And because these voltage drops are proportional to the rotor current, itself proportional to the torque for all practical purposes, it follows that speed changes are proportional to torque changes.

As a consequence, the mechanical (speed-torque) characteristics of the sub-synchronous cascade will be represented by a family of curves which, throughout the linear section of the slope, are practically parallel to one another. The family of speed-torque characteristics that may be obtained for various values of angle α are represented in Fig. 12.10. As can be seen, these characteristics are similar to those of the d.c. motor controlled by changing the voltage applied across the rotor circuit. Consid-

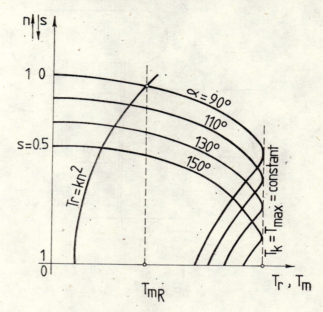

Fig. 12.10. Speed-torque characteristics of the static converter electric cascade (Scherbius cascade).

ering the effect of speed drop on the cooling process to be negligible, the asynchronous motor can be loaded over the whole range of the speed control required up to the nominal current; therefore, the sub-synchronous cascade can achieve a speed change at a constant torque.

Power balance sheet. The power balance sheet of the sub-synchronous cascade with a static converter is shown in Fig. 12.11. The iron and Joule stator losses are deducted from the total active power P_1 taken by the motor from the mains, while the remainder is transferred to the rotor circuit. The fraction $(1-s)$ of this electromagnetic power (denoted by P) is converted into mechanical power and the major part will generally be the useful power P_2 supplied to the machine. Slip power sP, less the Joule losses in the rotor circuit, is rectified, converted to frequency f_1 and fed back to the mains via the recovery transformer, less the transformer losses Σp_{TR}.

If the motor requirements equal the nominal power P_{1R} and if other losses are neglected, then $P_{1R} = P = T\Omega_1$; hence, $T = \dfrac{P_{1R}}{\Omega_1} =$ constant. However, the mechanical power developed is $P_2 = T\Omega_1(1-s)$, i.e. it varies with slip s which, in turn, is controlled by angle α. The range over which the motor can be used, with no need of being overloaded by the cascade arrangement, is shown in Fig. 12.11.

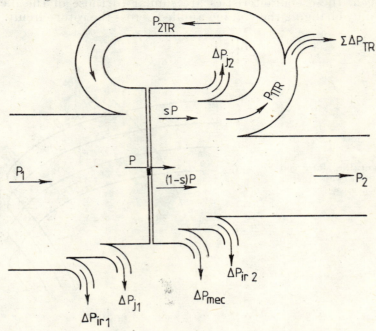

Fig. 12.11. Power balance diagram of the static converter electric cascade (Scherbius-cascade).

Design analysis of static components. The sub-synchronous static converter cascade arragement includes, as its main static components, the rectifier and the inverter. The power for which each component will be designed depends on the extent of the range over which the speed is to be controlled, Δn, and hence on the maximum slip, s_{0max}, corresponding to the no-load actuating state of the cascade arrangement. In what follows, we deal with each separate static component of such an arrangement.

The rectifier components are designed as a function of the mean value of the rectified current, I_d, and of the maximum value of the ideal rectified no-load voltage, U_{doi}. The usual relations below refer to silicon diodes in a non-controlled three-phase bridge arrangement.

Assuming the inductive voltage drop to be negligible ($d_x = 0$), the mean rectified current values can be found from the relation

$$I_d = 1.20 \sqrt{\frac{3}{2}} I_2 = 1.20 \times 1.22 \, I_2 \qquad (12.24)$$

where 1.20 is a safety coefficient allowing for the comparatively high leak reactance of the motor.

The maximum value of the ideal no-load rectified voltage is

$$U_{doi} = \frac{3\sqrt{2}}{\pi} s_{0\,max} U_{20} = 1.35\, s_{0\,max} U_{20} \qquad (12.25)$$

where U_{20} represents the phase voltage in the rotor circuit for the no-load locked rotor state.

To design the inverter components, the following data should be available: the mean value of the rectified current, I_d, from relation (12.24); and the ideal no-load voltage of the inverter, $U_{di\,IN}$, as derived from the maximum value of U_{doi} [see relation (12.25) above], corresponding to the minimum speed over the required control range. In the above analysis, it is assumed that the inverter incorporates a thyristor three-phase controlled bridge arrangement [12.12].

The ideal no-load voltage of the inverter $U_{di\,IN}$ can be approximated by two different relations, depending on the recovery network chosen. If recovery is through the mains, then

$$U_{di\,IN} = \frac{-1}{\cos\alpha_{max} - d_{IN}} U_{doi} \qquad (12.26)$$

where $\cos\alpha_{max}$ is the maximum ignition (control) angle and d_{IN}, the sum of voltage drops as referred to $U_{di\,IN}$. Because the recovery transformer in such a case is connected to the same network as the electric motor, voltage variations will be evenly distributed over the whole electric system and, as a consequence, they can be overlooked. If, however, the recovery is via another network, then

$$U_{di\,IN} = \frac{-1}{(1 - \Delta U_r)\cos\alpha_{max} - d_{IN}} U_{doi\,max} \qquad (12.27)$$

where ΔU_r is the voltage drop across the recovery network, $U_{doi\,max}$ the ideal no-load voltage across the d.c. intermediate circuit corresponding to minimum speed and maximum stator voltage, and the other symbols have the same meanings as in equation (12.26).

Example 12.1. **Design analysys of static components of a sub-synchronous static converter cascade.** It is required to design a variable-speed drive for a radial pump, based on a sub-synchronous static converter cascade. The data available are

(i) *operating parameters of the radial pump*: discharge head for a zero output, $H_0 = 73$ m; static head of pumping system, $H_s = 41$ m; pump power rating, $P_{pR} = 450$ kW; speed, $n = 1475$ rpm;

(ii) *operating parameters and rating of the wound-rotor asynchronous motor*: power, $P_R = 500$ kW; speed, $n_R = 1475$ r.p.m.; efficiency, $\eta_R = 94.3\%$; power factor, $\cos\varphi = 0.84$; stator voltage, $U_{1n} = 3300$ V; stator current, $I_{1R} = 114$ A; locked-rotor breakaway voltage in no-load actuating state, $U_{20} = 600$ V; and rotor current (for $P_R = 500$ kW), $I_{2R} = 510$ A;

(iii) *operating parameters of the electric network*: supply voltage, $U_{sR} = 3200$ V $\pm 5\%$; supply voltage frequency, $f_{sR} = 50$ Hz; voltage of the recovery network, $U_{rR} = 500$ V $\pm 5\%$; and frequency of the same, $f_{rR} = 50$ Hz.

Solution

(1) *Estimation of initial quantities*

Minimum pump speed over the required control range:

$$n_{min} = n_R - \sqrt{\frac{H_{st}}{H_0}} = 1475\sqrt{\frac{41}{73}} = 1{,}100 \text{ rpm}$$

Maximum rotor slip over the required control range

$$s_{max} = 1 - \frac{n_{min}}{n_0} = 1 - \frac{1{,}100}{1{,}500} = 0.266 = 26.60\%$$

Maximum phase-voltage across the rotor, corresponding to the minimum speed

$$U_2 = s_{max} U_{20} = 0.266 \times 600 = 160 \text{ V}$$

Maximum rotor current, corresponding to the maximum power and maximum speed

$$I_2 = \frac{P_{max}}{P_R} I_{2n} = \frac{450}{500} 510 = 459 \text{ A}$$

(2) *Rectifier design analysis.* The rectifier is designed on the assumption that it features a non-controlled three-phase bridge, using semi-conducting diodes throughout. In this case, the following data are needed:

— the ideal no-load voltage

$$U_{doi} = 1.35 \times U_2 = 1.35 \times 160 = 216 \text{ V}$$

— the maximum rectified current (theoretical value)

$$I_d = 1.20 \times 1.22 \times 459 = 672 \text{ A}$$

— the rated rectified current supplied by the rectifier

$$I_{dR} = k_1 I_d$$

In this last expression, k_1 can only be found if **two** other quantities are known

— the minimum slip corresponding to the maximum speed

$$s_{min} = 1 - \frac{n_{max}}{n_0} = 1 - \frac{1{,}468}{1{,}500} = 0.0213 = 2.13\%$$

— the frequency of the rotor voltage corresponding to the minimum slip

$$f_{2\,min} = s_{min} f_R = 0.0213 \times 50 \approx 1 \text{ Hz}$$

Having found these data, we look into the Brown Boveri catalogue and choose DSA 200 diodes. Thus, for $f = 1$ Hz and $\alpha = 120°$, the reduction coefficient will be 0.97. Therefore

$$k_1 = \frac{1}{0.97}$$

and

$$I_{dR} = \frac{1}{0.97} \times I_d = 1.03 \times 672 = 695 \text{ A}$$

From the same catalogue, we choose a standard rectifier featuring a non-controlled three-phase bridge arrangement, designed for a rated current of $I_{dR} = 780 \text{ A} > 695 \text{ A}$.

The diode off-state (reverse) voltage, neglecting the effect of the smoothing choke for $\pm 5\%$ change in the supply voltage and an ignition angle $\alpha = 30°$, has the peak value

$$\hat{U}_r \approx \frac{1.05}{\cos 30°} \times 1.05 \times U_{doi} = \frac{1.05^2}{0.866} \times 216 = 274 \text{ V}$$

Consequently, we choose diodes corresponding to a rated current of 200 A (split between two parallel-connected diodes) and to the rated off-state (reverse) voltage of 600 V > 274 V.

(3) *Inverter design analysis.* The design analysis below is carried out assuming that thyristors, connected in a controlled three-phase bridge arrangement, are used throughout, and that the inverter is connected to the recovery network via a recovery (matching) transformer. On this basis, the following data are needed:

— maximum rectified voltage for ideal no-load running

$$U_{doi\ max} = 1.05 U_{doi} = 1.05 \times 216 = 227 \text{ V}$$

— cosine of the maximum ignition angle for $\alpha_{max} = 155°$

$$\cos \alpha_{max} = -0.906$$

Summing up the resistive, d_{r1}, and the inductive, d_{x1}, voltage drops, we find the total voltage drop across the inverter, d_{IN}. For the rated current of the transformer (taken as equal to the current corresponding to the maximum speed), the voltage drops above will have the following values

$d_{r1} = 0.03$ (depending on the curent)
$d_{x1} = 0.04$ (depending on the current)
$d_c = 0.01$ (independent of the current)

Total $d_{IN} = 0.08$

When the asynchronous motor is used as a turbopump prime-mover, the current ratio in ideal conditions will be

$$i_d = \frac{\text{Current corresponding to the minimum speed}}{\text{Rated current of the recovery transformer}}$$

or

$$i_d = \left(\frac{n_{min}}{n_R}\right)^2 = \left(\frac{1 \times 100}{1 \times 468}\right)^2 = 0.56$$

from which

$(d_r + d_x) = (d_{r1} + d_{x1}) \times i_d$
$\qquad\qquad\ = 0.07 \times 0.56 \approx 0.04$
$d_c = \qquad\qquad\qquad 0.01$

Total $d_{IN} = \qquad\qquad 0.05$

With the total voltage drop found, we can now compute the ideal no-load rectified voltage of the inverter

$$U_{di\ IN} = \frac{-1}{(1-\Delta U_r)\cos \alpha_{max} - d_{IN}} U_{doi\ max} =$$

$$= \frac{-1}{(1-0.05)(-0.906) - 0.05} \times 227 = 250 \text{ V}$$

The nominal rectified current of the inverter is

$$I_{dR} = I_d = 672 \text{ A}$$

Having these data now available, we look into the Brown Boveri catalogue and select a standard inverter, featuring a controlled three-phase bridge arrangement, designed for a rated current

$$I_{dR} = 750 > 672 \text{ A}$$

Assuming a possible mains voltage surge of 5%, the maximum value of the thyristor off-state (reverse) voltage will be

$$\hat{U}_r = 1.05 \times 1.05 \times U_{di\ IN} = 1.05^2 \times 250 = 275 \text{ V}$$

The maximum value obtained for the off-state voltage, denoted by U_{DRM} in the positive direction and by U_{RRM} in the reverse direction, is multiplied by an overvoltage coefficient, taken as 2.2. Hence

$$U_{DRM} \text{ or } U_{RRM} = 2.2 \hat{U}_r = 2.2 \times 275 = 605 \text{ V}$$

From the same catalogue, we now choose the inverter thyristors as a function of the maximum mean on-state current $I_{TAV} = 200$ A (split between two parallel-connected thyristors), and of the cutt-off voltage U_{DRM} or $U_{RRM} = 800$ V > 605 V.

Remark : The notation used above, that is I_{TAV}, U_{DRM} and U_{RRM}, is recommended by the International Electricity Code.

Special arrangements. One of the operational peculiarities of an asynchronous motor with slip-rings, connected in a sub-synchronous static converter cascade arragement, is the reduction of the power factor (Fig. 12.12). This deterioration of the power factor arises because the

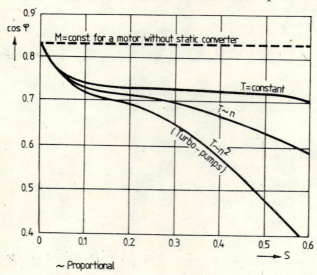

Fig. 12.12. Variation curves for the power factor as a function of slip for a cascade-connected asynchronous motor, driving various working machines :

curve $T =$ constant, for working machines with a constant resistive torque ; curve $T \sim n$, for working machines having a linear resistive torque ; curve $T \sim n^2$, for working machines with a parabolic resistive torque.

converter reactive power can only be taken via the motor, and consequently for the same torques the motor will have higher current requirements. As a consequence, the cascade arrangement loads the a.c. network with both the reactive power absorbed by the motor and that taken up by the inverter. However, it is worth noting that whereas the reactive power taken up by the motor depends only on the load (resistive) torque, that taken up by the inverter also depends on inverter modulation which is proportional to the extent of the required range of speed control. Therefore, we can say that the reactive power absorbed by the motor has a fixed value, whereas that absorbed by the inverter can be reduced, but only if special cascade arrangements are used for the purpose.

The power losses, undergone by the converters of a cascade arrangement, depend on both the maximum rotor voltage (which, in turn, corresponds to the minimum speed) and the maximum rotor current (which, in turn, corresponds to the maximum motor torque). Being machines with a parabolic speed-torque characteristic, i.e. their load (resistive) torque varies with the square of the speed, the asynchronous motors, used as adjustable speed drives for turbopumps, show high rotor voltage and law rotor current at low speeds, whereas, at high speeds, the rotor voltage is low and the rotor current is high. This being so, if special cascade arrangements are used, the voltage applied across the inverter ter-

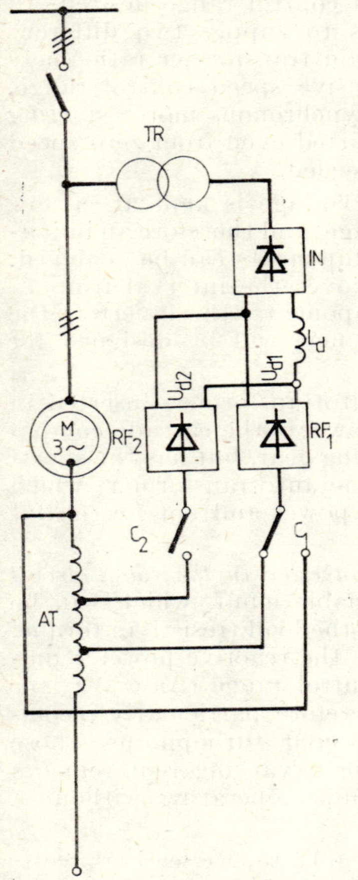

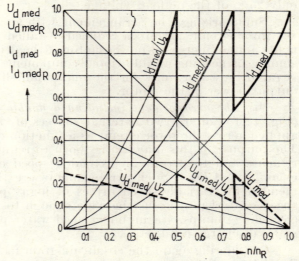

Fig. 12.13. Electric wiring of an asynchronous cascade with an autotransformer in the rotor circuit:

TR — transformer; M — motor; RF_1, RF_2 — rectifiers; IN — inverter; AT — autotransformer; C_1, C_2 — switches.

Fig. 12.14. Adapting current and voltage in the rotor circuit of a cascade-connected asynchronous motor with an autotransformer in the rotor circuit.

minals can be reduced at low speeds and hence a reduction in the converter power losses will result. Two special cascade arrangements for wound-rotor asynchronous motors, with autotransformer or a double converter in the rotor circuit, are described below.

(a) *The first arrangement* (Fig. 12.13) *features an autotransformer, inserted in the rotor circuit*, to reduce the voltage which still has high values at low speeds. Considering the parabolic characteristic of the turbopump resistive torque, the transformer ratio will be so chosen that the maximum current corresponds to the rated speed (Fig. 12.14).

The autotransformer should be designed for one or two output voltages, according to the extension of the speed control range desired. In Fig. 12.13 the autotransformer has two taps to supply two different output voltages, each corresponding to a certain transformer ratio (here 1 : 4 and 1 : 2). These ratios permit an extensive speed control range, the lower limit of which corresponds to the asynchronous motor starting speed; thus, the driving equipment may be started even from zero speed without any other starting auxiliaries being needed.

Owing to the autotransformer, the rotor voltage is kept at a low enough level (not above 25% of the no-load voltage) and therefore an important reduction in the power of the cascade components can be achieved. If, for instance, the speed is to be controlled over the interval from 75 to 100% of the rated speed, the cascade components (the inverter, the smoothing choke and the matching transformer) will be designed for only 25% of the no-load voltage.

Summarizing, the arrangement with an autotransformer, inserted in the rotor circuit, permits a reduced installed power of the cascade components and an increased power factor of the driving gear, but has the shortcoming of needing an additional component (the autotransformer) which is rated, however, only for the maximum slip power and not for the full output power of the motor.

(b) *The arrangement including a double converter in the rotor circuit* (Fig. 12.15) offers two external sources of variable e.m.f. which can be applied across the rotor circuit as a function of the load (resistive) torque. The purpose of this diagram is both to reduce the reactive power requirements and to permit an extended speed control range ($\Delta n = 1 : 7$ or even $\Delta n = 1 : \infty$). Such arrangements are therefore particularly recommended for variable-speed drives of high powered turbopumps whose speed control range must be wide enough to meet varying requirements (recycling pumps, steam boiler feed-water pumps, operating without a static discharge head, etc.).

Considering again the circuit diagram in Fig. 12.15, we see that rectifier *RF1* is directly connected to the rotor circuit, whereas rectifier *RF2* is connected to the same circuit via the circuit-breaker *I*. The matching transformer has three windings. One is the secondary, and the other two primary windings of the transformer are connected to inverters *IN1* and *IN2*. The cascade is tripped on via two starting resistors *Rs1* and *Rs2* in the d.c. intermediate circuits.

This arrangement is best suited for the variable-speed drives of machines with a parabolic speed-torque characteristic whose resistive torque is proportional to the square of the speed, i.e. turbopumps. The two converters may be either series or parallel-connected, depending on the resistive shaft torque developed by the turbopump, which is in turn a function of speed. As a consequence, the control is divided into two sections: a low speed range and a high speed range. For the lower section, where the resistive torque is much lower than its rated value, the converters are

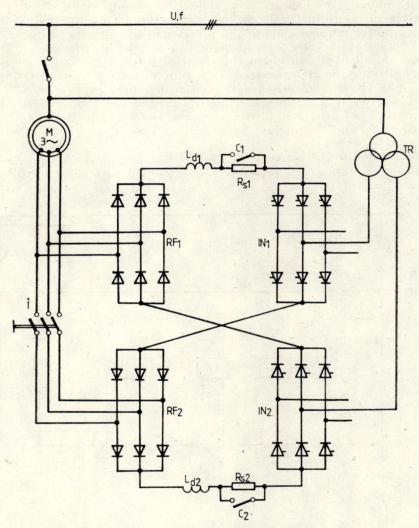

Fig. 12.15. Electric wiring of an asynchronous cascade with a double converter in the rotor circuit:

M — motor; TR — transformer; RF_1, RF_2 — rectifiers; IN_1, IN_2 — inverters; L_{d_1}, L_{d_2} — smoothing coils; C_1, C_2 — switches.

series-connected and, consequently, an additional e.m.f. will result which is twice as high as that obtained with other cascade arrangements using converters. For the higher section, where the resistive torque and the rotor current are substantially increased, the converters are parallel-connected and their power will consequently be halved when compared with that of the ordinary cascade arrangements.

In summary, we can say that the second arrangement, (*b*), of double converter in the rotor circuit, achieves the same level of reduction in the reactive power requirements of the inverter as (*a*).

A double-converter arrangement was built by Siemens (F.R. of Germany) and used to control the speed of three slip-ring asynchronous motors, driving the high-powered pumps of a steam generator feed-water plant in an electric power station. Each electric motor has $P = 8{,}200$ kW and $n = 1{,}490$ rpm. The speed control range ($\Delta n = 780 - 1{,}490$ rpm) of each motor was broken into two sections, the lower section $\Delta n_l = 750 - 1{,}125$ rpm. and the higher $\Delta n_h = 1{,}125 - 1{,}490$ rpm. [12.4], [12.18].

Connection to the network. To recover the slip power of the cascade asynchronous motor, the latter is connected with its rotor circuit to a recovery network, which can either be the mains or another recovery circuit.

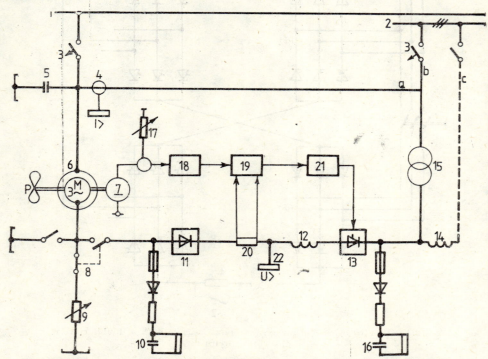

Fig. 12.16. Connection to the mains of a static converter sub-synchronous cascade :

1 — *HV* line; *2* — *LV* line; *3* — switch, protected against overloads and short-circuit hazards; *4* — protection against short-circuits; *5* — protection against excessive voltages and off-set of the reactive power; *6* — wound-rotor asynchronout motor; *7* — speed-voltage generator; *8* — switch; *9* — starting rheostat; *10* — protection against excessive voltages across the rectifier; *11* — rectifier; *12* — smoothing coil; *13* — inverter; *14* — a.c. coil; *15* — matching (recovery) transformer; *16* — protection against excessive voltages across the inverter; *17* — potentiometer; *18* — speed controller; *19* — current controller with limiting feature; *20* — shunt; *21* — ring counter; *22* — protection against excessive voltages across the d.c. intermediate circuit.

Adjustable-Speed Drives

There are several possible wiring diagrams for the cascade arrangement and the recovery network. Looking at Fig. 12.16, they can be described as follows:

(i) *the asynchronous motor 6* is supplied from the mains *1* (more often than not, a high-voltage network) while the slip power is recovered in the same network via a recovery matching transformer *15* (path *a* in Fig. 12.16);

(ii) *the asynchronous motor 6* is fed through the mains *1* (high-voltage network) while the slip power is recovered in another network *2* (in most cases, a low-voltage network) via a recovery (matching) transformer *15* (path *b* in Fig. 12.16);

(iii) *the asynchronous motor 6* is fed from the mains *1* (high-voltage network) while the slip power is directly recovered in another network *2* (low-voltage), with no recovery (matching) transformer (path *c* in Fig. 12.16).

Typical control diagram. The control diagram of a static converter sub-synchronous cascade arrangement (Fig. 12.17) includes a speed control stage, followed by a current control stage (a speed-current cascade diagram). The actual speed value is obtained via speed transducer T_n (a d.c. tacho-generator) which supplies a voltage proportional to speed that is subsequently compared with the voltage output of potentiometer *Pot*, which represents the selected value for speed. The real current value

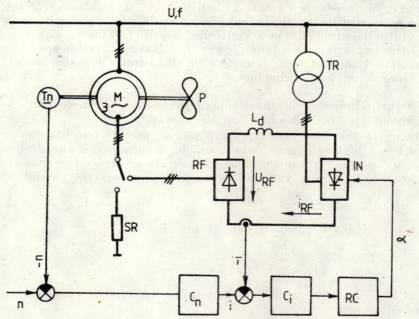

Fig. 12.17. Typical control arrangement for a static converter electric sub-synchronous cascade (Scherbius-cascade):

P — pump; M — wound-rotor asynchronous motor; RF — rectifier; IN — inverter; TR — matching (recovery) transformer; L_d — smoothing coil; SR — starting rheostat; T_n — speed-voltage generator (speed transducer); C_n — speed-controller; C_i — current controller; RC — ring counter.

is obtained via a current transformer connected in the converter d.c. intermediary circuit. The output of speed controller C_n supplies the set value for the current controller C_i; during speed-up and overload time periods, this C_i component also limits the current level to a value, suitable for the actuating end component IN. The speed of the asynchronous motor is controlled via angle α. The pulses across the thyristor gates are supplied by the ring counter RC, which features a grid control device, sychronised to the supply network of the asyncronous motor. Hence, the e.m.f. induced in the transformer is reflected in the rotor circuit of the motor at a level, controlled via the angle α.

In the case of a sub-synchronous cascade with static converter, speed control is achieved in the same way as for a d.c. motor whose commutator is fed from a controlled rectifier, except that the sub-synchronous cascade uses the current level of the converter d.c. intermediate circuit as a control variable instead of the current level of the supply circuit of the motor. These two driving systems show great similarity, since both use the current to set the level of the motor torque. There is, however, a difference: whereas with constant electromagnetic fields the motor torque increases more or less linearly with the rotor current in the case of a d.c. motor, in the case of the wound-rotor asynchronous motor — all other factors being the same — the motor torque increases more or less linearly with the active component of the rotor current, that is, with the current in the converter d.c. intermediate circuit. As can be seen, the static converter sub-synchronous cascade is a simplified arrangement, as is the d.c. motor whose commutator is supplied via a controlled rectifier. This is why the total cost of this arrangement is much lower than, say, a variable-speed control system based on frequency changes in the supply voltage, which is a much more complex arrangement.

Energetic indices. The major energetic indices of an electric cascade arrangement are the efficiency and the power factor.

Owing to the recovery of the slip power, the electric cascade arrangement has a high efficiency, which can be determined if the total losses across the cascade are known. These losses are given by the sum of constant losses $\Delta PC_{el\,c}$ and variable losses $\Delta PV_{el\,c}$ across the cascade

$$\Delta P_{el\,c} = \Delta PC_{el\,c} + \Delta PV_{el\,c} \tag{12.28}$$

The efficiency is given by

$$\eta_{el\,c} = \frac{\Delta P_{2\,el\,c}}{\Delta P_{2\,el\,c} + \Delta P_{el\,c}} \tag{12.29}$$

where $P_{2\,el\,c} = Tn/1{,}000$ is the useful power, delivered at the shaft of the driving equipment.

The constant losses $\Delta PC_{el\,c}$ are independent of the load of the asynchronous motor and include the mechanical losses due to friction (in the bearings and cooling fan), the iron losses in the stator core only, (through hysteresis and eddy currents), and other losses. These losses can be determined by summing up the constant losses of the asynchronous motor, operating at the rated parameters ΔPC_{mn} (plus 5%) and the constant losses in the recovery (matching) transformer in no-load conditions ΔPC_{TRo}

$$\Delta PC_{el\ c} = 1.05\ \Delta PC_{mR} + \Delta PC_{TRo} \qquad (12.30)$$

At rated speed, the constant losses of the asynchronous motor are given by

$$\Delta PC_{mR} = P_R \frac{1 - \eta_R}{\eta_n} - \Delta PV_{mR} \times 0.001 =$$

$$= P_R \frac{1 - \eta_R}{\eta_R} - 10\ T_R \omega_1 s_R \left(1 + \frac{r'_1}{r_2}\right) \times 0.01 \qquad (12.31)$$

where ΔPV_{mR} represents the variable losses of the motor at the rated speed.

In the case of a sub-synchronous cascade, the variable losses are substantially affected by the motor load and hence depend directly on the current requirements. In fact, these losses are proportional to the second power of the current and represent the winding losses (both of the stator and of the rotor). They may be expressed as

$$\Delta PV_{el\ c} = I_d^2 R'_{eqv} + 3 I_d \Delta U \qquad (12.32)$$

where R'_{eqv} is the equivalent resistance of the motor, which includes the referred resistance of stator winding r'_1.

The equivalent resistance R'_{eqv} depends on the inverter wiring. For the case of a three-phase bridge

$$R'_{eqv} = 2r_2 + 2r'_1 + r_{SC} + 2r_{TR} \qquad (12.33)$$

where r_2 is the resistance of the rotor winding, r'_1 — the referred resistance of the stator winding, r_{SC} — the resistance of the smoothing choke, r_{TR} — the reactance of transformer windings as referred to that of secondary windings.

The rectified current I_d depends on the motor load. It is

$$I_d = \frac{\sqrt{2}\ E_2}{2x_2} - \sqrt{\frac{E_2^2}{2x_2^2} - \frac{\sqrt{2}\ 10\ \omega_1\ T}{k_1 x_2}} \qquad (12.34)$$

The current, taken by an asynchronous motor from the mains, always lags in phase behind the voltage applied in symmetrical, balanced conditions. The physical explanation is the permanent need for an asynchronous motor to take up the reactive power, needed for its excitation. As a consequence, the motor operates as an induction coil and its power factor therefore always has an inductive, sub-unity value. In no-load conditions, the power factor is very low, of the order 0.2, since the active power requirements are comparatively low too (equal to machine losses), while the reactive power, needed for excitation is high enough — in fact the same as that required at full load. As the motor takes up load and the useful power developed increases, the active power, taken from the mains, increases too, while the reactive power stays practically at the same level, and therefore the power factor is improved, reaching a maximum value of 0.85—0.90 at about the useful rated power.

In all applications involving speed control of a cascade—connected asynchronous motor, the power factor of the motor deteriorates since the reactive power for the converters can be obtained only through the motor (i.e. the reactive power is taken up not only by the motor but by the inverter too). Thus, for the same torques, the current requirements of the motor will be greater.

The power factor can be estimated by starting from the phasor diagrams of a sub-synchronous cascade arrangement with a static converter (Fig. 12.18). We write

$$P_{el\ c} = P_m - P_{TR}$$
$$Q_{el\ c} = Q_m - Q_{TR}$$
$$\cos \varphi_{el\ c} = \nu_d \frac{P_m - P_{TR}}{\sqrt{(P_m - P_{TR})^2 + (Q_m + Q_{TR})^2}} \qquad (12.35)$$

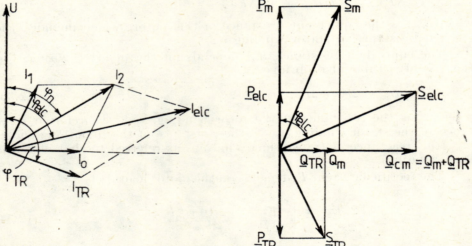

Fig. 12.18. Simplified phasor diagrams, used to derive the power factor of a static converter sub-synchronous cascade.

where $\nu'_d = 0.955$ is a coefficient, expressing the current characteristic degradation in both the stator and the transformer primary circuits;

$$P_m = \frac{T n_0}{955}; \quad Q_m = P_m \tan \varphi_m \qquad (12.36)$$

represents the active and the reactive power requirements, respectively, of the asynchronous motor;

$$P_{TR} = \frac{T}{955}\left(n_0 - \frac{n}{n_{sc}}\right); \quad Q_{TR} = \sqrt{S_{TR}^2 - P_{TR}^2} \qquad (12.37)$$

represent the active and reactive power developed, respectively, in the recovery (matching) transformer primary, or in the inverter circuit (if no recovery transformer is included in the arrangement);

$$S_{TR} = 0.815 \, m_{TR} I_d E_{2\,TR} \qquad (12.38)$$

is the apparent power of the transformer (for the arrangement with a three-phase bridge for the inverter).

The expression for the motor power factor in a sub-synchronous cascade arrangement with a static converter is

$$\cos\varphi_m = \cos\varphi_0 \cdot \cos\frac{\gamma}{2} \qquad (12.39)$$

where

$$\cos\varphi_0 = \frac{0.815\,I_d}{\sqrt{(0.815\,I_d)^2 + I_0^2 k_{TR}^2}} \qquad (12.40)$$

and

$$\cos\frac{\gamma}{2} = \sqrt{1 - \frac{2x_2 I_d}{2E_2}} \qquad (12.41)$$

and finally

$$\cos\varphi_{.n} = \frac{0.815}{\sqrt{2}\,E_2}\sqrt{\frac{2E_2^2 - \sqrt{2}\,x_2 E_2 I_d}{(0.815\,I_d)^2 + I_0^2 k_{TR}^2}} \qquad (12.42)$$

The low power factor, both at low loads and at full load, is an important drawback of the sub-synchronous static converter cascade. This factor should therefore be improved. For this purpose, condenser banks are used in most cases for high power variable-speed drives.

The power requirements of such a condenser bank are derived from the equation

$$Q_{comp} = Q_{el\,c} - Q'_{el\,c} \qquad (12.43)$$

and the power factor, obtained after compensation

$$\cos\varphi_{el\,c\,comp} = \nu_d \cdot \frac{P_m - P_{TR}}{(P_m - P_{TR})^2 + (Q_m + Q_{TR} - Q_{comp})^2} \qquad (12.44)$$

where $Q'_{el\,c}$ expresses the reactive power requirements, corresponding to the standard power factor, $\cos\varphi_{std}$, and may be written

$$Q'_{el\,c} = P_{el\,c} \tan\varphi_{std} \qquad (12.45)$$

Benefits and drawbacks. Benefits of the sub-synchronous static converter cascade include:

(i) *it uses an asynchronous motor with a wound-rotor*, which is more sturdy, more economical, and lighter than a d.c. motor for the same power and speed ratings;

(ii) *as a general rule, the power of the cascade arrangement is lower* than that of the asynchronous motor, being equal to the product of the power multiplied by the maximum slip required in operation, whereas for a d.c. motor the power of the rectifier circuit should be equal to the full output power of the motor, whatever the range of speed adjustment;

(iii) *it is highly efficient*, owing to the recovery of the slip power, developed by the rotor of the asynchronous motor.

Drawbacks include:

(i) *a low power factor*, as a result of high reactive power requirements, because not only the motor but also the converter takes up power;

(ii) *slightly reduced motor efficiency* at maximum speed, but only when the rotor is not short-circuited at the maximum speed;

(iii) *owing to the narrow speed control range*, the electric cascade needs additional starting means (an external starting resistor).

Scope. Because the power, and therefore the cost, of the components of a sub-synchronous static converter cascade depend on the extent of the required speed control range, this arrangement is generally economical in such applications as the variable-speed drives of turbopumps with ratings from 400 to 25,000 kW which require a narrow speed control range ($\Delta n = 1 : 1.2 - 1 : 2$). In most cases, such turbopumps are used in pumping systems with static heads (with back-pressure), examples being drinking and industrial water supplies, pressurized irrigation systems, etc. In some cases, where a wide speed control range is required, e.g. the variable-speed drives of high-powered turbopumps in pumping systems with no static head, special arrangements are used, such as the autotransformer or the double converter arrangement (see special arrangements). The latter case includes the turbopumps, used for recycling purposes or for feed-water supplies of steam boilers.

Compared with other variable-speed drives, the sub-synchronous static converter cascade is the arrangement most commonly used with turbopumps. This is confirmed by world-wide practice. For instance, Siemens (F. R. of Germany) has supplied a full range of sub-synchronous static converter cascade equipment from 1964 onwards, most of which were designed as variable-speed drives of turbopumps of various ratings. More details on this subject are to be found in Table 12.1.

12.2.2 The Asynchronous Electromechanical Cascade (the Krämer-cascade)

Circuit diagram (Fig. 12.19). A wound-rotor three-phase asynchronous motor AM supplies, via a rotor circuit with a rectifier RF in a three-phase bridge arrangement, an independent excitation to a d.c. motor DCM.

The latter is rigidly coupled to the shaft of the asynchronous motor through a geared reducer.

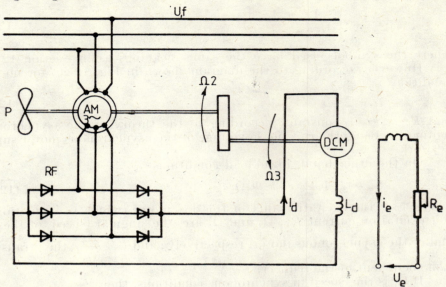

Fig. 12.19. Electric wiring of a static converter electromechanical cascade (Krämer):
P — pump; AM — wound-rotor asynchronous motor; DCM — d.c. motor; RF — rectifier; L_d — smoothing coil; R_e — excitation rheostat.

Mathematical analysis of operation. Speed control is obtained by changing the excitation current I_e of motor DCM. This is clear enough if we look at the main equivalent circuit of the rotor circuit of the asynchronous motor (Fig. 12.20) and assume that, when the system including the two rigidly coupled motors is in no-load conditions, the rotor phase current I_2 of the asynchronous motor and the rotor current I_a of the d.c. motor are both zero. The rms value of the phase-voltage at the input of the rectifier RF is U_2 and at the output U_a

$$U_2 = k_u U_a \qquad (12.46)$$

where coefficient k_u depends on the rectifier wiring diagram $\left(k_u = \dfrac{\pi}{3\sqrt{6}}\right)$ (for the case of the three-phase bridge). The rotor voltage U_2, referred to

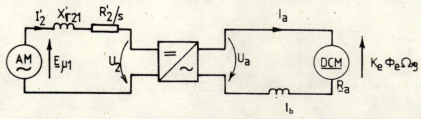

Fig. 12.20. Equivalent circuit of the static converter electromechanical cascade (Krämer-cascade).

591

the number of turns and to the stator frequency of the asynchronous motor is (according to the theory of electrical machines)

$$U'_2 = U_2 \frac{1}{s_0} \cdot \frac{w_1 k_{w1}}{w_2 k_{w2}} \qquad (12.47)$$

where the symbols used have the same meanings as in Section 12.1.1.

However, according to the diagram shown in Fig. 12.20, for no-load conditions

$$E_{\mu 1} = U'_2 \qquad (12.48)$$

and $E_{\mu 1} \approx U_1 =$ constant (according to the theory of the asynchronous motor), U_1 being the rms phase voltage of the asynchronous motor supply network.

On the other hand, in no-load conditions

$$U_a = k_e \Phi_e \Omega_3 = k_e \Phi_e \Omega_2 i \qquad (12.49)$$

where Φ_e is the excitation flux of the d.c. motor (which is a function of the excitation current I_e); Ω_3 and Ω_2 are the angular speed of the d.c. and of the asynchronous motor, respectively; and $i = \dfrac{\Omega_3}{\Omega_2}$, the transmission coefficient of the reducer.

If Ω_1 is the speed in synchronous conditions, then

$$\Omega_2 = \Omega_1(1 - s_0) \qquad (12.50)$$

according to the slip definition

$$\left(s = \frac{\Omega_1 - \Omega_2}{\Omega_1} \right)$$

From expressions (12.46) to (12.50), it follows that

$$E_{\mu 1} = U'_2 = U_2 \frac{1}{s_0} \times \frac{w_1 k_{w1}}{w_2 k_{w2}}$$

$$k_u = \frac{w_1 k_{w1}}{w_2 k_{w2}} \times \frac{1}{s_0} U_a = k_u k_e \Omega_1 i \frac{w_1 k_{w1}}{w_2 k_{w2}} \times \frac{s - s_0}{s_0} \times \Phi_e$$

or

$$s_0 = \frac{k_u k_e \Omega_1 i \dfrac{w_1 k_{w1}}{w_2 k_{w2}} \Phi_e}{E_{\mu 1} + k_u k_e \Omega_1 i \dfrac{w_1 k_{w1}}{w_2 k_{w2}} \Phi_e} \qquad (12.51)$$

Usually things are so arranged that, when $\Phi_e = \Phi_{eR}$, the two terms of the denominator of expression (12.51) are equal, and

$$s_0 = \frac{\Phi_e}{\Phi_e + \Phi_{eR}}$$

Adjustable-Speed Drives

In this case, for $\Phi_e = 0$, it follows that $s_0 = 0$ (the natural characteristic of the asynchronous motor, assuming that the motor resistance R_a, which is reflected in the rotor circuit of the asynchronous motor, is negligible).

For $\Phi_e = \Phi_{eR}$, $s_0 = 0.5$, which is the maximum value attainable because the rated flux cannot be exceeded without overloading the d.c. motor excitation circuit.

Mechanical (speed-torque) characteristics. The family of mechanical (speed-torque) characteristics of the system, built up with the two rigidly coupled motors on the same load shaft, is shown in Fig. 12.21, as a function of the excitation flux Φ_e. The abscissa is the total torque of the asynchronous motor, that is, the sum of the asynchronous motor torque T_{am}

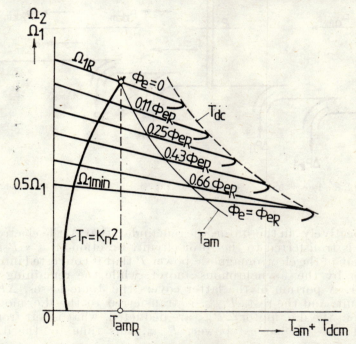

Fig. 12.21. Family of mechanical characteristics of the electromechanical cascade (Krämer-cascade).

and the d.c. motor torque iT_{dcm}, as referred to the asynchronous motor shaft. As can be seen from Fig. 12.21, with a fall of speed, the speed-torque characteristics and the overloading capacity of the motor increase (i.e. high torques are obtained at low speeds). For this reason, the electromechanical cascade is also called *"the constant power cascade"*. However, since the turbo-pump power not only changes but drops rather considerably with speed (because the mechanical power requirements of turbopumps vary in direct proportion to the third power of the speed)), this property of the electromechanical cascade is of no benefit in the case of variable-speed drives of centrifugal pumps; it is, however, rather useful in the case of variable-speed drives of volumetric pumps.

Power balance sheet. The power balance sheet of a Krämer cascade is represented in Fig. 12.22. If P_{1am} represents the power requirements of the asynchronous motor, while ΔP_{ir1} and ΔP_{j1} are the iron and Joule

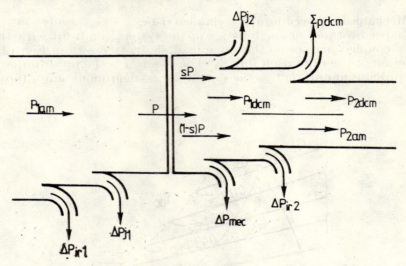

Fig. 12.22. Power balance diagram for the static converter electromechanical cascade (Krämer).

osses, respectively, in the stator, the remainder, that is, the electromagnetic power P, is transferred to the rotor circuit. Fraction $(1-s)P$ represents that portion of the electromagnetic power P that is converted into mechanical power by the asynchronous motor, while the remaining sP is the slip power. A portion of the latter covers the Joule losses, ΔP_{j2}, in the rotor circuit, and the rest, P_{1dcm}, is transferred to the d.c. motor. When the losses in the d.c. motor ΣP_{dcm} are deducted, what is left from P_{1dcm} is converted into mechanical power, P_{2dcm}, this time by the d.c. motor.

Ignoring the cascade losses and assuming that the motor takes up the rated power, then

$$P_{1R} \approx P_R = (1-s)P_R + sP_R$$

However, since $(1-s)P_R \approx M_{am}\Omega_2$ and $sP_R \approx T_{dcm}\Omega_3 = iT_{dcm}\Omega_2$, it follows that

$$P_{1R} \approx (T_{am} + iT_{dcm})\Omega_2$$

which demonstrates that for, a constant power fed to the input, the cascade supplies at the output a mechanical power practically constant, with high torques at low speeds and vice-versa, which is the basic feature of the Krämer cascade (see also Fig. 12.21).

Therefore, by increasing the field current of the d.c. motor by means of a simple field resistor R_c (see Fig. 12.19), speed Ω_2 of the cascade can be reduced down to about half Ω_1, i.e. half of the synchronism speed; moreover, this reduction is achieved smoothly and with practically no power losses. The d.c. motor power $P_{1dcm} \approx sP_{1R}$, where the slip s is taken at the maximum value reached by the cascade, that is 0.5. Hence, the power of the d.c. motor is about half that of an asynchronous motor. As a rule, slips in excess of 0.5 are not used, because the increased power requirements of the d.c. motor unduly increase its cost.

Typical control diagram. The control variable in any speed control arrangement is U_e, i.e. the field flux Φ_e. The control diagram of an electromechanical cascade is shown in Fig. 12.23. Like the electrical cascade,

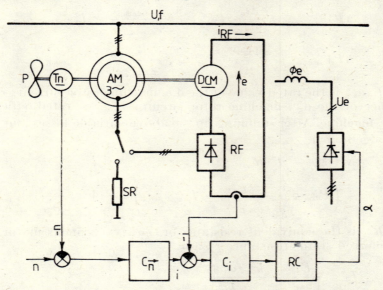

Fig. 12.23. Typical control arrangement for the static converter electromechanical cascade (Krämer-cascade):

P — pump; AM — asynchronous motor with a wound rotor; DCM — d.c. motor; RF — rectifier; SR — starting rheostat; Tn — speed-voltage generator (speed transducer); C_n — speed controller; C_i — current controller; RC — ring counter.

this arrangement includes controllers both for the rpm. and for the rectified current i_{RF}, which govern the control angle α of the controlled field rectifier.

The energetic indices, for an electromechanical cascade, are the efficiency and the power factor.

Efficiency is estimated as for the electrical cascade, and has the form

$$\eta_{em\ c} = \frac{\Delta P_{2em\ c}}{\Delta P_{2em\ c} + \Delta P_{em\ c}} \qquad (12.52)$$

where $\Delta P_{em\ c} = \Delta PC_{c\ em} + \Delta PV_{c\ em}$ is the total cascade loss.

The constant losses of the electromechanical cascade $\Delta PC_{c\ em}$ are equal to those of the asynchronous motor (mechanical + iron losses) plus 5% and are determined with equation (12.31).

The variable losses of the electromechanical cascade, $\Delta PV_{c\ em}$, include the losses dependent on the current requirements of the asynchronous motor, $I_d^2 R'_{eqv}$, and the losses induced by the d.c. machine, that is, the mechanical losses $\Delta PC_{mec\ dcm}$ and the iron losses $\Delta PV_{ir\ dcm}$.

The losses induced by the d.c. machine can be estimated accurately enough with the relations below

$$\Delta PV_{mec\ dcm} = \frac{\Omega_2}{2\Omega_{3n}} P_{dcm\ R} \frac{1 - \eta_{dcm\ R}}{\eta_{dcm\ R}} - I_{a\ dcmR}^2 \qquad (12.53)$$

$$\Delta PV_{ir\ dcm} = \frac{\Omega_2 \Phi^*}{2\Omega_{3S}} P_{dcm\ R} \frac{1 - \eta_{dcm\ R}}{\eta_{dcm\ R}} - I_{a\ dcm\ R}^2 r_a \qquad (12.54)$$

where $I_{a\ dcm\ R}$ is the rated value of the d.c. machine rotor current; r_a — total resistance of the d.c. machine rotor circuit; $\eta_{dcm\ R}$ — rated efficiency of the d.c. machine. After summing up the above variable losses and manipulating algebraically

$$\Delta PV_{em\ c} = I_d^2 R'_{eqv} +$$
$$+ \left(P_{dcm\ R} \frac{1 - \eta_{dcm\ R}}{\eta_{dcm\ R}} - I_{a\ dcm\ R}^2 r_a \right)(1 + \Phi^*) \frac{\Omega_2}{2\Omega_{3R}} \qquad (12.55)$$

where R'_{eqv} is the equivalent resistance of the asynchronous motor (including a referred stator resistance r_1, and

$$I_d = \frac{E_2}{\sqrt{2}\ x_2} + \frac{k_{dcm}\ \omega_1\ \Phi^*}{\sqrt{2}\ k_1 x_2} -$$
$$- \sqrt{\left(\frac{E_2}{2x_2} + \frac{k_{dcm}\ \omega \Phi}{2k_1 x_2} \right)^2 - \frac{10\sqrt{2}\ T\omega}{k_1\ x_2}} \qquad (12.56)$$

represents the rectified current. The coefficient $k_{dcm} = \dfrac{T_{dcm\ R}}{I_{aR}}$. The power factor of the electromechanical cascade is actually the asynchronous motor power factor, and hence $\cos \varphi_{em\ c} = \cos \varphi_{am}$. It is computed with equation (12.42).

Scope. The electromechanical cascade is applied to the variable-speed drives of pumps necessitating high torques at low working speeds (e.g.

volumetric pumps), which require only a narrow range of speed control ($\Delta n \leqslant 1 : 1.4$).

For the variable-speed drives of low-powered pumps ($P < 100$ kW), packaged electromagnetic cascades can be used. Making a longitudinal cut through a compact electromechanical cascade (Fig. 12.24), we can

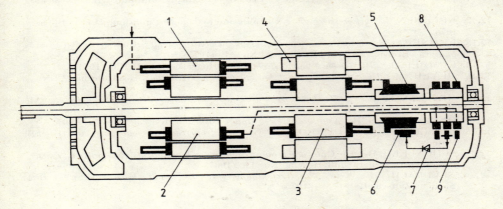

Fig. 12.24. Packaged electromechanical cascade (Krämer-cascade):
1, 2 — stator and rotor, respectively, of the wound-rotor asynchronous motor; *3, 4* — armature and inductor, respectively, of the d.c. motor; *5,6* — commutator and brushes, respectively, of the d.c. motor; *7* — semiconductor diodes of the rectifier; *8, 9* — rings and brushes, respectively, of the asynchronous motor.

see the asynchronous motor *2* and the armature *3* of the d.c. machine. Both are mounted on the same shaft and are housed, with the stator *1* of the asynchronous motor and the inductor *4* of the d.c. machine, in the same casing. The cascade rectifier is located on the front side of the casing and is built with silicon semiconducting diodes.

REFERENCES

12.1. ALBRECHT, S., GAHLEITNER, A., *Bemesung des Drehstrom Asynchronmotors in einer untersynchronen Stromrichterkaskade.* Siemens-Zeitschrift, No. 40 (1966).
12.2. BECKER, D., *Aufbau und Betriebsverhalten untersynchroner Stromrichterkaskaden.* Electro-anzeiger, No. 5, (1967).
12.3. ELGER, H., WEISS, M., *Untersynchrone Stromrichterkaskade als drehzalregelbarer Antrieb für Kesselspeisepumpen.* Siemens-Zeitschrift, No. 5, (1968).
12.4. ELGER, H., *Schaltungsvarianten des untersynchronen Stromrichterkaskade.* Siemens-Zeitschrift, No. 3, (1977).
12.5. FRANSUA, A., MĂGUREANU, R., CÎMPEANU, A., CONDRUC, M., TOCACI, M., *Mașini și sisteme de acționări electrice.* Technical Publishing House, Bucharest, 1978.
12.6. GLOBEL, O., *Drehstrom-Gleichstromkaskaden zum Antrieb von Wasserwerkspumpen.* Siemens-Zeitschrift, No. 10, (1962).
12.7. GRAF, K., *Entrainement des pompes à régulation de vitesse par convertisseur en cascade hyposyncrone pour l'approvisionnement en eau du Lac de Constance.* Revue Siemens, No. 6, (1973).

12.8. GRAF, K., *Wasserversorgung mit drehzahlgesteuerten Pumpen antrieben für die Stadt Barquismento (Venezuela)*. Siemens-Zeitschrift, No. 11, (1975).
12.9. GUNTER, S., RUDOLF, S., *Die grössten Kesselspeispumpen Antriebe mit Untersynchronerstromrichter-Kaskade*. Elektrizitätwirtschaft, No. 11, (1971).
12.10. IONEL, I. I., *Acționarea electrică a turbomașinilor*. Technical Publishing House, Bucharest, 1980.
12.11. IONEL, I. I., *Instalații de pompare reglabile*, Technical Publishing House, Bucharest, 1976.
12.12. KELEMEN, A., *Acționări electrice*. Didactic and Pedagogic Publishing House, Bucharest, 1976.
12.13. KUHNERT, S., *Antriebe mit Untersynchroner Stromrichter-Kaskade für Kesselspeispumpen*. Brown Boveri Nachrichtern, No. 11, (1975).
12.14. KUMMEL, F., *Elektrische Antriebstechnik*. Springer-Verlag, Berlin, 1971.
12.15. MEYER, M., *Uber die Untersynchrone Stromrichter-Kaskade*. Electrotechn. Z., No. 82, (1961).
12.16. MIKULASCHEK, F., *Die Ortskurven die Untersynchronen Stromrichter-Kaskade*. AEG-Mitt, No. 52 (1962).
12.17. ONISCENKO, G. B., *Asynhronnyi venilinnyi kaskad*. Energyia, Moscow, 1967.
12.18. ONISCENKO, G. B., IUNKOV, M. G., *Elektroprivod turbomehanizmov*. Energyia, Moscow, 1972.
12.19. SCHÖNFELD, R., *Die Untersynchrone Stromrichter-Kaskade als Regelantrieb*, Zeitschrift für Messen. Steuern und Regeln, No. 11, (1967).
12.20. SEELBACH, H., *Neue Wasserversorgung für die Stadt Maracaibo (Venezuela)*. Siemens-Zeitschrift, No. 10 (1975).

13

Electrical Drive of Turbopumps, Run by Means of Variable-speed Transmission Mechanisms

Generally, alternating-current motors are the most commonly used as pump drives, and they inherently operate at fixed speeds. Insertion of transmission elements into the drive train will allow desired adjustments of turbopump speed within certain limits. Transmission mechanisms used for this purpose are made of two members that are mechanically independent: (a) the input member, or the primary, connected to the motor, and (b) the output member, or the secondary, connected to the load (turbopump). By this arrangement the speed of the primary is kept constant and that of the secondary can be progressively and continuously varied. Two types of transmission mechanisms have been developed in the practice of turbopump speed control, namely, the eddy-current slip coupling and fluid coupling.

13.1 The Eddy-current Coupling

13.1.1 Description and Basic Principle

The eddy-current coupling is an electromechanical torque-transmitting device, installed between a constant-speed prime mover and a load (turbopump) in order to obtain adjustable speed operation. This is in fact an electromechanical torque-transmitting device which transforms torque T_1, speed n_1, and power P_1 from the primary into T_2, n_2, P_2 in the secondary, of different magnitudes from those of the primary. A typical self-contained air-cooled eddy-current coupling is shown in Fig. 13.1.

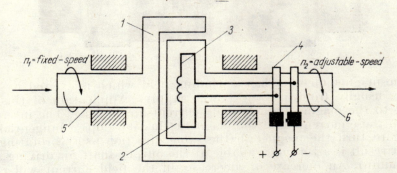

Fig. 13.1. Schematic arrangement of an eddy-current slip coupling.

Electric Drive by Transmission Mechanisms

The input and output members are mechanically independent, with the output magnet member revolving freely with the input ring or drum member. An air gap separates the two members, and a pair of antifriction bearings serve to maintain their proper relative position. The magnet member, placed in the exterior, is called the *armature circuit*. It is made of a solid iron cylinder *1*, made and of a driving shaft *5*, mechanically connected to the drive shaft *6*, which rotates it at constant speed. The input ring, placed inside, is called the *inductor* and is made of the rotor *2*, (laminated with grooves for the excitation winding *3*) and the led shaft (secondary) *6*, on which the slip rings *4* are fixed for electrical connection to the source of direct current.

Figure 13.2 shows a longitudinal section of such a coupling. We can see how the shafts are supported against the bearings, and the way in which Joule heating is eliminated.

Fig. 13.2. Cutaway diagram of an eddy-current slip coupling.

The magnet member has a winding field which is excited by direct current, usually from a static power supply. Application of this field current to the magnet induces eddy-currents in the ring. The interaction between these currents and the magnetic flux develops a tangential force tending to turn the magnet in the same direction as the rotating ring. The net result is a torque available at the output shaft for driving a load (turbopump). An increase or decrease in the field current will change the value of torque developed, thereby allowing adjustment of load

600

speed. As the inductor speed rises, the relative speed between inductor and armature falls, and with it the electromotive force induced. Thus, the driving torque, transmitted by the coupling, is diminished until it equals the load torque presented by the turbopump, and thus steady operation of the driving system is established. When the load torque rises, the rotation speed falls, the relative speed between the inductor and armature rises, and hence the sliding as well as the eletromotive force induced both rise. The increase of the induced electromotive force leads to growth of the driving torque, transmitted by the coupling, until the moment when the driving torque is again equal to the now increased load torque. Thus, steady operation is again established, but this time at a smaller rotation speed.

13.1.2 Computation of Eddy-current Coupling

Generally, two quantities should be determined for computation of an eddy-current coupling: the slip power, and the coupling efficiency.

The slip power. Since the input power P_{E-cci}, received from the primary shaft of the electromagnetic coupling, is proportional to the driving torque T_1, developed by the electric motor, and to the speed n_1 of the motor, it follows that

$$P_{E-cci} = \frac{T_1 n_1}{955} \tag{13.1}$$

The power output, transmitted to the secondary shaft P_{E-cco}, is proportional to the driving torque T_1, transmitted by the electromagnetic coupling, and to the speed of the secondary shaft n_2. Thus

$$P_{E-cco} = \frac{T_1 n_2}{955} \tag{13.2}$$

The difference between power input and output is the slip power of the eddy-current coupling P_{E-ccs}, which is lost in the coupling and is expressed by

$$P_{E-ccs} = P_{E-cci} - P_{E-cco} = \frac{T_1 n_1}{955} \left(\frac{n_1 - n_2}{n_1} \right) = \frac{T_1 n_1}{955} s_{E-cc} \tag{13.3}$$

where s_{E-cc} denotes the coupling slip.

Eddy-current couplings are produced in various types and dimensions. The selection of a coupling is made in accordance with the maximum driving torque, transmitted by the coupling, which corresponds to the maximum slip and maximum permitted speed of the coupling. The value of permitted slip loss for each type of coupling is limited (Table 13.1), and thus it should be checked by computation. The relative value of maximum slip loss in the coupling is about equal to the maximum slip loss occur-

TABLE 13.1 Characteristics of eddy-current couplings of IMS-type (U.S.S.R.)

Coupling type	Rated torque, in daN/m	Rated slip, in rev/min	Max. permitted speed, in rev/min	Permitted power losses in kW with respect to input speed, in rev/min							Tension of supply source and rated energizing current of coupling				Max. permitted total torque (GD^2), in daN·m²	
											eddy-current command by voltage regulator (connection in parallel with excitation winding)		eddy-current command by magnetic amplifier (connection in series with excitation winding)			
				3000	1500	1000	750	600	500		U,V	I,A	U,V	I,A	Armature	Inductor
IMS-7.5	7.5	100	3000	6.5	4.0	3	2.5	2	1.8		220	1.0	380	0.5	6.5	3.5
IMS-22	22.0	50	3000	15.0	9.0	7	5.5	5	4.0		220	1.4	380	0.7	15.0	7.5
IMS-40	40.0	50	1500	28.0	17.5	13	10.5	9	8.0		220	3.0	380	1.5	80.0	40.0
IMS-75	75.0	50	1500	65.0	40.0	30	24.5	21	18.0		220	3.0	380	1.5	189.5	65.0
IMS-100	100.0	50	1500	65.0	40.0	30	24.5	21	18.0		220	3.0	380	1.5	195.0	65.0
IMS-160	160.0	50	1500	108.0	66.0	50	41.0	35	30.0		220	5.0	380	2.5	330.0	140.0

ring in an asynchronous motor of the same power. Thus, in the case of a turbopump connected to a pumping system without static head, the relative value of slip power is about 15 — 18,5 per cent of the nominal power of the driving asynchronous motor and correspond to a speed $n = 0.61 n_0$.

Coupling efficiency. When we do not consider the losses, the efficiency is expressed by

$$\eta_{E-cc} = \frac{n_2}{n_1} \tag{13.4}$$

When the losses are considered (frictional losses occurring in bearings, ventilation losses, and the losses for excitation), which together represent some 2 per cent of the nominal power and which are proportional to the speed, the efficiency expression becomes

$$\eta_{E-cc} = \frac{n_2}{n_1} - 0.02 \frac{n_2}{n_1} = 0.89 \frac{n_2}{n_1} \tag{13.5}$$

The total efficiency of a driving installation, provided with an eddy-current coupling, $\eta_{di\,E-cc}$, is given by the equation

$$\eta_{di\,E-cc} = \eta_m \eta_{E-cc} \tag{13.6}$$

Since the useful power output at the driving motor shaft ($P_u^* = P_u/P_R$) corresponds to the power input at the primary shaft of the coupling P_{E-cci}, it follows that

$$P_u^* = \frac{P_{E-ccuo}}{P_R} \times \frac{1}{\eta_{E-cc}} \tag{13.7}$$

where P_{E-ccuo} is the useful power output at the coupling secondary shaft, and P_R the rated power of the electric motor.

From formula (13.7), we may derive the coupling efficiency

$$\eta_{E-cc} = \frac{P_{E-ccuo}}{P_R} \times \frac{1}{P_u^*} \tag{13.8}$$

Known as the motor efficiency is given by the equation [13.1]

$$\eta_m = \frac{P_{m.u}^*}{P_{m.u}^* + \frac{1}{2}\left(\frac{1}{\eta_{m.R}} - 1\right)(1 + P_u^2)} \tag{13.8 a}$$

By substituting equations (13.8) and (13.8a) into (13.6), we get the general equation of efficiency for a driving installation, provided with an eddy-current coupling

$$\eta_{di\,E-cc} = \frac{P_u^*}{P_u \dfrac{n_1}{0.98 n_2} + \dfrac{1}{2}\left(\dfrac{1}{\eta_{1R}} - 1\right)\left[1 + \left(P_R^* \dfrac{n}{0.98 n_2}\right)^2\right]} \tag{13.9}$$

Electric Drive by Transmission Mechanisms

It is to be noted that at small speeds n_2 and large slips s, the efficiency $\eta_{d\,i\,E-cc}$ of a driving installation, provided with eddy-current coupling, is rather low and is decreasing approximately linearly with speed. Thus, speed control of turbopumps by means of eddy-current couplings is not economical in energy terms.

Example 13.1. **Computation of a speed-controlled driving installation employing an eddy-current coupling.** Consider a turbopump driven at variable speed by means of an eddy-current coupling. Assuming as known operation parameters of the turbopump, $P = 38$ kW and $n = 930$ rpm, we want to determine: (1) the selection computation for the coupling, and (2) the efficiency computation of the driving installation.

Solution.

(1) *Computation for the selection of the coupling.* The rated value of the load torque, presented by the turbopump shaft, is computed by means of equation (9.4):

$$T_R = 955 \frac{38}{930} = 39.80 \text{ daN} \cdot \text{m}$$

The slip power of the coupling is

$$P_{s\,max} = 1.85\, P_R = 0.185 \times 38 = 7 \text{ kW}$$

From Table 13.1, we select a coupling with the rated torque of $T_{E-cc\,R} = 40$ daN.m and the maximum allowed slip power $P_{s\,max} = 13$ kW.

For turbopump driving by means of the coupling, we choose a squirrel-cage motor with $P_R = 55$ kW, $n_R = 930$ rev/min, and $\eta_R = 0.88$.

(2) *Computation of the installation efficiency.* The function $\eta = f(n)$ is computed for various values of speed from equation (13.9). The value of useful driving motor power is found from

$$P_u^* = \frac{T_R \left[0.1 + 0.9 \left(\frac{n}{n_R} \right)^2 \right] n}{955\, P_R} = \frac{39.8 \left[0.1 + 0.9 \left(\frac{n}{n_R} \right)^2 \right] n}{955 \times 55}$$

The other calculations have been organized in Table 13.2.

TABLE 13.2 **Efficiency computation for an eddy-current coupling**

n rev/min	$T^* = 0.1 + 0.9 \left(\frac{n}{n_R} \right)^2$	P^*	$\frac{P^* n_1}{0.98\, n}$	η_{E-cc}
930	1.00	0.690	0.740	0.82
800	0.77	0.460	0.570	0.70
600	0.48	0.210	0.350	0.50
400	0.27	0.083	0.210	0.30
200	0.14	0.021	0.105	0.12

Making use of the data presented in this Table, Fig. 13.3 shows the function $\eta = f(n)$. It is noted that the efficiency is substantially lower for small speeds.

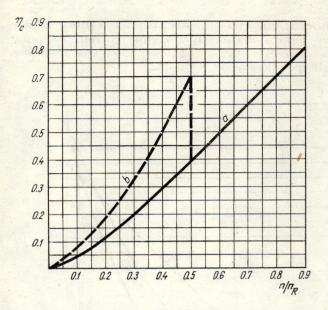

Fig. 13.3. Efficiency versus speed, for an eddy-current slip coupling:
a — one-speed motor; b — two-speed motor.

13.1.3 Load Characteristic Curves

Since the operation principle of an eddy-current coupling is similar to that of an asynchronous motor, their speed-torque characteristics are also similar, showing a reduced stability (Fig. 13.4).

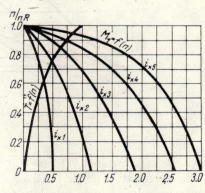

Fig. 13.4. Driving torque-speed characteristic curves for an eddy-current slip coupling.

Turbopumps are working machines with variable torque. Their torque rises proportionally to the square of their speed, and their power in proportion to the speed torque product. These rules are valid provided the pumping system where the turbopumps operate has only dynamic head (for instance, the case of recirculation pumps). In that case, the characteristic curves are as shown in Fig. 13.5.

This shows both the power curves of the pump and motor, and the efficiency curve of the eddy-current coupling, as well as the curve of

605

slip losses. We may conclude that, since the efficiency decreases substantially with speed reduction, the pumps driven by an eddy-current coupling should have a relatively flat characteristic curve; this shape is characteristic for turbopumps with small specific speed.

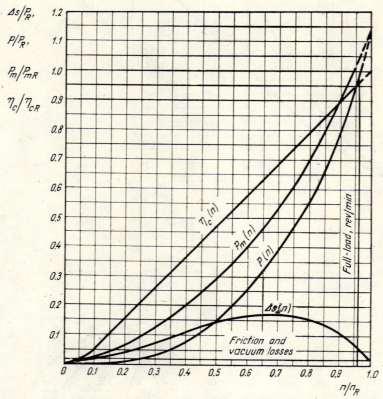

Fig. 13.5. Load characteristics of an eddy-current slip coupling for a friction-only pumping system:

$P(n)$ — load; $P_m(n)$ — motor power; $\eta_c(n)$ — efficiency, $\Delta s(n)$ — slip loss.

However, the great majority of pumps operate in pumping system with a static head. In that case, up to the speed $n = 0.8\,n_R$, the characteristic curves are similar to those of a pump, started with a closed discharge valve (Fig. 13.6). Thus, the curve OAB represents the power curve of the turbopump, started with a closed discharge valve, and the curve OAC, the case with an open discharge valve (point A corresponds to the moment at which the pump head equals the system pumping head). Due to the sudden rise of power within the AC interval, the slip losses are substantially power. Thus, their value does not exceed 10 per cent of the rated power. In general, this is the case for a pump connected to a pumping system with high static head.

606

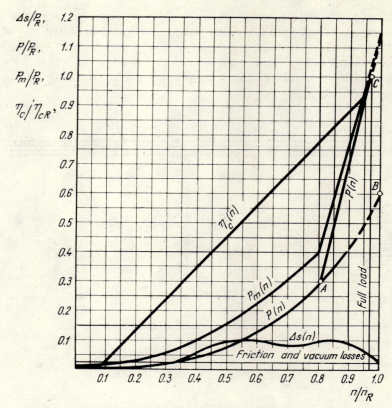

Fig. 13.6. Load characteristics of an eddy-current slip coupling for a pumping system with static and dynamic head:

$P(n)$ — load power; $P_m(n)$ — motor power; $\eta_c(n)$ — efficiency $\Delta s(n)$ — slip loss.

13.1.4 Speed Control

Most eddy-current couplings have load-speed control. An integral part of this system is the tachometer generator. It provides an indication of the exact output speed and enables the control of load speeds within relatively close tolerances, regardless of reasonable variations in load-torque requirements.

Precise speed control of an eddy-current coupling is obtained by the use of feedback circuitry. It consists of an output-speed sensing device which produces a voltage proportional to speed, a speed selector which establishes a command voltage, and a resultant voltage which is impressed on an amplifier supplying excitation to the coupling field. Figure 13.7 shows a simplified diagram.

Adjustment of the speed-control potentiometer creates an imbalance in the control circuit. This results in an excitation change, to regain balance by modification of the speed feedback signal to a suitable value at the new operation point. The response speed of the control is extremely high. The speed-control potentiometer may be set by mechanical or electrical means which are responsive to pressure or flow. Constant-pressure system controls are also frequently encountered in eddy-current coupling applications.

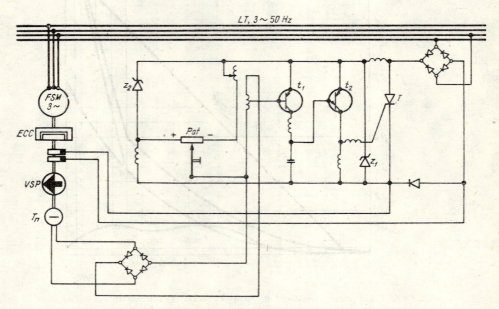

Fig. 13.7. Simplified diagram of an eddy-current coupling control:
FSM — fixed-speed motor; VSP — variable-speed pump; ECC — eddy-curent coupling; T_n — speed transducer.

13.1.5 Range of Utilization

The use of eddy-current coupling is generally limited to cases in which capital expenditure is more important than the running costs. The exact applicability limits could be determined by an optimization calculation. This calculation should consider the drive motor power, speed control range, and the number of working hours during a year.

Small power eddy-current couplings ($P \leqslant 100$ kW) have seen great development abroad (U.S.A., U.S.S.R.), being used in water supply installations. High power couplings have been applied only to research installations. For instance, the aerodynamic wind tunnels at the experimental installations of the Boeing Company (U.S.A.) used an eddy-current coupling of 13,000 kW power.

13.1.6 System Review

Driving installations, fitted with eddy-current couplings, have the following advantages and disadvantages:

Advantages:
— they have a simple construction, and hence their cost is reduced;
— they can drive turbopumps with high moments of inertia (developed at starting) without overheating the starting cage;
— they require low command powers, allowing at the same time easy coupling and decoupling of shafts;
— they secure the possibility of using an asynchronous motor for high power turbopumps.

Disadvantages:
— they have a low efficiency at small rotation speeds, due to the rise of slip losses;
— they require complex exploitation because the eddy-current coupling is an electrical machine with two rotating parts;
— their speed-torque characteristics show low rigidity and thus reduced stability.

13.2 Fluid Coupling

The term fluid coupling can be used loosely to describe any device, utilizing a fluid to transmit power. The fluid is invariably a natural or synthetic oil. This is due to the fact that oil is capable of transmitting power, is a lubricant, and is able to absorb and dissipate heat.

Fluid couplings hydraulically transform torque T_1, speeds n_1, powers P_1, from the primary into T_2, n_2, P_2 in the secondary, of different magnitudes from those of the primary. They also augment or reduce, as needed, these quantities between the primary and secondary.

13.2.1 Basic Principle

A fluid coupling is a hydraulic transformer, made of two main parts: the outer member (the primary) and the inner member (the secondary) (Figs. 13.8 and 13.9).

The outer member comprises a *pump impeller 1*, and a primary (driving) shaft *3*, mechanically connected to the driving motor axle. This member operates according to the radial-flow pump principle. The inner member is made of a *turbine impeller (runner) 2*, and a secondary (driven) shaft *4*, connected to a turbopump shaft. This part operates in an identical way to a hydraulic turbine.

The pump impeller transmits the hydraulic energy of the runner by means of an oil vortex, moving within a certain working circuit (Fig. 13.10).

Electric Drive by Transmission Mechanisms

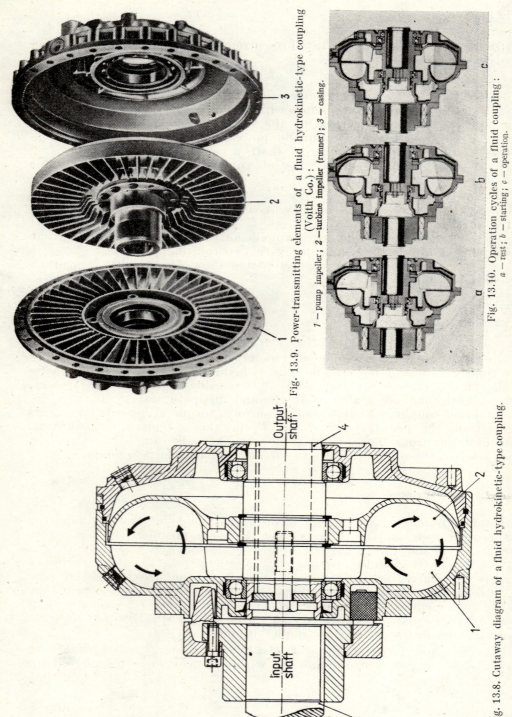

Fig. 13.9. Power-transmitting elements of a fluid hydrokinetic-type coupling (Voith Co.):
1 — pump impeller; 2 — turbine impeller (runner); 3 — casing.

Fig. 13.10. Operation cycles of a fluid coupling:
a — rest; b — starting; c — operation.

Fig. 13.8. Cutaway diagram of a fluid hydrokinetic-type coupling.

As the volume of liquid discharged into the working circuit is larger or smaller than that entering, the thickness of the liquid vortex falls or rises, and thus the circulated flow rate Q varies. Hence the magnitude of power transmitted varies proportionally to the $T_2 n_2$ product. The variation of liquid flow rate in the hydraulic circuit results in a speed change of the secondary (driven) shaft, and thus the fluid coupling transmits the torque $T_1 > T_2$ and the speed $n_2 < n_1$. The primary (driving) shaft has a fixed speed n_1, independent of the secondary shaft, which always has a variable speed n_2.

Energy is transmitted from the primary to the secondary shaft by varying the liquid motion during its circulation through the grooves of the pump and turbine impellers. The liquid motion in the coupling and the energy transmission take place only when the pump impeller rotates faster than the runner. The mechanical energy, transmitted from the driving motor, is turned in the pump impeller into hydraulic energy impressed on the liquid flux. The liquid directly penetrates the turbine impeller, where hydraulic energy is converted into mechanical energy, supplied to the secondary shaft. Then, the working liquid again enters the pump impeller and the cycle starts again.

Due to hydraulic and power losses, the power at the secondary shaft P_2 is always smaller than that at the primary shaft P_1. This is the reason for the occurrence of the slip effect during torque transmission between runner and pump impeller.

13.2.2 Fluid Coupling Computation

Since the secondary, with speed n_2, is lagging the primary with peed n_1, by slip s, we may write

$$P_1 = \frac{T_1 \omega_1}{102} = 0.0014\, T_1 n_1; \quad s = \frac{n_1 - n_2}{n_1} = 1 - \frac{n_2}{n_1}$$

$$P_2 = P_1 \eta (1 - s) = 0.0014\, T_2 n_2, \text{ with } \eta = \eta_1 \eta_2 = \eta_{h1} \eta_{h2} \eta_{m1} \eta_{m2}$$

When $s = 0$, $n_2 = n_1$ and $P_2 = P_1 \eta$; when $s = 1$, $n_2 = 0$ and $P_2 = 0$.

The computation relations are derived by introducing the double unitary quantities Q_1' and n_1' just as for turbopumps, making use of diameter D (the maximum inner diameter of the primary)

$$Q_1' = \frac{Q}{D^2 \sqrt{H}}; \quad Q = Q_1' D^2 \sqrt{H}; \quad n_1' = \frac{nD}{\sqrt{H}}; \quad H^{3/2} = \left(\frac{n}{n_1'}\right)^3 D^3,$$

$$P = \frac{\rho Q H}{102 \eta} = \frac{\rho}{102 \eta} Q_1' D^2 H^{3/2} = \frac{\rho Q_1'}{102 \eta n_1'^3} n^3 D^5 = \frac{T}{102} \frac{\pi n}{30} \quad (13.10)$$

For the primary, we obtain the relations

$$T_1 = c_1 \rho n_1^2 D^5; \quad P_1 = 0.0014 c_1 \rho n^3 D^5 \quad (13.11)$$

and analogously, for the secondary

$$T_2 = c_2 \rho n_1^2 D^5; \quad P_2 = 0.0014 c_2 \rho n_1^3 D^5 \tag{13.12}$$

Computation by similarity equations. The equations (13.11) and (13.12) hold for any turbopump, and hence for fluid couplings, too. Selecting an already built model (denoted by indices M), we have the similarity equations

$$\frac{P}{P_M} = \frac{c \rho n^3 D^5}{c_M \rho_M n_M^3 D_M^5}; \quad \frac{T}{T_M} = \frac{c \rho g n^2 D^5}{c_M \rho_M n_M^2 D_M^5} \tag{13.13}$$

For the same amount of working oil $\rho = \rho_M$ and the same type of fluid coupling ($n_s = n_{sM}$, thus with $c = c_M$), we get the scale factor of geometrical similarity λ and the characteristic diameter D:

$$\frac{D}{D_M} = \left(\frac{P}{P_M}\right)^{1/5} \left(\frac{n_M}{n}\right)^{3/5} = \lambda; \quad D = \left(\frac{P}{P_M}\right)^{1/5} \left(\frac{n_M}{n}\right)^{3/5};$$

$$D_M = \lambda D_M$$

For $\rho = \rho_M$ however, with n_s different from n_{sM}, but not by more than ± 5 per cent, the equation becomes

$$c/c_M \sim n_s/n_{sM}, \quad \text{and hence} \quad \lambda' = \frac{n_1}{n_{sM}} \left(\frac{P n_M^3}{P_M n^3}\right)^{1/5}; \quad D = \lambda' D_M \tag{13.14}$$

Computation by statistical equations. Given n_1, P_1 and ρ (for oil) from the design, the circulated flow rate Q and discharge head H of the primary are determined

$$P_1 = \rho \frac{QH}{102 \eta_1} \text{ [kW]}; \quad \eta_1 = \eta/\eta_2; \quad Q = \frac{85 \eta_1}{\rho H} P_1 \text{ [m}^3\text{/s]} \tag{13.15}$$

$$H = \frac{\eta_{h1}}{gk} c_{u2m} u_{2m} = 0.0534 \frac{\eta_{h1}}{k} D_{2m} c_{u2m} \tag{13.16}$$

The specific speed η_s, expressed in dimensionless form, is

$$\nu_s = \frac{n_s}{(2g)^{3/4} \rho^{1/2}}$$

and numerically

$$\nu_s = 0.1072 \frac{n_1 P_1^{0.5}}{\rho^{0.5} H^{1.25}}$$

The computation proceeds according to the statistical equations established by Proskura:

$$D_{2i} = (10.8 - 6\,\nu_s)\left(\frac{Q}{n_1}\right)^{1/3} \tag{13.17}$$

$$D_{1m} = 0.8\,\nu_s^{1/3} D_{2m}$$

$$\frac{D_{2e} - D_{2i} - 2\dfrac{\varepsilon}{1000}}{2D_{2m}} = 0.113\,\nu_s^{2/3}\,;\quad D_{2o} \sim D2_{2m} - D_{2i} \tag{13.18}$$

$\varepsilon = 0.05$ mm for $\nu_s = 0.25$ and $\varepsilon = 0.08$ mm for $\nu_s \geqslant 0.4$

$$D_{2m} = \frac{4.42}{\nu_s^{2/3}}\left[D_{2o} - (10.8 - 6\nu_s)\left(\frac{Q}{n_1}\right)^{1/3} - 0.002\,\varepsilon\right] \tag{13.19}$$

$$z = 8.65 + \frac{10 \cdot 8}{\nu_s^{1/3}}. \tag{13.20}$$

where z stands for the blade number.

Size determination by means of equations (13.12). From the previous power analysis (13.12), we get the outer diameter D_{2o}

$$P_1 = 0.0014\,c\rho\,n_1^3\,D_{2o}^5$$

or

$$D_{2o} = 3.72\sqrt[5]{\frac{P_1}{c\rho n_1^3}} \tag{13.21}$$

where the coefficient c is established empirically by measurements. The other dimensions are determined according to the model selected.

13.2.3 Characteristic Curves

The experimentally established characteristics may be represented in a graphical form: Fig. 13.11 shows two such graphs.

Figure 13.11,a shows, against secondary speed $n_2 = (1 - s)$, the torques T_2 of the secondary for $n_1 = 1\,000$ to $3\,000$ rev/min and the isoslip curves for constant $s = 10 - 5$ per cent.

Figure 13.11,b shows the variations of quantities T_2 and η with the n_2/n_1 ratio; $T_2 = f(n_2/n_1)$ and $\eta = f(n_2/n_1)$. The efficiency varies linearly for $n_2/n_1 = 1 - s < 0.5$, but the η curves are different for rising $1 - s$ values, as shown for $s = 3 - 10$ per cent.

Figure 13.12 shows the relation between slip s and the degree of filling (formation of oil vortex).

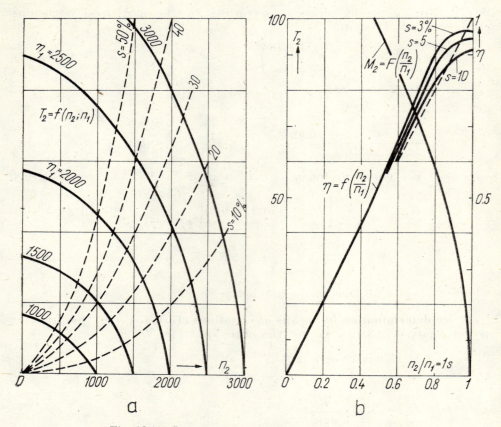

Fig. 13.11. Characteristic curves of a fluid coupling.

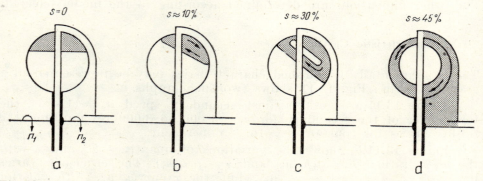

Fig. 13.12. Hydraulic coupling regulation by varying the oil vortex, with increasing slips:
$$s = 1 - \frac{n_2}{n_1}.$$

13.2.4 Description of Types

All fluid couplings may be divided into four categories as follows: hydrokinetic, hydrodynamic, hydroviscous and hydrostatic couplings.

Hydrokinetic coupling. While all types of fluid couplings are used in starting and controlling pumps, the most commonly used is the hydrokinetic machine (Figs. 13.13 and 13.14).

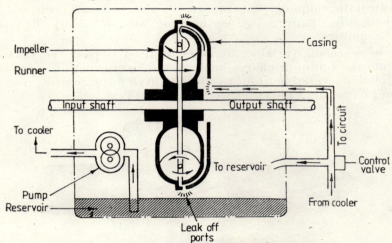

Fig. 13.13. Fluid coupling, hydrokinetic-type.

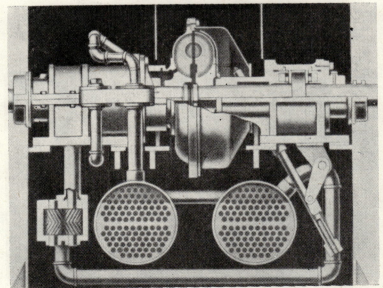

Fig. 13.14. Cutaway of a hydrokinetic coupling (American Standard).

Electric Drive by Transmission Mechanisms

Hydrodynamic coupling. This type of fluid coupling is occasionally used to drive pumping equipment, usually in the portable pump field (Fig. 13.15).

Hydroviscous coupling. This coupling is relatively new in commercial use. There are several manufacturers in the U.S.A. who are marketing this type of drive for a wide range of pump applications.

Hydrostatic coupling. This invariably makes use of positive-displacement hydraulic pumps in conjunction with a positive-displacement hydraulic motor.

In some cases, variable amounts of fluid bypass the normal flow route from the pump discharge back to the pump suction. This provides a controllable variable flow to the positive-displacement motor and, therefore, a variable output speed (Fig. 13.16).

Fig. 13.15. Fluid coupling between pump and motor.

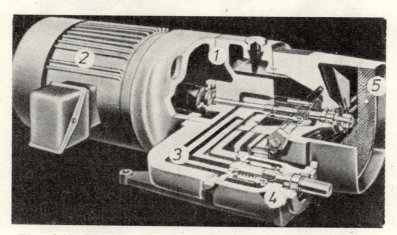

Fig. 13.16. Cutaway of a hydrostatic drive (Sperry Vickers Co.).

REFERENCES

13.1. IONEL, I. I., *Acționarea electrică a turbomașinilor*. Technical Publishing House, Bucharest, 1980.
13.2. KARASSIK, J. I., KRUTZSCH, C. W., FRASER, H. W., MESSINA, J. P, *Pump Handbook*. McGraw-Hill Book Co., New York, 1976.
13.3. ONISCENKO, G. B., IUNKOV, M. G., *Elektroprivod turbomechanizmov*. Energhia, Moscow, 1972.
13.4. PAVEL, D., ZAREA, S., *Turbine hidraulice și echipamente hidroenergetice*. Didactic and Pedagogic Publishing House, Bucharest, 1968.

14

Techniques for Matching Pumps to Economic Requirements

Introduction. In previous chapters we have presented the theory, computation methods, and phenomenology of the main equipment (pumps, valves, pipelines, motors, converters, etc.) which form the basis of pumping installations. The present chapter refers to the assembling of such equipment for the systematic arrangement of a pumping installation intended for a particular and typical application.

Each scheme presented here is accompanied by its working diagram, including the main characteristic curves of the pumps and pumping system. They are:

— the $H(Q)$ curve representing the pumping head (pressure secured by the pump);
— the $H_s(Q)$ curve ,, the pumping head required in the system (the necessary pressure);
— the $P(Q)$ curve ,, the effective power absorbed by the pump;
— the $\eta(Q)$ curve ,, the pump efficiency;
— the $e(Q)$ curve ,, the specific energy (energy consumed for pumping one cu.m. of water at a given pumping head).

These curves, integrated into a complete diagram, can be used for optimization considerations (maximum performance for minimum cost of electric power) for each type of installation studied to meet a particular application. The system head curve can be used for selecting the correct pump, depending on the physical phenomena of the pumping system.

A complete diagram also shows:

— medium global efficiency of pumping installation, $\eta_{i\,med}$, as a percentage;
— medium unitary power consumption for one cu.m of water, pumped at one meter head, $e_{u\,med}$, in Wh/cu.mm;
— electric power savings, obtained by using variable-speed pumps instead of fixed-speed types.

Classification. The pumping installations may be classified as follows:

1. *According to the facility they supply*
   1.1. Pumping installations for populated centres.
   1.2. Pumping installations for industrial zones.
   1.3. Pumping installations for land reclamation systems.

2. *According to their rotation speed*
   2.1. Pumping installations equipped with fixed-speed pumps. In this case, all pumps have fixed-speed drives which cannot be adjusted.
   2.2. Pumping installations, equipped with variable-speed pumps. In this case one or two pumps, out of the total number, have variable speed and make use of adjustable-speed drives.

14.1 Pump Application to Facilities of Populated Centres

Populated centres are great consumers of drinking water. As main consumers we find: individual multi-storey buildings (hotels, offices, hospitals, appartments, etc.), dwellings and housing settlements in residential districts, and town facilities. Drinking water is supplied to all these consumers by the following categories of pumping installations: (a) pressure-boosting installations, and (b) waterwork transmission and distribution pumping station.

14.1.1 Pressure-boosting Installations for Water Supply to Multi-storey Buildings

These are pumping installations, boosting the water pressure in an auxiliary distribution system, and also restoring the existing pressure in the main system. They are fed by the town low-pressure main, and discharge water into a high-pressure distribution system to which the facilities to be supplied are connected.

Pressure-boosting installations can be connected to the main in two ways:
— *directly*, in the case of supplying consumers with low flow rate and, implicitly, when the main diameter is large enough (for generating small velocity variations, Δv, in transitory working conditions), and when the available pressure on the suction side is at least 1.5 daN/sq.cm.;
— *indirectly*, by means of an intermediate break-pressure tank in the case of more important consumers (dwellings, housing settlements, etc.), where the town distribution system is less developed, in a zone having low pressure (under 1.0 daN/sq.cm.).

According to the geometry of buildings established by urban planning and architectural design, water supply to consumers is achieved in two ways:
— by high-rise water systems, in the case of water boosting to high-rise buildings (hotels, office headquarters, etc.);
— by long-line water systems, in the case of water supply to housing settlements (dwellings, appartments villas in spas, etc.).

Water supply to buildings by high-rise systems. The supply systems, used for high-rise buildings, are developed vertically and, therefore, are short and require small flow rates, associated with high pumping heads. The necessary pumping head depends on the geometric height of the relevant building. Since the maximum supply value should not exceed 6 daN/sq.cm for buildings higher than 30 to 45 m, the distribution system should be divided into several pressure zones with adequate pressure-boosting installations. Thus, depending on the geometrical heights of the respective buildings, we could have a pressure-boosting installation for one pressure zone, for two pressure zones, or for several pressure zones.

A. Pressure-boosting installations for multi-storey buildings with one pressure zone. These installations are fit for the water supply of buildings 30 to 45 m. high.

The pumping system of such an installation is generally characterized by:

— a small maximum required flow rate ($Q = 10$ to 90 cu.m/h) varying suddenly within a large flow range ($\Delta Q = 1:10 \ldots 1:5$) and remaining at minimum and mean values for long time intervals, and maximum values for short time intervals. The measurements taken in some buildings with one pressure zone [14.2, 14.13] show that $Q_{min} \approx$ $\approx 25\% Q_{max}$, lasting for about 7 to 14 hours a day;

— an average pumping head ($H = 45$ to 60 m) and an $H_s(Q)$ characteristic curve of flat shape (due to reduced length of the system) with a ratio $H_{st}/H_t \approx 0.8$.

The unitary power, required by an optimum number of pumps for the distribution of the maximum required flow rate is small ($P \leq 10$ kW).

In the above-mentioned conditions, one could use both fixed-speed pumps and variable-speed pumps in the pressure-boosting installation.

Arrangement of a pressure-boosting installation with fixed-speed pumps. Selection of a pumping installation depends on the required flow rate and on its distribution to one or more pumps. If the maximum consumption is low ($Q_{max} \leq 20$ cu.m/h.), the installation should be equipped with one single duty pump, doubled by a stand-by pump, each selected for a unitary flow rate of 110 per cent of the peak flow rate.

Figure 14.1 shows schematically a pressure-boosting installation with one single duty fixed-speed pump. It has simple pressure switch

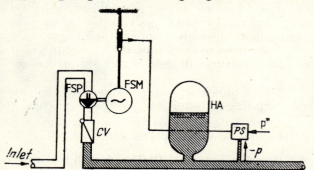

Fig. 14.1. Schematic arrangement of a pressure-boosting installation with one fixed-speed duty pump:

FSM, FSP — fixed-speed motor and pump; *HA* — hydraulic accumulator; *CV* — no-slam check valve; *PS* — pressure switch.

regulation (see Section 8.2.1) by a conventional pressure switch, introduced into the command circuit of an electric motor and a hydropneumatic

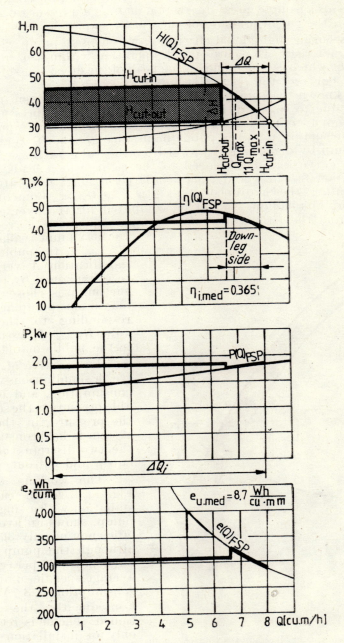

Fig. 14.2. Performance curves of a pressure-boosting installation with one fixed-speed duty pump.

tank on the pump discharge line. The stand-by pump command is also by means of a simple pressure switch, having a pressure 0.2 daN/sq.cm lower than the cut-in pressure of the duty pump.

For securing high global efficiency, we choose a pump with a skewed $H(Q)$ characteristic. It secures pump operation within a narrow flow-rate range, ΔQ_p, which can be found on the peak of the down-leg side of the $\eta(Q)$ efficiency curve of the respective pump (Fig. 14.2).

It is known that pressure changes within public systems during each 24 hours due to consumption variations. The suction pressure variation Δp_{sc} is about 1 to 1.5 daN/sq.cm. This variation (for direct pump connection to the suction side of a public main) changes the $H(Q)$ characteristic of the pump by displacing it upward by the value of Δp_{sc}. At the same time, the curve is also displaced to the right, which leads to a rise of pump unitary flow rate and thus to a larger average flow rate considered in computing the hydrophor tank ($Q_{avr1} > Q_{avr2}$). Consequently, the hydrophor tank volume should be computed for the average flow rate derived from the $H(Q)$ curve of the respective pump, corresponding to the maximum available pressure $H_{sc\,max}$ in the public main. The modification of average flow rate is explained by the fact that the pump is chosen for a total maximum pumping head corresponding to the minimum value $H_{sc\,min}$ of pressure available in the public main (Fig. 14.3).

Irrespective of the successive or simultaneous variations in consumption, and of the available pressure in the suction main, the pressure in the discharge system is held constant between the two discrete values of the cut-in and cut-out pressures.

The working diagram in Fig. 14.2, for the case of an installation with a single $Sadu_{50 \times 5}$ pump, shows an average global efficiency of only 36.5 per cent although the pump maximum efficiency is 47 per cent, and the average specific unitary energy consumption is 8.7 Wh/cu.m.m. Consequently, the above-mentioned scheme is recommended only for installations with small flow rates and a small number of pumps.

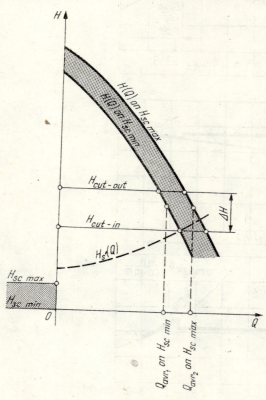

Fig. 14.3. Performance diagram showing the modification of head flow rate curve with variations in input pressure.

Example 14.1 **Computation of a pressure-boosting installation, equipped with a single fixed-speed pump.** Let us consider a tower building with a height of 30 m (ground floor and ten levels) located in a town whose water supply system in that zone has a minimum available pressure of 1 daN/sq·cm ($H_{sc\,min} = 10$ m). The suction line has a head loss of 2 m. We also know the working parameters of the high-pressure distribution system: maximum flow rate per hour $Q_{max} = 7.20$ cu.m/h; minimum flow rate per hour $Q_{min} = 0.72$ cu.m/h; discharge geodesic head $H_{gds} = 32$ m; utilization head at the sanitary facility of the top floor $H_{util} = 3$ m; and discharge head loss $h_e = 8$ m. For recovering the available pressure in the supply main, and for electric power savings, the pumping installation is directly connected, on its suction side, to this main. We have to choose: the pump, the electric motor, and the hydrophor tank.

Solution

(1) *Pump selection.* The pump unitary flow rate is determined

$$Q_{unit} = \frac{110}{100} \times 7.20 = 7.90 \text{ cu.m/h}$$

The total pumping head of the system is computed from (5.30)

$$H_t = 32 - 10 + \frac{0.3 - 0.0}{0.1} + 0 + 8 + 2 = 35 \text{ m}$$

We choose a pump with $Q = 8$ cu.m/h, $H = 35$ m, $\eta = 47$ per cent and $n = 2930$ rev/min.

(2) *Selection of electric motor.* Motor power is computed from (2.12)

$$P_m = \frac{8 \times 35}{367 \times 0.60} \times 1.50 = 1.91 \text{ kW}$$

We choose a squirrel-cage asynchronous motor with $P_m = 2.2$ kW, $n = 2850$ rev/min. $U = 380$ VY (for direct starting), $\eta_m = 82$ per cent.

(3) *Size of hydrophor tank.* The pump mean flow rate is computed from relation (8.12) and the cut-in flow rate and cut-out flow rate are computed from Fig. 14.2

$$Q_m = \frac{2}{3} \times \frac{(8.40 + 6.60)^2 - 8.4 \times 6.60}{(8.40 + 6.60)} = 7.5 \text{ cu.m/h}$$

Since the pumps are controlled by compensated pressure switch regulation, the net volume of the tank is computed from equation (8.21) after we have previously established graphically the values of H_{cut-in} and $H_{cut-out}$

$$V_{net} = \frac{1 \times 7.50 - (1-1) 7.50}{4 \times 6} \times \frac{45.50 + 10}{14} \times$$

$$\times \frac{31 + 10}{31 + 10 - (1-1)3 + 2 \times 3} = 1.07 \text{ cu.m}$$

Therefore, we choose a tank with a volume 1.60 cu.m. The dead volume, V_{dead}, is computed with formula (8.22) and the fact that the selected tank has a rated diameter $D_R = 1.0$ m, the dead head $H_{dead_1} = 0.29$, the bottom valve has a rated diameter $D_R = 80$ mm, and $H_{dead\,2} = 0.05$ m.

$$\Delta V_{dead} = 1 \times \left(0.05 + \frac{2}{3} \times 0.29\right) \times 0.785 = 0.19 \text{ c.m}$$

The total volume is computed from equation (8.4, a)

$$V_t = 1.07 + 0.19 = 1.26 \text{ cu.m} < 1.60 \text{ cu.m}.$$

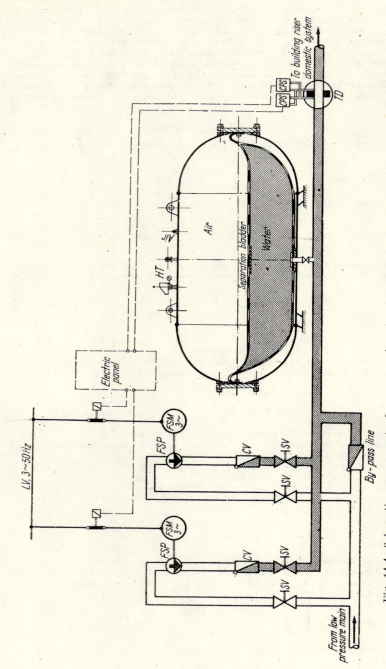

Fig. 14.4. Schematic arrangement of a pressure-boosting installation with two fixed-speed duty pumps:
FSM, FSP — fixed-speed motor and pump; *HT* — hydropneumatic tank with flexible separation bladder; *CV* — no-slam check valve; *CPS* — compensated pressure switch; *SV* — shut-off valve; *LV* — low voltage.

Pumps for Population Facilities

If the value of maximum demand is large (20 cu.m/h $< Q_{max} <$ < 90 cu.m/h), the pressure-boosting installation will be equipped with *two or more pumps*, each selected for a unitary flow rate of 55 per cent (in the case of two pumps) of the peak demand (Fig. 14.4).

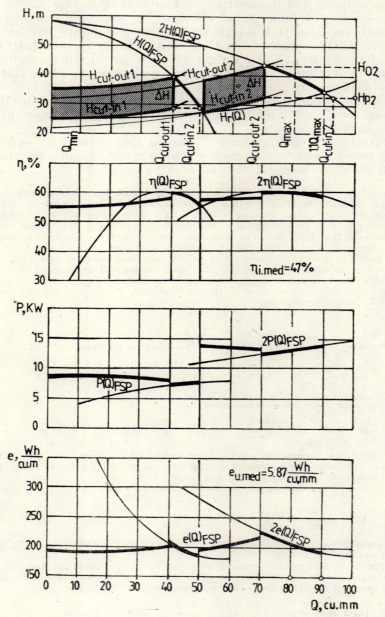

Fig. 14.5. Performance curves of a pressure-boosting installation with two fixed-speed duty pumps.

625

Compensated pressure switch regulation (see section 8.2.1) is recommended for automated control of pumps, by means of which we achieve pumps starting in a rising pressure cascade and, implicitly, the advantage of closer $H(Q)$ and $H_s(Q)$ characteristics, leading both to the reduction of losses and to electric power savings (Fig. 14.5).

For securing a high global efficiency of the pumping installation during automated control, it is neccesary that the pumps selected should have a flat $H(Q)$ characteristic, since that is closer to the $H_s(Q)$ characteristic of the system. Thus, we get a higher global efficiency than that shown in Fig. 14.2 and a reduced energy consumption. These statements could be checked by means of data shown in Fig. 14.5. For instance in the case of Lotru$_{65}$ type pumps, we have an average global efficiency of 47 per cent, and an average unitary energy consumption of 5.87 Wh/cu.m.m.

The stand-by pump automated control also employs a compensated pressure switch, with a starting pressure 0.2 daN/sq.cm lower than that at which the duty pump enters the working cycle.

Example 14.2 **Computation of a pressure-boosting installation, equipped with two fixed-speed pumps.** Consider a building 30 m high (ground floor and ten levels) requiring a maximum flow rate of 80 cu.m/h. The building is located in a town whose drinking water supply system secures a 10 m pressure head at the point of connection. The supply line of the building shows a 2 m head loss. The operation parameters of the distribution system are: maximum flow rate $Q_{max} = 80$ cu.m/h; minimum flow rate $Q_{min} = 8$ cu.m/h; geodesic discharge head $H_{g\,ds} = 32$ m; useful head at the top floor sanitary facility $H_{util} = 3$ m; and head loss on the discharge line $h_{ds} = 8$ m. Pressure is boosted by means of an installation, equipped with two fixed-speed pumps. For recovering available pressure in the supply main, the pumps will be directly connected on their suction side. We want to choose: the pumps, electric motor, and hydrophor tanks.

Solution.

(1) *Pump selection.* Unitary flow rate of the pumps is determined

$$Q_u = \frac{55}{100} \times 80 = 44 \text{ cu.m/h}$$

Pumping head of the system is computed from (5.30)

$$H_t = 32 - 10 + \frac{0.30 - 0}{0.10} + 0 + 8 + 2 = 35 \text{ m}$$

Two radial-flow pumps are selected, each having: $Q = 45$ cu.m/h; $H = 35$ m; $\eta = 60$ per cent; and $n = 2930$ rev/min. The total flow rate, resulting with the pumps mounted in parallel, is 90 cu.m/h > 80 cu.m/h (see Fig. 14.5).

(2) *Motor selection.* Driving motor power is computed from (2.12)

$$P_m = \frac{45 \times 35}{367 \times 0.60} \times 1.20 = 8.9 \text{ kW}$$

A squirrel-cage asynchronous motor is chosen, with $P_m = 10$ kW; $\eta_m = 87.5$ per cent; $n = 2930$ rev/min; $U = 380$ VΔ.

(3) *Hydrophore tank volume.* Pump average flow rate is computed by means of equation (8.12). The cut-in and cut-out flow rates are derived from Fig. 14.5 for the working conditions of the last pump on the $2H(Q)$ curve, but in accordance with the $H(Q)$ curve

$$Q_m/2 = \frac{2}{3 \times 2} \times \frac{(93 + 70)^2 - 93 \times 70}{93 + 70} = 41 \text{ cu.m/h}.$$

The net volume of the hydropneumatic tanks is computed from equation (8.21). The cut-in and cut-out heads from computation for the turbopump (the second pump) are also derived from Fig. 14.5. Thus

$$V_{net} = \frac{2 \times 41 - (2-1)41}{4 \times 6} \times \frac{43.50 + 10}{12} \times$$

$$\times \frac{33 + 10}{33 + 10 - (4.50 + 2.2)} = 9.60 \text{ cu.m}$$

This volume is distributed to two tanks of 5 cu.m each, one of them being filled only with air.

The dead volume is computed from equation (8.22) for the following geometrical dimensions of the tanks: $D_R = 1.40$ m, $\Delta H_{dead\,1} = 0.438$, and of the bottom valve, $D_R = 125$ mm; $\Delta H_{dead} = 0.10$ m. Thus

$$V_{dead} = 1 \times \left(0.10 + \frac{2}{3} \times 0.438\right) \times 1.54 = 0.61 \text{ cu.m}$$

and the total volume is obtained by applying relation (8.4. b)

$$V_t = 9.60 + 0.61 = 10.20 \text{ cu.m} \approx 10 \text{ cu.m}.$$

Arrangement of a pressure-boosting installation, equipped with variable-speed pumps. The flat shape of the $H_s(Q)$ characteristic curve of the pumping system requires two pumps, of which only one should have variable speed. Their unitary flow rate is chosen as 55 per cent of the peak demand in the distribution system of the respective building. For higher accuracy in the adjustment of pumping installation flow rate, it is recommended that the $H(Q)$ curves of the pumps should have a skew shape, so as to form perpendicular intersections between them and the $H_s(Q)$ curve of the pumping system.

Since unitary power, resulting from the distribution of maximum consumption between two pumps, has a small value ($P \approx 10$ kW), there are two possibilities for driving the variable-speed pump:

— by means of a squirrel-cage motor, fed by an a.c. voltage regulator with built-in triacks system, and having the semiconductors placed at a star joint of the respective motor (Fig. 14.6);

— by the same type of motor, but fed by a frequency converter with built-in power transistors (instead of thyristors) (Fig. 14.7 and 14.8).

Both schemes allow manual and automated speed control as well. Manual control is by a potentiometer, and automated control by a control loop with a continuous pressure transducer (Fig. 14.9).

Due to this latter control, the discharge line of the pumping installation preserves a constant pressure, irrespective of successive or simultaneous variations in consumption or inlet pressure.

Since the $H(Q)$ pump characteristic is similar to the $H_s(Q)$ system characteristic (Fig. 14.10) due to speed control, we thus also have a flow rate control with reduced losses, which increases the global efficiency of the pumping installation and leads to savings of electric power. This can be seen on the diagram shown in Fig. 14.10.

The diagram has been drawn for the parallel operation of two Lotru$_{65}$ type pumps, one of which has adjustable speed. Thus the global efficiency

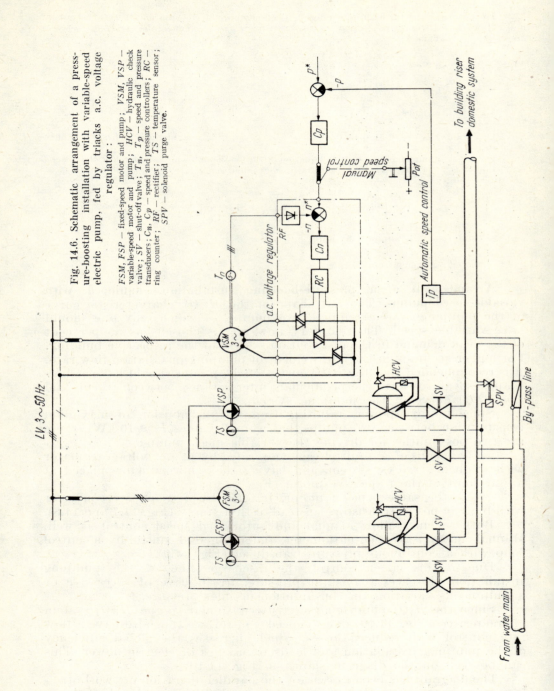

Fig. 14.6. Schematic arrangement of a pressure-boosting installation with variable-speed electric pump, fed by triacks a.c. voltage regulator:

FSM, FSP — fixed-speed motor and pump; VSM, VSP — variable-speed motor and pump; HCV — hydraulic check valve; SV — shut-off valve; T_n, T_p — speed and pressure transducers; C_n, C_p — speed and pressure controllers; RC — ring counter; RF — rectifier; TS — temperature sensor; SPV — solenoid purge valve.

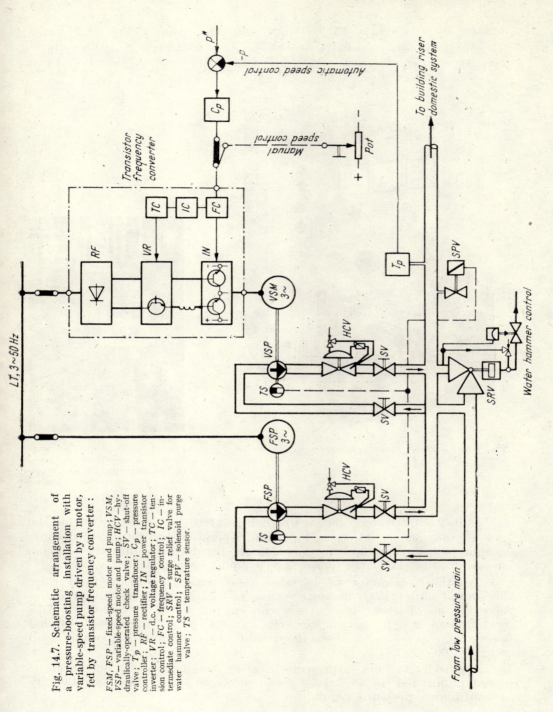

Fig. 14.7. Schematic arrangement of a pressure-boosting installation with variable-speed pump driven by a motor, fed by transistor frequency converter:

FSM, FSP — fixed-speed motor and pump; VSM, VSP — variable-speed motor and pump; HCV — hydraulically-operated check valve; SV — shut-off valve; Tp — pressure transducer; C_p — pressure controller; RF — rectifier; IN — power transistor inverter; VR — d.c. voltage regulator; TC — tension control; FC — frequency control; IC — intermediate control; SRV — surge relief valve for water hammer control; SPV — solenoid purge valve; TS — temperature sensor.

of the installation is 49 per cent, and the unitary specific energy 5.87 Wh/cu.m.m, compared with 36.5 per cent and 8.7 Wh/cu.m.m, respectively, in the case of a conventional installation (see Fig. 14.2).

Fig. 14.8. Power transistor frequency converter (Siemens Co.).

Fig. 14.9. Continuous pressure transducer (regulator), mounted on the discharge line of a pressure boosting pump (Danfoss Co.).

Note that the diagram shown in Fig. 14.10 has been drawn for the case of the scheme shown in Fig. 14.6, where electric losses by slipping cannot be recovered by an a.c. voltage regulator. Using the scheme shown in Fig. 14.7, the performance could be much better.

Another advantage of variable-speed pumps resulting from their horizontal outlet $H(Q)$ pressure characteristic, is the fact that pressure-boosting installations, using them can supply water to buildings (with one pressure zone) much higher ($H_{geom.\ max} = 45$ m) than the conventional ones ($H_{geom.\ max} = 30$ m), without exceeding the maximum allowable head range of 60 m.

Figure 14.11 shows a pressure-boosting installation with one variable-speed pump, fed by a package frequency converter, and controlled by a continuous pressure regulator.

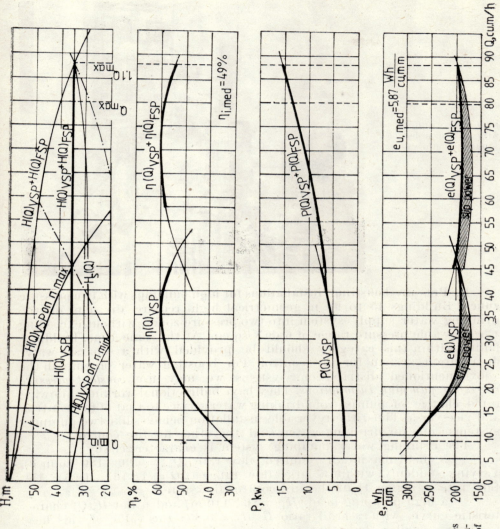

Fig. 14.10. Performance curves of a pressure-boosting installation with two pumps, one of which has variable-speed.

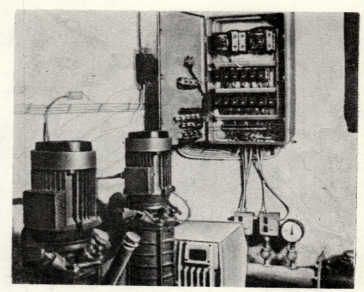

Fig. 14.11. Pumping set with one variable-speed pump (Danfoss Co.).

B. **Pressure-boosting installations for high buildings with two pressure zones.** Buildings 45 to 90 m geometrical high require division of the drinking water supply system into two pressure zones, without exceeding the allowable pressure range of 6 daN/sq.cm. At the same time, buildings falling within this category should be provided with a separate water distribution system for fire control. For boosting water pressure in the above-mentioned distribution systems, we may use *constant pressure reversible pumping installations*. They have bidirectional working, allowing rapid change of pumps from drinking water service to that of fire control, and vice-versa. Thus, the reversible installation behaves like two pumping systems: one for drinking water, and the other for fire control.

The drinking water pumping system is characterized by:
— a small maximum required flow rate ($Q_{max} = 90-180$ cu.m/h), varying suddenly within a large flow range ($\Delta Q = 1:10 \ldots 1:5$); this flow rate is higher than the fire control flow rate;
— a high pumping head ($H_t = 80-110$ m) and a flat $H_s(Q)$ characteristic curve, with a head ratio $H_s/H_t \approx 0.85$; the value of this head is always smaller than that for the fire control system, since for the latter we should add the head losses, produced in the water hose, and the pressure head, necessary for keeping a long jet of liquid.

Because of the distribution of maximum required flow rate between an optimum number of pumps, the maximum resulting unitary power of the pumps has a small value, $P = 40-55$ kW.

Arrangement of a reversible pumping installation with fixed-speed pumps. This installation is designed according to the scheme shown in Fig. 14.12. It is equipped with three identical fixed-speed pumps, of which two are duty pumps, each of them with a unitary flow rate of

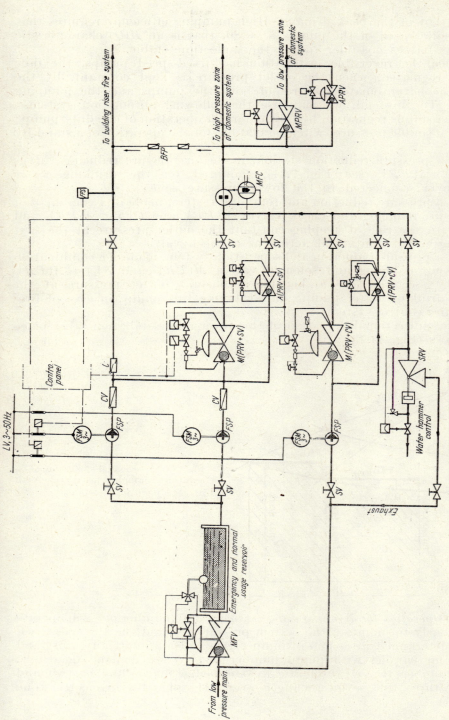

Fig. 14.12. Schematic piping arrangement of a reversible pumping installation with fixed-speed pumps:

FSP, FSM — fixed-speed pump and motor; *MFV* — modulating float valve; *MPRV, APRV* — main and auxiliary pressure-reducing valves; $M(PRV+SV)$, $A(PRV+SV)$ — main and auxiliary pressure-reducing and solenoid valve combinations; *SRV* — surge relief valve; $M(PRV+CV)$, $A(PRV+CV)$ — main and auxiliary pressure-reducing and check valve combinations; *BFP* — back-flow preventer; *CV* — check valve; *SV* — shut-off valve; *MFC* — monitor flow control; *FPS* — fire pressure switch.

55 per cent of the peak demand. High pumping efficiency requires that the specific speed of the pumps $n_s \approx 80$, that is an $H(Q)$ characteristic curve as flat as possible, and at the same time stable.

Since the reversible installation has fixed-speed pumps, the flow rate is regulated, when the outlet pressure is kept constant, by the simultaneous automated control of both the pumps and the pumping system. To this end, we may use the following automation systems:

— a simple regulation by flowmeter, for operation of one duty pump;

— a simple pressure switch regulation, for the operation of a stand-by pump;

— a pressure reduction station, made for pressure reducing valves $MPRV + APRV$ (see Fig. 7.31), mounted on the drinking water changeover, connected to the lower pressure zone;

— a pressure reduction and retaining system, made of a combination of pressure-reducing and check valves $M(RPV$ and $CV) + A(PRV$ and $CV)$, with the role of keeping constant the outlet pressure at the first drinking-water pump, and acting as a check valve;

— a pressure reduction and separation system, made of a combination of pressure-reducing and solenoid valves $M(PRV$ and $SV) + A(PRV$ and $SV)$ (see Fig. 7.28, b), keeping a constant outlet pressure at the second drinking water pump, and switching this pump (in case of fire) to the fire control system.

The bidirectional operation of this reversible installation takes place as follows (Fig. 14.13):

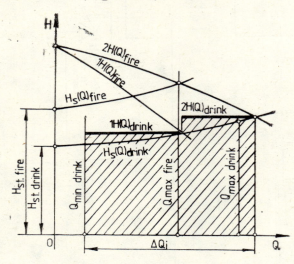

Fig. 14.13. General diagram of a reversible pumping installation with fixed-speed pumps.

(1) *Operation for drinking water service*. Pump number one will operate continuously to maintain the system pressure. Should the system exceed the predetermined rate, pump number two will automatically start and operate in parallel with pump number one. When demand decreases, pump number two is automatically stopped. Pumps number two and number three will be interchanged every 24 hours by an electric time

clock. Should pump number one fail to operate for any electrical reason, pumps number two and number three will automatically start, while a pilot warning red-light will be lit on the control panel. Pumps number two and number three will continue to operate until a "pump-number-one-failure-reset" button is pressed on the panel.

(2) *Operation for fire control service.* Pump number one in drinking water service is permanently keeping constant both the pressure in the drinking-water distribution system and that in the fire-control system, regardless of variations in consumption or inlet pressure. The connection between the two systems is by a backflow preventer[1] (Fig. 14.14).

Fig. 14.14. Pressure-reducing backflow preventer with protection zone (Singer Valve Co.).

This device secures definite protection against backflow, and is ideally suited particularly to conditions under which safe water might become contaminated. The unit consists of two check valves and an automatically controlled "protection zone", equipped with a special differential relief valve. Should the fire line pressure drop 0.6 daN/sq.cm below the design pressure for six seconds, both pumps number two and number three will automatically be started. The solenoid controls on the main valve and on the auxiliary combination of pressure-reducing and solenoid valves will close these valves, allowing full flow from both pumps into the fire standpipe. Pump number one will continue to supply water to the domestic system. When the main pumps are on fire service, an alarm bell rings and a red light is lit on the panel. The main pumps will continue to operate until a "fire-pump reset" button is pressed on the panel, thus switching the system back to normal operation.

Figure 14.15 shows the operation diagram of a reversible installation, provided with three $WKLv_{50\times 5}$ type pumps (KSB). The performances are: medium global efficiency $\eta_{i\,med} = 58$ per cent; unitary specific energy

---

[1] Certified to *AWWA* recommendation from Joint Committee Report No. 82101(1958) on "The Use of Backflow Preventers for Cross-Connection Control".

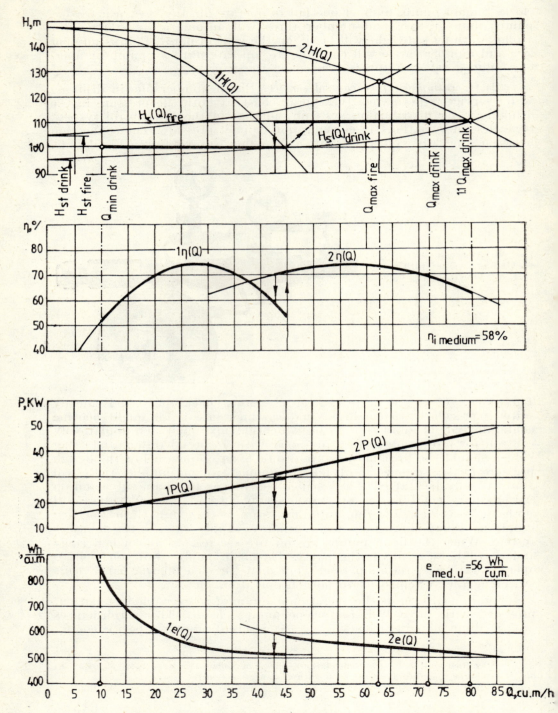

Fig. 14.15. Performance curves of a reversible pumping installation with fixed-speed pump.

consumption $e_{u\,med} = 5.60$ kW/cu.m.m. These performances are satisfactory for the working conditions mentioned above.

Figure 14.16 shows a pressure-boosting installation with pump regulation by flowmeter (see Fig. 8.30) and outlet pressure regulation by a combination of pressure-reducing and check valves (see Fig. 7.28, *a*).

Fig. 14.16. Pumping set to maintain constant head in tall buildings (Pacific Co.).

Arrangement of a pumping installation with variable-speed pumps[1]. According to the scheme shown in Fig. 14.17, this reversible installation is equipped with two duty pumps for drinking water, of which one has adjustable speed, and a stand-by fixed-speed pump for the fire-control system. Each of them has a unitary flow rate of 55 per cent of the peak demand.

Considering the flat shape of the system characteristic curve, the $H(Q)$ pump characteristic curves should have a skewed shape, leading to perpendicular intersections between the two types of characteristics. This results in a higher accuracy of regulation of the pumping installation flow rate.

The variable-speed pump should be driven by a squirrel-cage asynchronous motor, fed by a frequency converter with intermediate constant-tension circuit (Fig. 14.18).

Under these conditions, the operation of a reversible pumping installation for both systems (drinking water and fire control) requires the automated control of both pumps and pumping system. To this end, the following automation systems could be used:

— a simple pressure regulator (made of $T_p + C_p$ elements) with continuous action aiming to keep constant the duty pumps pressure VSP; this system is combined with a speed-control frequency converter;

---

[1] Author's original solution.

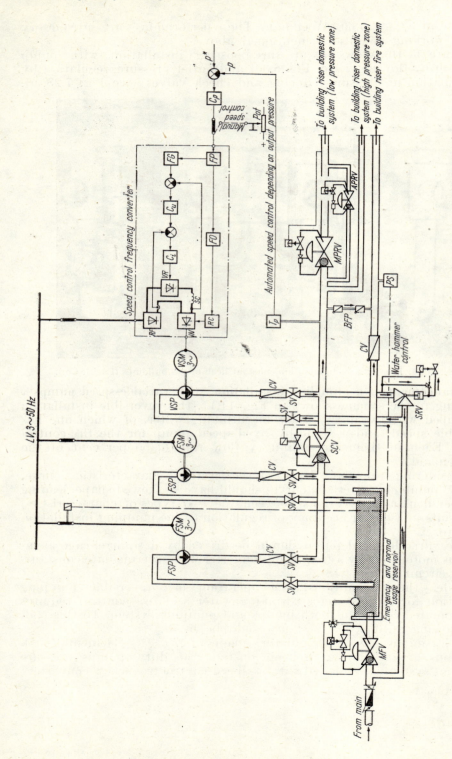

Fig. 14.17. Schematic arrangement of a reversible pumping installation with variable-speed pumps:

SP, *FSM* — fixed-speed pump and motor; *VSP*, *VSM* — variable-speed pump and motor; *MPRV*, *APRV* — main and auxiliary pressure-reducing valves; *SCV* — solenoid control valve; *SRV* — surge relief valve; *MFV* — modulating float valve; *SV* — shut-off valve; *CV* — no-slam check valve; *PS* — pressure switch; *Tp* — pressure transducer; *Cp* — pressure controller; *FG* — frequency generator; *Cu*, *Ci* — tension and current controller; *RF* — rectifier; *IN* — inverter; *VR* — voltage reguator; *FP* — frequency programme; *FO* — frequency oscillator; *SC* — smoothing coil, *RC* — ring counter; Pot. — potentiometer.

— a simple discontinuous pressure switch PS, for regulation of the fire-control pump FSP;

— a pressure reduction station, comprising the pressure-reducing valves $MPRV$ and $APRV$ (see Fig. 7.31), mounted on the drinking water pipe going to the lower pressure zone;

— a switching system, made up of the solenoid valve SCV, for switching the pump FSP from drinking water to fire control service;

— a backflow preventer (see Fig. 14.14) for permanently keeping the fire control system under pressure. It consists of two resilient-seated spring-loaded check valves, with a protection zone between them. This zone is freed to atmosphere by means of a differential control valve, acted upon by the difference between inlet and protection-zone pressures so as to keep a lower pressure in the protection zone;

— a water hammer control system, made of a surge relief valve SRV (see Fig. 7.83) for pipe-breaking prevention.

Bidirectional operation of the reversible pumping installation takes place as follows (Fig. 14.19):

(1) *Operation in drinking water service.* The pump VSP operates continuously with variable flow rate, for meeting minimum and average consumption, keeping its outlet pressure constant by an adjustable speed. When consumption exceeds the flow rate corresponding to maximum pumps speed, then, by means of a maximum voltage relay, the fixed-speed duty pump FSP is started and the variable-speed pump VSP lowers its speed in accordance with its reduced flow rate. When consumption goes on rising, the pump FSP keeps a fixed working point correspond to its own maximum efficiency, but the pump VSP raises its flow rate and keeps its outlet pressure constant by means of speed variation. Should consumption go lower, the reversible installation works vice-versa.

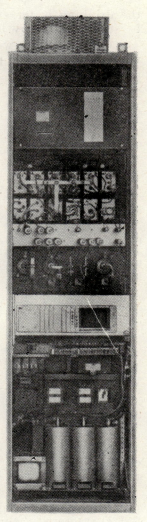

Fig. 14.18. Frequency converter of 50 kVA (Brown Boveri Co.).

(2) *Operation in fire control service.* Let us suppose the pumps VSP and FSP work to meet the drinking water consumption. At a given moment, a sudden pressure drop of a minimum 6 daN/sq.cm takes place for at least six seconds in the fire control system. This pressure drop is recorded by pressure switch PS, which orders the cut-in of pump FSP simultaneously with the closing of SCV. Given this condition, the whole liquid discharged by the two pumps working in parallel is directed only towards the fire control system, and the pump VSP remains to supply

the drinking water system. When the fire is over, the two pumps will be stopped only by hand.

Consequently, we may say that a reversible pumping installation with variable-speed pumps is able to keep a constant outlet pressure

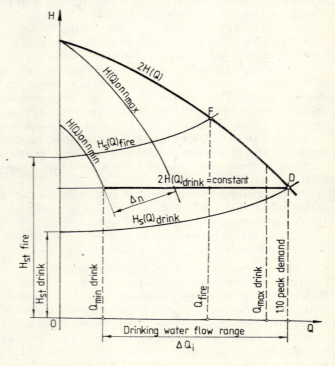

Fig. 14.19. General diagram of a reversible pumping installation with variable-speed pumps.

when supplying the drinking water system irrespective of variations in consumption or inlet pressure, securing at the same time the pressure in the fire control system.

Due to speed regulation, the artificial characteristic curves of the pumps have the shape of a horizontal line, which is very similar to the flat characteristic curve of the pumping system. Consequently, the flow regulation is achieved with reduced losses of pressure and slip energy, resulting in an increase of global efficiency of the installation, and saving of electric power. The advantages mentioned above are seen in Fig. 14.20.

This diagram corresponds to the scheme of Fig. 14.15 for the case of two duty pumps type $WKLv_{100 \times 5}(KSB)$. The graphs show rather high performances: medium global efficiency $\eta_{i\,med} = 71$ per cent; unitary medium specific power consumption $e_{u\,med} = 4.1$ Wh/cu.m.m. As compared to a reversible pumping installation with fixed-speed pumps, the installation with variable-speed pumps requires reduced running expenses, since it secures an average global efficiency of 71 per cent instead of 62 per cent, and a power consumption of only 450 Wh/cu.m instead of 580 Wh/cu.m.

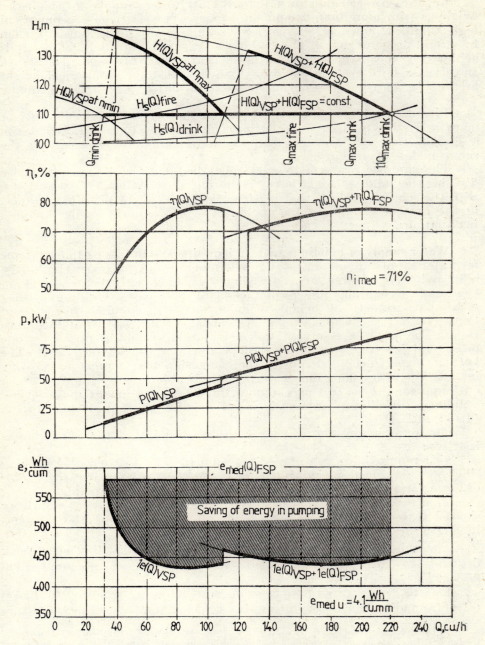

Fig. 14.20. Performance curves of a reversible pumping installation with variable-speed pumps.

C. Pressure-boosting installations for water supply to high-rise buildings with more than two pressure zones. The extremely geometrical high buildings known as *"Skyscrapers"* require a water distribution system in several pressure zones. Thus, the water supply of such buildings is achieved by means of several drinking water and fire control systems. In this case, both systems could be divided so that the geometrical height of the fire control pressure zone should be twice that of the drinking water zone.

Pressure could be boosted in these buildings by means of two or more reversible pumping installations with constant pressures, made in a similar way to those described at points A and B, and having the same modes of operation.

The number of reversible pumping installations is established by dividing the building height by a height unit of 60 or 90 m (in accordance with the scheme applied, that is with Figs. 14.12 or 14.17). For instance, in the case of a building 180 m. high, we use two reversible pumping installations.

Water supply of buildings with a long-line system. Long-line systems for water supply are met in cases of a common supply for several buildings, scattered over a large area (dwellings, housing settlements, villas in spas, etc.). Urban planning recommends, in these cases, buildings with a maximum geometrical height of 30 m or, in particular cases, 45 m. Consequently, the water supply system is not necessarily divided into zones. The water pressure in the distribution system is raised by one or several boosting installations, depending on the planned area. The author's experience [1] led him to the conclusion that a pressure-boosting installation with a long-line water system is able to supply water, under optimum conditions, to a maximum of 10,000 conventional flats in several buildings (dwellings, or housing settlements) [14.13; 14.14; 14.15; 14.16; 14.17].

The pumping system of a pressure-boosting installation is characterized by:

— a required flow rate of medium value ($Q = 50$ to 250 dm³/s) varying slowly within a medium range ($\Delta Q = 1:4\ldots1:3$), the moderating factor of these variations being the great number of consumption points;

— a medium pumping head ($H_t = 50\ldots60$ m) and a relatively skewed $H_s(Q)$ characteristic curve (due to the length of the system) having a head ratio $H_{st}/H_t = 0.66\ldots0.75$.

The unitary power of pumps, resulting from the distribution of peak demand to an optimum number of pumps, is of medium value ($P = 22\ldots160$ kW per pump).

Due to concentrated flow rates of large values ($Q = 50$ to 250 dm³/s) taken from urban mains by pressure-boosting installations, a break-pressure

---

[1] During the period 1964—1978 the author designed, within the "Proiect București" Design Institute, various pressure-boosting installations for water supply to some large housing settlements in Bucharest.

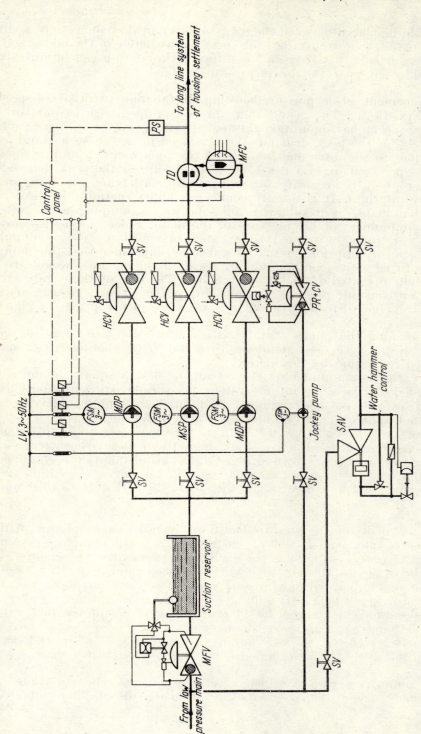

Fig. 14.21. Arrangement of a pressure-boosting installation with fixed-speed pumps for water supply of buildings with a long-line system:
MDP – main duty pump; *MSP* – main stand-by pump; *FSM* – fixed-speed motor; *HCV* – hydraulic check valve; *PR+CV* – combination of pressure reducer and check valve; *TD* – throttling device (diagrame of Venturi); *MFC* – monitor flow control; *SAV* – surge arrestor valve; *SV* – shut-off valve; *MFV* – modulating float valve; *PS* – pressure switch.

tank should be placed between the main pumps and the connection point to these mains. For recovering the pressure available in the mains and, implicitly, for energy savings, a small-flow jockey pump can be directly connected to by-pass the break-pressure tank.

Arrangement of a pressure-boosting installation with fixed-speed pumps (Fig. 14.21). The installation can be equipped with two or three duty pumps, each having unitary flow equal to 55 per cent or 37 per cent, respectively, of the peak demand. We also need in each case a stand-by pump, equal to the duty ones, and a jockey pump with a unitary flow rate of 25 per cent or 17.5 per cent, respectively, of the peak demand. This latter pump is necessary for economically satisfying minimum and night consumptions. It is usually connected, on its suction side, directly to the urban main, by-passing the tank, and with a pumping head chosen for the minimum available pressure in the main. Since the mains show pressure variations of some 1 to 1.5 daN/sq·cm above the minimum value due to consumption fluctuations, it is recommended that this pump should be fitted with a combination of pressure-reducing and check valve, hydraulically operated on the discharge side. The role of these valves is to maintain constant pressure at the pump outlet (below the maximum permitted value) irrespective of variations in consumption or inlet pressure.

For making the pump characteristic curves closer to those of the pumping system, we need pumps with a flat $H(Q)$ characteristic, and automated regulation by flowmeter.

Since the maximum allowed pressure of 6 daN/sq.cm should not be exceeded in the distribution system, pressure-boosting installations, equipped with fixed-speed pumps, are limited to buildings having a maximum height of 30 m (ground floor and 10 levels).

A few pressure-boosting installations for the water supply to 10,000 conventional flats of the Bucharest housing settlement have been achieved according to the scheme shown in Fig. 14.21. This installation is equipped with two main duty pumps, each having $Q = 140$ dm³/s, $H = 55$ m and $P = 125$ kW, and a jockey pump with $Q = 70$ dm³/s, $H = 40$ m and $P = 75$ kW.

Arrangement of a pressure-boosting installation, equipped with variable-speed pumps. In view of the skewed type of pumping system characteristic curve supplying the housing settlement, two schemes of pressure-boosting installation with variable-speed pumps could be used. Thus, for small required flow rates ($Q \leq 50$ dm³/s), we can make use of the scheme with a single duty pump (Figs. 14.22, a and b), doubled by a stand-by pump. The pumps are identical, having a unitary flow rate equal to 110 per cent of the peak demand.

The electric drive of both pumps is by means of a squirrel-cage motor, alternatively fed by a single frequency converter with a changeover switch.

The pump flow rate is adjusted, to meet consumption variations in the system, by a manual or automated variation of its speed. The automated

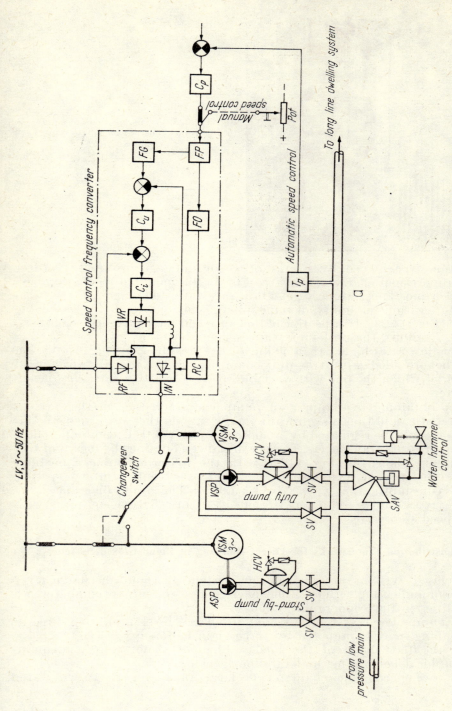

Fig. 14.22. *a*. Schematic arrangement of a pressure-boosting installation, equipped with variable-speed pumps:

VSP, *VSM* — variable-speed pump and motor; *HCV* — hydraulic check valve; *SV* — shut-off valve; *SAV* — surge anticipator valve; *Tp* — pressure transducer; *Cp* — pressure controller; *FP* — frequency programme; *FO* — frequency oscillator; *GF* — frequency generator; *Cu*, *Ci* — tension and current controllers; *RC* — ring counter; *RF* — rectifier; *IN* — inverter; *VR* — voltage regulator (chopper); *Pot.* — potentiometer.

Fig. 14.22, b. Variable-speed pumping set to maintain constant head in dwellings (Danfoss Co.).

regulation is achieved by means of a continuous pressure transducer with proportional characteristic (see Fig. 14.9), mounted on the common discharge line. Thus, we may get a flow rate variation range $\Delta Q = 0.25$ to $1.10\ Q_{max}$ within the limits of some isoefficiency curves with reasonable values (Fig. 14.23). Irrespective of flow rate and inlet pressure variations, ($\Delta H_{scmax} = 20$m), the outlet pressure is kept constant.

For flow rates higher than 50 dm³/s, the pressure-boosting installation could be provided with two duty pumps (Fig. 14.24, a, b), working in parallel, of which the one with variable speed is fed by a static frequency converter.

For obtaining overlapping of pump characteristics on the system characteristic, and thus a significant energy saving, the flow rate of the installation could be regulated by a more complex control, combining flow rate and pressure transducers. The performances, obtained by means of this system, are shown in Fig. 14.25 with the following values: medium overall efficiency of installation $\eta_{i\ med} = 78$ per cent; unitary specific energy consumption $e_{u\ med} = 37$ Wh/cu.m.m. The same diagram shows the energy saving as compared to that of an installation equipped with fixed-speed pumps.

14.1.2 Distribution Waterworks for Direct Boosting in Pump-Closed Systems

Pump-closed systems are distribution systems without discharge level tanks communicating with the atmosphere. These systems are characteristic of localities in flat topographical zones.

A pump-closed system (without discharge level tank), fed directly by pumping, may be imposed by three main factors:
— the flat shape of the land, which does not offer adequate topographical heights for underground level tanks;
— the uneconomic character of large tanks for the town water supply;

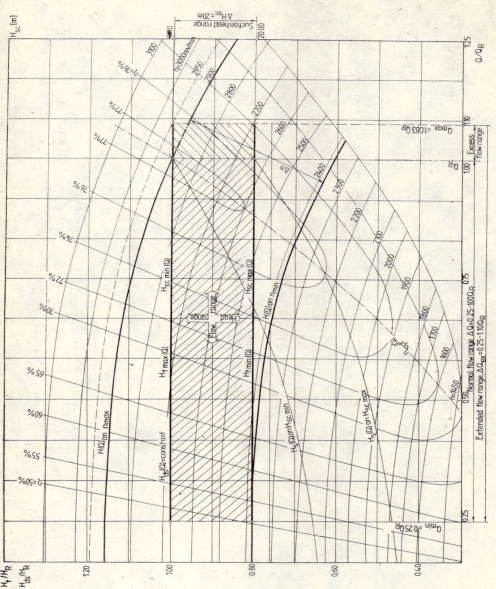

Fig. 14.23. Performance diagram of a pressure-boosting installation with variable-speed pumps for dwellings.

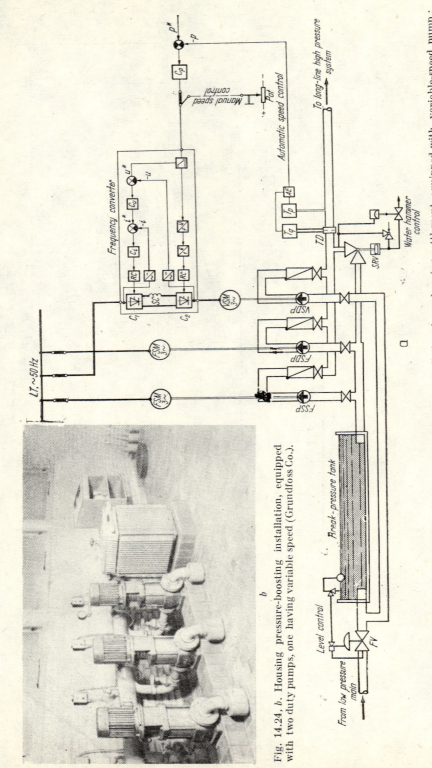

Fig. 14.24, b. Housing pressure-boosting installation, equipped with two duty pumps, one having variable speed (Grundfoss Co.).

Fig. 14.24, a. Schematic arrangement of a pressure-boosting installation for a housing settlement, equipped with variable-speed pump: FSP, FSM – fixed-speed pump and motor; VSP, VSM – variable-speed pump and motor; FV – float valve; SRV – surge relief valve; TD – throttling device; T_p, T_q – pressure and flow-rate transducers; AE – adding element; C_p, C_u, C_i – pressure, tension and current controllers; RC – ring counter; SC – smoothing choke; C_1, C_2 – controlled rectifier and inverter; $VSDP$ – variable-speed duty pump; $FSDP$ – fixed-speed duty pump $FSSP$ – fixed-speed stand-by pump.

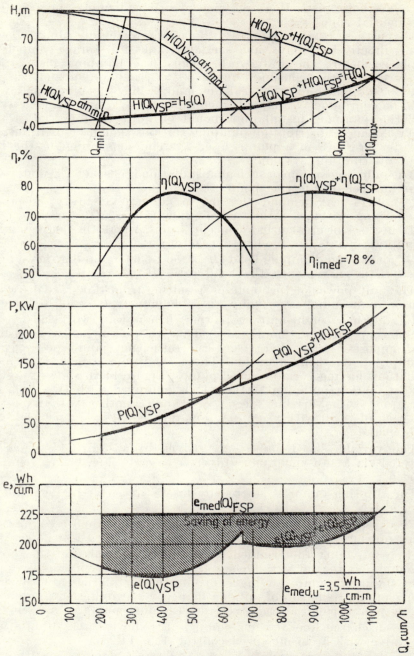

Fig. 14.25. Performance diagram of a pressure-boosting installation for a housing settlement, equipped with variable-speed pump.

— aesthetic restrictions, imposed by urban planning and architecture in residential zones.

Given these conditions, pressure boosting in the distribution systems is done directly by pumps, and the level tanks for storage compensation are placed before the waterwork. Thus, the pumping installation is directly taking the consumption peaks of the system and, therefore, should be designed for the peak demand, while the pumped flow rate should be continuously adapted to consumption variations in the system. Thus, this pumping installation should be quite flexible.

Two or more distribution pumping stations should be used for raising the water pressure in a pump-closed system, depending on the length of the distribution system.

The pumping system of such an installation is characterized, depending on the number of injection points, by:

— a high peak demand ($Q = 0.5$ to 5.0 cu.m/s), slowly and progressively varying within a rather narrow flow range ($\Delta Q = 1 : 3 \ldots 1 : 2.5$), mainly in cases of great urban concentrations, where the large number of consumption points is a moderating factor;

— an average pumping head ($H_t = 50$ to 60 m) and a skewed $H_s(Q)$ characteristic curve, with the head ratio $H_{st}/H_t \approx 0.7$.

The unit power of turbopumps, resulting from flow rate distribution by an optimum number of pumps, is high, namely $P = 400$ to $2,000$ kW.

Beside their high unitary powers, urban waterworks also require continuous operation (24 hours daily). Thus, they are big energy consumers, which imposes strict energy criteria (efficiency and specific energy consumption) on their operation. These conditions can only be met by installations having variable-speed pumps. Figure 14.26 shows the scheme of a distribution pumping station. It is equipped with two equal duty pumps, one with variable speed, *VSP*, and another with fixed-speed, *FSP*, each of them with a unitary flow rate, equal to 55 per cent of the peak demand.

The variable-speed pumps are driven by a slip-ring asynchronous motor, fed via a subsynchronous inverter cascade (Fig. 14.27), regulating its speed.

The slip-ring motor is started by means of a starting rheostat with electric drive.

The fixed-speed pump is driven by a synchronous motor, compensating the low-pressure factor of the asynchronous motor fed by the converter.

High reliability (determined by fire control regulations) is obtained by doubling both the variable-speed pump and the fixed-speed pump with a stand-by pump, having a driving motor of the same nature as the duty pump. By means of suitable changeover switch, the variable-speed stand-by pumps could alternatively use the duty pump converter.

The discharge line of each pump is fitted with a hydraulic needle check valve with a closing-speed control (Fig. 14.28).

For eliminating the adverse effects of water hammer resulting from power failure, the common discharge line is fitted with a surge anticipator

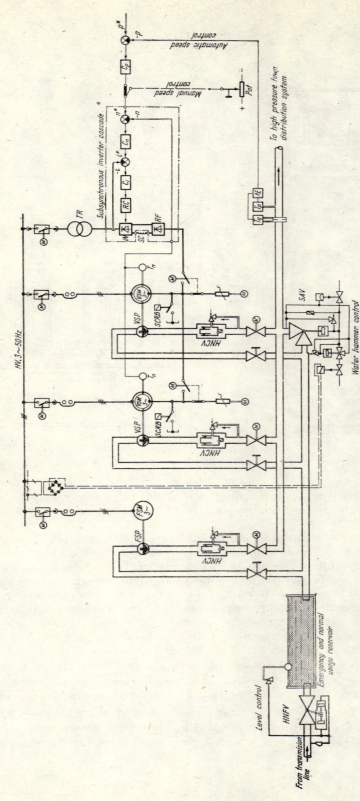

Fig. 14.26. Schematic arrangement of a variable-speed pump waterworks for direct boosting in a pump-closed town system:

FSP, FSM — fixed-speed pump and motor; VSP, VSM — variable-speed pump and motor; $HNFV$ — hydraulic needle float valve; $HNCV$ — hydraulic needle check valve; SAV — surge anticipator valve; $SCRB$ — short circuit and rising brushes device; T_p, T_Q, T_n — pressure, flow-rate and speed transducers; AE — adding element; C_p, C_n, C_i — pressure, speed and current controllers; RC — ring counter; IN — inverter; RF — rectifier; SC — smoothing choke; TR — transformer; $Pot.$ — potentiometer.

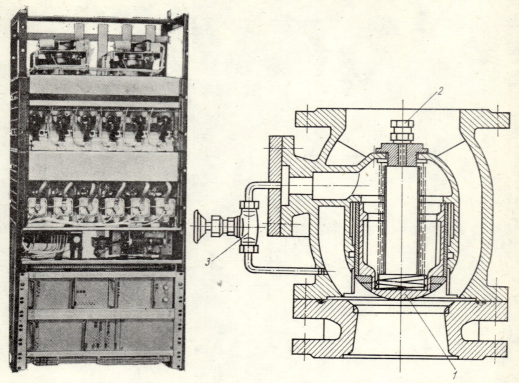

Fig. 14.27. Subsynchronous inverter cascade (Siemens Co.).

Fig. 14.28. Hydraulic needle check valve (Erhard Co.)

valve. This surge valve is provided with two control pilots, one of which is responsive to power failure, and the other to overpressure (see Fig. 7.102).

Rotation speed regulation may be effected both manually, by means of a potentiometer, and automatically. The automated regulation of outlet pressure implies the use of two transducers (one for pressure, T_p, and the other for flow rate, T_q), connected to a pressure controller by means of an adding element. Suitable selection of these elements results in a parabolic outlet pressure curve [an $H(Q)$ parabola with downward convexity], exactly superposed over the $H_s(Q)$ characteristic curve of the pumping system. Thus, flow rate regulation is achieved without losses, obtaining in this way a high global efficiency of the installation and, implicitly, a reduced consumption of electric energy. Let us draw, for instance, the working diagram for a distribution pumping installation as described above and fitted with $24NDS$-type pumps (Fig. 14.29). It shows the following performances (significantly higher than those of a conventional distribution pumping installation with fixed-speed pumps): medium overall efficiency $\eta_{imed} = 92$ per cent; medium specific energy

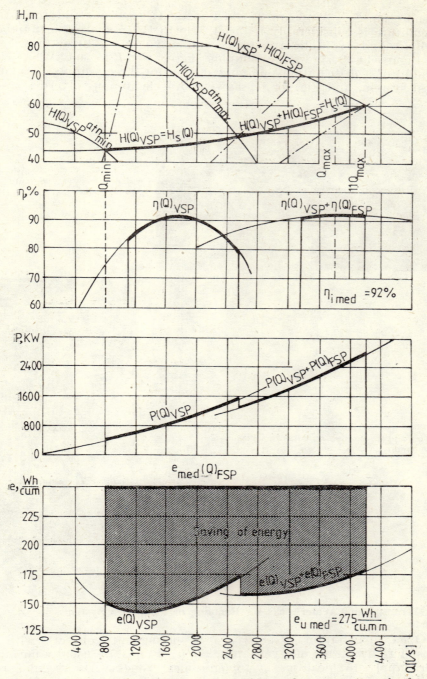

Fig. 14.29. Performance diagram of a variable-speed pump waterworks for boosting in a pump-closed system.

consumption $e_{umed} = 165$ to 275 Wh/cu.m. At the same time, investment expenditure is considerably reduced, due to the reduction in number of machines and hydraulic apparatus, and in volume of structures.

Figure 14.30 shows the pumping aggregates ($P = 500$ kW, $n = 1475$ rev/min) of a distribution pumping station for direct supply to a pump-closed system, and Fig. 14.31 shows a subsynchronous inverter cascade.

Fig. 14.30. Variable-speed pump waterworks for boosting in a pump-closed system (Brown Boveri Co.).

Fig. 14.31. Subsynchronous inverter cascade for the supply of the slip-ring motor shown in Figure 14.30 (Brown Boveri Co.)

14.1.3 Transmission Waterworks for Pumping in Tank-open Systems

Open distribution systems are in permanent contact with the atmosphere by means of level tanks. Such systems are usually meant for water supply to localities in topographical zones with irregular relief (mountain or hill regions). In such conditions, level tanks may be placed at favourable topographical height, between the distribution system and the pumps, in order to compensate the flow rate and to take over the peaks of consumption. Thus, pump flow rate is reduced, and the waterworks should be sized only for the maximum value of daily flow rate, which is below that of flow rate per hour.

The required tank compensation volume results from the superposition of the pumped flow rate graph on that of consumption. The pumped flow rate shape is influenced by the working conditions of the waterworks. When the waterworks operates with a constant flow rate (Fig. 14.32)

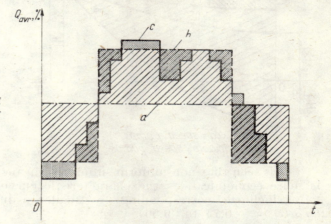

Fig. 14.32. Variation curve of consumption in a town system during 24 hours.

the tank has a uniform supply, and thus the pumped flow rate curve, $Q = f(t)$, is a horizontal line (curve a). We may note that this curve shows large deviations from the consumption curve (curve c) and this leads to a high value of compensation volume, about 8—16 per cent of the daily peak demand. If the flow rate pumped to the tank can be varied (for example, by means of variable-speed pumps), then the pumped flow rate graph will have the shape of the broken straight line (curve b). It will be noticed that the deviations of this curve from the consumption curve (curve b) are less pronounced, and therefore lead to a much smaller compensation volume, that is, 2.5—5 per cent of the daily peak demand, smaller by some 30 per cent than in the case of uniform feeding. Moreover, non-uniform tank supply (that is, a variable flow rate in the main, from pumping station to tank) has another advantage, namely, the energy savings. This may be shown from the fact that head losses on the discharge pipeline (the connection line between tank and waterworks) has a variation, proportional to the square of the flow rate. Thus, when water is pumped through long transmission lines with large head losses, that is, with inclined or very inclined $H_s(Q)$ characteristic curves, the flow rate decrease leads to a substantial fall of pumping head, which in its turn is proportional to the power absorbed by the pumps. Since waterworks for water supply to localities have a permanent operation (24 hours each day), the decrease of flow rate and of pumping head leads to an obvious energy saving.

The graph shown in Fig. 14.33 refers to situations in which a non-uniform supply (variable flow rate) is required. The graph shows many $H_s(Q)$ characteristic curves corresponding to different values of the H_{st}/H_t head ratio.

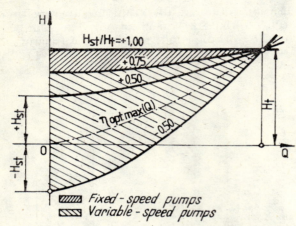

Fig. 14.33. Shape of pumping system characteristic curve $H(Q)$, versus head ratio H_s/H_t.

It is seen that non-uniform supply, by means of variable-speed pumps, is more economical, as the characteristic curve of transmission lines is more inclined, corresponding to smaller values of the H_{st}/H_t ratio ($H_{st}/H_t = +0.75$ to -0.50).

The pumping system of a transmission pumping station is characterized by a high required flow rate and pumping head, that is, $Q = 0.5$ to 10 cu.m/s, and $H_t = 70$ to 120 m, respectively.

The unitary power resulting from the distribution of peak demand to several pumps is high, namely $P = 160$ to 3,000 kW.

Arrangement of a transmission pumping station, equipped with fixed-speed pumps. For high values of the head ratio $H_{st}/H_t \geqslant +0.75$ on a transmission line, we may use fixed-speed pumps for the waterworks, provided with three or four pumps, each of them having a unitary flow rate, equal to 36—27 per cent of daily peak demand. These pumps are driven by asynchronous motors, or synchronous motor in the case of high unitary powers.

Arrangement of a transmission pumping station, equipped with variable-speed pumps. For values of $+0.75 > H_{st}/H_t > -0.50$ (see Fig. 14.33), important energy and cost savings can be obtained by making use of variable-speed pumps, according to the above arguments. The number of pumps depends on the value of this ratio: for small values ($H_{st}/H_t = +0.25 \ldots -0.50$), a.e. for a very steep system-head curve, a single variable-speed pump would be sufficient.

Let us analyze two cases: one with two pumps, and the other with a single pump. For the first case, the transmission pumping installation can be designed according to the scheme shown in Fig. 14.34. It shows

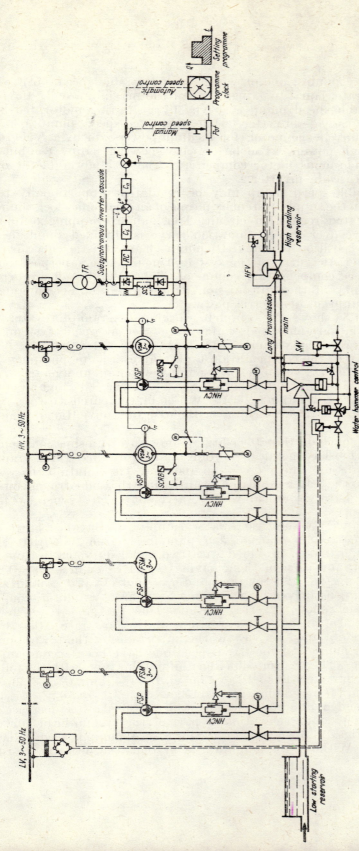

Fig. 14.34. Schematic arrangement of a variable-speed pump waterworks for pumping in a tank-open system:

FSP, FSM — fixed speed pump and motor; *VSP, VSM* — variable-speed pump and motor; *HFV* — hydraulic float valve; *SAV* — surge anticipator valve; *HNCV* — hydraulic needle check valve; C_n, C_i — speed and current controllers; I_n — speed transducer; I_m — current and rising brushes device; RC — ring counter; IN — inverter; RF — rectifier; SC — smoothing choke; $SCRB$ — short circuit and rising brushes device; TR — transformer; $Pot.$ — potentiometer.

two equal duty pumps, one of which has adjustable speed, both with a unitary flow rate, equal to 55 per cent of the maximum daily flow rate.

The variable-speed pump is driven by a slip-ring motor, fed by an asynchronous cascade inverter, while the fixed-speed pump is driven either by a squirrel-cage motor (for medium powers), or by a synchronous motor (for high powers). The latter case shows a happy combination, since the synchronous motor compensates the low power factor of the synchronous cascade inverter.

The variable-speed pump may be regulated proportionally to the consumption in the supply system by means of an electronic clock connected to the automated regulation circuit. Thus, an overlapping of pumped flow rate curve over the consumption curve is obtained simultaneously with the superposition of the $H(Q)$ pump characteristic curve over the $H_s(Q)$ characteristic curve of the system. Due to these superpositions, we obtain a reduced compensation volume of the level tank and an increased overall efficiency of the transmission pumping station, accompanied by a lower consumption of electric power.

Figure 14.35 shows the complete working diagram of a waterworks, provided with variable-speed pumps for aetank-open system. The system requires a maximum daily flow ratr of 1.8 cu.m/s. A non-uniform supply to the compensation tank requies a 65 m pumping head. Making use of $20NDS$ variable-speed pumps, weobtain an average overalle efficiency of the pumping installation $\eta_{i\,aver} = 91$ per cent, and an average specific energy consumption $e_{av} = 167$ Wh/cu.m, compared with $\eta_{i\,aver} = 61$ per cent and $e_{aver} = 2.45$ Wh/cu.m, in the case of conventional pumping installation.

We may conclude that the variable-flow rate transmission pumping station is the most economical solution for long-ditance water transport in long transmission lines ($L = 5$ to 100 km). Tiss solution has severa applications to the water supply of localities in Arh sand African countries as well as in Latin America, where water source are generally found, at great distances.

For instance [14.9], the water supply of Barquismento town (Venezuela) required a transmission pumping station like that shown in Figs. 14.36, a and b, provided with two equal duty pumps, one with variable speed, and each of them having $Q = 1.75$ cu.m/s, $H = 115$ m, $P = 2,200$ kW, $n = 1,190$ rev/min, and $U = 6.6$ kV at $f = 60$ Hz. The adjustable pump is driven by a slip-ring asynchronous motor, connected with a subsynchronous inverter cascade. Due to the cascade feeding, a speed range $\Delta n = 1000-1190$ rev/min and a flow rate $\Delta Q = 1300-1750$ dm³/s are obtained. In view of the importance of the town, both the variable-speed pump and the fixed-speed one have been accompanied by stand-by pumps. The transmission line is of unifilar type, with the followng construction characteristics: $D_r = 1.50$ m, $L = 55$ km., $Q = 3.50$ cu.i/s, $H_{st} = +65$ m, $H_{dy} = 50$ m, $H_t = 115$ m, $H_{st}/H_t = 0.56$.

In open-ended pumping systems which also show a decrease from inlet elevation to outlet elevation, a part of the system-head curve will be negative (Figs. 14.37, a and b). In this case a pump is used to increase gravity flow. Without a pump in the system, the negative resistance

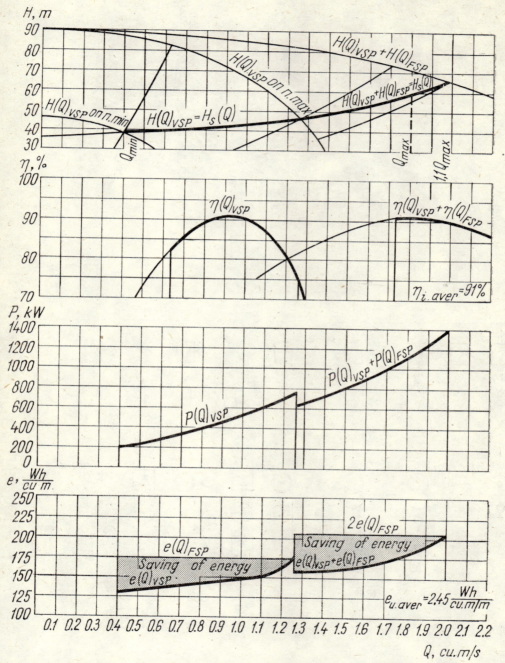

Fig. 14.35. Performance diagram of a variable-speed pump waterworks for pumping in a tank-open system.

Matching Pumps to Requirements

Fig. 14.36, *a*. Variable-speed pump waterworks in Barquismento for pumping in a tank-open system.

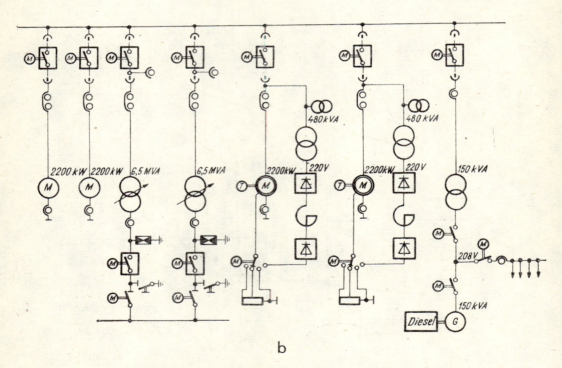

Fig. 14.36, *b*. Schematic wiring diagram of a **variable-speed pump** waterworks in Barquismento.

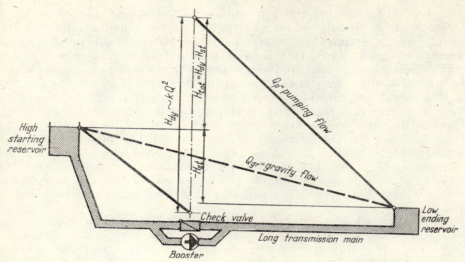

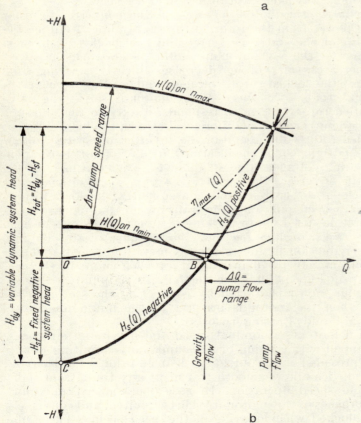

Fig. 14.37. *a*. Schematic diagram of a full variable-speed waterworks for pumping in a tank-open system:

VSP, *VSM* — variable-speed pump and motor; *HNCV* — hydraulic needle check valve; *SAV* — surge anticipator valve; *SCRB* — short circuit and rising brushes; *HNCV* — hydraulic needle check valve; *TR* — transformer; *IN* — inverter; *RE* — rectifier; *SC* — smoothing choke; *RC* — ring counter; *C*, *C* — current and speed controllers.

Fig. 14.37, *b*. Performance diagram of a full variable-speed waterworks for pumping in a long transmission line with very inclined characteristic.

661

or static head is the driving head ($-H_{st}$) which moves the liquid in the system. Steady state gravity flow is sustained at the flow rate corresponding to zero total system head (negative static head plus system resistance equal zero). If a flow is required at any rate greater than that which gravity can produce, a pump is necessary to overcome the additional system resistance. Significant power savings can be obtained in this case from variable speed pumping. This is especially true where there is a large flow rate variation and significant dynamic head (friction). The use of variable-speed pumps reduces the total power, consumed in dynamic head (friction) for any 24-hour period. The power reduction is proportional to the flow rate, multiplied by the head losses. The permanent adjustment of pump flow rate to water demand (due to speed variation) results not only in a considerable power reduction, but also in reduction of the reservoir compensating capacity and of the transmission main line section.

A transmission station will generally comprise two adjustable-speed pumps, each capable of handling the peak flow rate. These pumps are driven by asynchronous slip-ring motors, fed alternatively by means of a single subsynchronous inverter cascade (Fig. 14.38 a).

During speed control the motor operates on the converter in order to re-inject the slip power from the rotor to the supply line. Re-injection is achieved by means of the static converter and feed-back transformer.

The rotor voltage of the asynchronous motor depends on the speed. When the speed increases, this voltage decreases linearly, starting from the maximum value (the locked-rotor voltage), and reaches zero at the synchronous speed. Actually, the speed of the motor is set at the value for which the rectified rotor voltage corresponds exactly to the opposing voltage supplied by the inverter. The difference between the instantaneous values of the rectified rotor voltage and of the inverter opposing voltage is absorbed by the smoothing coil in the intermediate d.c. circuit. Permanent control of the opposing-voltage value is achieved by variation of the priming angle of the inverter. Practically, the opposing voltage value is independent of the rotor current and consequently the load torque.

Relative to fixed-speed pumping, the transmission pumping station, equipped with a variable-speed pump working in a very steep system-head curve, composed entirely of line friction, shows substantially reduced running costs. This reduction is due to approximately 50 per cent lower energy consumption, namely to 70 instead of 140 Wh/cu.m in the case of the graphs shown in Fig. 14.39, b for the RV_{120} type pump.

The asynchronous motor is started by connecting the stator to the high voltage line; then the motor is brought to the minimum control rate by means of a liquid starting rheostat.

For operation of motors at nominal speed and maximum efficiency, each motor is equipped with a rising-brushes and short-circuiting device. This device is also provided on the outside with an additional short-circuiting contactor. The device allows smooth transit at any time, from short-circuit to adjustable speed operation. In case operation at maximum speed persists, the brushes of the motor are short-circuited, and conversely, short-circuiting is blocked when lowering of the speed occurs. The rising-

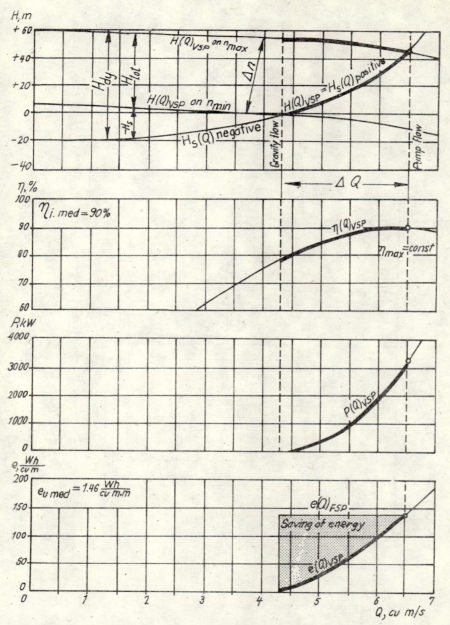

Fig. 14.38,b. Performance diagram of a full variable-speed pump waterworks for pumping into transmission main line with negative $H_s(Q)$ curve.

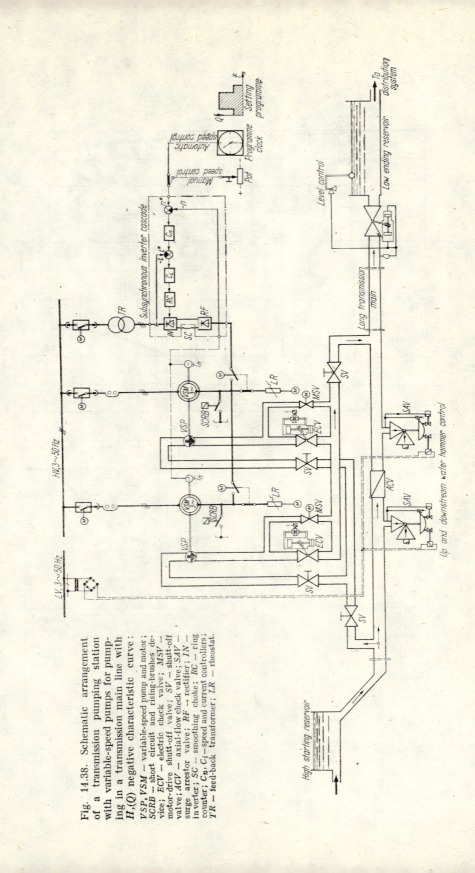

Fig. 14.38. Schematic arrangement of a transmission pumping station with variable-speed pumps for pumping in a transmission main line with $H_s(Q)$ negative characteristic curve: VSP, VSM — variable-speed pump and motor; $SCRB$ — short circuit and rising-brushes device; ECV — electric check valve; MSV — motor-drive shutt-off valve; SV — shutt-off valve; ACV — axial-flow check valve; SAV — surge arrestor valve; RF — rectifier; IN — inverter; SC — smoothing choke; RC — ring counter; C_n, C_i — speed and current controllers; TR — feed-back transformer; LR — rheostat.

brushes device mechanism ensures operation of the motor at maximum speed without losses in the converter or the feed-back transformer, and without losses due to friction of the brushes that inherently occur in the case of cascade connection.

The surges due to the normal starting or stopping of a pump are eliminated by use of a slow-operating electric check valve. It is most often a hydraulically-operated butterfly valve. Opening and closing times can be adjusted separately by means of the throttle valves which are fitted to the power piston cylinders. For normal shut-down, a button

Fig. 14.39.a Variable-speed transmission pumping station in Stuttgart (Siemens Co.).

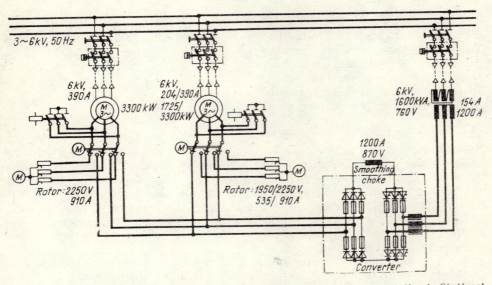

Fig. 14.39, b. Schematic wiring diagram of full variable−speed transmission station in Stuttgart.

would be pushed to stop the pump, but the pump does not stop at once. Instead, the pump check valve begins closing very slowly, gradually reducing the velocity of the water column. When the electric check valve is nearly closed, a limit switch on the valve de-energizes the pump, which coasts to a stop at the same time that the valve reaches its seat with a minimum of pressure fluctuations.

For power failure protection, the surge anticipator (arrestor) type valve is used (see Fig. 7. 103). This valve offers three-way protection, by underpressure, overpressure and power failure pilots, anticipating surges and opening before the surge at the valve. The anticipator valve will open during the coast-down phase of the pump following the power failure, and once the valve opens, it will remain open by reason of a hydraulic time-delay pilot until reversal of the surge valve occurs. After the specified time delay, the valve will begin to close at a very slow rate, bringing the water column to a gradual stop.

Water supply in localities around Stuttgart (F. R. of Germany) [14.8] has been achieved by means of a transmission pumping installation, similar to that shown in Fig. 14.38,a.

This installation (Figs. 14.39, a and b) includes two pumps, of which one is the duty pump, each of them having: $Q = 4.5$ cu.m/s, $H = 67.50$ m., $P = 3,300$ kW, $n = 595$ rev/min, $U = 6$ kW. Flow rate regulation has been achieved within the range $\Delta Q = 3.0 - 4.5$ cu.m/s, to which a speed adjustment $\Delta n = 425 - 595$ rev/min corresponds. Water transport to long distances is done by means of a long transmission line with $D_R = 1.60$, $L = 111$ km, $H_{st} = -110$ m, $H_t = 55$ m, $H_{st}/H_t = -0.50$.

14.2 Pumps Applications to Industrial Zones Facilities

The main consumers of industrial water are thermal and cooling circuits. Water pumping to these circuits is achieved by means of the following categories of pumping installations:
— Boiler-fed pumping plants
— Condenser-cooling pumping plants
— Circulating pumping installations for thermal systems

14.2.1 Boiler-fed Pumping Plants

Steam boilers are heat generators, feeding steam into the turbines of steam power plants. They are fed with water by means of boiler-fed pumps. These pumps increase the water pressure from the thermal circuit up to the necessary value in the boiler. Feed pumps are high-pressure pumps (about 200 daN/sq·cm), working at high temperatures (140—180 C°). Due to their function and to the amount of absorbed power per unit (1,000—15,000 kW), these pumps are the most important internal consumers at a steam power plant.

The pumping system including boiler-fed pumps is characterized by an average flow rate ($Q = 200 - 2,000$ t/h), varying slowly within a rather large flow range ($\Delta Q = 1 : 4$), and by an extremely high pumping head

($H = 1{,}000 - 6{,}000$ m), placed on a skewed-type $H_s(Q)$ curve and having a ratio $H_{st}/H_t = 0.66$.

For safe operation under any conditions, the pumping system peak demand is divided between several turbopumps as follows (Fig. 14.40) [14.21)]:

— a pump with $Q = 105\% Q_{max}$ driven by a steam turbine, or two partial-flow pumps with $Q = 55\% Q_{max}$, driven by variable-speed electric motors (Fig. 14.40, a);

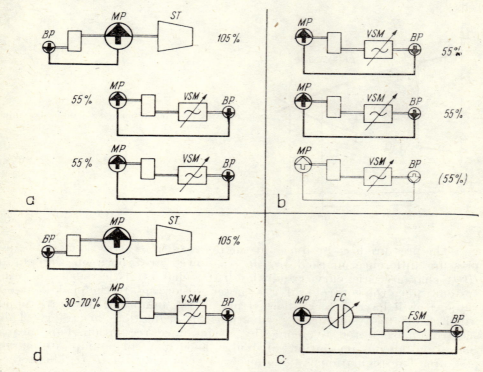

Fig. 14.40. Flow rate distribution possibilities in boiler-fed pumping plants:

a — one pump driven by a steam turbine coping with the whole flow rate, or two partial-flow pumps, driven by electric motors with adjustable speed; b — two or three partial-flow pumps, each driven by a variable-speed electric motor; c — two or three partial-flow pumps, each driven by a fixed-speed electric motor with a fluid coupling; d — one pump, driven by a steam turbine, meeting the whole flow rate, and one partial-flow pump, driven by a variable-speed electric motor. MP — main pump; BP — booster pump; FC — fluid coupling; FSM — fixed-speed motor; VSM — variable-speed motor; ST — steam turbine.

— two partial-flow pumps with $Q = 55\% Q_{max}$ each, driven either by a variable-speed electric motor (Fig. 14.40, b), or by a fixed-speed electric motor connected by a fluid coupling (Fig. 14.40, c);

— a pump with $Q = 105\% Q_{max}$, driven by a steam turbine, and a partial-flow pump with $Q = 70\% Q_{max}$, driven by a variable-speed electric motor (Fig. 14.40, d).

Due to continuous steam off-take from the boiler, boiler-fed pumps operate at constant pressure in any operation point, although the discharge level is variable. This discharge level variation is generated by the constant change of steam flow rate absorbed by the turbine, due to its shaft load variation.

The simplest method for regulating the feed-flow rate is by throttling. This is achieved by means of a control valve mounted on the boiler-feed pipe and acting according to the level. It works as follows (Fig. 14.41): When the flow rate falls, the pump pressure on the $H(Q)$ curve rises, and, at the same time, the required boiler pressure marked on the $H_s(Q)$ curve falls.

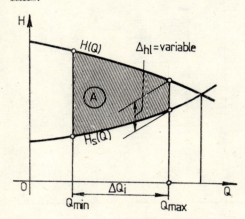

Fig. 14.41. Regulation of feed pump flow rate by throttling the fluid stream.

The pressure loss, produced at the float control valve, is equal to the pressure difference, marked by the shadowed area between the pump $H(Q)$ characteristic curve, and the $H_s(Q)$ characteristic curve of the system. Since the control valve is chosen according to the peak demand value, it will have a very short stroke in the minimum flow rate zone, which leads to a significant reduction of regulation accuracy. This shortcoming is often overcome by a combined regulation. This means the simultaneous regulation of boiler level and of pump discharge pressure. The regulation is done either by changing the $H_s(Q)$ characteristic curve of the system, when it works with fixed-speed pumps, or by changing the $H(Q)$ characteristic curve of the feeding pumps, when they are of variable-speed type. We may note that regulation of the $H_s(Q)$ characteristic curve is rarely done, and then only for very small flow rates ($Q < 180$ t/h).

Arrangement of a boiler-fed pumping installation, equipped with fixed speed pumps. A pumping installation, equipped with fixed-speed pumps, is shown in Fig. 14.42.

The scheme includes a fixed-speed pump, FSP, driven by a motor FSM, usually asynchronous. It also includes a regulation system based on differential pressure. This system makes use of a control valve CV_{dp}, working in accordance with the differential pressure produced in the level control valve MFV. The valve CV_{dp} is closed when differential pressure

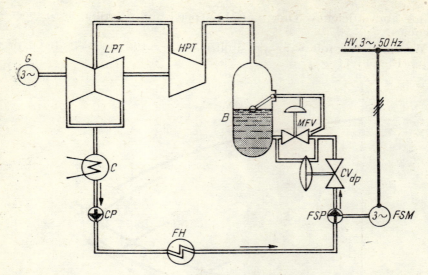

Fig. 14.42. Schematic piping arrangement of a boiler-fed pumping installation, equipped with fixed-speed pumps:

FSP — fixed-speed pump; FSM — fixed-speed motor; MFV — modulating float valve; CV_{dp} — control valve for differential pressure B — boiler tank; G — generator; C — condenser; CP — condensate pump; FH — feedwater heater; LPT, HPT — low and high pressure turbines.

rises, and vice-versa. The system maintains a constant differential pressure and a discharge pressure with parabolic variation. It follows that the $H(Q)'$ curve is parallel to the $H_s(Q)$ curve of the system (Fig. 14.43).

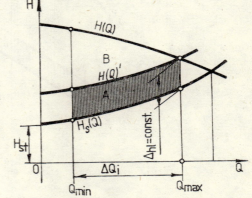

Fig. 14.43. Performance diagram of a boiler-fed pumping installation, equipped with fixed-speed pumps.

Arrangement of a boiler-fed pumping installation, equipped with variable-speed pumps. Steam power plant turbines of high power are often used under variable pressure. Under such conditions, the steam pressure varies proportionally to the turbine load, which requires a corresponding variation of discharge head and of feed pump flow rate. The large power turbines in steam power plants require high performance of the pumps, driving the installation. The performance refers to their regulation range, long life and efficiency. Only variable-speed pumps are able to meet such exigencies.

We have the following possibilities for variable-speed drive of feed pumps (Fig. 14.44):

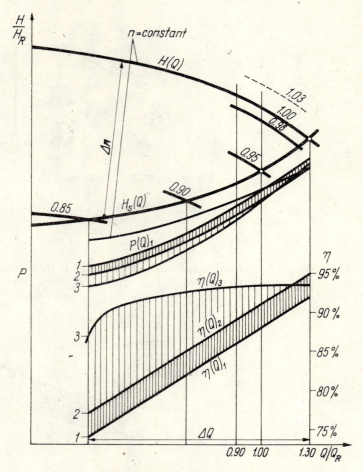

Fig. 14.44. Comparison of economic properties of several methods of feed pumps regulation by electric drive:

1 — by fluid coupling; *2* — by regulating rheostat; *3* — by subsynchronous inverter cascade.

— adjustable-speed electric drives with squirrel-cage motor, or with synchronous motor, connected by fluid coupling to the pump. These drives are affected by the losses occurring in the couplings;

— adjustable-speed electric drives with slip-ring motor and speed-adjusting rheostat. These drives are also affected by losses due to the Joule heating effect;

Pumps for Industrial Facilities

— adjustable-speed electric drives with subsynchronous inverter cascade. These drives are not affected by losses.

This latter solution is the most economical, since it secures a variable flow rate under necessary pressure without significant additional losses. Another advantage of this solution lies in the fact that the slip-ring motor requires a much lower starting power than the squirrel-cage type. This helps the design of auxiliary transformers, since the feed pumps are the major consumers of electric steam power.

According to unitary power of the feed pumps, two schemes of subsynchronous inverter cascade may be used: (a) with simple converter, and (b) with double converter.

The simple converter scheme (Fig. 14.45) is fit for high unitary powers ($P \leqslant 4,000$ kW). This scheme has been used by the Brown Boveri Company (Switzerland) for adjustable-speed electric drive of three pumps,

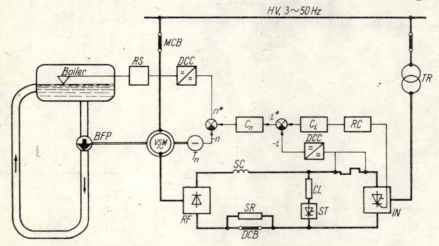

Fig. 14.45. Schematic arrangement of variable-speed drive with slip-ring motor, connected in simple inverter cascade to the feed waterpump of a boiling-water reactor:

BFP — boiler-fed pump; *VSM* — variable-speed motor; *RS* — rheostat for setting desired speed; *DCC* — D.C. converter; *RF* — rectifier with diodes; *IN* — inverter with thyristors; *ST* — switching thyristor; *SR* — starting resistor; *CL* — current limiter; *SC* — smoothing choke; *DCB* — D.C. breaker; *RC* — ring counter (pulse control set); C_n, C_i — speed and current regulators (controllers); *TR* — transformer; *MCB* — mains circuit-breaker; T_n — tachometer (speed transducer).

feeding water to the boilers of a reactor at Mühleberg nuclear power station (F.R. of Germany). Each pump is driven by a slip-ring asynchronous motor of $P = 3,300$ kW, having rotation speed $n = 2980$ rev/min, speed control range $\Delta n = 2,300 - 2,980$ rev/min, and supply voltage $U = 6,000$ V.

The double converter scheme (Fig. 14.46) is meant for feeding pumps with extremely high unit powers ($4,000 < P < 25,000$ kW). It introduces two converters in subsynchronous cascade into the rotor circuit of the motor (Fig. 14.47). These two converters introduce two sources of variable electromotive voltage into the rotor circuit. The double converter (see section 12.2.1) has the advantage of a reduced power consump-

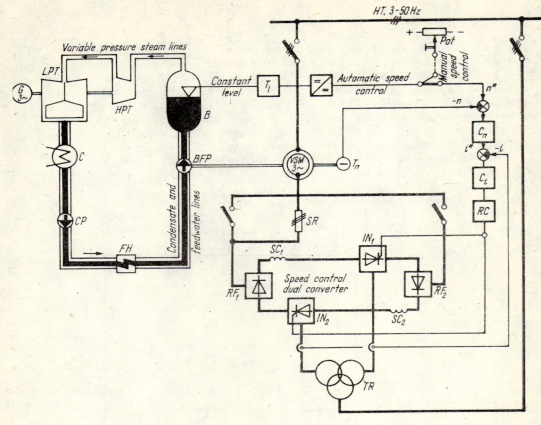

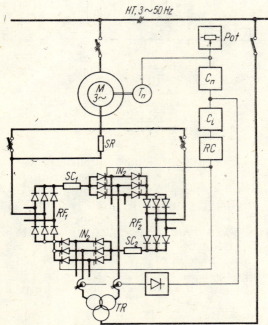

Fig. 14.46. Schematic arangement of a variable-speed drive with slip-ring motor, connected in double inverter cascade to the feedwater pump of a steam power plant:

G — generator; LPT, HPT — low and high pressure turbines; B — boiler; BFP — boiler-fed pump; C — condenser; CP — condensate pump; FH — feedwater heater; VSM — variable-speed motor; T_n — tachometer (speed transducer); IN_1, IN_2 — inverters; RF_1, RF_2 — rectifiers; SC_1, SC_2 — smoothing chokes; TR — transformer; SR — starting resistor; RC — ring counte (pulse control set); C_n, C_i — speed and curren regulators (controllers); T_l — level transducer

Fig. 14.47. Schematic wiring diagram of a double converter subsynchronous cascade:

RF_1, RF_2 — rectifiers; IN_1, IN_2 — inverters; SC_1, SC_2 — smoothing chokes; TR — transformer; SR — starting resistor; M — induction motor; T_n — tachometer (speed transducer); RC — ring counter (pulse control set); C_n, C_i — speed and current controllers; Pot — potentiometer.

tion, as well as of a large range of speed control ($\Delta n = 1:7$). This range is divided into two sub-ranges — an inferior one, and a superior one (Fig. 14.48).

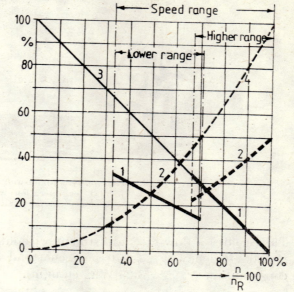

Fig. 14.48. Working diagram of a double converter subsynchronous cascade:

1 — tension variation; *2* — d.c. variation; *3* — rotor tension variation; *4* — rotor current variation.

A double converter scheme has been provided by the Siemens Company (F. R. of Germany) for driving some high-power feed pumps. Each pump is driven by an electric motor (Fig. 14.49) having $P = 8200$ kW, $U = 6300$ V, $n = 1490$ rev/min and speed variation range $\Delta n = 750 - 1490$ rev/min. Its two sub-ranges are divided as follows: the inferior range, $n_{inf} = 750 - 1125$ rev/min, and the superior one, $n_{sup} = 1125 - 1490$ rev/min.

The progress achieved during the last few years in electronic circuits and electric machine insulation has allowed the use of cascade driving installations. Thus, the Standinger steam power plant (F. R. of Germany) is fed by three pumps (Fig. 14.50) with a unitary flow rate of 55 per cent of the total flow rate ($Q = 1610$ t/h). In this case, pump rotation speed is very high, namely 4963 rev/min. To this end, each pump has been coupled to its driving motor by means of a transmission system with a 3.4:1 ratio. The driving motor has a power of 15,200 kW and a speed of 1450 rev/min, varying within a speed range of 725—1450 rev/min. A 330 kW booster pump is coupled to the back side of the motor.

In conclusion, pump speed control can lead to an artificial $H(Q)$ characteristic curve with a downward convexity, which is approximately superposed on the $H_s(Q)$ characteristic curve of the system (Fig. 14.51).

Fig. 14.49. High-power phase-wound electric motor ($P = 8,200$ kW), built for operation with double converter subsynchronous cascade (Siemens Co.).

This secures a flow rate adjustment without losses, which leads to a high global efficiency ($\eta_{i\,med} = 70$ per cent) and a low unitary electric power consumption ($e_{u\,med} = 3.25$ Wh/ cu.m.m).

As compared to a pumping installation, equipped with fixed-speed pumps, the pumping installation provided with variable-speed pumps shows a substantially lower energy consumption. For instance, in the case shown in Fig. 14.51, the consumption is 5,700 Wh/cu.m pumped water, instead of 8,000 Wh/cu.m.

Fig. 14.50. Feed pumping installation consisting of one duty pump and a high-power phase-wound motor ($P = 15,200$ kW), operating with double converter in subsynchronous inverter cascade (Brown Boveri Co.).

14.2.2 Condenser-cooling Pumping Plants

In general, condensers of steam power stations are cooled with water. The necessary water flow rates are extremely large, reaching 130—190 cu.m/h per MW, that is 3—4 cu.m/s for 100 MW. This high consumption is boosted by the high temperature of the environment and by the vacuum value to be maintained in the condenser system. For intance, cooling-water consumption in winter time is only 70 per cent of summer time consumption. The steam power stations built at present require cooling water flow rates that only large rivers can meet. When such raw water cooling flow rates are not available, water is cooled in closed-circuit cooling towers.

Thus, due to the high value of cooling flow rates, steam power plant location is not only dependent on its position relative to a thermal energy source (coal basin, etc.), but also, to a greater extent, on its position relative to the

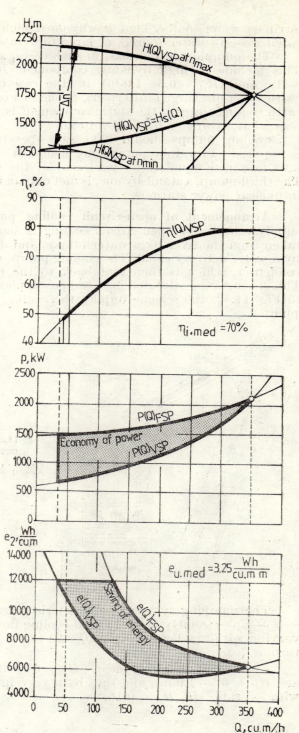

Fig. 14.51. Performance diagram of a boiler-fed pumping installation, equipped with variable-speed pumps.

auxiliary water source. That is why most high power steam plants are located close to water sources of large flow rates (large rivers, lakes, etc.).

The pumping system has, in such cases, a small total pumping head. This includes a high dynamic component $H_{dy} = 6-8$ m) and a small static one ($H_{st} = 0.5-1.0$ m) in the case of open-circuit installations (with water discharge into a river). In the case of a closed-circuit installation (with cooling towers), the static component is about equal to the dynamic one, that is $H_{st} = H_{dy} = 8-10$ m.

Cooling pumps should have safe operation, and no stand-by pump is required. Two duty pumps are generally used, each of them having unit flow rate equal to 55 per cent of the maximum total flow rate. The third pump, a stand-by one, is met only in the case of condenser cooling in nuclear power stations.

Arrangement of open-circuit cooling pumping plants. The water supply scheme for open-circuit cooling is shown in Fig. 14.52. Water is taken from the river by a water intake, and directed towards the pumps by a water delivery canal. The cooling pumps discharge water through the condenser, which is then sent back to the river by a discharge canal. This scheme is mostly used in steam power plants of high power. According to Fig. 14.52, the scheme implies the cooling of each turbine by its own pump.

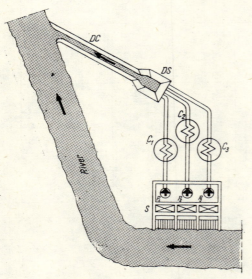

Fig. 14.52. Schematic diagram of an open-circuit cooling pumping plant:
P_1, P_2, P_3 — water circulating pumps; C_1, C_2, C_3 — condensers; S — strainers; DS — discharge pump; DC — discharge channel.

The pumping system is characterized by:
— an extremely high required cooling flow rate ($Q = 3.5-7.0$ cu.m/s), varying within a large range ($\Delta Q = 1:10$) in accordance with the steam turbine load;
— a small pumping head ($H_t = 6-8$ m, of which $H_{dy} = 6-7$ m and $H_{st} = 0.5-1.0$ m), and thus by a very inclined characteristic curve, with a ratio $H_{st}/H_t = 0.16$.

The unitary power resulting from the distribution of the maximum required cooling flow rate to a single pump is about 250—500 kW.

Consequently, the pumping system of a single cooling pump has a small pumping head, but a high dynamic component, which makes its $H_s(Q)$ characteristic curve take a parabolic shape, and closely follow the curve of maximum efficiency of a variable-speed pump.

For the conditions of a high required flow rate and a small pumping head, we recommend axial-flow pumps with vertical axle. There are two possibilities for flow-rate regulation of these pumps: by altering the position of blades, and variation of rotation speed. In both cases, the turbopump topogram will be necessary. This topogram will show the flow range, determined by steam turbine load modification, as well as the head range, resulting from variation of river water level. The area included between the four extreme points of operation represents the geometric range of all duty points. This area is also necessary for the correct selection of variation limits of impeller blades rotation angle, in the case of fixed-speed pumps, or rotation speed, in the case of variable-speed pumps.

Figure 14.53 shows the topogram of an axial-flow pump with adjustment by altering the position of impeller blades (pitch control).

The area between points P, P_1, P_2, P_3, represents the geometric range of all operation points of the pump. The diagram also shows that the higher the static head of the pumping system, the better its regulation efficiency.

The main parameters (flow rate and pumping head) of an axial-flow pump may be also adjusted by rotation speed variation. In this case, due to the high level variations at the source, we have to cope with many shortcomings which are not present in the case of radial-flow or mixed-flow pumps. They consist either of unstable pump operation in the area of the $H_s(Q)$ characteristic curve corresponding to minimum level (Fig. 14.54), or of uneconomical operation with low efficiency (although speed is adjustable) on the $H_s(Q)_{max}$ characteristic curve corresponding to the maximum level.

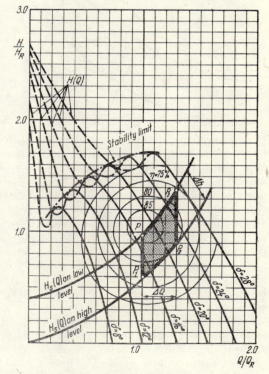

Fig. 14.53. Operation diagram of a cooling pump in the case of flow rate regulation by altering the position of impeller blades.

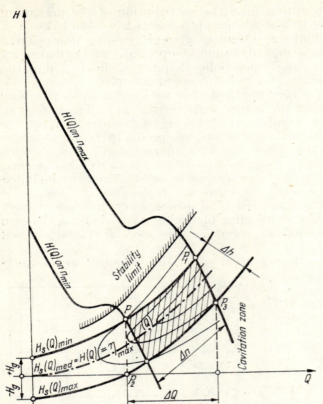

Fig. 14.54. Operation diagram of a cooling pump in the case of flow rate regulation by speed adjustment.

Therefore, the rational operation zone of an axial-flow pump is rather narrow (distributed on both sides of the maximum efficiency area, $\eta_{max}(Q)$). For obtaining a larger zone in cases of source level variations, we may use a combined adjustment, changing simultaneously rotation speed and position of impeller blades. Thus, unlike radial-flow or mixed-flow pumps, the axial-flow type should be regulated with great care, requiring a previous analysis of all possible operation conditions. Consequently, we may say that the problem of adjustable electric drive of high-power axial-flow pumps is insufficiently studied so far.

In view of the fact that cooling pumps in open-circuit systems require a rather large range of speed regulation, a slip-ring motor could be used connected in asynchronous inverter cascade mode and provided with an autotransformer in its rotor circuit (Fig. 14.55).

The autotransformer secures two values of outlet voltage (each value corresponding to a certain transformation ratio, for instance 1 : 4 and 1 : 2), as well as a wide range of speed control, so that the lower limit of speed control range corresponds to the cut-out speed of the asynchronous motor. Moreover, due to introduction of the autotransformer into the circuit, the driving installation can be started even from zero speed, without the additional starting apparatus needed for conventional schemes.

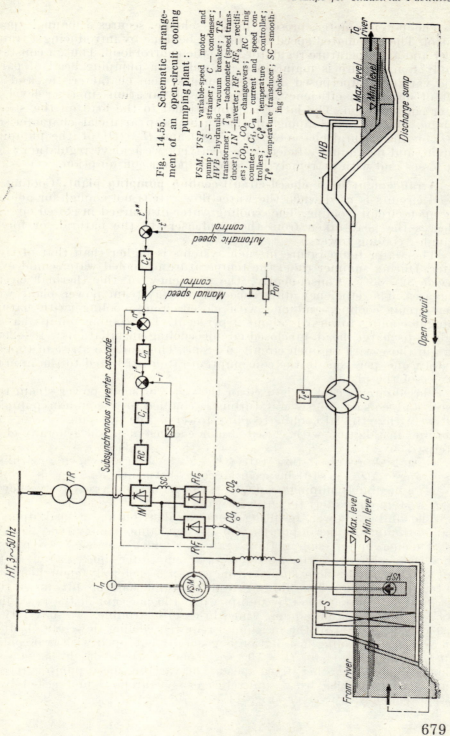

Fig. 14.55. Schematic arrangement of an open-circuit cooling pumping plant:

VSM, VSP — variable-speed motor and pump; S — strainer; C — condenser; HVB — hydraulic vacuum breaker; TR — transformer; T_n — tachometer (speed transducer); IN — inverter; RF_1, RF_2 — rectifiers; CO_1, CO_2 — changeovers; RC — ring counter; C_i, C_n — current and speed controllers; C_t^o — temperature controller; T_t^o — temperature transducer; SC — smoothing choke.

The potentiometer, present in the scheme, secures a manual speed control. The automated speed control can be done by introducing a transducer and a temperature regulator into the control loop. Thus, a constant supply temperature is maintained, as well as a pumping head of parabolic variation superposed on the natural system characteristic, and on that of maximum efficiency, irrespective of variations in river level or water consumption. The latter is proportional to the load on the steam turbine. Due to overlapping of the turbopump artificial characteristic $H(Q)$ on the natural characteristics $H_s(Q)$ and $H_{max}(Q)$ of the pumping system, and maximum efficiency, respectively, we achieve regulation without losses and with extremely low electric power consumption.

Arrangement of a closed-circuit cooling pumping plant. Cooling in closed circuits is used when the water flow rate is not enough for achieving open-circuit cooling. The cooling water discharged in closed circuits releases the heat taken from the condensers to the natural, or forced draught, cooling towers.

The water temperature in such systems is higher than that of river water. During summer time, the temperature of cooled water could even exceed $32-33C°$. Consequently, the efficiency of the thermal circuit decreases. The efficiency difference between a steam power plant with open-circuit cooling, and one with closed-circuit cooling (with natural draught cooling towers) is about 2 per cent. The cooling pumps have a higher discharge head in closed-circuit cooling installations (18—20m) than in those with an open circuit (6—8 m). This leads to a rise of approximately one per cent in the consumption rate distributed to the internal service of the plant.

Closed-circuit cooling is used in the case of small power steam turbines, and consists of several turbines, coupled to a common pumping system with artificial cooling (cooling towers). The pumping system of a pumping installation with closed-circuit cooling pumps is generally characterized by:

— a high required flow rate ($Q = 2-3$ cu.m/s) varying within a large range ($\Delta Q = 1:10$) in accordance with turbine load;

— a medium pumping head ($H_t = 18-20$ m) and a relatively skewed $H_s(Q)$ system characteristic curve, with a ratio $H_{st}/H_t = 0.5$.

The unitary power, resulting from the distribution of peak demand to an optimum number of pumps, is of average value, namely $160-250$ kW.

The maximum cooling flow rate is divided between two pumps. Since the cooling water flow rate, necessary in winter time, is about 70 per cent of that required in summer, one pump is generally enough for the winter consumption. This flow rate distribution has the advantage that in winter we have a stand-by pump, with 100 per cent of the peak flow rate demand, while in summer time, in case of trouble, we may rely at least on 70 per cent of the maximum required cooling flow rate. That is, a closed-circuit cooling installation is generally equipped with two cooling pumps, each selected for $Q = 70\% Q_{max}$, one of which should have variable speed, and the other, fixed speed. The variable-speed pump is recommended to be driven by a squirrel-cage asynchronous motor, fed by a static frequency converter (Fig. 14.56).

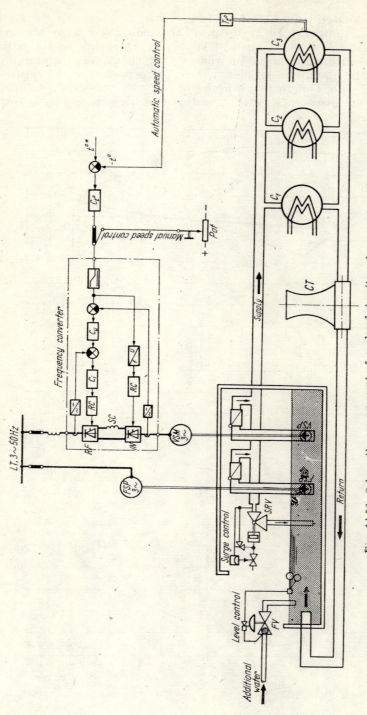

Fig. 14.56. Schematic arrangement of a closed-circuit cooler pumping plant:

VSM, VSP — variable-speed motor and pump; FSM, FSP — fixed-speed motor and pump; SRV — surge relief valve; FV — float valve; CT — cooling tower; C_1, C_2, C_3 — condensers; RF — rectifier; IN — inverter; RC — ring counter; C_i, C_u, C_t^0 — current, tension, temperature controllers; Tt_0 — temperature transducer; SC — smoothing coil.

The automated control of cooling water temperature is done by introducing a temperature transducer and controller into the control loop. This system keeps a constant supply temperature and a parabolic discharge pressure, superposed on the system characteristic pressure and located near the maximum efficiency curve, irrespective of the return temperature variation. Due to overlapping of the two pressures, we obtain a control without losses and with a reduced electric power consumption.

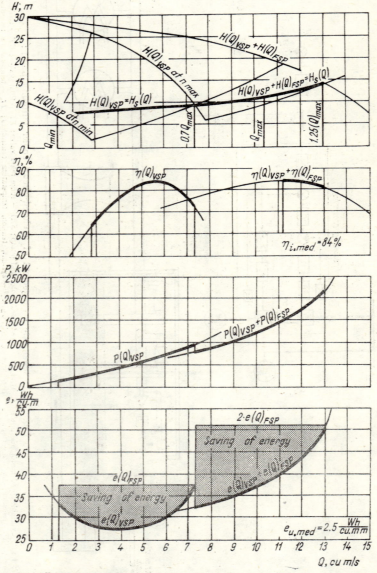

Fig. 14.57. Performance curves for a closed-circuit cooling pumping plant.

For instance, in the case of a pumping installation with two MV_{1401} type pumps, we get the operation diagram shown in Fig. 14.57.

This diagram shows that the unitary medium specific consumption of electric power is $e_{u\,med} = 2.5$ Wh/cu.m.m and the medium global efficiency $\eta_{i\,med} = 84$ per cent. The shadowed area stands for the energy saving, obtained by the use of one variable-speed pump instead of two fixed-speed pumps, that is 35 Wh/cu.m, instead of 45 Wh/cu.m.

Figure 14.58 shows a pumping cooling plant of small flow rate, equipped with variable-speed pumps.

Fig. 14.58. Closed-circuit cooling pumping plant with variable-speed pump (Danfoss Co.).

14.2.3 Circulating Pumping Installations for Thermal Systems

The thermal agent for urban heating may be:
— warm water, having a supply-temperature of 90°C, and a return temperature of at maximum 70°C;
— hot water, with a supply-temperature of 130—180°C, and a return at 65—70°C.

The first case is limited to very short systems (for instance, for a part of a residential district), since the flow rate circulated is high. Heating with hot water is, at present, the most efficient and widely applied method.

The related recirculation pumps have a pumping head made up only of a dynamic component, since these pumps overcome only linear and local head losses, occurring as hydraulic resistances which are proportional to the square of flow rate.

The $H_s(Q)$ characteristic curve of the system passes through the origin of $H-Q$ co-ordinate axes, and through the duty point, found

Matching Pumps to Requirements

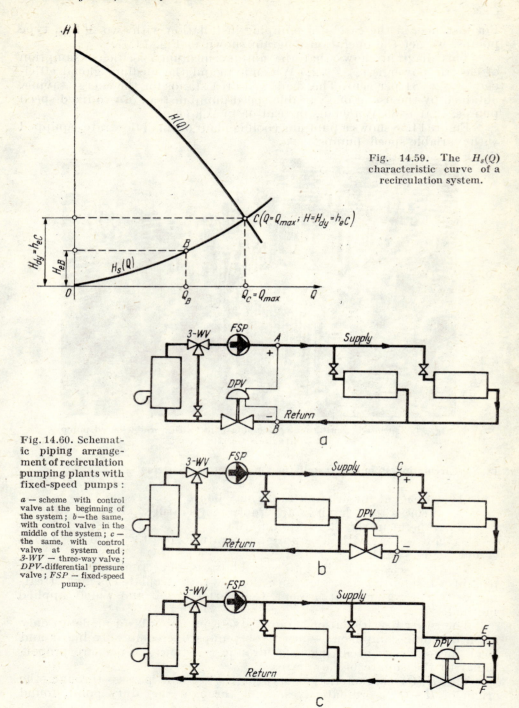

Fig. 14.59. The $H_s(Q)$ characteristic curve of a recirculation system.

Fig. 14.60. Schematic piping arrangement of recirculation pumping plants with fixed-speed pumps:

a — scheme with control valve at the beginning of the system; *b* — the same, with control valve in the middle of the system; *c* — the same, with control valve at system end; *3-WV* — three-way valve; *DPV* — differential pressure valve; *FSP* — fixed-speed pump.

at the intersection of this characteristic with the $H(Q)$ characteristic of the recirculation pump. These two points are not enough for drawing the $H_s(Q)$ curve, and thus we also have to determine the intermediate points B (Fig. 14.59).

These points may be determined by means of the following relation

$$h_{lB} = h_{lC}\left(\frac{Q_B}{Q_C}\right)^2 = k_2 Q_B^2$$

Recirculation pumps have to accelerate circulation of hot water in the system by producing differential pressure between two points, located on the outward and return pipes. To this end, two categories of pumps are used: (a) fixed-speed pumps, and (b) variable-speed pumps.

Arrangement of a recirculation pumping plant with fixed-speed pumps. As shown in Fig. 14.60, the two main parts of the installation are the pump FSP and a control valve DPV.

The location of the control valve, necessary for differential pressure control in the system, has a great influence on installation pressure and on operating costs. Diagrams shown in Fig. 14.61 show the pressure con-

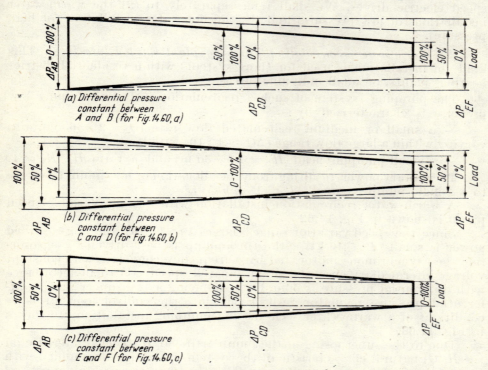

Fig. 14.61. Pressure conditions corresponding to control valve location between different points in the system:
a — at the beginning; b — in the middle; c — at the end of the system.

685

ditions corresponding to control valve location between the points AB, CD, or EF.

When the valve DPV is placed between the points AB, that is close to the pump FSP (Fig. 14.60, a), the installation works with a too high differential pressure (Fig. 14.61, a) when the load is smaller than 100 per cent. Consequently, for partial loads, we will notice a higher electric power consumption than necessary.

When the valve DPV is placed between the points CD, that is, in the middle of the system (Fig. 14.60, b), we obtain the smallest variation of differential pressure (Fig. 14.61, b).

When the installation is working under load for a long time, it is recommended to place the valve DPV between the points EF, in the remotest point of the system (Fig. 14.60, c), since in that case, we obtain the most economical operation (Fig. 14.61, c).

Arrangement of a recirculation pumping plant with variable-speed pumps. Since the $H_s(Q)$ characteristic curve of the recirculation system is exactly superposed on that of maximum efficiency of a variable-speed pump, it is recommended to provide these installations with adjustable-speed electric drives. We shall refer separately to (a) the warm-water installation (of low pressure) and (b) the hot-water installation (of high pressure).

(a) *Warm-water recirculating pumping plant* (*of low pressure*). This kind of installation is used for short systems within residential district heating plants, or small individual areas.

The pumping system of such a recirculation installation under low pressure is characterized by:

— a small or medium recirculated flow rate ($Q = 2-360$ cu.m/h) varying within a large flow range ($\Delta Q = 1:10$) with load;

— a small pumping head ($H_t = 1.5-30$ m) and a ratio $H_{dy}/H_t = 1$.

The unitary power, resulting from the allocation of maximum flow rate to a single pump, is small ($P = 1-40$ kW).

A warm-water recirculation installation with a single variable-speed pump is shown in Fig. 14.62.

Since the speed variation range is large ($\Delta n \approx 1:10$), and the absorbed power is small ($P \leqslant 40$ kW), the turbopump is best driven by a squirrel-cage asynchronous motor, fed by a frequency converter with constant voltage intermediary circuit. The automated speed control in accordance with differential pressure is done by introducing a transducer and a differential pressure regulator, respectively, into the control loop. Pressure conditions of a warm-water recirculating pumping installation are shown in Fig. 14.63.

Due to the superposing of the pump artificial $H(Q)$ characteristic on the $H_s(Q)$ natural characteristic of the system, which is compounded with that of maximum efficiency of a variable-speed pump, we obtain a control without losses, which leads to the lawest consumption of electric power (Fig. 14.64, a).

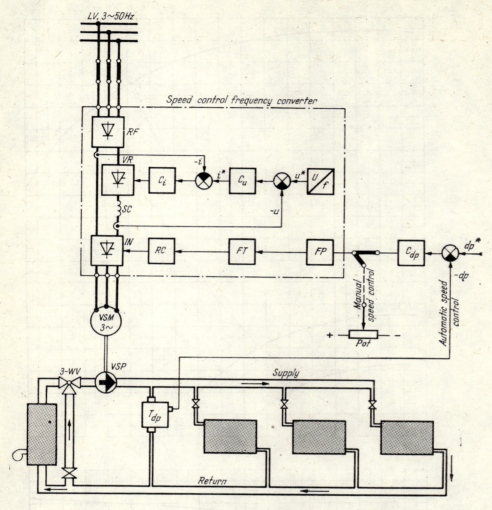

Fig. 14.62. Schematic arrangement of a warm-water pumping recirculation installation with a variable-speed pump :

VSM, VSP — variable-speed motor and pump; 3-WV — three-way valve; T_{dp} — differential pressure transducer; RF — rectifier; IN — inverter; VR — voltage regulator; SC — smoothing coil; C_i, C_u — current and tension controllers; RC — ring counter; FP — frequency programme; FT — frequency transmitter; C_{dp} — differential pressure controller.

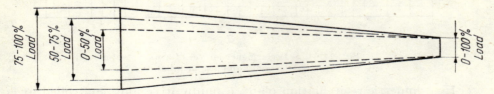

Fig. 14.63. Pressure condition of a warm-water pumping recirculation installation with variable-speed pumps.

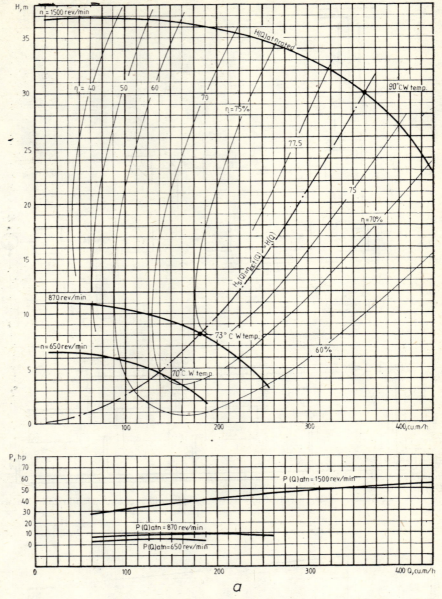

Fig. 14.64, *a*. Performance curves of a warm-water recirculation pumping installation.

In a domestic recirculation plant, a pump drive unit controlled by a microprocessor (Fig. 14.64, *b*) is used, which gives a high degree of flexibility.

Pumps for Industrial Facilities

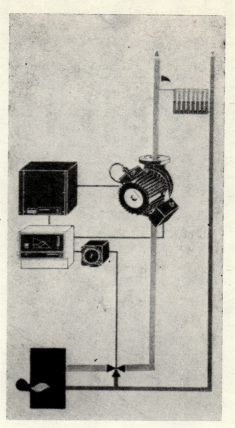

Fig. 14.64, *b*. Microprocessor-controlled recirculation plant (Grundfoss Co.).

(b) *Hot-water recirculation pumping plant* (*under high pressure*). This kind of installation is used for long line heating systems in towns and localities. A high-pressure heating system is characterized by:
— a high required maximum flow rate ($Q = 0.5-3.0$ cu.m/s) varying within a very large range ($\Delta Q = 1 : 25$);
— a medium pumping head ($H_t = 100 \div 150$ m) and a ratio $H_{dy}/H_t = 1$.

The unitary power, resulting from the allocation of maximum required flow rate to a single pump, is high ($Q = 500-3,000$ kW).

Figure 14.65 shows the scheme of a hot-water recirculation plant with a single pump.

Since the flow rate and speed have a very large variation range and the unitary power is high, the hot-water regulated pumps are best driven either by a d.c. motor (Fig. 14.66, *a*), or a synchronous motor with static converter. This latter motor is fed by a static converter, controlled by pulses coming from the motor (Fig. 14.66, *b*).

The pump supplying the installation is of special construction, having its stuffing box provided with cooling sleeves, coupled to a hydraulic circuit permanently containing a cooling medium.

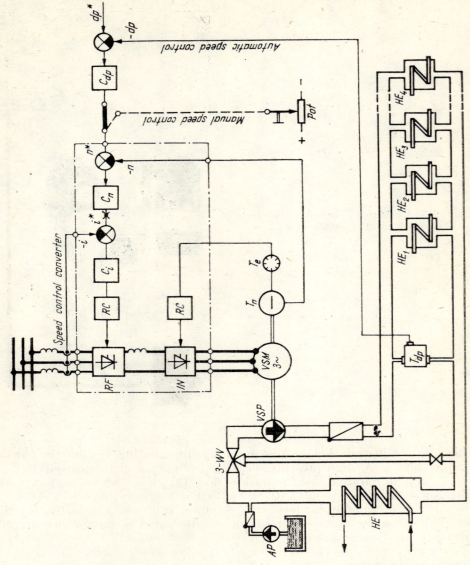

Fig. 14.65. Schematic arrangement of a hot-water recirculation pumping installation (under high pressure):

VSM, VSP — variable-speed motor and pump; AP — additional pump; HE — heat exchanger; $3\text{-}WV$ — three-way valve; RF — rectifier; IN — inverter; RC — ring counter; C_n, C_i — speed and current controllers; C_{dp} — differential pressure controller; T_{dp} — differential pressure transducer; T_n, T_e — speed and angle transducer; SC — smoothing choke.

Pumps for Industrial Facilities

Fig. 14.66, *a*. Variable-speed pump drive by d.c. motor.

Fig. 14.66, *b*. Static converter feeding a d.c. motor without collector (Siemens Co.).

Since the turbopump artificial $H(Q)$ characteristic curve may be superposed on the $H_s(Q)$ natural characteristic curve of the system, as well as on that of maximum efficiency relating to a variable-speed pump, we obtain a very economical control with a reduced electric power consumption. For instance, the diagram of Fig. 14.67 shows a reduced unitary power consumption ($e_{u\ med} = 1.48$ Wh/cu.m.m), and an extremely high global efficiency ($\eta_{i\ med} = 90$ per cent). As compared to the installation with one fixed-speed pump, the one mentioned above requires much lower operating costs due to its reduced energy consumption, namely 200 Wh/cu.m instead of 530 Wh/cu.m.m.

The Siemens Company of West Germany achieved a recirculation installation provided with a d.c. static converter motor (without commutator). In that case, the pump was driven by a motor having $P = 1,500$ kW, and $n = 1000$ rev/min.

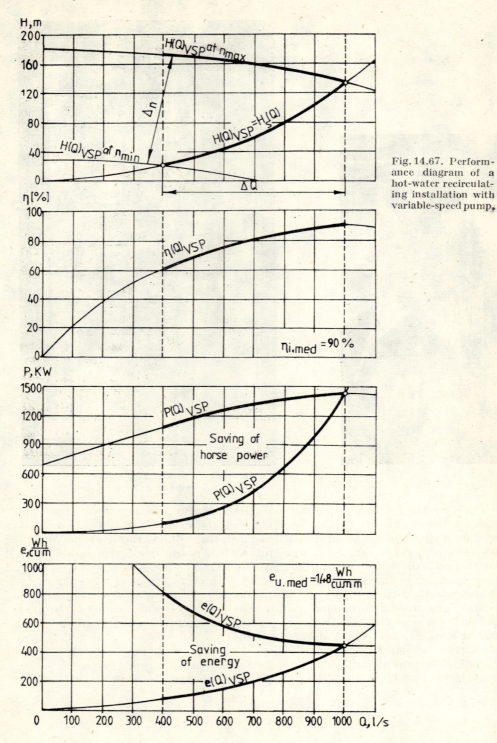

Fig. 14.67. Performance diagram of a hot-water recirculating installation with variable-speed pump.

14.3 Applications of Pumps to Land Reclamation Systems

Depending on the water pumping sense, land pumping stations may be: (1) for irrigation, when the water flow is secured by pumping from a source (river, lake, etc.), or (2) for drainage, when excess water in the soil is eliminated by pumping it from the drainage system to an emissary (river, see, etc.).

14.3.1 Land Irrigation Pumping Installations

Raising the water level in irrigation canals is done by:
— proper pumping stations, also called supply stations, located at the intersection of the main trunk ditch with the source. These stations are working under variable-suction level conditions ($\Delta L = 4-8$ m);
— boosting stations, located in one or several intermediate points of the main trunk ditch, working under constant-suction level conditions.

The pumping system of such installations is characterized by:
— an extremely large required flow rate ($Q = 1-100$ cu.m/s), varying slowly within a narrow flow range ($\Delta Q = 1:4$);
— a small pumping head ($H_t = 4-20$ m) in the case of irrigation works from overflow meadows, and a medium pumping head ($H_t = 30-60$ m) in the case of irrigation works from high platforms;
— a flat $H_s(Q)$ characteristic curve in the meadows, and a skewed one ($H_{st}/H_t \approx 0.8$) in high platforms.

The unitary power resulting from the allocation of maximum required flow rate to an optimum number of pumps is of a high value ($P = 500 - 8{,}000$ kW).

Arrangement of a land irrigation pumping installation, equipped with fixed-speed pumps. This pumping installation is generally provided only with duty pumps, without stand-by facilities. The peak demand is distributed to 4—8 pumps. Each pump-required flow rate is chosen as 27.5—13.75 per cent of the peak demand (the first value corresponds to four pumps used, and the second, to eight pumps). These pumps are generally driven by synchronous motors.

Axial-flow pumps are used in the case of small pumping heads ($H_t = 4-8$m), while in the case of medium pumping heads ($H_t = 12-40$m) mixed-flow single or multi-stage pumps are recommended. The flow rate of these pumps may be regulated by altering the position of blades at axial-flow pumps (Fig. 14.68), and by adjusting the inlet guide vanes in the case of mixed-flow pumps (Fig. 14.69). In both cases, we have to use the turbopump topogram. It will show the flow range determined by water consumption variation in the channel system, as well as the head range resulting from variations in suction level. The area between the four extreme points is the geometric range of all operation points.

This area has to be known both when designing and when using such installations, since we have to know pump behaviour so that suction level variation, or consumption variation, should not make the P, P_1, P_2, P_3 points pass into the unstable or cavitation zones.

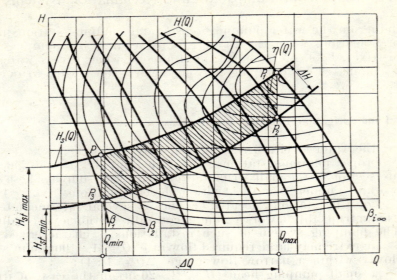

Fig. 14.68. Possible operation range of an axial-flow pump with flow-rate regulation by altering position of impeller blades.

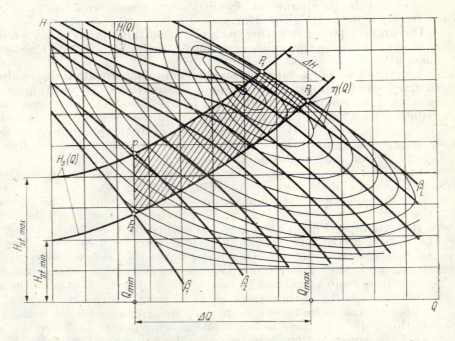

Fig. 14.69. Possible operation range of a mixed-flow pump with flow-rate regulation by adjusting inlet guide vanes.

Study of the topogram is also necessary in the case of pumps discharging into pipes which exhaust into discharge basins by siphoning. In this case, pump operation will be checked for the whole flow range both with a primed siphon and with a broken one. In the latter case, the pumping head is higher than the normal one.

Arrangement of a land irrigation pumping installation, equipped with variable-speed pumps. When the required flow rate is large and the pumping head small ($H_t = 4-6$ m), the pumping installation is equipped with axial-flow pumps, preferably with vertical axis. Under the same flow rate conditions but with medium pumping heads ($H_t = 12-40$ m), the mixed- or radial-flow pumps are recommended. In both cases, the maximum required flow rate could be distributed among one, two (the normal case), or three pumps, of which only one has variable speed. For each case, the pump will have a unitary flow rate, equal to 105 per cent, 55 per cent, and 35 per cent, respectively, of the maximum flow rate. From the design stage, we should study the control possibilities of variable-speed pumps in various conditions of flow rate and suction level variations, making use of the turbopump topogram. When the flow rate and suction level are varying simultaneously, the pump operation range is represented by the area limited by the four extreme points, P, P_1, P_2, P_3 (Fig. 14.70). This area, is the geometric range of all operation points.

A different situation arises in the case of regulation of axial-flow pumps. In that case a rigorous operation range should be established, so that the extreme points do not fall into unstable or cavitation zones.

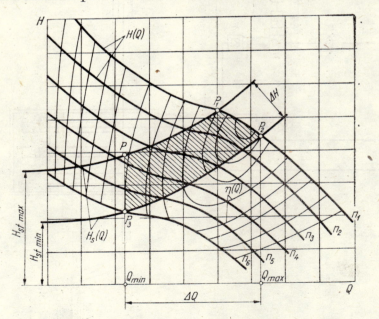

Fig. 14.70. Possible operation range of a mixed- or radial-flow pump with flow-rate regulation by rotation speed variation.

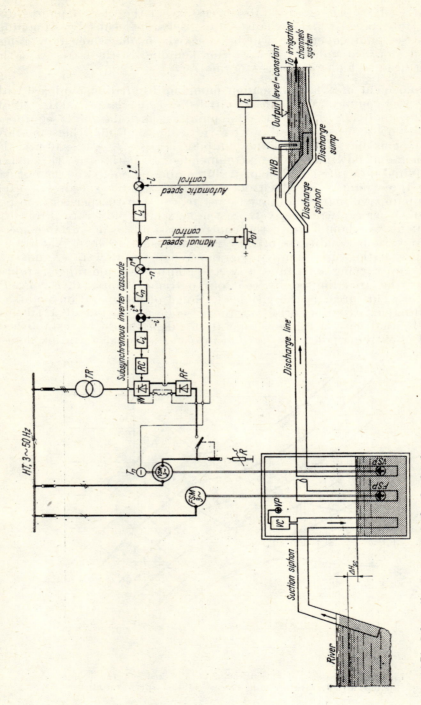

Fig. 14.71. Schematic arrangement of a land irrigation pumping installation, equipped with a variable-speed pump and a fixed-speed one: SM, VSP — variable-speed motor and pump; FSM, FSP — fixed-speed motor and pump; HVB — hydraulic vacuum breaker; TR — transformer; R — rheostat; RF — rectifier; IN — inverter; RC — ring counter; C_i, C_n, C_l — current, speed and level controllers; T_n, T_l — speed and level transducers.

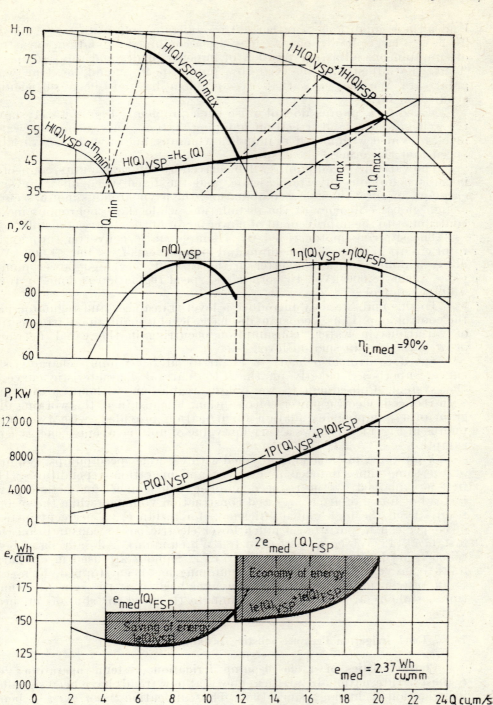

Fig. 14.72. Performance curves of a land irrigation pumping installation, equipped with variable- and fixed-speed radial-flow pumps.

Under these conditions, a rather narrow range is obtained ($\Delta Q \geqslant 0.5\, Q_R$). To have a larger flow range, we may use a combined regulation, consisting of simultaneous adjusting position of impeller blades and rotation speed. Thus, the adjustment of axial-flow pump speed should be done with great care. We may say that this problem has not been sufficiently studied so far.

Mixed-flow pumps do not raise such problems. Figure 14.71 shows the scheme for a land irrigation pumping station for raising the water level, provided with mixed- or radial-flow pumps, one of which has variable speed and the others, fixed speeds.

The variable-speed pumps will be driven by slip-ring motors, connected in subsynchronous inverter cascade, and the fixed-speed ones, by synchronous motors. In such a combination, the asynchronous motor secures a high global efficiency of the installation, while the synchronous motor contributes to the improvement of power factor of the electrical installation.

The slip-ring motor is fed by a simple static converter through its rotor circuit in the case of medium and high powers ($P = 400-4{,}000$ kW), or by a double static converter in the case of very high power pumps ($P = 4{,}000-8{,}000$ kW). In both cases, speed regulation is done by rotor voltage control.

To this end, we may introduce a level transducer and regulator into the control loop. Thus, a constant discharge level is maintained, irrespective of variations in water consumption occurring in the canal network, or of variations in suction level.

Speed control also leads to an artificial $H(Q)$ pump characteristic curve, superposed exactly on the $H_s(Q)$ natural characteristic curve of the system. Consequently, we obtain regulation without losses and with reduced electric power consumption. For instance, the working diagram of a pumping installation for irrigation provided, with two $56B_{17}$ pumps in Fig. 14.72, shows a very low value of unitary average energy consumption ($e_{u\,av} = 2.37$ Wh/cu.m.m).

A pumping installation, provided with variable-speed pumps, requires at the same time, reduced investment and operation expenditures. The first is explained by the reduced number of machines (for instance, two instead of four), of armatures and construction volumes, while the second results from the high global efficiency (for instance, 90 per cent instead of 65 per cent), and from a much lower electric power consumption (for instance, 142 Wh/cu.m instead of 180 Wh/cu.m.m.). All data have been taken from diagrams shown in Fig. 14.72. Siphons are used in discharge lines to prevent reverse flow after pumping has been stopped, by means of an automatic hydraulic vacuum breaker located at the summit. Sometimes, siphons are used only to eliminate the need for shut-off or check valves.

14.3.2 Land Irrigation Pressure-boosting Installations

Characteristics of a closed-pump irrigation system, operating "on request". International statistics show that new trends have been signalled in irrigation technology since 1955. Thus, irrigation *"on request"* (beneficiary's request — human, soil, plant, climatic, industrial, economic and social conditions, etc.), by "spraying", with water volume recording), has completely and rapidly replaced the "surface" irrigation. The older systems have been transformed and adapted to spraying in various countries, among which France, holds the first place, followed by Italy [14.1].

Thus, the distribution system is of closed type with "on request" operation. Water distribution "on request" means that each beneficiary has the possibility of using water when he needs it, in accordance with agricultural or other requirements, but within the designed limits of flow rates.

Delivery on request, obtained by automation and remote control, is an excellent solution. The irrigation system is very long, with many branches. The branches carry water intakes with irrigation hydrants having several automation functions: flow rate controller, flow rate limiter, and volume integrator.

The available pressure at the extremity of a closed system should at least be equal to the maximum required pressure at the irrigation hydrants. Consequently, the characteristic curve of the irigation system is the envelope of maximum pressure values necessary and sufficient for the convenient feeding of all irrigation hydrants in their completely open positions, for each value of the required flow rate. We notice, at the first branches from the pumping station, that the envelope wanted, irrespective of the required value of flow rate, is placed at the level of the pumping-system characteristic pressure ($H_s = H_t$), corresponding to the maximum required flow rate. Having in view the operation conditions of irrigation hydrants, the pumping system characteristic curve is represented by a horizontal line, parallel to the flow rate axis, with a constant ordinate, equalling the system characteristic pressure, corresponding to the maximum required flow rate.

A pumping installation operating on request is characterized by:
— a high required flow rate ($Q = 200 - 1,500$ dm³/s), showing a large daily variation within a wide flow range ($\Delta Q = 1 : 20$);
— a high pumping head ($H_t = 70 - 100$ m), located on a horizontal characteristic ($H_{st}/H_t = 1$).

Since an irrigation pressure-boosting station has to meet variable flow rate requirements, it has to include a great number of fixed-speed pumps for securing flexible operation and a high efficiency, the more so as the installation flow rate is high. When variable-speed pumps are used, their number is substantially lower. Since installation costs depend greatly on the number of pumps, we have to find an optimum number by comparing additional power costs with corresponding savings in capital investment. The unit power, resulting from the distribution of the required maximum flow rate to an optimum number of pumps, is of medium value ($P = 100 - 500$ kW) in the case of fixed-speed pumps, and of high value ($P = 200 - 1,000$ kW), in the case of variable-speed pumps.

In view of the great number of irrigation pressure-boosting stations present in large irrigation complexes (for instance, 167 such stations in Sadova irrigation complex of the Socialist Republic of Romania, covering a surface of 80,000 ha), variable-speed pumps are chosen for reduction of the number of machines.

Arrangement of a land irrigation pressure-boosting installation with fixed-speed pumps. Depending on the pump automated control, two kinds of stations are generally used: (1) with regulation by flowmeter, and (2) with combined regulation by pressure switch and flowmeter.

(1) *Pressure-boosting installations with regulation by flowmeter.* Due to the wide regulation flow range, the station should be provided with 3—6 duty pumps and a small jockey pump, the latter having half the flow rate of the former. The unitary flow rate of the turbopumps depends on their number and is determined by means of Table 14.1.

TABLE 14.1 **Selection of unit capacity of pumps**

Pump numbers	Flow rate of main pump, in % of peak demand	Flow rate of jockey pump (pilot lead pump), in % of peak demand
3	36.60	19.30
4	27.50	13.75
5	22.00	11.00
6	18.30	19.15

A high global efficiency is obtained by making use of pumps with a flat $H(Q)$ characteristic, as stable as possible in the zone of small flow rates. We find such characteristics at medium specific speed pumps ($n_s > 80-125$).

Automated control of pumps is effected as follows: (a) by flowmeter for duty pumps, choosing a flowmeter with a number of contacts, equal to the number of main pumps, and (b) by pressure switch for the jockey pumps, making use of a hydrophor tank. The jockey-pump flow rate determines the cut-in flow rate of the first duty pump, as well as the width of the flow range the flowmeter could secure.

The hydrophor tank will have a small volume if jockey pumps, too, have a small flow rate, but this implies a small value of the cut-in flow rate of the first duty pump, which leads to the selection of a flowmeter with a large flow range. It means that the flowmeter should have a throttling device (diaphragm or Venturi tube) with a high differential pressure (Table 14.2), and this leads to a high pressure loss in the device. For avoiding the above-mentioned difficulties, we may use an electromagnetic flowmeter, but it is very expensive.

Operation depends on the demands of the pumping plants as the jockey pump operates discontinuously, collaborating with the hydrophor to maintain the required pressure in the system during periods of low flow demand. When the system demand increases to a set flow, the flowmeter shuts off the jockey pump and turns on the main pump number *1*. If system demand goes on rising, the flowmeter will turn on successively the next main pumps, to operate simultaneously with the main pump number *1*. As

TABLE 14.2 **Selection of differential pressure at orifice of flowmeter**

Pump numbers	Differential pressure, in m
3	2.5
4	5.0
5	10.0
6	15.0

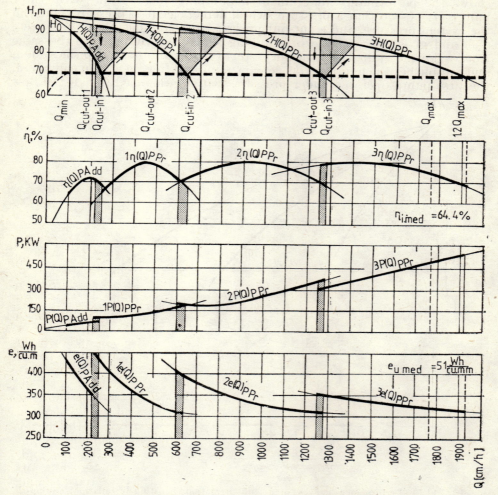

Fig. 14.73. Performance curves of a land irrigation pressure-boosting installation, equipped with fixed-speed pumps, regulated by flowmeter.

system demand decreases, the flowmeter will shut off successively the main pumps, and the jockey pump will be turned on. A manual transfer switch allows manual interchange of main pumps.

Figure 14.73 shows the operation diagram of a pressure-boosting installation, regulated by a flowmeter. In this case, the installation is equipped with three main VDF_{300} pumps and one VDF_{200} jockey pump. The diagram shows a relatively high global efficiency ($\eta_{i\ med} = 64.4$ per cent) and a medium specific power consumption ($e_{u\ med} = 5.11$ Wh/cu.m.m).

(2) *Pressure-boosting installation with regulation by associated pressure switch and flowmeter.* In this case the number of pumps should be maximum 3—5 for keeping a simple electric installation. Figure 14.74 shows the performance diagram of a pumping installation with three pumps. It is seen that a pressure switch detects a cut-in pressure, equal to the system pressure, and a cut-out pressure, much higher than the system one, and gives the starting or stopping signal accordingly. At the same time, a flowmeter analyzes the variation of required flow rate in the system and establishes the number of necessary pumps. After the cut-in command, the first pump operates under pressure switch guidance. When the flow

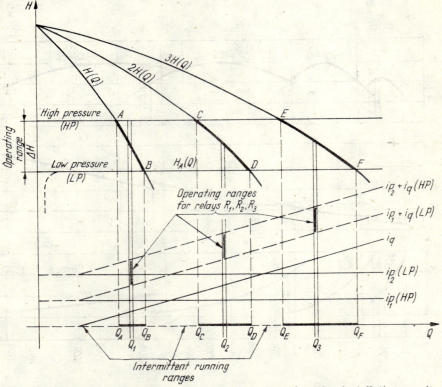

Fig. 14.74. Performance curves of a land irrigation boosting-installation, equipped with fixed-speed pumps, regulated by associated pressure switch and flowmeter (*CdC* method).

rate exceeds the flow rate of one pump, the pressure falls and the pressure switch gives the starting command. In other words, the starting or stopping order is prepared by the flowmeter and confirmed by the pressure switch.

In practice, this command is achieved by the summation of two electric currents;

— an "i_p" current, resulting from a circuit where resistances are connected, or not, in series, depending on the pressure switch position; the i_p current varies within the limits i_{p_1} and i_{p_2};

— an "i_q" current, resulting from the flowmeter and amplified; i_q is varying linearly with flow rate.

The function $i_p + i_q$ is represented by two lines, corresponding to high pressure (HP) and to low pressure (LP). The sum of the two currents passes through circuit relays R_1, R_2, R_3 associated with pumps P_1, P_2, P_3.

The relay operation zones are located between the two lines ($i_{p_1} + i_q$) and ($i_{p_2} + i_q$). In the Q_0-Q_A, Q_B-Q_C, Q_D-Q_E intervals, the operation range is discontinuous, while in the Q_A-Q_B, Q_C-Q_D, Q_E-Q_F intervals, it is stable.

Arrangement of a land irrigation pressure-boosting installation, equipped with variable-speed pumps. The horizontal shape of the system $H_s(Q)$ characteristic requires the presence of at least two pumps and a maximum of three, of which only one has variable speed. The use of a single pump is not economical, since the horizontal $H_s(Q)$ characteristic would then cross the smaller efficiency curve in the minimum flow rates zone on the topogram of the variable-speed pump.

A higher accuracy of flow-rate regulation is obtained when the $H(Q)$ characteristic curve of the pumps is inclined. The more inclined the $H(Q)$ curves, the more perpendicular the intersections between pump and system curves.

Since pump unitary values range between 200 and 1,000 kW, variable-speed pump driving is done as follows:

— with a squirrel-cage motor, fed via its stator by means of a frequency converter when unitary power is smaller than 400 kW (Fig. 14.75);

— with a slip-ring motor, fed via its rotor by a subsynchronous inverter cascade, for unitary powers higher than 400 kW (Fig.14.76).

Both schemes (see Figs. 14.75 and 14.76) show both manual and automated regulation of rotation speed. The automated regulation is done by introducing a continuous pressure transducer and controller into the control loop. In this way, pressure is held constant regardless of variations in consumption within the closed irrigation system.

Due to the superposition of the $H(Q)$ pump artificial characteristic curve on the $H_s(Q)$ natural characteristic curve of the system, the flow rate is regulated without hydraulic and electric losses, and thus global efficiency of the pumping installation rises, finally leading to a reduced electric power consumption. Figure 14.77 shows the diagram of an irrigation boosting-station, equipped with three identical *14NDS* type pumps, one of which has variable speed.

The diagram shows a medium global efficiency $\eta_{i\,med} = 89$ per cent, and a unitary medium power consumption $e_{u\,med} = 3.32$ Wh/cu.m.m.

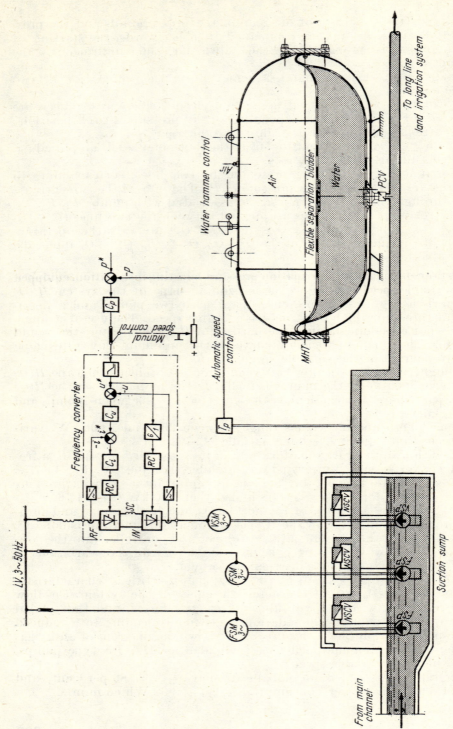

Fig. 14.75. Schematic arrangement of a land irrigation pressure-boosting installation with variable-speed pumps, driven by a squirrel-cage motor, fed via a frequency converter:

VSM, VSP — variable-speed motor and pump; FSM, FSP — fixed-speed motor and pump; $NSCV$ — non-slam check valve; MHT — membrane hydropneumatic tank with separation bladder; RF — rectifier; IN — inverter; SC — smoothing coil; RC — ring counter; C_i, C_u, C_p — current, tension and pressure controllers; T_p — pressure transducer.

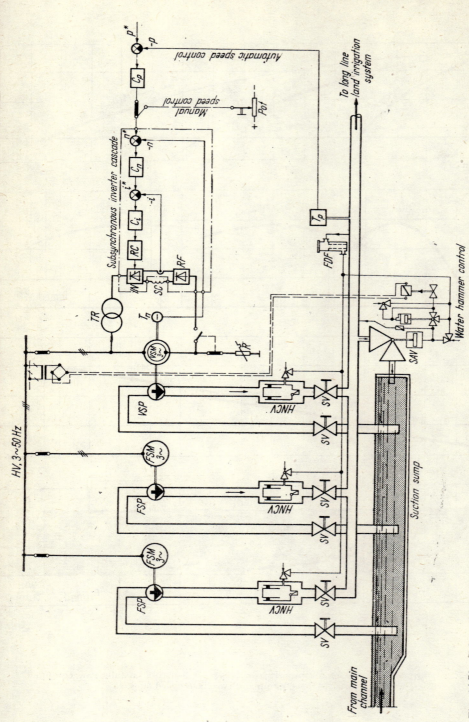

Fig. 14.76. Schematic arrangement of a land irrigation pressure-boosting installation with variable-speed driving by a slip-ring motor, fed via a subsynchronous inverter cascade:

VSM, VSP — variable-speed motor and pump; $HNCV$ — hydraulic-operated needle check valve; SV — shut-off valve; SAV — surge anticipator valve; FDF — fixed-drum filter (cartridge filter); TR — transformer; IN — inverter; RF — rectifier; SC — smoothing coil; RC — ring counter; C_n, C_i, C_p — speed, current and pressure controllers; T_n, T_p — speed and pressure transducers

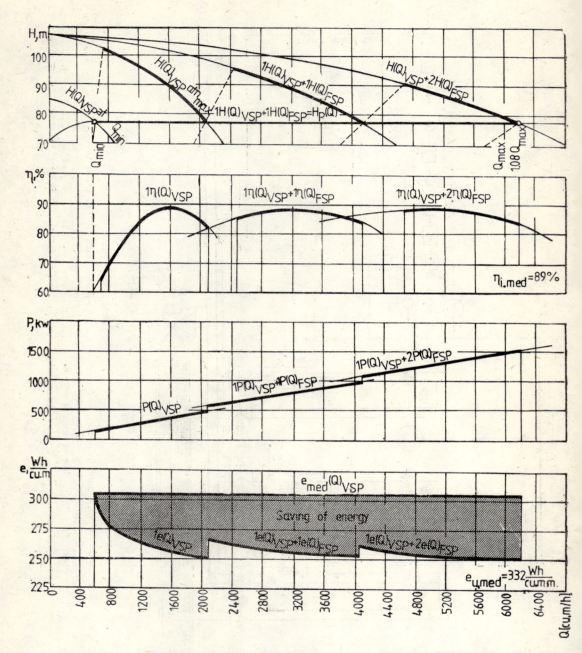

Fig. 14.77. Performance curves of a land irrigation pressure-boosting installation, equipped with one variable-speed pump and two fixed-speed ones.

Relative to an installation equipped with fixed-speed pumps, the one mentioned above requires much reduced running expenses due to its performance: $\eta_{i\,med} = 89$ per cent as against 76.7 per cent in the first case, and $e_{u\,med} = 260$ Wh/cu.m as against 305 Wh/cu.m in the first case (values taken from Fig. 14.77).

Water hammer, occurring due to power failure, can be controlled in two ways:
— by making use of hydropneumatic tanks, fitted with rubber separation bladder (see Fig. 14.75) in the case of pumping installations with small flow rates and short distribution systems;
— by using surge anticipator valves (see Fig. 14.76) in the case of pumping installations with large flow rates and long distribution lines.

In both cases, slamming of the valves is avoided by using no-slam hydraulic check valves.

If we use hydraulically-operated check valves or a surge anticipator valve, a supply of clean water, connected to a fixed drum filter, is compulsory.

A concrete example of a land irrigation boosting-installation with variable-speed pumps is the Lagruère installation [14.74] within the irrigation system of Mas d'Agenais, Lot-et-Garonne (France), equipped in 1970 with four pumps, including one with variable speed (Figs. 14.78, a

Fig. 14.78, a. A land irrigation pressure-boosting installation, equipped with one variable-speed pump and three fixed-speed ones (Lagruère — France).

Fig. 14.78, b. Subsynchronous inverter cascade (Siemens Co.).

and b). Each pump had $Q = 200$ dm³/s, $H = 90$ m, $P = 250$ kW, and $n = 1450$ rev/min. This installation had a flow rate of 800 l/s and covered 1,400 ha.

14.3.3 Land Drainage Pumping Installations

Since ancient times, drainage has been used to maintain a high productivity of irrigated lands. Mormons were the first who practised irrigation in America, and considered drainage as an essential part of irrigation works.

Statistics mention hundreds of millions of hectares, requiring drainage before irrigation, for preventive goals, and after irrigation, for curative purposes.

At present all irrigation programmes in the U.S.A. also include the necessary investments for drainage works. This situation derives from the observation that various lands which, before irrigation having a 6—30 m ground water level, became swamped after irrigation and then witnessed a secondary salination due to excessive increase of ground water level.

The component elements of a drainage system can be grouped as follows: agricultural land to be drained, drainage regulation system, collection-exhaustion system, trunk sewer, drainage pumping station, and the outlet element (emissary).

A pumping station, located between the trunk sewer and the emissary, should meet the following conditions:
— to hold a constant level in the trunk sewer irrespective of the amount of liquid flow towards the station;
— to exhaust water in conditions of variable level to the emissary.

The pumping system for a drainage installation is characterized by:
— a large required flow rate ($Q = 2 - 20$ cu.m/s) varying within a narrow range ($\Delta Q = 1 : 2.5$);
— a very small pumping head ($H_t = 3 - 7$ m) and a flat $H_s(Q)$ characteristic ($H_{st}/H_t = 0.75$).

The unitary power, resulting from the maximum required flow rate distribution among an optimum number of pumps, is of medium value ($P = 40 - 400$ kW).

Under the above-mentioned conditions, only axial-flow pumps are suitable for use.

Arrangement of a land drainage pumping installation, equipped with fixed-speed pumps. The pumping installation should be provided only with duty pumps, without any stand-by pump. The maximum collected flow rate is distributed to 3 or 4 pumps, each with a unitary flow rate of 35 or 26 per cent of the maximum flow rate directed towards the station (the first value corresponds to three- and the second to four-pumps in use). The flow rate of the pumping station can be regulated in two ways:
— by variation of the number of pumps;
— by adjusting the position of impeller blades for one or two pumps.

In the first case, pump operation can be automatized in accordance with the level, by introducing a level meter in each cut-in/cut-out pump

Pumps for Land Systems

Fig. 14.79. Vertical spindle auto-controlled variable pitch axial pump. The electrically operated servomotor is controlled in its turn by the efficient liquid level float (Vickers Co.).

Fig. 14.80. The pump performance curves for a unit with four differing pitch settings of the impeller:
a — pump-head curves; b — pump-power curves; c — pump-efficiency curves.

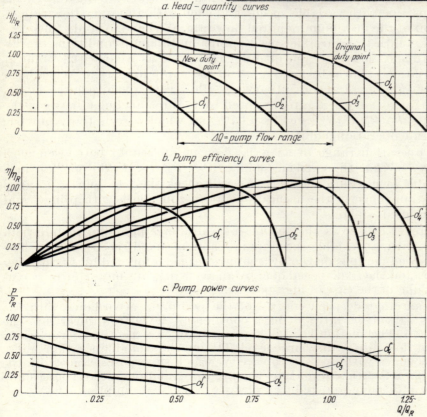

circuit. Thus, when the suction flow rate rises, we obtain starting of successive pumps in ascending level cascade. The suction tank is used as an auxiliary element for limiting the maximum starting frequency of the pumps, since its volume acts as a flow-rate compensator. It accumulates the water volume and is used at least for the starting of the pumping installation, or for compensation of minimum flow rate, as well as at any time when the effluent flow rate exceeds the operation points of the pumps.

In the second case, we make use of axial-flow pumps with variable-pitch impellers, capable of automated pitch adjustement (Fig. 14.79). Operation can be completely automatic, with the output from the pumps responding to the variation in flow of the drainage effluent reaching the suction pumps. The electrically operated pitch-changing control output can be seen mounted on top of the main pump motor. The high pump efficiency is maintained over a wide range of quality handled (Fig. 14.80).

Arrangement of a land drainage pumping installation, equipped with variable-speed pumps. The maximum effluent flow is divided between two pumps, one with variable speed, and each with a unitary flow rate of 55 per cent of the maximum required flow rate. We have to study the operation possibilities of variable-speed pumps in different conditions (variation of effluent flow rate and of discharge level, as well as the simultaneous variation of both) by making use of the pump topogram. When both parameters are simultaneously varying (flow rate and discharge level), the pump operation range is represented by the area, included between the four points P_1, P_2, P_3, P_4 (see Fig. 14.70) and which also is the geometric range of all operation points.

The operation range of an axial-flow pump should be chosen with great care, so that the extreme points P_1, P_2, P_3, P_4 do not fall into the unstable or cavitation zones. We will check the possibilities to keep the operation range in the natural operation zone, both with primed and with unprimed siphons.

Figure 14.81 shows the scheme of a drainage pumping installation, equipped with a variable-speed pump and a fixed-speed pump. The variable-speed pump is driven by a squirrel-cage motor, fed by a frequency converter, and the fixed-speed one, by an asynchronous motor or by a synchronous one in the case of high powers.

The automated speed control of the pump, in accordance with the level, is attained by introducing a continuous level transducer into the control loop. The suction level is kept constant, irrespective of variations in effluent flow rate or discharge level. The envelope of all operation points is an $H(Q)$ curve with upward concavity, which superposes exactly on the system characteristic curve, thus leading to regulation without losses and to a high efficiency and reduced electric power consumption.

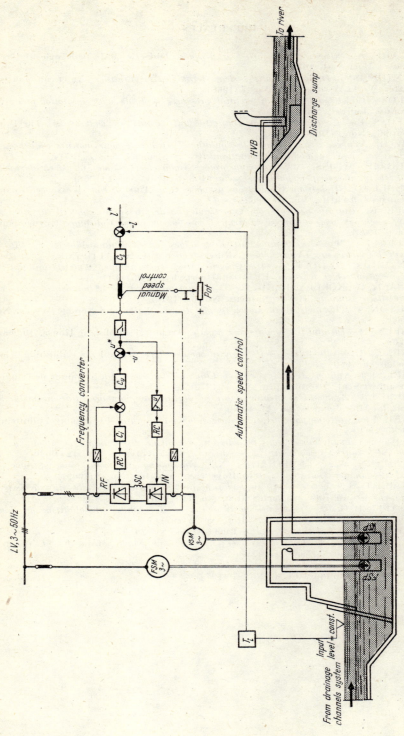

Fig. 14.81. Schematic arrangement of land drainage pumping installation, equipped with a variable-speed pump and a fixed-speed one:
FSP, FSM = fixed-speed pump and motor; VSP, VSM = variable-speed pump and motor; HVB = hydraulic vacuum breaker; T_l, C_l = level transducer and level controller; C_i, C_u, C_l — current, tension, and level controllers; Pot — potentiometer. RF — rectifier; IN — inverter; SC — smoothing choke; RC — ring counter;

REFERENCES

14.1. BLIDARU, V., *Sisteme de irigații și drenaje*. Didactic and Pedagogic Publishing House, Bucharest, 1976.
14.2. BRAXTON, S.Y., *Design of water systems for high-rise buildings*. American Journal of Water Works Associations, No. 7 (1966).
14.3. DANFOSS, Co., *Regulation of differential pressure in plants with circulation pumps*. The Danfoss Journal, No. 3 (1976).
14.4. DANFOSS, Co., *Controlling pumps connected in parallel in waterworks*. The Danfoss Journal, No. 1 (1976).
14.5. DIRR, R., *Nouveaux équipements de régulation électronique pour moteurs asynchrones de grande puissance a double alimentation*. Revue Siemens, No. 7 (1971).
14.6. ELGER, H., WEISS, M., *Untersynchrone Stromrichterkaskade als drehzahlregelbarer Antrieb für Kesselspeisepumpen*. Siemens-Zeitschrift, No. 4 (1968).
14.7. GLOEBEL, O., *Drehstrome Gleichstromkaskaden zum Antrieb von Wasserwerkspumpen*. Siemens-Zeitschrift, No. 10 (1962).
14.8. GRAFF, K., *Entrainement de pompes a régulation de vitesse par convertiseur en cascade hyposynchrone pour l'approvisionnement en eau du Lac de Constance*. Revue Siemens, No. 6 (1973).
14.9. GRAFF, K., *Wasserversorgung mit drehzahlgesteuerten Pumpenantrieben für die Stadt Barquismento (Venezuela)*. Siemens-Zeitschrift, No. 11 (1975).
14.10. GUNTER, S., RUDOLF, S., *Die Grössten Kesselspeisepumpen Antriebe mit untersynchronerstromrichter Kaskade*. Elektrizitätswirtschaft, No. 11 (1971).
14.11. HABÄCK, A., KÖLLEVSPERGER, D., *Application et perfectionnement du moteur à convertiseur statique*. Revue Siemens, No. 6 (1971).
14.12. HORTON, L., *Variable speed sewage pump drives*. Water and Sewage Works, No. 3 (1963).
14.13. IONEL, I. I., *Instalații de pompare reglabile*. Technical Publishing House, Bucharest, 1976.
14.14. IONEL, I. I., *Acționarea electrică a turbomașinilor*. Technical Publishing House, Bucharest, 1980.
14.15. IONEL, I. I., *Stations de pompage à régulation débitmetrique*, La technique de l'eau et de l'assainissement. Hidrotehnica, No. 1 (1972).
14.16. IONEL, I. I., *Zgomotul și atenuarea lui în stațiile de pompare din ansamblurile de locuințe*. Hidrotehnica, No. 11 (1968).
14.17. IONEL, I.I., *Stații de pompare cu reglare automată în regim discontinuu*. Hidrotehnica, No. 3 (1969).
14.18. JOST, J., *Alimentation en eau potable d'un réseau de distribution par pompage a la demande, sans réservoir en charge sur le réseau*. Travaux, No. 12 (1969).
14.19. KLEIMANN, W., STAMBOLIDIS, A., *Anwendung untersynchroner Stromrichterkaskaden in der Industrie*. Brown Bovery Nachrichten, No. 2 (1969).
14.20. KULMERT, S., *Antriebe mit untersynchroner Stromrichterkaskade für Kesselspeispumpen*. Brown Bowery Nachrichten. No. 11 (1975).
14.21. PEDOTTI, P. G., *Les problèmes posés à l'étude de grandes installations de pompes alimentaires*. Revue technique Sulzer, No. 2 (1971).
14.22. RICHTER, H., *Boiler feed pumps*. Combustion, No. 2 (1970).
14.23. SEELBACH, H., *Neue Wasserversorgung für die Stadt Maracaibo (Venezuela)*. Siemens-Zeitschrift, No. 10 (1975).
14.24. * * * *La vitesse variable remplace le chateau d'eau du réseau d'irrigation du Mas d'Agenais*. Techniques CEM, No. 12 (1973).

Subject Index

Absolute
  pressure, 72
  roughness, 192
Alarm rotameter, 451
Altitude valve, 375
Asychronous cascade, 563
Atenuation of cavitational destruction, 181
Atmospheric pressure, 72

Backflow preventer, 635
Backward waves, 216
Bladder accumulator, 421
Broken siphon, 491

Cascade
  with autotransformer in rotor circuit, 581
  with double converter in rotor circuit, 582
Cavitation
  bubbles, 165
  characteristic curves, 184
  coefficient, 161, 185
  destruction, 178
Casing, 71
Characteristic line, 287
Collapse of bubbles, 174
Compensated pressure switch, 440
Constant voltage d.c. link converter, 551
Continuous pressure regulator, 463
Controls for surge prevention, 384
Controls for surge dissipation, 385

Control ratio, 332

Dead volume, 430
Detecting cavitation, 181
D.C. link converter, 549
Distribution pumping station, 646
Duty point, 128
Dynamic head, 235
Direct waves, 246

Eddy-current coupling, 601
Effective closing time, 252
Electric check-valve, 387
Electric recirculation valve, 392
Fixed cavitation, 161
Flow-rate switch, 449
Fluidity, 191
Friction coefficient, 194, 199

Gaseous cavitation, 169
Geodesic suction-head, 239
Global efficiency, 157

Head loss, 200, 201
Hot-water recirculation pumping plant, 689
Hyamat package pressure boosting installation, 423
Hydrocel membrane type accumulator, 419

Subject Index

Impeller, 71
Implosion of vapour bubbles, 344
Incipient cavitation, 161
Inlet guide vanes, 153
Intrinsic characteristic, 330

Joule effect, 525

K_{vs} coefficient, 322
Krämer-cascade, 590

Leading edge, 61
Local resistance coefficient, 216—234

Manometric suction head, 237
Maximum pressure head, 261
Membrane-type accumulator, 418
Microprocessor-controlled
   recirculation plant, 689
Mobile cavitation, 161
Motor with electronic comutator, 517

Non-modulating float
   valve, 371
Non-overload, power characteristic, 120
Nucleation centres, 167

Overlapping angle, 569
Overload power characteristic, 120

Pacomonitor flow control, 450
Package constant pressure
   system, 450
Parabolic pressure regulation, 466
Parallel-connection, 135
Pressure-boosting installations :
   for multistorey buildings
      with one pressure-zone, 620
   for building with two pressure-zones, 632
Pressure-reducing controls, 353
Pressure-sustaining controls, 359
Prewhirl, 154

Reaction coefficient, 38

Reflection time, 267
Relative
   pressure, 72
   roughness, 192
Riemann invariances, 299

Scherbius cascade, 571
Series-connection, 137
Shifting the duty point, 142
Similarity equations, 477
Slip power, 525
Sonic wave, 246
Specific
   energy, 155
   flow, 27
   head, 27
   resistance S, 196
   speed, 27
   speed n_q, 84
   speed n_s, 81
Speed
   control coefficient, 148
   control range, 147
   torque characteristic, 470
Static converter synchronous motor, 517
Supercavitation, 161
Surge anticipator valve, 404
Surge relief valve, 399
Swirling cavitation, 161

Time delay pilot, 409
Time of travel, 249
Total head, 235
Trailing edge, 61
Transmission main line
   with negative characteristic curve, 663
Transmission pumping stations, 654
Transvector control, 560
Turbopump starting, 481

Unit frictional loss, 203
Universal characteristic, 110

Vacuummetric suction head, 238
Vaporous cavitation, 169
Variable voltage, d.c. link converter, 554
Vena contracta, 327
Vibratory cavitation, 161

Wall roughness, 192

Warm-water recirculating pumping plant, 686
Water supply to buildings by high-rise systems, 619
Water supply of buildings with long-line system, 642
Wave drag, 249
Wave speed, 247
Working characteristic, 109